Bacterial Lipopolysaccharides

Bacterial Lipopolysaccharides

The Chemistry, Biology,
and Clinical Significance
of Endotoxins

Edited by Edward H. Kass
and Sheldon M. Wolff

1973

The University of Chicago Press
Chicago and London

This volume is based on the proceedings of a conference on bacterial endotoxins held at Airlie House, Warrenton, Virginia, May 31–June 2, 1972. The chairmen of the conference were Edward H. Kass and Sheldon M. Wolff. The papers were initially published as a supplement to volume 128, no. 1, of *The Journal of Infectious Diseases,* July, 1973.

The support of the following for the publication of these proceedings is gratefully acknowledged: Burroughs Wellcome Company; Ciba Pharmaceutical Company; Hoffmann-La Roche Incorporated; Lilly Laboratory for Clinical Research; Merck Sharp and Dohme Research Laboratories; the National Institute of Allergy and Infectious Diseases, National Institutes of Health; Schering Corporation; The Upjohn Company; Wallace Laboratories; Warner-Chilcott Laboratories; and Wyeth Laboratories.

The University of Chicago Press, Chicago 60637
The University of Chicago Press, Ltd., London

International Standard Book Number: 0-226-42564-9
Library of Congress Catalog Card Number: 73-81482

Contents

Session III: Biological Responses

Session IV: Models of Endotoxic Activity

Session V: Clinical Aspects

SESSION I: CHEMISTRY

Molecular Organization of the Rigid Layer and the Cell Wall of *Escherichia coli*

Volkmar Braun

From the Max-Planck-Institut für Molekulare Genetik, Berlin-Dahlem, Germany

A model for the supermolecular structure of the rigid layer (murein-lipoprotein complex) of the cell wall of *Escherichia coli* was devised, based on chemical structure analysis. According to these data, 250,000 lipoprotein molecules of known amino-acid sequence are evenly distributed over a one-layered murein net. The sequence and conformation of the lipoprotein is unusual. The murein net consists of polysaccharide chains cross-linked by short peptides. Since one disaccharide unit of the polysaccharide chains is probably 10.3 A long, the lipoprotein molecules, hooked statistically to every 10th–12th disaccharide unit, are spaced 100–120 A apart. The distance from center to center of neighboring polysaccharide chains is 12–13 A. From this rather precise view of the distribution of the lipoprotein molecules on the net, a model can be presented for the basic arrangement of the wall. Immunologic studies show that the lipoprotein protrudes from the innermost part of the wall towards the surface, and it is likely that it plays mainly a structural role. Due to the hydrophilic and hydrophobic properties of the lipoprotein, the other main components of the wall (e.g., lipopolysaccharide, phospholipids, proteins) could be organized around the lipoprotein molecules so that these could form the covalently anchored core of subunits of which the cell wall is built.

This paper presents a tentative view of the molecular organization of the cell wall of *Escherichia coli, Salmonella,* and closely related species of which the lipopolysaccharide (endotoxin) is a major component.

Membranes are considered to be liquid-cristalline bilayers into which proteins are integrated as functional units [1–3]. Many functions are attributed to membranes, and much data about their overall chemical composition (proteins, phospholipids, glycolipids, glycoproteins, polysaccharides) is available. However, little is known about the molecular organization of any membrane with respect to the spatial arrangement of single, well-defined macromolecules. Electron microscopy, a principal tool in the study of intact membranes, resolves membranes into layers and globules, but in most cases it is difficult to correlate these observations at the supermolecular level with functions and structures known from enzymatic and chemical studies. Most other physical methods require that the structures have a long-range order that cannot be expected from a liquid-cristalline structure. Membranes can be taken apart by chemical methods, but often only under denaturing conditions and with irreversible loss of the in-vivo arrangement of their constituent parts.

In this paper, the structure and molecular arrangement of a lipoprotein in the outer membrane of *E. coli* will be described [4].

The Rigid Layer and the Murein Net

These studies became feasible after it was shown that the lipoprotein is a part of the so-called rigid

This work was done in collaboration with V. Bosch, K. Hantke, and H. Wolff of this institute; with H. Mayer and S. Schlecht of the Max-Planck-Institut für Immunbiologie, Freiburg; and with H. Hagenmaier of the University of Tübingen. Part of the work was supported by the Deutsche Forschungsgemeinschaft.

Please address requests for reprints to Dr. Volkmar Braun, Max-Planck-Institut für Molekulare Genetik, Berlin-Dahlem, Germany.

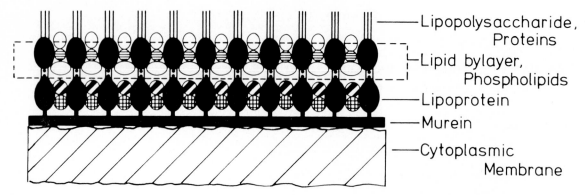

Lipopolysaccharide, Proteins

Lipid bylayer, Phospholipids

Lipoprotein

Murein

Cytoplasmic Membrane

Figure 1. Section through the *Escherichia coli* cell envelope, which consists of the outer membrane, the cell wall, and the inner, cytoplasmic membrane. For the wall, hypothetical subunits are drawn to represent the major building blocks. It is proposed that the pattern of distribution of the lipoprotein molecules over the murein (see figure 6) is representative of the arrangement of the subunits. Other major components of the wall, although suggested in this model, do not necessarily have to be in a one-to-one relationship to the number of lipoprotein molecules. The lipid at the N-terminal end of the lipoprotein is drawn in such a way that it is interacting with the lipid A of lipopolysaccharide in a lipid bilayer, of which both components besides phospholipids could be a part.

layer [5]. The rigid layer is the innermost layer of the outer membrane (cell wall) of *E. coli* (figure 1) [6–9]. Lipoprotein molecules are distributed over the murein (or, as it has also been called, the peptidoglycan or mucopeptide), which is a bag-shaped net consisting of polysaccharide chains cross-linked by short peptide bridges (figure 2). The net, with around 250,000 covalently linked lipoprotein molecules [10], encloses the whole cell as one giant macromolecule (molecular weight, $1–4 \times 10^9$ daltons). Since it consists only of covalent bonds, the structure is preserved even under vigorous conditions of isolation, such as treatment with 4% boiling dodecylsulfate. The detergent solubilizes all other constituents of the outer membrane (i.e., lipopolysaccharide, proteins, and phospholipids), which make up 80% of the outer membrane; only the rigid layer, or the murein-lipoprotein complex, remains undissolved and can thus be isolated in pure form.

To get an idea about the distribution of lipoprotein molecules over the murein net, we first had to learn more about the supermolecular structure of murein. In murein, N-acetyl-muramic acid and N-acetyl-glucosamine are linked by β-1,4-glycosidic bonds (figure 3); when one compares this structure to X-ray analysis of similar polysaccharide chains in cellulose and chitin, one can deduce that the disaccharide units of murein are also 10.3 A long. The polysaccharide chains, built of such subunits, are quite rigid rods. They are cross-

linked by a peptide bond between the meso-diaminopimelic acid and the D-alanine of neighboring peptide side chains (figure 4). In addition to the chemical structure of the basic subunit and the mode of cross-linkage, one knows roughly that the thickness of the rigid layer is 15–30 A [6, 7]. Whether the lipoprotein contributes to the thickness observed in electron microscopy is unknown. In an unfolded conformation, penetrating deeply into the cell wall, it probably would not be observed by this technique. Nothing else is known for application to the construction of a murein model; X-ray analysis revealed no interpretable details [11]. Therefore, we added as another parameter the number of murein repeating units per unit of surface of the wall.[1] The number of cell envelopes and their surface area were determined by electron microscopy, and the number of disaccharide units was determined by amino-acid analysis, based on the murein-specific components muramic acid and diaminopimelic acid. In addition, the number of disaccharide units was determined by incorporation of radioactive diaminopimelic acid into a double mutant unable to synthesize and to decarboxylate diaminopimelic acid. With four different strains, we found that for one

---

[1] V. Braun, H. Gnirke, U. Henning, and K. Rehn, "A Model for the Structure of the Shape Maintaining Layer of the *E. coli* Cell Envelope," manuscript in preparation.

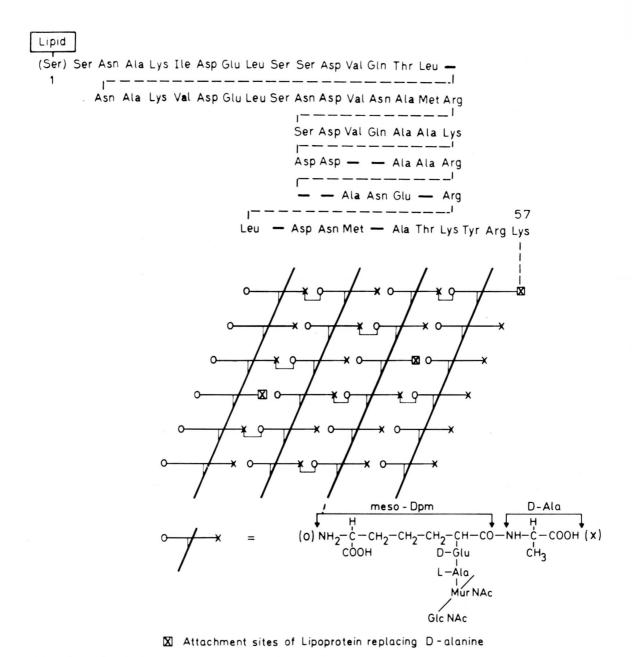

Figure 2. Proposed model for the murein net on which only one lipoprotein molecule is drawn. Underlying murein net: the parallel heavy lines symbolize the polysaccharide chains. They are cross-linked by the T-like peptide side chains. Nothing is known about the conformation of the peptide side chains. Here they are drawn to allow a long-range covalent fixation of the murein, which is a necessity for *Escherichia coli* and other gram-negative bacteria. The length of the T-bar corresponds well with the average distance between polysaccharide chains (12.4 A) calculated for a one-layered murein from the number of repeating units per unit of cell surface. The extended aliphatic chain of meso-Dpm-D-Ala is about 13 A long. For the folding of the L-Ala-D-Glu, an arrangement with several hydrogen bonds to the sugar backbone can be chosen (e.g., as proposed in [29]). Since *E. coli* is cross-linked only to the variable extent of 15%–30% [12], some links between meso-Dpm (o) and D-Ala (x) were left open. But despite the lack of some cross-linking peptide bonds, the conformation of the peptide side chain is considered to be the same for all. Penicillin inhibits the cross-linking reaction [14], and the cell lyses. Attachment site of the lipoprotein: to account for the fact that, on the average, one lipoprotein is covalently linked to every 10th–12th disaccharide unit of the murein, four attachment sites are indicated in the murein net. But for the sake of clarity, only the sequence of one lipoprotein molecule is drawn. The lipoprotein is linked by the ε-amino group of its C-terminal lysine to the carboxyl group of meso-diaminopimelic acid of the murein and replaces there the D-alanine [30]. The sequence is presented in a way that emphasizes its repetitive design.

4

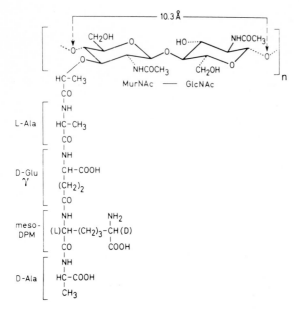

Figure 3. Structure of the repeating disaccharide unit of the *Escherichia coli* murein. Length of the disaccharide unit N-acetyl-muramyl-N-acetyl-glucosamine (10.3 A) is deduced from the X-ray data on cellulose and chitin [19]. The peptide side chain L-alanyl-γ-D-glutaminyl-meso-diaminopimelyl-D-alanine is linked to the lactyl group of muramic acid. No stereo structure, comparable to that of the disaccharide unit, is known for the peptide side chain.

disaccharide unit a surface area of 128 A² is available. Since one disaccharide unit is 10.3 A long, 12.4 A are left for the breadth. This would mean that, for a murein consisting of one layer, the distance from center to center of parallel polysaccharide chains is 12.4 A.

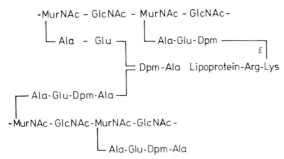

Figure 4. Cross-linkage between two neighboring polysaccharide chains, which are polymers of the basic disaccharide unit N-acetylmuramyl-N-acetyl-glucosamine (MurNAc-GluNAC). The peptide side chains cross-link these by a peptide bond between the amino group of the optical D-center of diaminopimelic acid (Dpm) and the carboxyl group of the terminal D-alanine (D-Ala) (see also figure 3). The attachment site of the lipoprotein is also indicated.

The data are still not sufficient to construct an unambiguous murein model. Four additional arguments can be added.

(*1*) For the *E. coli* murein, a covalent, long-range fixation of the structure in both directions of the plane must be claimed since, even in boiling concentrated dodecylsulfate, the murein retains its structure and shape. Rod-shaped *E. coli* cells yield rod-shaped murein, and spherical mutants yield spherical murein [12, 13].

(*2*) From building of models, it is obvious that within a polysaccharide chain the peptide side chains always extend from the side and in the direction of muramic acid.

(*3*) It is also likely that all peptide side chains have the same spatial configuration and are fixed in this configuration by hydrogen, ionic, and other bonds. The cross-linking enzyme, transpeptidase [14], only closes a peptide bond between two peptide side chains fixed in the appropriate configuration and only strengthens by a covalent bond a configuration already present. In *E. coli* the extent of cross-linkage is quite variable, ranging between about 15% and 30%, which favors the above view. The enzyme, probably itself fixed in a membrane, presumably is also unable to recognize peptide side chains in different spatial orientations and certainly is unable to bring together moving peptide side chains.

(*4*) Points (*1*) and (*2*) imply that the peptide side chains must be outside the plane of the polysaccharide chains. These four arguments pose fairly narrow restrictions on construction of models and the result is presented in figure 2. The structure of murein is assumed to be regular, as is the case with most biological structures that are built of subunits, e.g., viruses [15], surface proteins of *Spirillum* [16], *Bacillus brevis* [17], *Bacillus polymyxa* [18], cellulose, or chitin [19].

This model does not imply that the whole murein is built of such highly ordered, crystalline-like areas. They could be interrupted by less dense areas forming links between the highly ordered ones, although there is no evidence for this. Depending on the orientation of the plane of the sugar relative to the cell surface, the polysaccharide chains cover 40%–90% of the cell surface, leaving enough space for the passage of molecules or even macromolecules like alkaline phosphatase [20], other periplasmic enzymes [21], or phospholipases [22].

We cannot decide whether a monolayer or a multiple layer exists. On the basis of the number of murein repeating units found, we calculated that we could construct at most three layers, but then it would be impossible to obtain a two-dimensional, covalent fixation. So if the one-layered *E. coli* murein were the basic structure for gram-positive bacteria as well, the murein of *Bacillus subtilis* would consist of 40 layers. Forty layers would just account for the thickness of around 400 A observed under the electron microscope [23]. This fact favors the one-layered arrangement chosen for *E. coli*.

The Lipoprotein

What is the structure of the lipoprotein, how is it fixed to the murein, and how is it distributed over the murein net shown in figure 2? This unusual protein contains at the N-terminal end a covalently linked lipid [24] and at the C-terminal end the covalently bound murein. The interesting sequence of the lipoprotein is known [25], as is one important feature of its conformation (V. Braun and H. Hagenmaier, unpublished observations). The sequence is presented in figure 2 to emphasize its repetitive design.

Starting from the third amino acid, two consecutive homologous oligopeptides of 14–15 amino acids have nearly an identical amino-acid sequence, with only very few and very conservative amino-acid replacements [isoleucyl (Ile)/valyl (Val); seryl (Ser)/asparaginyl (Asn); glutaminyl (Gln)/Asn; leucyl (Leu)/methionyl (Met)]. The C-terminal half of the original gene, coding apparently for 15 amino acids, is then probably fused four times with the duplicated portion. Only the attachment sites of the lipid (lipid-(Ser)-Ser-) and the murein (-tyrosyl (Tyr)-Arg-Lys-murein) extend from the repetitive block, and these latter portions do not fit into the repetitive design when another striking feature of the sequence is considered.

Again commencing with the third amino acid, every following fourth (or third, since the formation of the amino acids is alternating, i.e., 1, 2, 3, 4, 1, 2, 3, 1, 2, 3, 4, etc.) amino acid has a hydrophobic side chain [Ile, Leu, Val, Leu, Val, Leu, Val, Met, Val, alanyl (Ala), Ala, Ala, Leu, Met, (Tyr)] that is, on the average, one of every 3.5 amino-acid residues. Since 3.6 amino-acid residues

comprise one turn of an α-helix, the hydrophobic side chains would all occur on the same side of the α-helix, provided the lipoprotein has an α-helical conformation. Measurements of the circular dichroism (figure 5) showed indeed about 70% α-helical content, calculated on the basis that polylysine is 100% helical at *p*H 11.75. The astonishing fact is that the lipoprotein went through treatment with boiling 4% dodecylsulfate, and thus the helical content measured above is that of the renatured protein. Studies are in progress for the isolation of the protein under less denaturing conditions; we want to determine whether the whole repetitive part is α-helical. There are no amino acids like proline in the sequence that would break

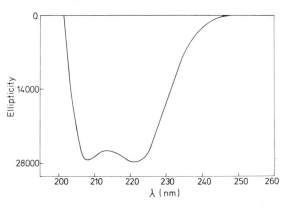

Figure 5. Conformation of the lipoprotein in aqueous solution. Circular-dichroism spectrum of lipoprotein released by lysozyme digestion of murein contains two disaccharide units of murein and is aggregated. This spectrum is typical for an α-helix like that of polylysine at *p*H 11.75. A random-coil conformation would show no negative ellipticity between 210 nm and 250 nm, and a β-structure exhibits only one negative peak at 217 nm. The analysis of this curve according to [31] indicates more than 70% α-helix, no β-structure, and the rest random coil. Aggregation of the lipoprotein molecules could cause light scattering and differential absorption dampening, which result in a decrease of the observed α-helical content [32]. In addition, the spectrum is that of the renatured lipoprotein; thus the actual helical content of the protein in vivo is probably still higher. The molar ellipticities, θ (degree $\times$ cm²/decimole), were calculated from the equation $\theta = \theta/(d \cdot m)$, where θ is the ellipticity in mdegrees, d is the path length in mm, and m is the molarity of amino-acid residues obtained by division of the protein concentration in mg/ml by 110, the average residue molecular weight (see figure 2). The instrument used was the Dichrograph of Joan-Russel, France. The measurement was done at *p*H 7.2 and 22 C.

helical structures. A protein in a membrane does not, for energy reasons, have to fold and to bury hydrophobic side chains in the hydrophobic interior of the protein globule, as water-soluble proteins necessarily do [26]. The hydrophobic side chains could interact with other hydrophobic components of the cell wall in such a way that an α-helix with a hydrophobic and a hydrophilic face could exist.

Conformational studies were also important to explain an intriguing observation. Antibodies made in rabbits against the denatured lipoprotein bound to the native lipoprotein in the cell wall. We now know that the antibodies were produced against a renatured, apparently mostly native lipoprotein. We did immunologic studies to localize the lipoprotein in the membrane system of *E. coli*. We were interested in determining whether the lipoprotein molecules extend from the murein towards the inside of the cell or the outside or both. By cleavage of the bond between the lipoprotein molecules and the murein with trypsin, a structural distortion of the wall is brought about, which apparently weakens the binding of the cytoplasmic membrane to the wall [4, 10]. The question is whether the lipoprotein is directly or indirectly involved in the binding of the cytoplasmic membrane to the wall. To decide whether the lipoprotein is built into the cell wall or whether it is primarily interacting with the cytoplasmic membrane, we cleaved the underlying murein net with lysozyme, and, after separation of the wall and membrane according to the method of Osborn [27], the amount of lipoprotein was determined immunologically in both membranes. We found, by passive hemagglutination, 20–30 times more lipoprotein in the cell wall than in the cytoplasmic membrane (V. Bosch and V. Braun, unpublished observations). If it were interacting primarily with the cytoplasmic membrane, the lipoprotein would have been found there after release from its anchorage, the murein net.

Having localized the lipoprotein in the cell wall, we wanted to determine how far it extends towards the surface of the cell. From heat-inactivated strains of *E. coli* and *Salmonella* enveloped with polysaccharide capsules, strains without capsules, strains with a complete lipopolysaccharide R-core but deficient in O-specific side chains, and strains with incomplete R-core antisera were produced in rabbits and tested with lipoprotein-coated human

Table 1. Reaction of antiserum from heat-inactivated *Escherichia coli* or *Salmonella* with lipoprotein.

Strains	Titer (PHA)*
S-form with capsule	10–40
S-form without capsule	40–160
R-forms, complete R-core	160–1,280
R-forms, incomplete R-core	640–2,560

* PHA = phytohemagglutinin; S = smooth; R = rough.

erythrocytes for passive hemagglutination.[2] As shown in table 1, antisera produced from encapsulated strains contain nearly no antibodies against lipoprotein, and in smooth strains without capsules, only a low titer was found. R-forms, especially those with incomplete R-core, showed increasingly high titers. That this was not mainly due to the different responses of the rabbit immune system towards encapsulated S- and R-forms was shown by experiments with antiserum produced against the isolated lipoprotein and tested with living bacteria as antigens.[3] The results agree qualitatively with those listed in table 1. This demonstrates that the lipoprotein is directed towards the cell surface, but that it is buried in the cell wall.

Supporting chemical findings, where in closely related strains of *E. coli* this lipoprotein was found with only few amino acid differences [10], the immunologic cross-reaction of lipoprotein antisera shows that, for this group of Enterobacteriaceae, the lipoprotein is a common antigen.

Is it possible to obtain a more precise idea about the distribution of lipoprotein molecules in the cell wall? About 250,000 molecules are present in one cell wall. Since no patches can be observed when the isolated rigid layer is studied with the electron microscope, and since the number of molecules is very high, an even distribution is very likely. We know from chemical studies that, on the average, one lipoprotein molecule is linked to every 10th–12th repeating disaccharide unit of the murein. One repeating unit is 10.3 A in length; therefore, the lipoprotein molecules are probably spaced 100–120 A apart along the polysaccharide chains. The polysaccharide chains, according to the results mentioned, are in a one-layered murein 12–

[2] V. Braun, H. Mayer, and S. Schlecht, "Immunologic Localisation of the Murein-Lipoprotein in the Cell Wall of *E. coli* and *Salmonella*," manuscript in preparation.
[3] See footnote 2.

13 A apart. A detailed view of the distribution pattern of the lipoprotein molecules follows and is schematically represented in figure 6.

The lipoprotein is covalently fixed with its C-terminal end on the murein, and the lipid is bound at the N-terminal serine. This lipid, containing mainly palmitic acid bound to an unknown compound, perhaps an α-aminopolyhydroxy carbonic acid, could reach quite far towards the cell surface and interact there with the lipid A of the lipopolysaccharide [28], phospholipids, and other hydrophobic components. These lipids could be a part of a lipid bilayer. Electron microscopy reveals two typical membrane layers in the cell wall [6–8]. The hydrophobic and hydrophilic properties of the lipoprotein (also reflected in its α-helical polypeptide chain) make it quite suitable for a variety of polar and nonpolar interactions with other cell-wall components. It is probably a structural membrane protein. Its pattern of distribution could be representative of the molecular organization of the whole cell wall, if the lipoprotein molecules form the covalently anchored core of subunits of which the cell wall is built. The other main components of the wall would be organized around the lipoprotein by noncovalent forces. This view is schematically represented in figure 1, in which the known components are drawn in black. This picture leaves space for structural heterogeneities to account for the various functions of the wall. The idea underlying this model can be checked when more is known about other cell-wall constituents, especially proteins, with respect to their number and stoichiometric relation to the murein-lipoprotein.

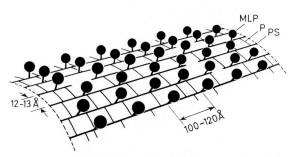

Figure 6. Tentative distribution of the lipoprotein molecules over the murein net. The murein is drawn without the details of figure 2, since the emphasis lies on the netlike structure over which the lipoprotein molecules are distributed. It is proposed that the spacing of the lipoprotein molecules along the polysaccharide chains is 100–120 A and that the polysaccharide chains are 12–13 A apart. The direction in which the polysaccharide chains run in this section of the rigid layer is not known. They could run parallel to the *Escherichia coli* rod as drawn here but also in any other direction.

References

1. Singer, S. J., Nicolson, G. L. The fluid mosaic model of the structure of cell membranes. Science 175: 720–731, 1972.
2. Overath, P., Schairer, H.-U., Hill, F., Lamnek-Hirsch, I. Structure and function of hydrocarbon chains in bacterial phospholipids. *In* D. F. Holzl-Wallach and H. Fischer [ed.] 22. Colloquium der Gesellschaft für Biologische Chemie, Mosbach. Springer, Berlin, 1971, p. 149–164.
3. McConnell, H. M. Molecular motion in biological membranes. *In* F. O. Schmitt [ed.] The neurosciences: second study program. Rockefeller University Press, New York, 1970, p. 697–706.
4. Braun, V., Rehn, K. Chemical characterization, spatial distribution and function of a lipoprotein (murein-lipoprotein) of the *E. coli* cell wall. The specific effect of trypsin on the membrane structure. Eur. J. Biochem. 10:426–438, 1969.
5. Weidel, W., Pelzer, H. Bagshaped macromolecules —a new outlook on bacterial cell walls. Adv. Enzymol. 26:193–232, 1964.
6. Murray, R. G. E., Steed, P., Elson, H. E. The location of the mucopeptide in sections of the cell wall of *Escherichia coli* and other gram-negative bacteria. Can. J. Microbiol. 11:547–560, 1965.
7. de Petris, S. Ultrastructure of the cell wall of *Escherichia coli* and chemical nature of its constituent layers. J. Ultrastruct. Res. 19:45–83, 1967.
8. Frank, H., Dekegel, D. Electron microscopical studies on the localisation of the different components of cell walls of gram-negative bacteria. Folia Microbiol. (Praha) 12:227–233, 1967.
9. Martin, H. H. Biochemistry of bacterial cell walls. Annu. Rev. Biochem. 35:457–484, 1966.
10. Braun, V., Rehn, K., Wolff, H. Supramolecular structure of rigid layer of the cell wall of *Salmonella, Serratia, Proteus* and *Pseudomonas fluorescens*. Number of lipoprotein molecules in a membrane layer. Biochemistry 9:5041–5049, 1970.
11. Balyuzi, H. H. M., Reaveley, D. A., Burge, R. E. X-ray diffraction studies of cell walls and peptidoglycans from gram-positive bacteria. Nature [New Biol.] 235:252–253, 1972.
12. Schwarz, U., Leutgeb, W. Morphogenetic aspects of murein structure and biosynthesis. J. Bacteriol. 106:588–595, 1971.
13. Henning, U., Rehn, K., Braun, V., Höhn, B., Schwarz, U. Cell envelope and shape of *Escherichia coli* K12. Eur. J. Biochem. 26:570–586, 1972.

8

14. Izaki, K., Matsuhashi, M., Strominger, J. L. Biosynthesis of the peptidoglycan of bacterial cell walls. 8. Peptidoglycan transpeptidase and D-alanine carboxypeptidase: penicillin-sensitive enzymatic reaction in strains of *Escherichia coli.* J. Biol. Chem. 243:3180–3192, 1968.

15. Caspar, D. L. D., Klug, A. Physical principles in the construction of regular viruses. Cold Spring Harbor Symp. Quant. Biol. 27:1–24, 1962.

16. Houwink, A. L. A macromolecular mono-layer in the cell wall of Spirillum spec. Biochim. Biophys. Acta 10:360–366, 1953.

17. Brinton, C. C., Jr., McNary, J. C., Carnahan, J. Purification and in vitro assembly of a curved network of identical protein subunits from the outer surface of a bacillus (abstract). Bacteriol. Proc. 69:48, 1969.

18. Nermut, M. V., Murray, R. G. E. Ultrastructure of the cell wall of *Bacillus polymyxa.* J. Bacteriol. 93:1949–1965, 1967.

19. Carlström, D. The polysaccharide chain of chitin. Biochim. Biophys. Acta 59:361–364, 1962.

20. Malamy, M. H., Horecker, B. L. Release of alkaline phosphatase from cells of *Escherichia coli* upon lysozyme spheroplast formation. Biochemistry 3:1889–1893, 1964.

21. Brockman, R. W., Heppel, L. A. On the localization of alkaline phosphatase and cyclic phosphodiesterase in *Escherichia coli.* Biochemistry 7:2554–2562, 1968.

22. Bell, R. M., Mavis, R. D., Osborn, M. J., Vagelos, P. R. Enzymes of phospholipid metabolism: localization in the cytoplasmic and outer membrane of the cell envelope of *Escherichia coli* and *Salmonella typhimurium.* Biochim. Biophys. Acta 249:628–635, 1971.

23. Kelemen, M. V., Rogers, H. J. Three-dimensional molecular models of bacterial cell wall muco-peptides (peptidoglycans). Proc. Natl. Acad. Sci. U.S.A. 68:992–996, 1971.

24. Braun, V., Hantke, K. Evidence for a covalent bond between lipid and a membrane protein (rigid layer of the cell wall of *E. coli*). *In* H. Peeters [ed.] 19th Proceedings protides of the biological fluids. Academic Press, New York, 1972, p. 221–224.

25. Braun, V., Bosch, V. Repetitive sequences in the murein-lipoprotein of the cell wall of *Escherichia coli.* Proc. Natl. Acad. Sci. U.S.A. 69:970–974, 1972.

26. Perutz, M. F., Kendrew, J. C., Watson, H. C. Structure and function of haemoglobin. Some relations between polypeptide chain configuration and amino acid sequence. J. Mol. Biol. 13:669–678, 1965.

27. Osborn, M. J., Gander, J. E., Parisi, E., Carson, J. Mechanism of assembly of the outer membrane of *Salmonella typhimurium.* Isolation and characterisation of cytoplasmic and outer membrane. J. Biol. Chem. 247:3962–3972, 1972.

28. Westphal, O., Lüderitz, O. Chemische Erforschung von Lipopolysacchariden gramnegativer Bakterien. Angew. Chem. 66:407–417, 1954.

29. Higgins, M. L., Shockman, G. D. Procaryotic cell division with respect to wall and membranes. Critical Reviews in Microbiology 1:29–72, 1971.

30. Braun, V., Sieglin, U. The covalent murein-lipoprotein structure of the *Escherichia coli* cell wall. The attachment site of the lipoprotein on the murein. Eur. J. Biochem. 13:336–346, 1970.

31. Greenfield, N., Fasman, G. D. Computed circular dichroism spectra for the evaluation of protein conformation. Biochemistry 8:4108–4116, 1969.

32. Urry, D. W. Protein conformation in biomembranes: optical rotation and absorption of membrane suspensions. Biochim. Biophys. Acta 265:115–168, 1972.

Lipid A: Chemical Structure and Biological Activity

Otto Lüderitz, Chris Galanos, Volker Lehmann,
Marjatta Nurminen, Ernst T. Rietschel,
Günter Rosenfelder, Markus Simon, and
Otto Westphal

*From the Max-Planck-Institut für Immunbiologie,
Freiburg, Germany*

Further details of the chemical structure of lipid A of *Salmonella* have been evaluated. It was found that pyrophosphate bridges interlink β-1,6-linked glucosamine disaccharide units, which are esterified by lauric, palmitic, and 3-D-myristoxymyristic acid and which are substituted at the amino groups by 3-D-hydroxymyristic acid. Lipid A is the biologically active component of lipopolysaccharides. Free lipid A, solubilized by complexing with bovine serum albumin, exhibits strong endotoxic activity in a number of biological tests. Lipid A exhibits immunogenic properties when suitably exposed on the bacterial surface. Immunization of rabbits with lipid A leads to the production of specific antibodies to lipid A that are capable of cross-reacting with a large variety of lipopolysaccharides.

Lipopolysaccharides are located in the cell wall of gram-negative bacteria where they form, complexed with lipids and proteins, the outer membrane of the cell. They represent the O antigens and the endotoxins of these organisms (for reviews see [1]).Lipopolysaccharides consist of three regions of contrasting chemical and biological properties. The O-specific polysaccharide (region I), carrying the main serologic specificity, is linked to the core polysaccharide (region II), which is common to groups of bacteria. The core is linked through a 2-keto-3-deoxyoctonate (KDO) trisaccharide to the lipid component (region III) termed lipid A [2]. Biochemical and genetic investigations on endotoxins performed in the last decades have led to an understanding of the detailed structure, the biosynthesis, and the underlying genetics of the polysaccharide component of salmonella lipopolysaccharides. These results were obtained by the use of so-called rough (R) mutants that are defective in the synthesis of the complete lipopolysaccharide. Thus, mutants were isolated from strains of *Salmonella, Escherichia coli, Shigella flexneri,* and other bacteria whose lipopolysaccharides lack the O-specific chains or additional parts of the core. However, knowledge about the chemistry of the lipid A component of lipopolysaccha-

rides was rather fragmentary until recent efforts in various laboratories led to elucidation of the structure and biosynthesis of this part of endotoxins. Simultaneously, biological studies on lipid A were resumed; these had led 15 years ago to strong indications that lipid A played a role in endotoxicity. This concept, however, was not generally accepted at that time [3].

The present paper summarizes recent results obtained in our laboratory on the chemistry and biology of lipid A. Most studies were performed on *Salmonella,* but comparison will be made with other bacterial groups.

Chemical Structure of Lipid A

Isolation of lipid A. In principle there are two sources of lipid A. A number of mutant strains are known that are devoid of heptose (Hep) and therefore synthesize a defective lipopolysaccharide containing only lipid A and KDO. These natural products, generally designated as glycolipids, proved to represent good starting materials for investigations on lipid A [4].

Alternatively, lipid A may be isolated from any lipopolysaccharide by hydrolysis with mild acid [5, 6]. This treatment cleaves the ketosidic linkage of KDO to lipid A. In this way water-insoluble preparations of lipid A are obtained. Due to the acidic conditions of isolation, these preparations may be degraded to some extent. In the case of

Please address requests for reprints to Dr. Otto Lüderitz, Max-Planck-Institut für Immunbiologie, D-78 Freiburg/Br, Stübeweg 51, Germany.

lipopolysaccharides containing more stable linkages to lipid A, more drastic conditions of hydrolysis are necessary, and some destruction is inevitable (G. Rosenfelder, unpublished observations).

Lipopolysaccharides are generally extracted in our laboratory with phenol and water (in the case of smooth (S) forms [7]) and with a phenol-chloroform-petroleum ether mixture (in the case of R forms [8]). After ultracentrifugation, the preparations may be purified further by electrodialysis (C. Galanos, unpublished observations). Metal ions and amines like ethanolamine (EtN), spermidine, and putrescine, which contaminate lipopolysaccharides, are thus removed. With some S and R lipopolysaccharides, this treatment leads to their deposition as an insoluble gel on the anode membrane, from which they can be collected. The material can then be solubilized by conversion to the sodium salt.

Structure of the lipid A backbone. It is known that salmonella lipid A contains glucosamine, phosphate, and long-chain fatty acids [5, 9]. Ethanolamine, which is present in the core polysaccharide (where it is linked to heptose through a pyrophosphate and to KDO through a phosphate bridge), also occurs as a constituent in the lipid A region linked in a still unknown way.

Regarding the structure of lipid A, two proposals were made that, at first sight, seemed to exclude each other. Nowotny [10] concluded from his findings that glucosamine residues in lipid A were linked together by 1, 4 bridges of phosphate diester groups. On the other hand, results of Burton and Carter [11] suggested the existence of glycosidic linkages between glucosamine (GlcN) units. As will be seen below, both kinds of linkages probably occur in lipid A.

Most of the more recent structural studies were performed on the glycolipids of Re mutants of *Salmonella* containing KDO and lipid A or of Rd$_2$ mutants containing additional heptose. Removal of the long-chain fatty acids by hydrazinolysis led to the recognition of the backbone structure of lipid A units [4, 12]. Pentasaccharides of the formula (KDO)$_3$-2,3-GlcN-β1,6-GlcN, phosphorylated in position 4′ and/or 1, were isolated. These oligosaccharides could be converted by mild hydrolysis to the corresponding KDO-free phosphorylated glucosamine disaccharides.

Studies on model substances like RNA, cephalin, uridine diphosphate-N-acetyl glucosamine (UDP-GlcNAc), or Hep-Hep-(-P-P-EtN)KDO revealed that, under the conditions used, treatment with hydrazine for 10 hr led to a cleavage of pyrophosphate bridges, while phosphate bridges were attacked much more slowly. Hydrazinolysis was therefore performed under milder conditions (2 hr) with the glycolipid of an Rd$_2$ mutant (V. Lehmann, unpublished observations). After chromatography of the hydrazinolysate on Sephadex, three fractions (A–C) were isolated. Their structures were identified by application of hydrolysis with mild acid, treatment with phosphatase, and deamination with nitrous acid. Fraction A (figure 1) consisted of phosphorylated oligosaccharides corresponding to those mentioned above. Fraction B, which contained heptose, KDO, glucosamine, and phosphate in a ratio of 2:6:4:3 (or 4), could be identified as a dimer of fraction A interlinked by a pyrophosphate bridge. Fraction C, having the same molar ratio of constituents as B, represents an oligomer of several repeating units. In addition, fraction C contains 1 mole of amide-linked 3-hydroxymyristic acid per 2 moles of glucosamine.

These studies also confirmed our previous findings [13] that the three residues of KDO detec-

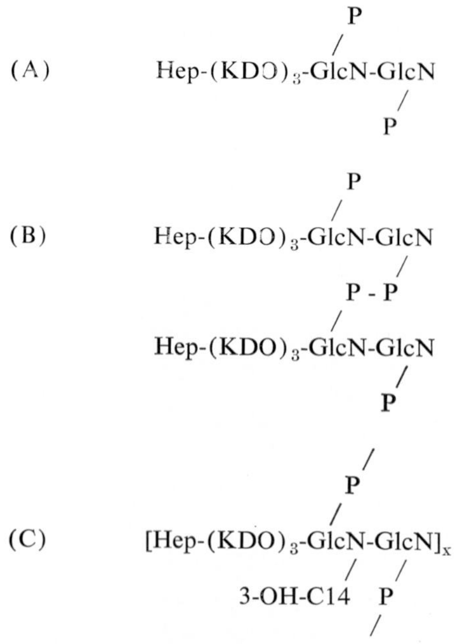

Figure 1. Three fragments isolated after limited hydrazinolysis of the glycolipid of a *Salmonella minnesota* Rd$_2$ mutant. Key to abbreviations used in figures: Hep = heptose; KDO = 2-keto-3-deoxyoctonate; GlcN = glucosamine; P = phosphate; EtN = ethanolamine; 3-OH-C14 = 3-D-hydroxymyristoyl.

table in glycolipids form a trisaccharide. When fraction A (or the corresponding heptoseless pentasaccharide from a Re mutant) was successively treated with nitrous acid (deamination, formation of substituted anhydromannose), and alkali (β-elimination of the oligosaccharide from position 3′), the oligosaccharides Hep-(KDO)$_3$ and (KDO)$_3$, respectively, could be isolated and identified after reduction with NaBH$_4$ (V. Lehmann, unpublished observations, see figure 2). It is the first time that KDO oligosaccharides were isolated from lipopolysaccharides.

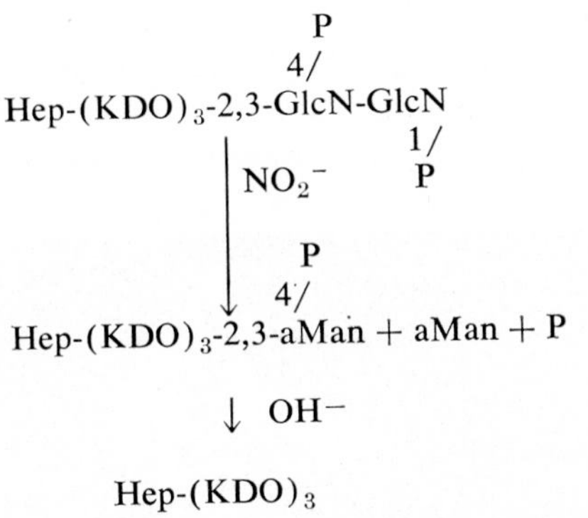

Recently, Adams and Singh have shown that, in the lipopolysaccharide of *Serratia marcescens,* the lipid A backbone also contains β1,6-linked glucosamine units [14, 15], while in *E. coli* 086 and *S. flexneri,* glucosamine oligomers with β1,4 linkages are present [16].

Nature and linkages of fatty acids. Lauric, myristic, palmitic, and 3-D-(—) hydroxymyristic acid represent the long-chain fatty-acid constituents in lipid A of *Salmonella.* These fatty acids are released on acid hydrolysis, while treatment with alkali liberates, in addition, Δ²-tetradecenoic acid (α,β-unsaturated myristic acid [17]). It was concluded that this unsaturated acid was an artifact derived by β-elimination from O-acylated 3-hydroxymyristic acid ester. The substituent of the hydroxy fatty acid was then identified as myristic acid. Finally, it could be shown that after treatment of the glycolipid with mild alkali, 3-myristoxymyristic acid was liberated.

These findings, and the fact that the amino groups of the glucosamine units are substituted by 3-hydroxymyristic acid residues, led to the formula of a lipid A unit represented in figure 3 [17].

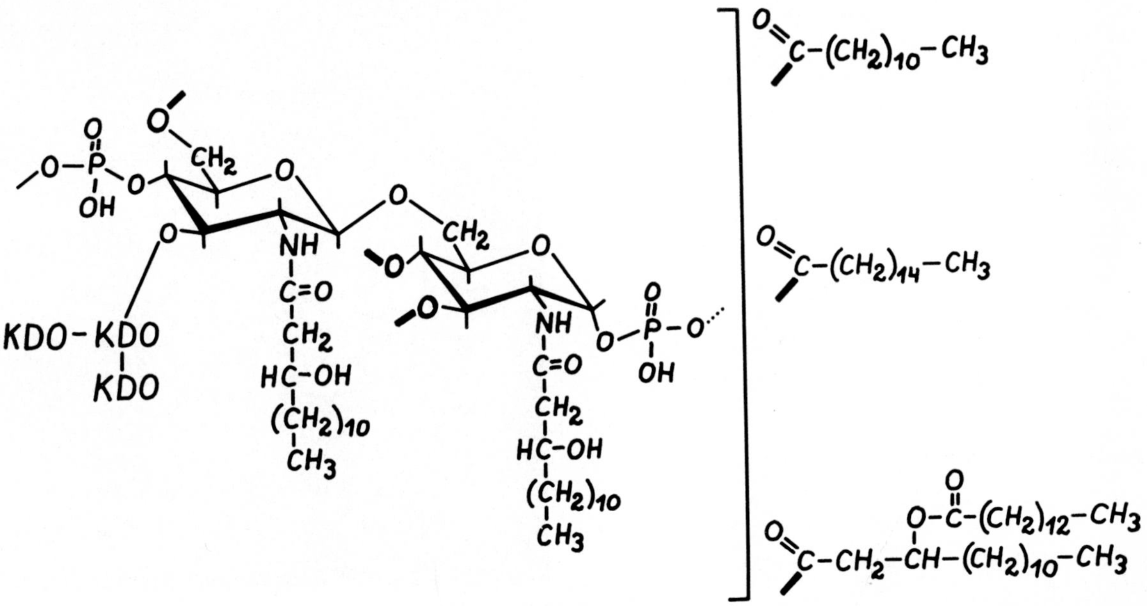

$$
\begin{array}{c}
\text{P} \\
4/ \\
\text{Hep-(KDO)}_3\text{-2,3-GlcN-GlcN} \\
\qquad\qquad\qquad 1/ \\
\Big\downarrow \text{NO}_2^- \qquad \text{P} \\
\\
\text{P} \\
4/ \\
\text{Hep-(KDO)}_3\text{-2,3-aMan} + \text{aMan} + \text{P} \\
\\
\downarrow \text{ OH}^- \\
\\
\text{Hep-(KDO)}_3
\end{array}
$$

Figure 2. Isolation of a Hep-(KDO)$_3$ oligosaccharide from a *Salmonella minnesota* Rd$_2$ glycolipid. Similarly, a KDO trisaccharide was isolated from a Re glycolipid. (aMan = 2,5-anhydromannose.)

Figure 3. Proposed structure of a lipid A unit of the *Salmonella minnesota* R595 glycolipid with an attached KDO trisaccharide. The three fatty-acid residues shown are linked in an unknown distribution to the hydroxyl groups of the glucosamine residues available at positions 3, 4, and 6′. Additional small amounts of myristic acid and unsubstituted 3-hydroxymyristic acid may occur as esters linked to glucosamine.

12

The three available hydroxyl groups in positions 6', 3, and 4 of the glucosamine disaccharides are esterified by about equal amounts of lauric, palmitic, and 3-myristoxymyristic acid (and small amounts of myristic and 3-hydroxymyristic acid). The amino groups are substituted exclusively with 3-hydroxymyristic acid residues, whose hydroxyl group is free.

The investigation of lipopolysaccharides of genera other than *Salmonella* clearly revealed that the nature and distribution of fatty acids may vary [18]. Thus, different or no hydroxy fatty acids may be present in the lipid A structure from other bacterial groups (table 1).

It is well known that the removal of ester-linked fatty acids by treatment with mild alkali drastically reduces the biological activity of lipopolysaccharides; this indicates that fatty-acid residues are important for endotoxicity [19]. However, it seems that not only the quantity but also the nature of the ester residues is of importance, and there are some preliminary indications that the presence or absence of acylated hydroxy fatty-acid esters may determine the degree of endotoxicity of a lipopolysaccharide (E. T. Rietschel, unpublished observations).

On the basis of our present knowledge of the structure of lipid A, an atomic model of an Ra lipopolysaccharide has been built (figure 4). It can be seen that the oligosaccharide extends in one direction, while the fatty acids can easily be directed to the other. A corresponding formula of the inner-core lipid A region of salmonella lipo-

Table 1. Summarized data on fatty-acid composition of lipid A from various sources [17, 18].

Fatty acid*	Source
C12, C14, C16 **3-OH-C14**	*Salmonella* [2, 41–44] *Escherichia coli* [45–51] *Proteus mirabilis* [52] *Aerobacter aerogenes* [53, 54] *Bordetella pertussis* [48]
C12, C14, C16 3-OH-C12 3-OH-C14	*Serratia marcescens* [55–57] *Neisseria perflava* [58]
C12 (C16) 3-OH-C10 (2-OH-C12) **3-OH-C12†**	*Pseudomonas aeruginosa* [59–61] *Pseudomonas alcaligenes* [62] *Azotobacter agilis* [63]
C12 3-OH-C10 3-OH-C14	*Rhodopseudomonas capsulata* [64]
C13 **3-OH-C13** **3-OH-C15**	*Veillonella* [65, 66]
(C16), C15,‡ C17‡ 3-OH-C16 **3-OH-C15‡** **3-OH-C17‡**	*Myxococcus fulvus* [67] *Polyangium* [67] *Flexibacter* [67] *Cytophaga* [67]
C16, C17, C18	*Brucella melitensis* [68] *Brucella abortus* [69]
C12, C16, C18, C22 3-OH-C14	*Anacystis nidulans* [70]

* The number that follows the letter C indicates the number of carbon atoms in each fatty acid. Thus, C14 is myristic acid; 3-OH-C14 is 3-hydroxymyristic acid.
† Bold-face type indicates amide-linked fatty acids.
‡ Iso-branched fatty acids.

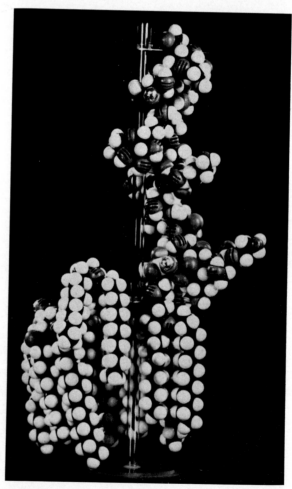

Figure 4. Atomic model of the partial structure of a salmonella Ra lipopolysaccharide. Shown is an acylated diglucosamine unit (lipid A unit) carrying the Ra oligosaccharide at position 3', and a second acylated diglucosamine unit linked at position 1 through a pyrophosphate bridge.

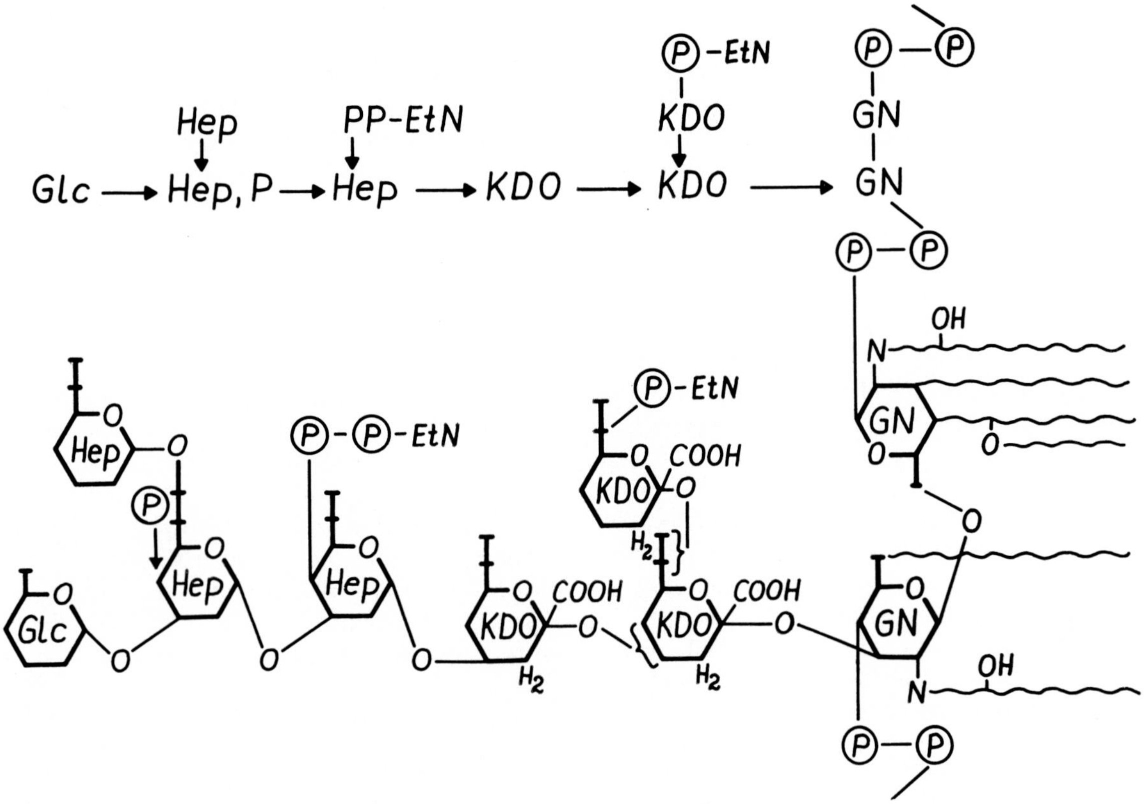

Figure 5. Structure of the inner core and lipid A of a salmonella Rc glycolipid. The presence of a third heptose unit and a phosphate group on heptose II was identified recently in mutants of higher R classes (V. Lehmann, unpublished observations). 4-Amino-arabinose, which occurs in salmonella glycolipids in an unknown linkage, is not shown [2].

polysaccharides is shown in figure 5. A series of partial degradation products of the lipid A part of salmonella glycolipids, obtained after treatment with sodium hydroxide, hydroxylamine, or hydrazine, has been identified. The changes in structure after such treatments are given in table 2. From these compounds the corresponding sugar-free lipid A fragments can be obtained by treatment with acetic acid; this results in a cleavage of the KDO linkages and an additional partial cleavage of pyrophosphate linkages.

Possible linkage of lipopolysaccharide to other cell-wall constituents. The fact that defective lipopolysaccharides (glycolipids) from R mutants can be extracted with a mixture of phenol-chloroform-petroleum ether (PCP) under very mild conditions (4 C) indicates that glycolipids are not covalently linked to the bacterial cell wall [8]. On the other hand, S-form lipopolysaccharides, which are not soluble in PCP, are often extracted with

Table 2. Series of available degradation products of lipid A derived from a glycolipid of *Salmonella*.

Treatment of glycolipid	Changes in structure
None	See formula, figures 3 and 6
0.25 N NaOH, 56 C, 5 min	Loss of (mainly) myristoxy-myristic acid [17]
0.25 N NaOH, 56 C, 60 F	Partial (about 50%) loss of ester-linked fatty acids [17, 19]
Hydroxylamine	Loss of ester-linked fatty acids, and, possibly, cleavage of pyrophosphate linkages [17]
Hydrazine (2 hr)	Loss of ester-linked and partial loss of amide-linked fatty acids (figure 1C)
Hydrazine (12 hr)	Loss of all fatty acids and cleavage of pyrophosphate bridges (figure 1A)

phenol and water at 68 C [7]; in this case linkages may be cleaved. Other, mostly milder, methods of extraction often yield preparations containing lipopolysaccharide, protein, and cephalin (trichloroacetic acid [20], ethylene glycol [21]). Although it has been assumed that these mixtures represent complexes formed by hydrophobic interactions, the existence of covalent linkages between the components could not be excluded.

Recently, Wober and Alaupović [22, 23] and Tarcsay et al.[1] have studied such "complexes." From their results with *S. marcescens,* the authors concluded that the lipopolysaccharide is covalently linked to protein through lipid A, probably by a N-glycosidic linkage of reducing-terminal glucosamine residues to asparagine.

On the basis of these findings, it seems possible that lipopolysaccharide is synthesized on protein-linked lipid A. In subsequent steps the linkage to protein would be cleaved and the S or R lipopolysaccharide transported to the outer wall. In this case only a small fraction of lipopolysaccharide would be found to be linked to membrane protein. Alternatively, one could assume that lipopolysaccharide is synthesized in the membrane and is not attached to protein. After transport to the surface, only completed S lipopolysaccharide is linked to protein.

Studies analogous to those of Wober, Tarcsay, and Alaupović [22, 23][2] are presently being done on R mutants.

Biosynthesis of lipid A. While the biosynthesis and genetics of the O-specific chains and the core of salmonella lipopolysaccharides have been studied intensively, not much is known about how lipid A is synthesized. Heath et al. [24] have demonstrated the transfer of KDO from cytidine monophosphate-KDO to a fragment of lipid A identified as an oligomer of N-3-hydroxymyristoyl-glucosamine. These results may suggest that addition of KDO occurs at an early stage in synthesis of lipid A, before incorporation of ester-linked fatty acids. Later, Taylor and Heath [25] studied the enzymatic incorporation of 3-hydroxy fatty acids from the corresponding acyl-carrier-protein derivative into a phospholipid of *E. coli* B. The authors discuss the possibility that this phospho-

lipid may function as an acyl donor for glucosamine residues of lipid A.

Very recently, Humphreys and Meadow [18, 26] demonstrated the incorporation of 3-hydroxylauric acid from 3-hydroxylauryl-coenzyme A into an EDTA preparation of *Pseudomonas alcaligenes* in whose lipid A component this is the main hydroxy fatty acid. On the basis of their results, obtained with the radioactive lipopolysaccharide isolated from the incubation mixture, the authors concluded that the ^{14}C-hydroxy fatty acid was incorporated into lipid A.

We have shown recently that an enzyme preparation of a mutant of *Salmonella typhimurium* defective in glucosamine-6-phosphate deaminase [27] catalyzes the synthesis of uridine diphosphate-glucosamine from uridine triphosphate, adenosine triphosphate, and glucosamine (M. Simon, unpublished observations). We are investigating the question of whether UDP-GlcN may function as a precursor for the synthesis of the lipid A backbone.

Neither KDO- nor lipid A-defective mutants have been isolated so far. It seems that the inner regions of lipopolysaccharides have functions important for the cells and are necessary for their survival. N-acetylglucosamine–negative mutants have been described; they are defective in glucosamine-6-phosphate deaminase [27]. Recently, mutants lacking glucosamine-6-phosphate synthetase were identified [28, 29]. These mutants should be useful for studying early steps in the biosynthesis of lipid A and for investigating the question of whether a lipid A pool exists in normal, growing cells.

Biological Activity of Lipid A

Anticomplementary activity of lipopolysaccharides. It has long been known that lipopolysaccharides can interact with complement in vitro in the absence of added antibody, resulting in loss of hemolytic activity. It was also suggested that the inactivation of complement by lipopolysaccharide might be a measure of its endotoxicity, because certain biological activities that are manifested in vivo during infection by gram-negative bacteria were also shown to follow the in-vitro interaction of complement with lipopolysaccharide [32].

[1] L. Tarcsay, C. S. Wang, and P. Alaupović, manuscript in preparation.

[2] See footnote 1.

We have tested the anticomplementary activity of a large number of S and R lipopolysaccharides derived from *Salmonella, E. coli,* and *Arizona* ([33] and C. Galanos, unpublished observations). Only about one third of these proved to be highly anticomplementary (figure 6). The others exhibited only low anticomplementary activity, no matter how much lipopolysaccharide was used. In fact, after a maximum was reached with a given amount of lipopolysaccharide, addition of more resulted in fixation of less complement; (figure 6, curve B). Therefore, the difference between "active" and "inactive" lipopolysaccharides is qualitative and not quantitative. The physicochemical state of a lipopolysaccharide, the presence of divalent ions and basic amines, and the batch of guinea-pig serum used may influence the anticomplementary activity of a lipopolysaccharide. Further studies on these phenomena are under way.

We investigated the presence of natural antibodies in the guinea-pig serum used as a source of complement; these antibodies may be partly involved in the interaction of complement and lipopolysaccharide. This was done by tests of the hemolytic activity of the guinea-pig serum with ovine erythrocytes that were passively sensitized with various alkali-treated lipopolysaccharides. It is known that treatment with alkali abolishes the anticomplementary activity of a lipopolysaccharide [33]. Titers ranging from 1:8 to 1:128 were found against 52 of 54 lipopolysaccharide preparations obtained from *Salmonella, Arizona,* and *E. coli.* It is evident that natural antibody activity in guinea-pig serum is at least partly involved in the interaction of complement with lipopolysaccharide (C. Galanos, unpublished observations). It was shown that free solubilized lipid A (see below) obtained by acid hydrolysis of various lipopolysaccharides (in complex form with solubilizing carriers) was highly anticomplementary, even when lipid A was obtained from lipopolysaccharides which themselves were inactive [33]. Therefore, the region in the lipopolysaccharide molecule responsible for anticomplementary activity is the lipid A.

Endotoxic activity of lipid A. We used two approaches to identify the biologically active center of lipopolysaccharides. First, we modified chemically the KDO residues in an Re glycolipid to determine whether or not the oligo(poly)saccharide portion of lipopolysaccharides plays a specific role in endotoxicity. Secondly, free lipid A was tested directly for endotoxic activity.

Table 3 shows the results obtained with chem-

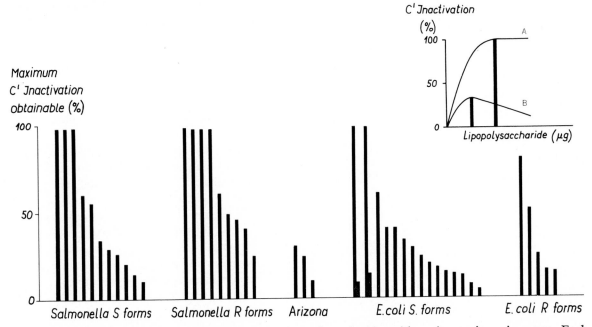

Figure 6. Anticomplementary activity of various lipopolysaccharides with a given guinea-pig serum. Each column denotes maximal activity obtainable with each preparation that cannot be surpassed by higher concentrations of lipopolysaccharide [33]. The two curves are meant to illustrate this.

Table 3. Biological properties of unaltered and modified Re glycolipid of *Salmonella minnesota*.

Glycolipid	HAI (μg/ml)*	50% inactivation of complement (μg)	Mouse lethality LD_{50}[†] (μg)	Chick-embryo lethality LD_{50} (μg)[‡]	Pyrogenicity LD-3[§] (μg/kg)
Unaltered	8	4	15	0.006	0.0035
N-dinitrophenylated	2	5	20	0.001	ND‖
O- and N-succinylated	>250 (16)**	22	12	ND	0.002
Carboxymethylated	>250 (8)**	5	80	ND	ND

* Minimal amount of inhibitor that prevented hemagglutination.
† Mouse lethality tests were performed at 37 C.
‡ Tests performed in Dr. A. Nowotny's laboratory.
§ Test performed in laboratory of Drs. D. W. Watson and Y. B. Kim. LD-3 = minimal pyrogenic dose after 3 hr.
‖ ND = not done.
** Preparation tested after treatment with alkali.

ically modified glycolipid [30]. Amino groups (present in the form of KDO-linked phosphoryl-ethanolamine residues) were dinitrophenylated, hydroxyl groups (and amino groups) of KDO (and ethanolamine) were succinylated, and the carboxyl groups of KDO were converted into their methyl esters. It was found that these modifications did not significantly change the activity of the glycolipid when tested for lethality to mice or chick embryos, pyrogenicity, or inactivation of complement. However, the serologic activity of the parent glycolipid was completely lost after succinylation and carboxymethylation. That this was due to the substituents introduced into the molecule could be shown by treatment with alkali, which simultaneously removed the substituents and restored full serologic activity [30]. These results showed that chemical modification of the oligosaccharide portion, even to a degree where the original serologic specificity was completely lost, did not result in a significant alteration of the endotoxic activity of the glycolipid. It is concluded that the polysaccharide part of lipopolysaccharides is not specifically involved in endotoxicity.

A new approach was used to demonstrate directly the role of lipid A as the endotoxic principle. This made it possible to obtain free lipid A in a stable, soluble form [3, 31, 33]. Lipid A as obtained from lipopolysaccharides (R form) after hydrolysis with mild acid is insoluble in water and, therefore, biologically inactive. It was found that lipid A can be solubilized by complexing with certain carriers that are water-soluble and nontoxic by themselves. Complexes of lipid

A with alkali-treated lipopolysaccharides, alkali-treated lipid A, or serum albumin were prepared [33]. For demonstration of their biological activity, inactivation of complement was used as a convenient in-vitro test. The complexes mentioned above were found to be highly active in this test. Subsequent investigations with complexes of lipid A and serum albumin revealed endotoxic activity in a number of in-vivo and in-vitro tests [33, 34]. Table 4 shows that the following assays were performed: mouse lethality, pyrogenicity [34], bone-marrow necrosis,[3] Limulus gelation [35], and complement inactivation [34]. Although the lipid A complexes were found to be in general somewhat less active than the corresponding parent lipopolysaccharides, the results strongly indicate that lipid A represents the biologically active center in endotoxic lipopolysaccharides and that the polysaccharide component acts as a solubilizing carrier that can be replaced by other, noncovalently bound carriers like serum albumin.

Recently, the toxicity of lipopolysaccharide from an Rb mutant of *Salmonella minnesota* and the serum-albumin complex of its lipid A was measured in adrenalectomized mice; LD_{50} values of 0.075 μg and 0.02 μg, respectively, were obtained (C. Galanos and M. Parant, unpublished observations).

Preparation and properties of antiserum to lipid A. Immunologic studies on the lipopolysaccharides (O antigens) of gram-negative bacteria have

[3] M. Yoshida, C. Galanos, E. T. Rietschel, and O. Lüderitz, manuscript in preparation.

Table 4. Biological properties of lipid A-bovine serum albumin complexes as compared with the parent lipopolysaccharides (glycolipids).

Preparation		Mouse lethality LD_{50} (μg)	Pyrogenicity LD-3[*] ($\mu g/kg$)	Bone-marrow reaction ED_{50}[†] (μg)	Complement inactivation[‡] (%)	Limulus-lysate gelation [35] ($\mu g/ml$)
Salmonella minnesota R60	LPS§	300	. . .	. . .	30‖	
	LA/BSA	1,700	0.003	18.8	86	
S. minnesota R345	LPS	300	. . .	6.2	5‖	
	LA/BSA	1,700	0.013	7.2	85	10^{-8}
S. minnesota R5	LPS	650	. . .	. . .	70	10^{-12}
	LA/BSA	4,000	0.007	10.9	85	
S. minnesota R595	LPS	1,750	0.004	3.9	80	10^{-12}
	LA/BSA	4,000	0.007	11.0	86	
Escherichia coli EH100	LPS	. . .	. . .	. . .	15‖	
	LA/BSA	1,700	. . .	. . .	80	
BSA control		>10,000	>100	>100	0	$>10^{-3}$

[*] LD-3 = minimal pyrogenic dose after 3 hr; data from [34].
[†] ED_{50} = 50% effective dose.
[‡] With 10 μg of preparation [33].
§ LPS = lipopolysaccharide; LA = lipid A; BSA = bovine serum albumin.
‖ Larger amounts of these preparations do not lead to inactivation of more complement; in fact, smaller amounts are inactivated.

revealed that distinct structures in the O-specific polysaccharide chains function as the serologic determinants against which the O antibodies are directed. In incomplete lipopolysaccharides of R mutants, oligosaccharides of the basal core represent the determinant regions [2]. In addition to these specific O and R antigens, a number of antigens present on the enterobacterial cell have been described that, under certain conditions, exhibit common specificities, thus causing cross-reactions between species [36]. The best known example is the common antigen of Kunin [37], which is present on the surface of all Enterobacteriaceae. However, this antigen is detectable only with antisera obtained with a limited number of species that carry the common antigen in an immunogenic form. In most strains, its immunogenicity is not expressed due to the presence of lipopolysaccharide [38].

A similar situation seems to exist with regard to lipid A, which is a common component of all lipopolysaccharides and is structurally identical or similar in all Enterobacteriaceae [3]. The formation of antibodies against this common structure of lipopolysaccharides might be expected, but such antibodies were not described previously. In fact, results indicate that the formation of antibodies to lipid A is suppressed to a large extent when immunization is performed with lipopolysaccharides or bacterial cells, in which lipid A carries long polysaccharide chains (S forms) or oligosaccharide units (R mutants). Only recently it was shown that lipid A, when presented to the animal (rabbit) in a suitable form, can act as an immunogen. It was found that immunization with intact *S. minnesota* R595 bacteria, which have the glycolipid KDO-lipid A on their cell surface, resulted in low levels of antibodies to lipid A. Improved titers were obtained by immunization with the same bacteria from which KDO had been cleaved by hydrolysis with mild acid (figure 7). Such hydrolyzed bacteria contain free lipid A in their outer cell membranes. These hydrolyzed bacteria, when coated with additional lipid A that was isolated from a lipopolysaccharide by hydrolysis with mild acid and used for immunization, elicited the highest titers of antibody to lipid A so far obtained (figure 7). Formation of antibody shows a time course resembling that of antibody formation against the polysaccharide structure of lipopolysaccharides [39]. The results show that lipid A, when sufficiently exposed on the bacterial cell, stimulates the formation of specific antibodies in the animal.

18

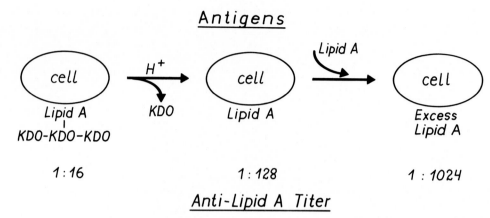

Antigens

1:16   1:128   1:1024

Anti-Lipid A Titer

Figure 7. Preparation of immunogenic lipid A. Modification of Re mutant cells and titers of antibody to lipid A obtained with unaltered cells, hydrolyzed cells, and hydrolyzed cells complexed with lipid A [36].

For determination of anti-lipid A activity in such antisera, the passive-hemolysis test with human red blood cells sensitized to lipid A was used. The sensitivity of the test can be increased significantly by use of ovine erythrocytes (and the antisera absorbed with these cells). Another specific assay for determining anti-lipid A activity involved liposomal model-membranes [39]. Lipid A-sensitized erythrocytes were also used for the absorption of antibodies to lipid A from immune sera and their reisolation. The specificity of the immune sera for lipid A could thus be demonstrated.

Preparations of lipid A obtained from one *E. coli* and a number of salmonella lipopolysaccharides exhibit strong serologic cross-reactions with different antisera to lipid A. All lipopolysaccharides tested cross-react with the antiserum to lipid A, some to a high degree. Similarly, many antibacterial antisera and "normal sera" exhibit activity against lipid A. This means that although lipid A is not always sufficiently exposed on the surface of the bacteria for induction of antibodies, antibodies to lipid A are capable of reacting with lipopolysaccharides, even with those of S forms. It is well known that antibodies directed against surface structures of the bacterial cell play an important role in defense mechanisms against infection [40]. The finding that antibodies to lipid A are capable of reacting with lipopolysaccharides indicated that they might function as opsonins and promote phagocytosis of bacteria. One system has been studied so far. *E. coli* 0111 were treated with antiserum to lipid A, and their fate in the peritoneum of the mouse was followed. It was found that antibodies to lipid A were effective at high

dilutions in sensitizing these bacteria for subsequent phagocytosis and killing [36]. Recent studies have shown that antibodies to lipid A also enhance the blood clearance of heat-killed *Salmonella enteriditis* (C. Galanos and M. Parant, unpublished observations).

Thus, antibodies to lipid A are capable of reacting with lipopolysaccharides on the cell surface of S-form bacteria. How far this is true for Enterobacteriaceae in general is a very important question. Since all enterobacterial cell walls contain lipid A, highly specific lipid A antiserum may be found to be reactive universally with a great number of otherwise serologically distinct species. Studies are under way to find out whether anti-lipid A antibodies can afford protection to enterobacterial infections.

Possible interactions of antibodies to lipid A with the endotoxic properties of lipopolysaccharides or lipid A are another important subject for investigation.

We have preliminary indications that amide-linked 3-hydroxymyristic acid functions as an immunodominant group in lipid A.

References

1. Kadis, S., Weinbaum, G., Ajl, S. J. [ed.] Microbial toxins. Vol. 5. Academic Press, N.Y., 1971. 507 p.
2. Lüderitz, O., Westphal, O., Staub, A. M., Nikaido, H. Isolation and chemical and immunological characterization of bacterial lipopolysaccharides. *In* G. Weinbaum, S. Kadis, S. J. Ajl [ed.] Microbial toxins. Vol. 4. Academic Press, New York, 1971, p. 145–233.
3. Lüderitz, O., Jann, K., Wheat, R. Somatic and capsular antigens of gram-negative bacteria. *In*

M. Florkin and E. H. Stotz [ed.] Comprehensive biochemistry. Vol. 26A. Elsevier, New York, 1968, p. 105–228.

4. Gmeiner, J., Lüderitz, O., Westphal, O. Biochemical studies on lipopolysaccharides of *Salmonella* R mutants. 6. Investigations on the structure of the lipid A component. Eur. J. Biochem. 7:370–379, 1969.

5. Westphal, O., Lüderitz, O. Chemische Erforschung von Lipopolysacchariden gramnegativer Bakterien. Angew. Chemie 66:407–417, 1954.

6. Grollman, A. P., Osborn, M. J. O-phosphorylethanolamine: a component of lipopolysaccharide in certain gram-negative bacteria. Biochemistry 3: 1571–1574, 1964.

7. Westphal, O., Lüderitz, O., Bister, F. Über die Extraktion von Bakterien mit Phenol/Wasser. Z. Naturforsch. 7b:148–155, 1952.

8. Galanos, C., Lüderitz, O., Westphal, O. A new method for the extraction of R lipopolysaccharides. Eur. J. Biochem. 9:245–249, 1969.

9. Ikawa, M., Koepfli, J. B., Mudd, S. G., Niemann, C. An agent from *E. coli* causing hemorrhage and regression of an experimental mouse tumor. III. The component of fatty acids of the phospholipide moiety. J. Am. Chem. Soc. 75:1035–1038, 1953.

10. Nowotny, A. Chemical structure of a phosphomucolipid and its occurrence in some strains of *Salmonella*. J. Am. Chem. Soc. 83:501–503, 1961.

11. Burton, A. J., Carter, H. E. Purification and characterization of the lipid A component of the lipopolysaccharides from *Escherichia coli*. Biochemistry 3:411–418, 1964.

12. Gmeiner, J., Simon, M., Lüderitz, O. The linkage of phosphate groups and of 2-keto-3-deoxyoctonate to the lipid A component in a *Salmonella minnesota* lipopolysaccharide. Eur. J. Biochem. 21:355–356, 1971.

13. Dröge, W., Lehmann, V., Lüderitz, O., Westphal, O. Structural investigations on the 2-keto-3-deoxyoctonate region of lipopolysaccharides. Eur. J. Biochem. 14:175–184, 1970.

14. Adams, G. A., Singh, P. P. The chemical constitution of lipid A from *Serratia marcescens*. Canad. J. Biochem. 48:55–62, 1970.

15. Bundle, D., Shaw, N. Synthesis of 2-acetamido-6-O-(2-acetamido-2-deoxy-β-D-glucopyranosyl)-2-deoxy-D-glucopyranose. Carbohydrate Res. 21: 211–217, 1972.

16. Adams, G. A., Singh, P. P. Structural features of lipid A preparations isolated from *Escherichia coli* and *Shigella flexneri*. Biochim. Biophys. Acta 202:553–555, 1970.

17. Rietschel, E. T., Gottert, H., Lüderitz, O., Westphal, O. Nature and linkages of the fatty acids present in the lipid A component of *Salmonella* lipopolysaccharides. Eur. J. Biochem. 28:166–173, 1972.

18. **Humphreys, G. O. Studies on *Pseudomonal* lipopolysaccharides. Thesis, University of London, 1971.**

19. Neter, E., Westphal, O., Lüderitz, O., Gorzynski, E. A., Eichenberger, E. Studies on enterobacterial lipopolysaccharides. Effects of heat and chemicals on erythrocyte-modifying antigenic, toxic and pyrogenic properties. J. Immunol. 76: 377–385, 1956.

20. Boivin, A., Mesrobeanu, L. Recherches sur les antigènes somatiques et sur les endotoxines des bactéries. I. Considérations générales et exposé des techniques utilisées. Rev. Imumnol. (Paris) 1:553–569, 1935.

21. Morgan, W. T. J. Studies in immuno-chemistry. II. The isolation and properties of a specific antigenic substance from *B. dysenteriae* (*Shiga*). Biochem. J. 31:2003–2021, 1937.

22. Wober, W., Alaupović, P. Studies on the protein moiety of endotoxin from gram-negative bacteria. Characterization of the moiety isolated by phenol treatment of endotoxin from *Serratia marcescens* 08 and *Escherichia coli* 0141: K85 (B). Eur. J. Biochem. 19:340–356, 1971.

23. Wober, W., Alaupović, P. Studies on the protein moiety of endotoxin from gram-negative bacteria. Characterization of the protein moiety isolated by acetic acid hydrolysis of endotoxin from *Serratia marcescens* 08. Eur. J. Biochem. 19:357–367, 1971.

24. Heath, E. C., Mayer, R. M., Edstrom, R. D., Beaudreau, C. A. Structure and biosynthesis of the cell wall lipopolysaccharide of *Escherichia coli*. Ann. N.Y. Acad. Sci. 133:315–333, 1966.

25. Taylor, S. S., Heath, E. C. The incorporation of β-hydroxy fatty acids into a phospholipid of *Escherichia coli* B. J. Biol. Chem. 244:6605–6616, 1969.

26. Humphreys, G. O., Meadow, P. M. The biosynthesis of lipid A in *Pseudomonads* (abstract). J. Gen. Microbiol. 68:V–VI, 1971.

27. White, R. J. Control of amino sugar metabolism in *Escherichia coli* and isolation of mutants unable to degrade amino sugars. Biochem. J. 106:847–858, 1968.

28. Wu, H. C., Wu, T. C. Isolation and characterization of a glucosamine-requiring mutant of *Escherichia coli* K-12 defective in glucosamine-6-phosphate synthetase. J. Bacteriol. 105:455–466, 1971.

29. Sarvas, M. Mutant of *Escherichia coli* K-12 defective in D-glucosamine biosynthesis. J. Bacteriol. 105:467–471, 1971.

30. Rietschel, E. T., Galanos, C., Tanaka, A., Ruschmann, E., Lüderitz, O., Westphal, O. Biological activities of chemically modified endotoxins. Eur. J. Biochem. 22:218–224, 1971.

31. Westphal, O., Nowotny, A., Lüderitz, O., Hurni, H., Eichenberger, E., Schönholzer, G. Die Bedeutung der Lipoid Komponente (Lipoid A) für die biologischen Wirkungen bakterieller Endotoxine (Lipopolysaccharide). Pharm. Acta Helv. 33: 401–411, 1958.

32. Gewurz, H., Snyderman, R., Mergenhagen, S. E., Shin, H. S. Effects of endotoxic lipopolysac-

charides on the complement system. *In* S. Kadis, G. Weinbaum, and S. J. Ajl [ed.] Microbial toxins. Vol. 4. Academic Press, New York, 1971, p. 127–149.

33. Galanos, C., Rietschel, E. T., Lüderitz, O., Westphal, O. Interaction of lipopolysaccharides and lipid A with complement. Eur. J. Biochem. 19: 143–152, 1971.

34. Galanos, C., Rietschel, E. T., Lüderitz, O., Westphal, O., Kim, Y. B., Watson, D. W. Biological activities of lipid A complexed to bovine serum albumin. Eur. J. Biochem. 31:230–233, 1972.

35. Yin, E. T., Galanos, C., Kinsky, S., Bradshaw, R. A., Wessler, S., Lüderitz, O., Sarmiento, M. E. Picrogram-sensitive assay for endotoxin: gelation of Limulus polyphemus blood cell lysate induced by purified lipopolysaccharides and lipid A from gram-negative bacteria. Biochim. Biophys. Acta 261:284–289, 1972.

36. Galanos, C., Lüderitz, O., Westphal, O. Preparation and properties of antisera against the lipid-A component of bacterial lipopolysaccharides. Eur. J. Biochem. 24:116–122, 1971.

37. Kunin, C. M. Separation, characterization, and biological significance of a common antigen in Enterobacteriaceae. J. Exp. Med. 118:565–586, 1963.

38. Whang, H. Y., Neter, E. Immunosuppression by endotoxin and its lipoid A component. Proc. Soc. Exp. Biol. Med. 124:919–924, 1967.

39. Kataoka, T., Inoue, K., Galanos, C., Kinsky, S. Detection and specificity of lipid A antibodies using liposomes sensitized with lipid A and bacterial lipopolysaccharides. Eur. J. Biochem. 24: 123–127, 1971.

40. Whitby, J. L., Rowley, D. The role of macrophages in the elimination of bacteria from the mouse peritoneum. Brit. J. Exp. Pathol. 40:358–370, 1959.

41. Nowotny, A., Kasai, N., Tripodi, D. The use of mutants for the study of the relationship between structure and function in endotoxins. *In* L. Chedid [ed.] Structure et effets biologiques des produits bactériens provenants de germes gram négatifs. Coll. Int. du CNRS no. 174, Paris, 1967, 174:79–94, 1969.

42. Kasai, N., Nowotny, A. Endotoxic glycolipid from a heptoseless mutant of *Salmonella minnesota*. J. Bacteriol. 94:1824–1836, 1967.

43. Romeo, D., Girard, A., Rothfield, L. Reconstitution of a functional membrane enzyme system in a monomolecular film. I. Formation of a mixed monolayer of lipopolysaccharide and phospholipid. J. Mol. Biol. 53:475–490, 1970.

44. Creach, O., Entressangles, B., Colobert, L. Les lipides de la paroi ectoplasmique d'*Eberthella typhi* I. Acides gras libres et lipides neutres de l'endotoxine. Biochim. Biophys. Acta 116:80–94, 1966.

45. Burton, A. J., Carter, H. E. Purification and char-

acterization of the lipid A component of the lipopolysaccharides from *Escherichia coli*. Biochemistry 3:411–418, 1964.

46. Ikawa, M., Koepfli, J. B., Mudd, S. G., Niemann, C. An agent from *E. coli* causing hemorrhage and regression of an experimental mouse tumor. III. The component fatty acids of the phospholipide moiety. J. Am. Chem. Soc. 75:1035–1038, 1953.

47. Kasai, N., Yamano, A. Studies on the lipids of endotoxins—thin-layer chromatography of lipid fractions. Jap. J. Exp. Med. 34:329–344, 1964.

48. Kasai, N. Chemical studies on the lipid component of endotoxin, with special emphasis on its relation to biological activities. Ann. N.Y. Acad. Sci. 133:486–507, 1966.

49. McIntire, F. C., Sievert, H. W., Barlow, G. H., Finley, R. A., Lee, A. Y. Chemical, physical, and biological properties of a lipopolysaccharide from *Escherichia coli* K-235. Biochemistry 6:2363–2372, 1967.

50. Rooney, S. A., Goldfine, H. An extractable form of lipid A formed in a temperature-sensitive mutant of *E. coli* CR34. Fed. Proc. 30:1173, 1971.

51. Taylor, A., Knox, K. W., Work, E. Chemical and biological properties of an extracellular lipopolysaccharide from *Escherichia coli* grown under lysine-limiting conditions. Biochem. J. 99:53–61, 1966.

52. Nesbitt, J. A., III, Lennarz, W. J. Comparison of lipids and lipopolysaccharide from the bacillary and L forms of *Proteus* P18. J. Bacteriol. 89: 1020–1025, 1965.

53. Gallin, J. I., O'Leary, W. M. Lipoidal components of bacterial lipopolysaccharides: nature and distribution of fatty acids in *Aerobacter aerogenes*. J. Bacteriol. 96:660–664, 1968.

54. Koeltzow, D. E., Conrad, H. E. Structural heterogeneity in the lipopolysaccharide of *Aerobacter aerogenes* NCTC 243. Biochemistry 10:214–224, 1971.

55. Adams, G. A., Singh, P. P. The chemical constitution of lipid A from *Serratia marcescens*. Can. J. Biochem. 48:55–62, 1970.

56. Alaupović, P., Olson, A. C., Tsang, J. Studies on the characterization of lipopolysaccharides from two strains of *Serratia marcescens*. Ann. N.Y. Acad. Sci. 133:546–565, 1966.

57. Bishop, D. G., Still, J. L. Fatty acid metabolism in *Serratia marcescens:* III. The constituent fatty acids of the cell. J. Lipid Res. 4:81–86, 1963.

58. Adams, G. A., Kates, M., Shaw, D. H., Yaguchi, M. Studies on the chemical constitution of cell-wall lipopolysaccharides from *Neisseria perflava*. Can. J. Biochem. 46:1175–1184, 1968.

59. Fensom, A. H., Gray, G. W. The chemical composition of the lipopolysaccharide of *Pseudomonas aeruginosa*. Biochem. J. 114:185–196, 1969.

60. Hancock, I. C., Humphreys, G. O., Meadow, P. M. Characterization of the hydroxy acids of *Pseudomonas aeruginosa* 8602. Biochim. Biophys. Acta 202:389–391, 1970.

61. Roberts, N. A., Gray, G. W., Wilkinson, S. G. Release of lipopolysaccharide during the preparation of cell walls of *Pseudomonas aeruginosa*. Biochim. Biophys. Acta 135:1068–1071, 1967.

62. Key, B. A., Gray, G. W., Wilkinson, S. G. The purification and chemical composition of the lipopolysaccharide of *Pseudomonas alcaligenes*. Biochem. J. 120:559–566, 1970.

63. Kaneshiro, T., Marr, A. G. Hydroxy fatty acids of *Azotobacter agilis*. Biochim. Biophys. Acta 70:271–277, 1963.

64. Weckesser, J., Drews, G., Fromme, I. Chemical analysis of and degradation studies on the cell wall lipopolysaccharide of *Rhodopseudomonas capsulata*. J. Bacteriol. 109:1106–1113, 1972.

65. Bishop, D. G., Hewett, M. J., Knox, K. W. Occurrence of 3-hydroxytridecanoic and 3-hydroxypentadecanoic acids in the lipopolysaccharides of *Veillonella*. Biochim. Biophys. Acta 231:274–276, 1971.

66. Hewett, M. J., Knox, K. W., Bishop, D. G. Biochemical studies on lipopolysaccharides of *Veillonella*. Eur. J. Biochem. 19:169–175, 1971.

67. Rosenfelder, G. Identifizierung der langkettigen Fettsäuren in Lipopolysacchariden einiger Myxobakterien. Eur. J. Biochem. 1972 (in press).

68. Lacave, C., Asselineau, J., Serre, A., Roux, J. Comparaison de la composition chimique d'une fraction lipopolysaccharidique et d'une fraction polysaccharidique isolées de *Brucella melitensis*. Eur. J. Biochem. 9:189–198, 1969.

69. Berger, F. M., Fukui, G. M., Ludwig, B. J., Rosselet, J. P. Increased host resistance to infection elicited by lipopolysaccharides from *Brucella abortus*. Proc. Soc. Exp. Biol. Med. 131:1376–1381, 1969.

70. Weise, G., Drews, G., Jann, B., Jann, K. Identification and analysis of a lipopolysaccharide in cell walls of the blue-green alga *Anacystis nidulans*. Arch. Mikrobiol. 71:89–98, 1970.

Permeability of Model Membranes Containing Phospholipids and Lipopolysaccharides: Some Preliminary Results

Hiroshi Nikaido and Taiji Nakae

From the Department of Bacteriology and Immunology, University of California, Berkeley, California

The outer membrane layer of the cell walls of gram-negative bacteria appears to be highly permeable for small, hydrophilic molecules. The hypothesis that this permeability is due to the presence of lipopolysaccharides was tested by the use of model bilayer membranes consisting of lipopolysaccharides and phospholipids. When closed vesicles bounded by such bilayer structures were tested, it was found that the presence of lipopolysaccharide did not markedly increase their permeability to uncharged organic molecules.

The cell walls of gram-negative bacteria are often rather complex in construction and seem to contain at least two layers (peptidoglycan and "outer membrane") [1]. The inner, peptidoglycan layer appears as an extremely thin, structureless layer when thin sections of enteric bacteria such as *Escherichia coli* or salmonella are examined under the electron microscope. The outer layer (i.e., outer membrane) shows typical features of a unit membrane, with the dark-light-dark trilaminar image about 75 A thick. Furthermore, the outer membrane consists of the typical components of biologic or unit membranes, i.e., proteins and phospholipids, except that it contains an additional and unique constituent, lipopolysaccharide (LPS) [2–4].

These observations suggest that the outer membrane layer is a true unit membrane with a hydrophobic interior sandwiched between hydrophilic surfaces. Such phospholipid-bilayer structures are known to be essentially impermeable to alkali metal ions and to hydrophilic organic molecules of the size of hexoses or larger [5], but the outer membrane seems to show a high degree of nonspecific permeability. For example, for the transport of β-galactosides in *E. coli,* the inner, cytoplasmic membrane is known to act both as the main barrier to permeability and as the site of the active transport system [6]. Only under very spe-

cial conditions can the passive diffusion of β-galactosides through the outer membrane become rate-limiting [7].

Since LPS is a unique component of the outer membrane, we can hypothesize that the presence of this compound might be related to the high permeability of the outer membrane. This hypothesis was examined by generation of "liposomes" (i.e., closed vesicles bounded by lipid bilayers [8], from phospholipids and LPS [9] and by measurement of the rate of diffusion of various hydrophilic molecules across LPS-containing, bilayer structures.

Materials and Methods

LPS. LPS was extracted from SR (SL901) and Rc (LT2M1) mutants of *Salmonella typhimurium* LT2 by the procedure of Galanos et al. [10]. The structure of these LPS preparations has been described [11]. LPS was quantitated by a colorimetric assay for heptose [12].

Phospholipids. *S. typhimurium* LT2 was harvested in the late exponential phase of growth in L broth and was extracted according to Folch [13]. The lipid extract was washed twice with Folch's "theoretical-upper-phase" mixture, dried in a rotary evaporator, and kept at −70 C as a solution in chloroform. Phospholipids were quantitated by assay for total phosphate [14].

Other chemicals. ^{3}H-inulin (methoxy-^{3}H), ^{3}H-dextran, and ^{14}C-glucose were obtained from ICN, New England Nuclear, and Schwarz, and had specific activities of 125 mCi/g, 764 mCi/g, and 240 mCi/mmole, respectively.

This study was supported in part by grant no. AI-09644-03 from the U.S. Public Health Service.

Please address requests for reprints to Dr. Hiroshi Nikaido, Department of Bacteriology and Immunology, University of California, Berkeley, California 94720.

Preparation of liposomes. Three methods were used. In all cases, the tubes containing the dried films were put in a desiccator containing silica gel, and the desiccator was exhaustively evacuated with a vacuum pump for 1 hr before resuspension of the film was attempted.

(*A*). Phospholipids in chloroform were added to a centrifuge tube, and chloroform was removed by a stream of N_2. Phospholipid films were resuspended in a salt solution that in some cases contained LPS. Glass beads were added, and the tubes were shaken with a Vortex mixer. The suspension was kept at room temperature (24 C) for 1 hr before use [9].

(*B*). An aqueous suspension of phospholipids alone or of phospholipids plus LPS was kept at 60 C for 1 hr and was then slowly cooled; this process is called annealing [15]. The suspension was then dried (as a film at the bottom of a centrifuge tube) under a stream of N_2 in a water bath at 40 C, and the film was resuspended in an appropriate aqueous solution. The tube was shaken with a Vortex mixer until the suspension became homogeneous (about 20 min) and then was left at room temperature for 1 hr before use.

(*C*). An aqueous suspension of phospholipids or of phospholipids plus LPS was dried up as a film under a stream of N_2, and the film was resuspended in an appropriate aqueous solution with the aid of a Vortex mixer. The suspension was heated at 60 C for 30 min and then slowly cooled to room temperature (about 2 hr); this annealing presumably promoted the formation of mixed bilayers [15].

Results

Liposomes bounded by mixed bilayers containing phospholipids and LPS. Rothfield and his associates [15] have shown that LPS molecules can become inserted between phospholipid molecules comprising either monolayers or bilayers. Thus, if an aqueous suspension containing both LPS and phosphatidylethanolamine is heated and slowly cooled, an aggregate containing both LPS and the phospholipid is formed [15]. Phosphatidylethanolamine exists as concentric lamellae of bilayers in aqueous solutions [15], and electron microscopy suggested the insertion of LPS molecules into these phospholipid bilayers [15].

More recently, liposomes, or closed vesicles bounded by the LPS-phospholipid mixed bilayer, were generated by Kataoka et al. [9]. These workers suspended dried films of phospholipids in aqueous suspensions of LPS and showed by an immunologic technique that most liposomes contained LPS molecules in the bilayer structure.

We used both the Rothfield annealing technique and the technique of Kataoka et al. to generate liposomes bounded by LPS-phospholipid mixed bilayers. The formation of liposomes was followed by the entrapment of ^{3}H-labeled polysaccharide in the intravesicular space. In a typical experiment, liposomes were generated by method *B* (see above) from phospholipids or from phospholipids plus LPS in 0.2 ml of 0.2 M NaCl containing ^{3}H-dextran (2 μCi). The suspensions were applied to columns (1.25 × 13 cm) of Sepharose 4B, which were then eluted with 0.2 M NaCl. This procedure separated vesicles from free ^{3}H-dextran in the outside medium; from the amounts of ^{3}H-dextran trapped inside the vesicles, we could calculate the intravesicular space. In this experiment, 0.24 μmole of phospholipids trapped 0.72 μl of fluid. This value (i.e., 3 μl trapped per 1 μmole) is comparable to those reported in the literature (e.g., 3 μl trapped per 1 μmole when phosphatidylcholine containing 4% phosphatidic acid was suspended in 0.145 M KCl [5]). When a mixture of 0.24 μmole of phospholipids and an Rc LPS (preparation 5/8/72, free of phospholipids, 0.24 μmole of heptose) was used, a much larger amount (1.6 μl) of the fluid was sequestered in the intravesicular space. This result is at least consistent with the idea that LPS-phospholipid mixed bilayers are formed and that LPS increases the bilayer surface area and/or increases interlamellar distance through electrostatic repulsion caused by the presence of many anionic groups in its "inner-core" region [11].

Permeability of phospholipid-LPS liposomes. Permeability of these liposomes was examined first by the osmotic-swelling method [5]. Liposomes were made in the presence of fairly large molecules (i.e., raffinose, which was assumed to be practically incapable of permeating the bilayer structure) and then were diluted into isotonic solutions of various substances. If these substances did not penetrate, they acted as osmotic stabilizers, and liposomes did not shrink or swell. However, if these substances diffused into the intravesicular space, the osmotic pressure exerted by the intravesicular raffinose molecules produced an influx of water, and the resultant swelling of the liposomes was detected as a decrease in optical density of the suspension. From the results (figure 1), it is clear that the permeability of phospholipid-LPS liposomes was not markedly higher than that of the phospholipid liposomes; liposomes of both kinds were essentially impermeable to glucose and were moderately permeable to erythritol.

The permeability of liposomes toward glucose was also measured by a more direct method. Liposomes were generated from Rc LPS (preparation 8/13/71, 1.2 µmoles of heptose) and phospholipids (1.5 µmoles of organic phosphate) by method C (see above) in 0.5 ml of 0.2 M NaCl-0.01 M MgCl$_2$ containing ^{3}H-inulin and ^{14}C-glucose. The suspension was applied to a Sepharose 4B column (1.25 × 11.5 cm), the column was eluted with 0.2 M NaCl-0.01 M MgCl$_2$, and fractions containing liposomes but not free inulin or glucose were collected.

These fractions were pooled and left at room temperature; at indicated times portions were withdrawn and fractionated on a Sepharose 4B column for determination of the release of ^{3}H-inulin and ^{14}C-glucose into the outside medium. In contrast to the experiment of figure 1, this experiment was carried out with a more defective LPS (Rc) and in the presence of Mg^{++}. Yet it

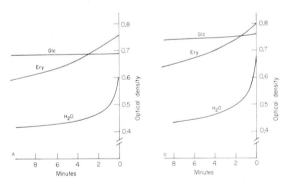

Figure 1. Osmotic swelling of liposomes. Liposomes were prepared from (**A**) phospholipids (6.4 µmoles of organic P) and from (**B**) phospholipids plus lipopolysaccharide from an SR mutant (0.2 µmole of heptose) by method A (see text). They were suspended in 1.0 ml of 80 mM raffinose-5 mM NaCl at 40 C; after 1 hr at this temperature, 0.1-ml portions were diluted in 0.9 ml of water, 80 mM glucose (Glc), or 80 mM erythritol (Ery) (each containing 5 mM NaCl) and kept at 40 C. Optical density at 500 mµ was recorded by use of a Hitachi 122 spectrophotometer.

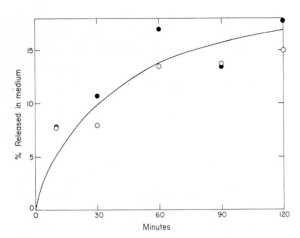

Figure 2. Release of ^{3}H-inulin (○) and ^{14}C-glucose (●) for phospholipid-lipopolysaccharide liposomes. Liposomes were made (see text) and were left at room temperature. At times indicated, portions were fractionated on Sepharose 4B columns, which had been equilibrated with 0.2 M NaCl-0.01 M MgCl$_2$.

is clear from figure 2 that the vesicles released glucose very slowly into the medium, at about the same rate at which inulin appeared in the medium, probably as a result of the disintegration of vesicles.

Permeability of "LPS" vesicles. It proved to be difficult to generate closed vesicles from LPS preparations free of phospholipids (authors' unpublished observations). One preparation of LPS from strain LT2Ml (preparation 8/13/71) consistently produced vesicles when dried as a film and resuspended (by shaking with a Vortex mixer for 20 min) in 0.2 M NaCl-0.01 M MgCl$_2$; this preparation, however, was heavily contaminated with phospholipids (3 moles of phospholipids per 1 mole of LPS heptose). Vesicles containing ^{3}H-inulin and ^{14}C-glucose were generated from this "LPS" preparation (0.24 µmole of heptose) by suspension in 0.5 ml of 0.2 M NaCl-0.01 M MgCl$_2$ and, after 1 hr at room temperature, were passed through a column of Sepharose 4B (1.25 × 10 cm). Fractions containing the liposomes were pooled and were left at room temperature; at various times 0.25-ml portions were applied to Sepharose columns, and the radioactivity present in the vesicles and in the extravesicular fluid was determined. The results (figure 3) show that the vesicles are essentially impermeable to glucose and that the apparent release of glucose is probably brought about by the slow disintegration of

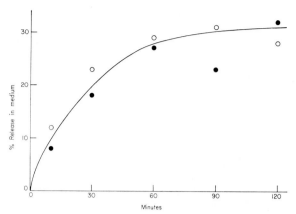

Figure 3. Release of ^{3}H-inulin (○) and ^{14}C-glucose (●) from "lipopolysaccharide" vesicles. "Lipopolysaccharide" (contaminated with phospholipids) from an Rc mutant (preparation 8/13/71) was treated as described in the text, and the release of radioactive compounds was followed at room temperature.

the vesicles, which results in the release of ^{3}H-inulin.

Discussion

The results appear to indicate that the incorporation of LPS molecules into the phospholipid-bilayer structure does not make the mixed bilayer markedly more permeable to small, hydrophilic molecules such as glucose. This conclusion, however, is dependent on the assumptions (*1*) that most of the LPS added has been inserted into the phospholipid-bilayer structure, (*2*) that LPS occupies a significant portion of the surface area of the bilayer, and (*3*) that the LPS is distributed more or less evenly among the population of vesicles. At least the first assumption seems well substantiated because filtration on Sepharose gel showed that most LPS became incorporated into large (phospholipid) aggregates (i.e., vesicles), at least when Rc LPS preparations were treated according to method *C* (authors' observations); and LPS markedly increased the amount of fluid entrapped by phospholipid liposomes at moderate ionic strength (0.2 M NaCl), presumably by increasing both the surface area (through insertion among phospholipid molecules) and the interlamellar distance (through increased electrostatic repulsion). In addition, we used the annealing technique, which was shown to be optimal for the production of LPS-phospholipid mixed bi-

layers [15]; under these conditions it seems likely that both conditions (*1*) and (*3*) were fulfilled, and yet the vesicles were still essentially impermeable to glucose (figure 2). Finally, in the phospholipid-contaminated Rc-LPS preparation (8/13/71), LPS-free phospholipid vesicles or even LPS-free regions of the bilayer are unlikely to exist because of the low ratio of phospholipid to LPS. In fact, the ratio is not far from the minimal one capable of producing a stable monolayer, according to Romeo et al. [16]. We can thus expect the presence of a very homogeneous mixed bilayer, in which approximately 25% of the surface area is calculated (on the basis of constants given by Romeo et al. [16]) to be occupied by LPS molecules. This obviously satisfies assumption (*2*), as well as assumptions (*1*) and (*3*). The impermeability of these "LPS" vesicles (figure 3) thus is strong evidence against our original hypothesis that LPS may increase the permeability of phospholipid bilayers.

It seems possible, however, that some areas of the outer membrane have a structure different from that of the phospholipid-LPS bilayer. Since we have not been able to produce closed vesicles bounded by pure LPS and Mg^{++}, more studies are necessary in this direction.

References

1. Glauert, A. M., Thornley, M. J. The topography of the bacterial cell wall. Annu. Rev. Microbiol. 23:159–198, 1969.
2. Miura, T., Mizushima, S. Separation and properties of outer and cytoplasmic membranes in *Escherichia coli*. Biochim. Biophys. Acta 193:268–276, 1969.
3. Schnaitman, C. A. Effect of ethylenediaminetetraacetic acid, Triton X-100, and lysozyme on the morphology and chemical composition of isolated cell walls of *Escherichia coli*. J. Bacteriol. 108:553–563, 1971.
4. DePamphilis, M. L., Adler, J. Attachment of flagellar basal bodies to the cell envelope: specific attachment to the outer, lipopolysaccharide membrane and the cytoplasmic membrane. J. Bacteriol. 105:396–407, 1971.
5. Bangham, A. D., deGier, J., Greville, G. D. Osmotic properties and water permeability of phospholipid liquid crystals. Chem. Phys. Lipids 1:225–246, 1967.
6. Sistrom, W. R. On the physical state of the intracellularly accumulated substrates of β-galactoside-permease in *Escherichia coli*. Biochim. Biophys. Acta 29:579–587, 1958.

7. Robbie, J. P., Wilson, T. H. Transmembrane effects of beta-galactosides on thiomethyl-beta-galactoside transport in *Escherichia coli*. Biochim. Biophys. Acta 173:234–244, 1969.

8. Sessa, G., Weissmann, G. Phospholipid spherules (liposomes) as a model for biological membranes. J. Lipid Res. 9:310–318, 1968.

9. Kataoka, T., Inoue, K., Lüderitz, O., Kinsky, S. C. Antibody- and complement-dependent damage to liposomes prepared with bacterial lipopolysaccharides. Eur. J. Biochem. 21:80–85, 1971.

10. Galanos, C., Lüderitz, O., Westphal, O. A new method for the extraction of R lipopolysaccharides. Eur. J. Biochem. 9:245–249, 1969.

11. Lüderitz, O., Westphal, O., Staub, A. M., Nikaido, H. Isolation and chemical and immunological characterization of bacterial lipopolysaccharides. *In* G. Weinbaum, S. Kadis, and S. J. Ajl [ed.]. Microbial toxins. Vol. 4. Academic Press, New York, 1971, p. 145–233.

12. Osborn, M. J. Studies on the gram-negative cell wall. I. Evidence for the role of 2-keto-3-deoxy-octonate in the lipopolysaccharide of *Salmonella typhimurium*. Proc. Nat. Acad. Sci. U.S.A. 50: 499–506, 1963.

13. Folch, J., Lees, M., Stanley, G. H. S. A simple method for the isolation and purification of total lipides from animal tissues. J. Biol. Chem. 226: 497–509, 1957.

14. Ames, B. N., Dubin, D. T. The role of polyamines in the neutralization of bacteriophage deoxyribonucleic acid. J. Biol. Chem. 235:769–775, 1960.

15. Rothfield, L., Horne, R. W. Reassociation of purified lipopolysaccharide and phospholipid of the bacterial cell envelope: electron microscopic and monolayer studies. J. Bacteriol. 93:1705–1721, 1967.

16. Romeo, D., Girard, A., Rothfield, L. Reconstitution of a functional membrane enzyme system in a monomolecular film. I. Formation of a mixed monolayer of lipopolysaccharide and phospholipid. J. Mol. Biol. 53:475–490, 1970.

A Common Antigenic Constituent in Various Purified Salmonella Endotoxins

M. Raynaud, B. Kouznetzova, M. J. Navarro,
J. C. Chermann, M. Digeon, and
A. Petitprez

From Institut Pasteur, Garches, France; Institut National de la Santé et de la Recherche Médicale (INSERM), Paris, France; and Institut Pasteur, Lille, France

Methods for extraction of endotoxin (phenol-water; phenol-chloroform-petroleum ether; 1 M sodium chloride-0.1 M sodium citrate) were investigated. Artificial hybrid macromolecules, which are formed in the phenol-water procedure, are not observed with the "hypertonic method." Toxicity of products obtained by different techniques varies. With the chick-embryo method and the pyrogenic test, endotoxins extracted with hypertonic solutions were more toxic than lipopolysaccharides (LPS) extracted with phenol-water. Immunologic studies of subfractions derived from purified endotoxins (NCE) and from purified LPS reveal important differences; an antigenic subfraction common to hypertonic endotoxins extracted from *Salmonella typhi* O901, *S. typhi* rough R_2, and *Salmonella minnesota* R595, has been demonstrated. It was not possible to detect this subfraction of NCE in subfractions of LPS.

Our intention was to compare the properties of endotoxins extracted from various enterobacteria (*Salmonella typhi* O901, *S. typhi* R_2, and *Salmonella minnesota* R595) by 1 M sodium chloride-0.1 M sodium citrate [1] with the properties of lipopolysaccharides (LPS) prepared from the same bacteria either by the phenol-water method [2] (for the O901 smooth form) or by the method of Galanos and al. [3] (for the rough and extremely rough forms).

Use of the phenol-water method results in reorganization of external components of the cell wall. This makes it possible to prepare hybrid LPS from a mixture of different bacteria [4, 5], or by the action of sodium deoxycholate on a mixture of two different types of LPS [6].

We prepared hybrid LPS by mixing different bacteria, e.g., *S. typhi* R_2 + *S. typhi* O901; *S. typhi* O901 + *Escherichia coli* O14; *S. typhi* O901 + smooth *Salmonella typhimurium* 1406

This work was supported by INSERM contract no. 72. 4016.

We thank Miss A. Carlin, Mrs. J. Riza, and Mr. F. Garcia for their excellent technical assistance. We also thank Mr. L. Muller for the bacterial culture used.

Please address requests for reprints to Dr. M. Raynaud, Institut Pasteur, 92380 Garches, France.

S [7]. We showed that the NaCl-sodium citrate method used under the conditions similar to those in the above two methods on mixtures of different bacteria does not induce formation of hybrids. Furthermore, the NaCl-citrate method may be used for smooth, rough, and extremely rough forms and for all Enterobacteriaceae [8].

Materials and Methods

Bacteria were obtained by culture in Chemap and van Doorn fermentors, the latter of which conforms to the standards of van Hemert [9]. Details concerning the use of van Doorn fermentors for culture of *Salmonella* have been mentioned previously [10]. Briefly, the medium was composed of casein hydrolyzate, yeast extract, and glucose; after culture for 6 hr at 37 C, bacteria were harvested by centrifugation in a Sharples centrifuge and washed twice in 0.9% NaCl. They were then extracted by two techniques. (*1*) To 5 mg of bacterial nitrogen was added 1 ml of 1 M sodium chloride-0.1 M sodium citrate solution, *p*H 7.0 [1, 8]. This method was used for all three strains. (*2*) To obtain LPS, we used a modification of the method of Westphal [2] (the duration of contact at 65 C was 5 min) for *S. typhi* O901 and the method of Galanos and al. [3] for *S. typhi* R$_2$ and *S. minnesota* R595.

Endotoxin contained in the crude hypertonic extract was then precipitated by 25% (w/w) polyethylene glycol [11]; the O901 and R$_2$ endotoxins were purified by pronase digestion [12]. *S. minnesota* R595 crude endotoxin was not subjected to action by pronase, since this partially destroys it.

The preparations were then purified by gel filtration on Sepharose 4B (Pharmacia, Uppsala) equilibrated in NaCl (1 g per liter) [13]. The purified O901 and R$_2$ endotoxins (or heavy endotoxins) eluted with the exclusion volume corresponding to a molecular weight higher than 4,000,000; on the other hand, *S. minnesota* R595 endotoxin eluted after the exclusion volume in a zone corresponding to a molecular weight of about 1,000,000 (figure 1). The effluents corresponding to the peaks were collected, dialyzed against water (four or five days against 10 volumes of water changed twice a day), and then lyophilized.

Treatment by Triton X100. Endotoxin was put in solution in distilled water containing 5% Triton X100 (Rohm and Haas, Philadelphia, Pa.). Contact was maintained for at least 1 hr at room temperature (23 C). Triton X100 was

previously used by Weiser and Rothfield for biosynthesis of LPS [14].

Nitrogen was measured by Kjeldhal's method [15]; phosphorus was measured according to the method of Fiske and Subbarow [16] as modified by Horecker [17]. Hexoses were determined by the technique of Dubois et al. [18]. Optical density was read at 280 nm on a Beckman DB spectrophotometer.

Immunodiffusion. Immunodiffusion was performed by Ouchterlony's method [19]. We observed that if sodium deoxycholate or sodium dodecylsulfate was added to the gel, lines of precipitation did not appear as they do under normal conditions. On the other hand, it was possible to detect numerous precipitation lines when Triton X100 was added to the gel at a final concentration of 1%.

Sera. The sera used have been described [13, 20]. Their characteristics are shown in table 1.

Toxicity. (1) Chick embryos. Toxicity to chick embryos was determined by injection of 0.1 ml of endotoxin iv into Leghorn embryos on the 10th day of incubation at 39.5 C. Mortality was observed after 24 hr [21–23]. The LD$_{50}$ was determined by the log-probit method.

(2) Minimal pyrogenic dose. New Zealand white rabbits weighing 2.5–3.5 kg were used. The average normal temperature of each animal was previously determined with an electric recorder thermometer (Carrieri, type Z 8P). Only rabbits whose temperature remained between 39 C and

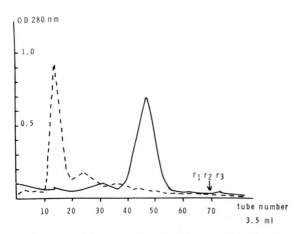

Figure 1. Elution profile of heavy endotoxins obtained by the 1 M NaCl-0.1 M sodium citrate method (NCE) on Sepharose 4B gel filtration (column K25-45, Pharmacia, Uppsala). Bed height was 42.5 cm. Sample applied for gel filtration was 5 ml of crude extract concentrated with polyethylene glycol. Flow rate was 1 ml/min. Buffer was 0.1% NaCl. Broken line = elution profile for the crude extract from *Salmonella typhi* O901 or from *S. typhi* R$_2$. Solid line = elution profile for the crude extract from *Salmonella minnesota* R595.

Table 1. Properties of the sera used in immunodiffusion experiments with endotoxins of *Salmonella typhi* O901, *S. typhi* R$_2$, and *Salmonella minnesota* R595.

Serum	Antibodies against
13475	Heterologous antigens r$_1$, r$_2$, r$_3$ Endotoxin from *Salmonella typhi* R$_2$
90193	Endotoxin from *Salmonella typhimurium* TV 119
90319	Endotoxin from *S. typhi* R$_2$
13525	Flagella Proteins of the wall Endotoxin from *S. typhi* R$_2$
483	Endotoxin from *S. typhi* O901
90307	Endotoxin from *S. typhi* O901
90516 90517 90564	Endotoxin from *Salmonella minnesota* R 595

39.5 C were selected. The substance to be studied was dissolved in pyrogenfree saline and injected into the marginal vein of the ear. A gradual rise in temperature for each animal was noted, and the maximal temperature for each animal was observed after 2–4 hr. Rise in temperature from the initial value was termed ΔT, and the curve ΔT was drawn as a function of the log of the dose injected. The minimal pyrogenic dose was that which induced a rise in temperature of 0.6 C/kg of body weight.

Electron microscopy. Heavy endotoxins obtained by the NaCl-citrate method (NCE) and LPS, some treated and some not treated with Triton X100, were ultracentrifuged in polyallomere tubes (International Equipment Ultracentrifuge, rotor SB405) for 12 hr at about 405,000 g. A translucent pellet, clearly visible, was obtained with NCE, even after treatment with Triton X100; however, LPS did not give any deposit after such treatment. After removal of the supernatant, the pellet was fixed with 25% gluteraldehyde dissolved in a 0.2 M sodium cacodylate buffer, *pH* 7.2, and was left for 4–18 hr at 4 C. The pellet, which adhered at the bottom of the tube, was washed twice with 0.1 M sodium cacodylate, *pH* 7.2, then stained by 2% osmic acid in 0.1 M sodium cacodylate for 1 hr. The supernatant was removed, and the pellet was dehydrated by addition of 70% and then 90% aqueous acetone; pure acetone was then added. The pellet was embedded in araldite overnight at 4

C [24]. Thin sections were obtained with a Porter-Blum MT 1 ultramicrotome and were collected on 200-mesh copper grids. The preparations were examined in a Hitachi HU 12 electron microscope at 75 kV.

Results

(1) Toxicity. Table 2 shows that the NCE were clearly more toxic per unit of weight than the corresponding LPS. This difference was observed both in tests of toxicity for chick embryos and in pyrogenic activity of *S. typhi* O901 and R_2. Results concerning toxicity were similar for *S. minnesota*.

(2) Antigenic analysis of purified endotoxins and LPS. We previously showed [25] that, with R_2 endotoxin prepared by gel filtration, one or two lines of precipitation could be detected in immunodiffusion, depending on the sera used. In the latter case, the two lines (serum 13 525) joined with the single line given by other sera (90 193, 90 319, and 13 475). The purified, heavy O antigen (NCE, *S. typhi* O901 extract) showed two lines of precipitation [13] whatever serum was used (90 307 or 483). On the other hand, the NCE extracted from *S. minnesota* R 595 showed a single line of precipitation with all sera employed (90 516, 90 517, and 90 564) (figure 2).

After the action of Triton X100 on the three endotoxins, lines of precipitation obtained by gel

Table 2. Chemical and biological properties of various endotoxins.

Product		Form	Chemical properties			Toxicity	
			N(%)	P(%)	Hexoses*	MPD† (µg/kg)	LD$_{50}$ (µg)
Salmonella typhi O901	Boivin		2.1	0.7	++	1×10^{-6}	1.7×10^{-3}
	NCE‡	Smooth	3.6	2.9	+	2×10^{-6}	0.6×10^{-3}
	LPS§		1.05	0.8	+	5×10^{-4}	1×10^{-2}
S. typhi R_2	NCE	R II	5.25	2.2	+	1×10^{-5}	1.8×10^{-3}
	LPS		...	...	+	$<1 \times 10^{-3}$	Not tested
Salmonella minnesota R595	NCE	ERR‖	4.75	2.9	0	1×10^{-3}	3×10^{-2}
	LPS		1.06	0.8	±#	Not tested	1×10^{-2}

* + = mentioned carbohydrate present; 0 = mentioned carbohydrate absent.

† The minimal pyrogenic dose (MPD) is the dose giving an increase in rectal temperature of 0.6 C for rabbits weighing 2–2.5 kg.

‡ NCE = purified heavy endotoxins obtained by 1 M NaCl-0.1 M sodium citrate.

§ LPS = lipopolysaccharides.

‖ ERR = extremely rough.

± = traces.

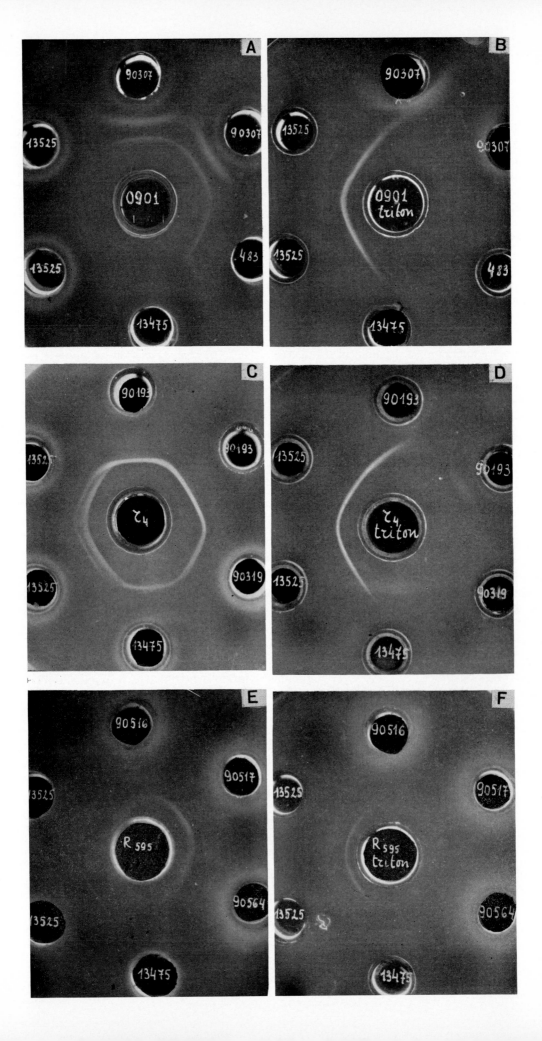

Figure 2. **(A)** Immunodiffusion in saline medium shows the purified heavy endotoxin obtained by the 1 M NaCl-0.1 M sodium citrate method (NCE) from *Salmonella typhi* O901. **(B)** Action of Triton X100 on this NCE. Note rescue of a new subfraction revealed by 13 525 serum (see text). **(C, E)** Same phenomenon as **A**. **(D, F)** Same phenomenon as **B**.

immunodiffusion containing 1% Triton X100 were greatly modified. The one or two lines initially observed were replaced by numerous lines that corresponded to antigenically distinct subfractions. That the position of these lines was farther from the wells containing the antigens indicated that these subfractions have a molecular weight lower than that of the NCE.

In addition, in all three cases, a new line of precipitation appeared with serum 13 525 obtained from a horse (no. 127) immunized for 10 months with an endotoxin extracted from *S. typhi* R₂ and degradated at a low temperature by alkaline alcohol [26].

Figure 3 shows that the antigenic configuration

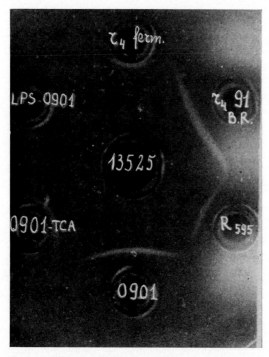

Figure 3. Antigenic configuration of three heavy endotoxins (*Salmonella typhi* O901, *S. typhi* R₂, and *Salmonella minnesota* R595) revealed by treatment with Triton X100 (outer wells). Center well contains 13 525 antiserum. R₄ form = *S. typhi* R₂ purified heavy endotoxin obtained by the 1 M NaCl-0.1 M sodium citrate method (NCE); R595 = *S. minnesota* purified NCE; O901 = *S. typhi* O901 purified NCE; O901 TCA = Boivin's extract from *S. typhi* O901; LPS = Westphal's extract from *S. typhi* O901.

revealed by treatment with Triton X100 was identical for the three NCE studied. This was not the case for LPS extracted from the same bacteria (figure 3). This configuration corresponded to the subfraction previously described [7, 13, 20].

The electron microscope brought to light morphologic differences between each pair (the NCE and the same endotoxin treated with Triton X100) (figure 4, A-F). The action of Triton X100 induced fragmentation with loss of the lamellar aspect of NCE.

After embedding and thin-sectioning of the lipopolysaccharides that had not been treated with Triton, a picture was found (figure 4, G-I) that was similar to that described by Shands et al. after negative staining [27, 28], by Bladen and Mergenhagen [29], and by Mergenhagen et al. [30, 31]. After treatment with Triton X100, the LPS were completely solubilized and could no longer be sedimented by ultracentrifugation under the conditions described above (see Methods).

Discussion

These results lead to a number of open questions. Important differences exist between NCE and LPS extracted from the same bacteria. These differences were observed in several tests. (*1*) NCE are, by unit of weight, more toxic than LPS ([13] and table 2) and do not have the same overall chemical composition [20]. (*2*) Triton X100 brings to light an antigenic subfraction common to the three forms of bacteria studied ([13] and figure 3) from which was extracted NCE. (*3*) This subfraction is also present in endotoxins extracted from smooth forms by Boivin's method, as we have previously shown [7]. (*4*) This subfraction cannot be detected in LPS. (*5*) Under the electron microscope, the NCE have a morphologic form clearly different from that of LPS [27-32] and of Boivin's antigens, as described by Spielvogel [33]. (*6*) NCE are also different from the granular forms that we have obtained by the action of 1 M NaCl-0.1 M sodium citrate solution on bacteria previously treated by sodium dodecylsulfate [34-36].

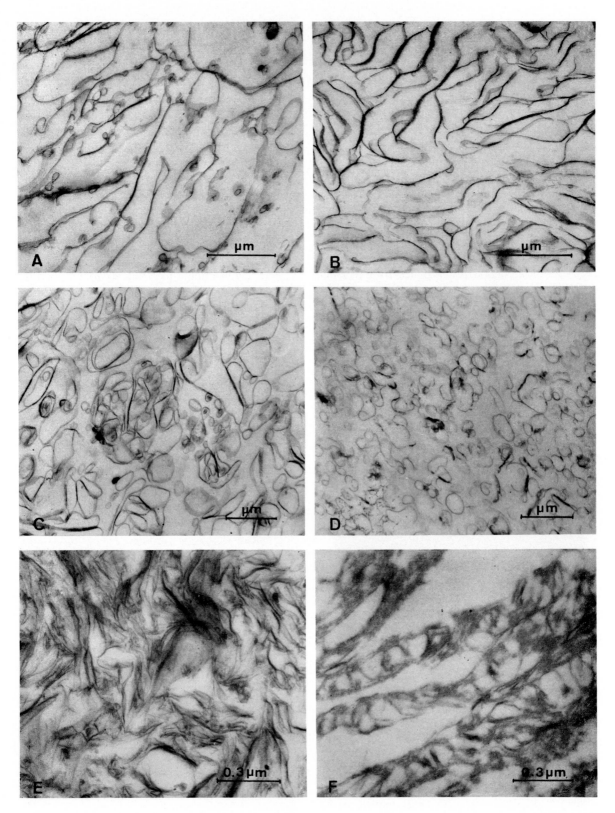

Figure 4.

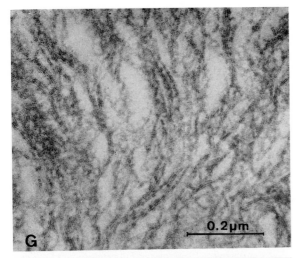

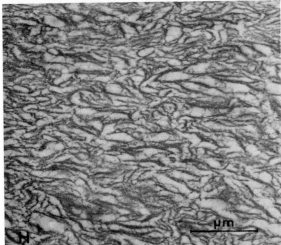

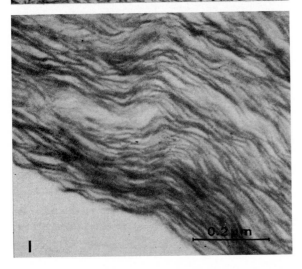

Figure 4. Morphologic differences between untreated purified heavy endotoxin obtained by the 1 M NaCl-0.1 M sodium citrate method (NCE), NCE treated with Triton X100, and lipopolysaccharide

References

1. Raynaud, M., Digeon, M. Sur une nouvelle toxine du bacille typhique extraite des formes rough. C. R. Acad. Sci. (Paris) 229:564–566, 1949.
2. Westphal, O., Lüderitz, O., Bister, F. Über die Extraktion von Bakterien mit Phenol/Wasser. Zeit. Naturforsch. 7 B:148–155, 1952.
3. Galanos, C., Lüderitz, O., Westphal, O. A new method for the extraction of R lipopolysaccharides. Eur. J. Biochem. 9:245–249, 1969.
4. Sarvas, M., Lüderitz, O., Westphal, O. Immunochemical studies on T1, S hybrids of *Salmonella paratyphi-B*. Ann. Med. Exp. Fenn. 45:117–126, 1967.
5. Chedid, L., Parant, M., Parant, F. Hybridization of endotoxins extracted from rough and smooth strains of salmonellae. *In* International symposium on enterobacterial vaccines, Bern, 1969 (Symp. Series Immunobiol. Standard v. 15). Karger, Basel, 1971, p. 103–114.
6. Rudbach, J. A., Milner, K. C., Ribi, E. Hybrid formation between bacterial endotoxins. J. Exp. Med. 126:63–79, 1967.
7. Chermann, J. -C. Les antigènes des formes rough des Enterobacteriaceae. Séparation, nature chimique, propriétés toxiques et immunologiques. Thèse Sciences, C. N. R. S. no. AO 1512. vol. 1, 1967. 210 p.
8. Digeon, M., Raynaud, M., Lavergne, M., Chermann, J. -C. Etude des divers antigènes élaborés par *Salmonella typhi* rough (souche R_2). I. Etude qualitative. Ann. Inst. Pasteur (Paris) 109(Suppl.): 66–84, 1965.
9. van Hemert, P. Vaccine production as a unit process. Drukkerij Clinkwijk, Utrecht, 1971. 175 p.
10. Mendiola, L. R., Kouznetzova, B., Chermann, J. -C., Sinoussi, F., Digeon, M., Raynaud, M. Purification of bacterial endotoxins by zonal centrifugation. Infec. Immun. 6:27–31, 1972.
11. Chermann, J. -C., Digeon, M., Raynaud, M. Une nouvelle méthode de purification des endotoxines:

(LPS). **(A, B)** *Salmonella typhi* O901 NCE. **A** is untreated and **B** is treated. Note small granules before treatment. **(C, D)** *S. typhi* R_2 NCE. **C** is untreated, **D** is treated. Note modification after treatment and appearance of fragments. **(E, F)** *Salmonella minnesota* R595 NCE. **E** is untreated and **F** is treated. Note lamellar aspect of NCE before treatment and a complete disorganization after treatment. **(G)** LPS extracted by the phenol-water method from *S. typhi* O901. Compare with **A**. **(H)** LPS from *S. typhi* R_2 extracted by the method of Galanos. Compare with **C**. **(I)** LPS from *Salmonella minnesota* R595 extracted by the method of Galanos. Compare with **E**.

34

la précipitation par le polyéthylène-glycol. C. R. Acad. Sci. [D] (Paris) 265:1251–1252, 1967.

12. Chermann, J. -C., Raynaud, M., Digeon, M. Etude des divers antigènes élaborés par *Salmonella typhi* rough (souche R₂). II. Séparation et propriétés. Ann. Inst. Pasteur (Paris) 113:375–398, 1967.

13. Chermann, J. -C., Digeon, M., Kouznetzova, B., Raynaud, M. Smooth and rough heavy endotoxins and lipopolysaccharides: preparation and properties. *In* International symposium on enterobacterial vaccines, Bern, 1969 (Symp. Series Immunobiol. Standard. v. 15). Karger, Basel, 1971, p. 123–130.

14. Weiser, M. M., Rothfield, L. The reassociation of lipopolysaccharide, phospholipid, and transferase enzymes of the bacterial cell envelope. Isolation of binary and ternary complexes. J. Biol. Chem. 243:1320–1328, 1968.

15. Kjeldahl, J. Sur une nouvelle méthode de dosage de l'azote dans les substances organiques. C. R. Carlsberg Labor. (Copenhagen) 2:1–12, 1888.

16. Fiske, C. H., Subbarow, Y. The colorimetric determination of phosphorus. J. Biol. Chem. 66:375–400, 1926.

17. Horecker, B. L., Ma, T. S., Haas, E. Note on the determination of microquantities of organic phosphorus. J. Biol. Chem. 136:775–776, 1940.

18. Dubois, M., Gilles, K. A., Hamilton, J. K., Rebers, P. A., Smith, F. Colorimetric method for determination of sugars and related substances. Anal. Chem. 28:350–356, 1956.

19. Raynaud, M., Digeon, M., Nauciel, C. Studies on the endotoxin and antigens of a rough strain of *Salmonella typhi*. *In* M. Landy and W. Braun [ed.] Bacterial endotoxins. Rutgers University Press, New Brunswick, N.J., 1964, p. 326–344.

20. Raynaud, M., Chermann, J. -C., Digeon, M. Etudes sur la structure chimique et immunologique des endotoxines Ret O de *S. typhi*. *In* L. Chedid [ed.] (Colloque International C. N. R. S. No. 174). La structure et les effets biologiques des produits bactériens provenant de germes gram-négatifs. Paris, 1969, p. 47–67.

21. Smith, R. T., Thomas, L. The lethal effect of endotoxins on the chick embryo. J. Exp. Med. 104:217–231, 1956.

22. Digeon, M., Raynaud, M. Toxicité et virulence comparées des formes smooth et rough de *Salmonella typhi* sur embryon de poulet. Ann. Inst. Pasteur (Paris) 100:344–351, 1961.

23. Finkelstein, R. A. Observations on mode of action of endotoxin in chick embryos. Proc. Soc. Exp. Biol. Med. 115:702–707, 1964.

24. Richardson, K. C., Jarett, L., Finke, E. H. Embedding in epoxy resins for ultrathin sectioning in electron microscopy. Stain. Technol. 35:313–323, 1960.

25. Digeon, M., Chermann, J. -C., Raynaud, M. Multiplicité des déterminants présents sur la molécule de l'antigène rough lourd des salmonelles. Ann. Inst. Pasteur (Paris) 113:843–856, 1967.

26. Bizzini, B., Digeon, M., Raynaud, M. Etudes sur la toxine R₂ du bacille typhique. III. Préparation d'une toxine soluble. Ann. Inst. Pasteur (Paris) 87:724–726, 1954.

27. Shands, J. W., Jr., Graham, J. A., Nath, K. The morphologic structure of isolated bacterial lipopolysaccharide. J. Mol. Biol. 25:15–21, 1967.

28. Shands, J. W., Jr., Graham, J. A. Morphologic structure and biological activity of bacterial lipopolysaccharides. *In* L. Chedid [ed.] (Colloque International C. N. R. S. No. 174). La structure et les effets biologiques des produits bactériens provenant de germes gram-négatifs. Paris, 1969, p. 25–34.

29. Bladen, H. A., Mergenhagen, S. E. Ultrastructure of *Veillonella* and morphological correlation of an outer membrane with particles associated with endotoxic activity. J. Bacteriol. 88:1482–1492, 1964.

30. Mergenhagen, S. E., Bladen, H. A., Hsu, K. C. Electron microscopic localization of endotoxic lipopolysaccharide in gram-negative organisms. Ann. N. Y. Acad. Sci. 133:279–291, 1966.

31. Mergenhagen, S. E., Gewurz, H., Bladen, H. A., Nowotny, A., Kasai, N., Lüderitz, O. Interactions of the complement system with endotoxins from a *Salmonella minnesota* mutant deficient in O-polysaccharide and heptose. J. Immunol. 100:227–229, 1968.

32. Boivin, A., Mesrobeanu, I., Mesrobeanu, L. Extraction d'un complexe toxique et antigénique à partir du bacille d'Aertrycke. C. R. Soc. Biol. (Paris) 114:307–310, 1933.

33. Spielvogel, A. R. An ultrastructural study of the mechanisms of platelet-endotoxin interaction. J. Exp. Med. 126:235–250, 1967.

34. Raynaud, M., Digeon, M., Chermann, J. -C., Giuntini, J. Sur une nouvelle forme d'endotoxines: les endotoxines "granulaires." C. R. Acad. Sci. [D] (Paris) 262:722–724, 1966.

35. Chermann, J. -C., Digeon, M., Raynaud, M., Giuntini, J. Etude de l'extraction des antigènes somatiques (R et O) à partir des parois bactériennes de salmonelles. Préparation d'endotoxines granulaires. Ann. Inst. Pasteur (Paris) 111(Suppl.): 59–69, 1966.

36. Digeon, M., Chermann, J. -C., Guyot-Jeannin, N., Raynaud, M. Les endotoxines granulaires rough. Propriétés toxiques. Ann. Inst. Pasteur (Paris) 111(Suppl.):70–77, 1966.

Heterogeneity and Biological Activity of Endotoxic Glycolipid from *Salmonella minnesota* R595

C. H. Chen, A. G. Johnson, N. Kasai,
B. A. Key,* J. Levin, and A. Nowotny

From the Department of Microbiology and Immunology, Temple University School of Medicine, Philadelphia, Pennsylvania; the Department of Microbiology, University of Michigan, Ann Arbor, Michigan; the School of Pharmaceutical Science, Showa University, Tokyo, Japan; and the Department of Medicine, Johns Hopkins University School of Medicine and Hospital, Baltimore, Maryland

The endotoxic glycolipid of *Salmonella minnesota* R595 was extracted either by the phenol-water method (GL) or directly with chloroform-methanol (GL-DE). Both glycolipids were soluble in chloroform-methanol (4:1). The yield of GL-DE was higher than that of GL. Examination of both materials by thin-layer chromatography (TLC) and column chromatography demonstrated that GL and GL-DE are chromatographically similar and that both are heterogeneous. Four fractions isolated from GL-DE with preparative TLC showed equal endotoxic activities in Shwartzman skin reaction, *Limulus* lysate gelation activity, and chick-embryo lethality. Chemical analyses indicate that all four are glycolipid in nature. The chromatographic behavior and the similar molar ratios of chemical constituents in these four components suggest that they may be chromatographically different due to polymerism. More detailed studies of some biological parameters have been done with GL and GL-DE by comparison with an endotoxin obtained from a smooth strain of *Serratia marcescens*. It was found that glycolipid showed activities similar to or higher than those observed for endotoxin.

Endotoxic lipopolysaccharides consist of three major components, i.e., lipids, polysaccharides, and some bound amino acids. One of the major problems inherent in structural studies of these complexes is the lack of specific methods for the cleavage of the various portions, i.e., lipid from polysaccharide from amino acids. Mild acidic hydrolysis has been used, but this has been shown to cause extensive heterogeneity of the resulting polysaccharide and lipid fractions [1–4].

The smooth-to-rough conversion of certain bacterial strains is characterized by partial or complete loss of the polysaccharide. One such mutant, *Salmonella minnesota* R595, which has been isolated by O. Lüderitz and associates [5], was found to contain endotoxic glycolipid lacking heptose in addition to the other core and species-specific sugars. Its composition has been shown to be mainly 2-keto-3-deoxy-octonic acid (KDO) and the lipid moiety of the parent, smooth-strain endotoxin. This glycolipid serves as a better starting material than smooth LPS for studies of the relationship between structure and function.

Another major problem in the study of bacterial endotoxins is their purity. It has been adequately demonstrated [6, 7] that, irrespective of the method used for isolation, endotoxins are extremely heterogeneous and that, therefore, meaningful structural studies cannot be carried out without prior isolation of homogeneous fractions.

This research was supported by grants no. AI-05581 and HL-01601 from the U.S. Public Health Service and by a grant from the New York Cancer Research Institute.

The authors wish to thank Mrs. Anne Nowotny, Mrs. Gloria Polin, and Mr. Chung-Ming Chang for their excellent assistance.

Please address requests for reprints to Dr. A. Nowotny, Department of Microbiology and Immunology, Temple University School of Medicine, Philadelphia, Pa. 19140.

* Present address: BHD Chemicals, Ltd., Poole, Dorset, England.

The purpose of this investigation was twofold. We aimed (*1*) to expand the knowledge of the biological activities of the glycolipid and thus to ensure that it merits the term endotoxic and (*2*) to investigate its heterogeneity with the aim of purifying it.

Materials and Methods

Bacterial strain. The heptoseless mutant strain *S. minnesota* R595 used throughout this work was obtained through the courtesy of Dr. Otto Lüderitz, Max Planck Institute for Immuno-biology, Freiburg, West Germany.

Cultivation of bacteria. The bacterial cells were cultivated as described earlier [8], except that the medium contained 1% glucose and 0.02% $MgSO_4 \cdot 7H_2O$. The cells were harvested at late log phase and kept frozen. The yield of the wet-packed cells was about 8 g/liter.

Phenol-water extraction of the R595 cells. The endotoxic glycolipid was extracted from R595 according to the method of Kasai and Nowotny [8], except that the step involving 2% sodium dodecyl sulfate extraction was omitted.

Chloroform-methanol extraction of the R595 cells. The possibility of extracting the endotoxic glycolipid directly with a chloroform-methanol (C-M) mixture was examined. The cells were washed with distilled water three times and lyophilized. The lyophilized cells (30 g) were extracted with a C-M 4:1 mixture (450 ml) by refluxing at 50 C for 6 hr. The residue after filtration was extracted two more times with C-M 4:1 (each time with 200 ml). The combined filtrate was concentrated in vacuo to dryness, dissolved again in C-M 4:1, then precipitated with methanol until the final concentration was C-M 1:2. The precipitate was dissolved in C-M 4:1 again, and this procedure was repeated two more times. The final soluble fraction in C-M 4:1 was termed directly extracted glycolipid (GL-DE) (figure 1).

Preparation of samples for biological assays. The samples of glycolipid used in biological assays were suspensions in physiological saline. Aliquots of glycolipid samples were evaporated to dryness under nitrogen, 90% of the final volume was added as pyrogenfree water, and the samples were sonicated at 1.7 A for 1 min; 10%

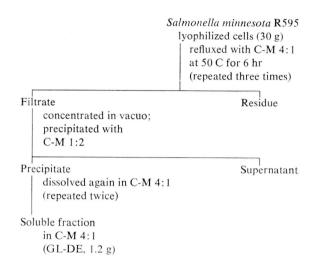

Figure 1. Direct extraction of endotoxic glycolipid (GL-DE) from *Salmonella minnesota* R595 cells with chloroform-methanol (C-M).

of the final volume was then added as 8.7% pyrogenfree sodium chloride.

Tumors. Ehrlich ascites tumor (EAT) was provided by Dr. R. Baserga of Temple University School of Medicine and maintained in the peritoneal cavities of C57B1/10 mice. Sarcoma 37 (S-37) was provided by Dr. H. F. Havas of Temple University School of Medicine and was maintained intraperitoneally in ICR mice. All mice used were males weighing 18–20 g at the time of inoculation of tumors.

Shwartzman reaction. Shwartzman skin tests were done according to the routine assay in this laboratory [9].

Limulus lysate assay. The *Limulus* lysate gelation assay was carried out as previously described [10]. A mixture of amebocyte lysate (0.05 ml) and material being tested (0.05 ml) was incubated in a water bath at 37 C for 4 hr and then at room temperature (24 C). The samples were observed periodically for 24 hr. The activity of the material being tested was expressed as the lowest concentration needed to form a solid gel.

Lethality to mice. Lethality to mice was determined in the usual manner [9] except that all injections were iv.

Tumor-hemorrhage assay. This assay was done by the method of Shear et al. [11] on ICR mice bearing S-37. Inocula of 6.5×10^6 S-37 cells were implanted subcutaneously seven days before iv administration of the endotoxic material. Hemorrhage was observed 24 hr after challenge by iv injection of 0.2-ml quantities into the lateral tail vein.

Chick-embryo lethality. Chick-embryo lethality was determined by the method of Smith and Thomas [12] as modified by Finkelstein [13].

Fifty percent end point. In assays of lethality and tumor hemorrhage, the 50% end point was determined by the Spearman-Karber method. Details of the procedure and calculations have been described elsewhere [9].

Adjuvant effect. Four- to six-week-old BALB mice, male and female, were used in studies of the adjuvant effect of GL-DE. Both bovine gamma globulin (BGG) and ovine red blood cells (oRBC) were used as antigens. Either 500 µg of BGG or 10^7 oRBC were injected intraperitoneally, either alone or in combination with amounts of GL-DE ranging from 0.1 µg to 100 µg, as indicated. Mice were bled at eight- and 10-day intervals, and titer was determined by standard active or passive hemagglutination methods. Enhancement of immunologic response by GL-DE was also determined by the Jerne plaque method [14].

Chemical analysis. The dry weight of the glycolipid samples used for biological assays and for chemical analysis was determined by microprocedure on a Cahn gram-electrobalance after the samples were dried under an infrared lamp. A new procedure has been established for the determination of molar ratios among hexosamine, fatty acids, phosphorus, and KDO in silicic-acid scrapings from thin-layer chromatographic plates without elution of the separated material. Phosphorus content was measured by the micromethod of Chen et al. [15]. Hexosamine was determined according to the method of Rondle and Morgan [9, 16]. Total fatty acids were determined by the procedure of Snyder and Stephens [17] after transesterification with boron trifluoride-methanol reagent [9]. KDO was determined based on the method of Weissbach and Hurwitz [18] as modified by Osborn [19]. Details of the methods will be published elsewhere.

Chromatography on a silicic acid column. A silicic acid column (1.5 × 95 cm) was prepared with a slurry of silicic acid (BioSil A, 200–325 mesh) in chloroform-methanol-water-ammonium hydroxide (100:50:8:4). The same solvent was used as the eluent. The flow rate was about 0.7 ml/min, and fractions of 4 ml were collected.

Thin-layer chromatography. All TLC was done with silicic acid BioSil A (2–10 µm) or Warner-Chilcott silica gel (plain) and with plates 20 × 20 cm in size and 0.25 cm thick. Thin layers of the preparations were made with either distilled water, distilled water containing ammonium hydroxide, or a phosphate buffer. The phosphate buffer [4] used throughout this work was prepared by adjustment of 0.1 M KH_2PO_4 solution to *p*H 7.8 with 0.1 N NaOH. To 30 g of silicic acid, 60 ml of H_2O, H_2O containing 4 ml of NH_4OH, or phosphate buffer was added, and, in the latter case, the *p*H of the slurry was again adjusted to 7.8. The plates were air-dried, activated at 120 C for 1 hr in a vacuum oven, then cooled in a desiccator.

The solvent systems used in these experiments were (*1*) chloroform-methanol-water (65:25:4); and (*2*) chloroform-methanol-water-ammonium hydroxide (100:50:8:4).

The spray reagents used, in addition to a dichromate-sulfuric acid reagent, were 0.2% ninhydrin in acetone for amino compounds and the molybdenum spray of Dittmer [20] for phosphorus-containing components.

Gas-liquid chromatography (GLC). Without elution, samples on the scraped silicic acid were transesterified with boron trifluoride-methanol reagent [9] and examined for fatty-acid esters by gas chromatography with use of a F and M 609 instrument equipped with hydrogen-flame ionization detector. The components on the chromatograms were identified by cochromatography with the authentic fatty-acid methyl esters.

High-voltage paper electrophoresis (HVE). Samples hydrolyzed with 6.1 N HCl for 4 hr at 100 C were examined for amino compounds in the usual manner [9].

Paper chromatography for carbohydrates. The water-soluble portion of a glycolipid hydrolysate (2 N aqueous HCl for 4 hr at 100 C) was examined for reducing compounds after the paper was developed in pyridine-ethyl acetate-water (5:2:5) (upper organic layer).

Ultraviolet spectra. Samples were examined as a 0.1% solution with a Beckman-DB spectrophotometer.

38

Results

Isolation of glycolipid. The yield of the endotoxic glycolipid (GL) extracted by the modified phenol-water extraction method [8] was 1% of the cells on a dry-weight basis. By the direct-extraction method with chloroform-methanol, 4% of the yield was obtained for GL-DE.

Examination of GL and GL-DE by paper chromatography and UV-absorption analysis showed that glycolipid was free of proteins, nucleic acids, and reducing sugars (other than glucosamine). TLC and HVE indicated the presence of phosphatidylethanolamine, spermidine, and putrescine. The same observation has been reported by Gmeiner et al. [21].

When GL was examined on BioSil A 2–10-μm TLC plates developed in solvent system A, a diffuse single component was obtained [8]. However, of numerous combinations of silicic acid and solvent system that were tried, some were able to separate GL into several components. One example of the separation of GL and GL-DE by TLC is shown in figure 2. The components shown were detected by spraying with dichromate-sulfuric acid. All of the major compounds were phosphate-positive when the plates were sprayed with molybdenum reagent. It is evident, therefore, that the endotoxic glycolipids are heterogeneous. By all systems examined, GL and GL-DE displayed a very similar pattern; the only difference was that GL contained more components at the solvent front.

In an attempt to fractionate the components of GL on a preparative scale, we applied 20 mg of GL-DE on a silicic acid column and eluted with solvent B. TLC has previously shown that all the constituents of GL-DE contain phosphorus, and, therefore, the column was monitored by analysis of the fractions from every alternate tube for their phosphorus content. The result is shown in figure 3. There were two major peaks, and each still contained more than one component. This pattern is similar to that produced on TLC plates with use of the same solvent system (figure 2). When the pooled fractions from each major peak were examined on TLC, the first peak corresponded to fraction *e* and the second to fractions *a, b, c,* and *d* of figure 2.

The system shown in figure 2 was chosen for the purification of components from GL. About 1 mg of GL-DE was applied on a plate, and the components were detected by spraying with water after the plate was developed. Five fractions, *a, b, c, d,* and *e,* were scraped from the plates and eluted on a medium glass Büchner funnel with chloroform-methanol mixture. The filtrates were filtered through an ultrafine glass funnel for removal of some of the silicic acid contamination. The filtrates were evaporated to dryness under nitrogen and dissolved again in C-M 3:1.

When the isolated fractions were examined on

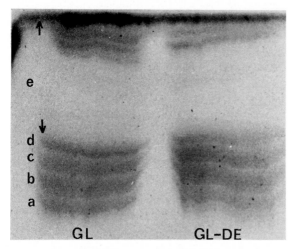

Figure 2. Thin-layer chromatography of glycolipid with silicic acid BioSil A (2–10 μm) and solvent system chloroform-methanol-water-ammonium hydroxide (100:50:8:4). Silicic acid slurry was prepared with distilled water-ammonium hydroxide. R_f: *a* = 0.30; *b* = 0.37; *c* = 0.46; *d* = 0.51.

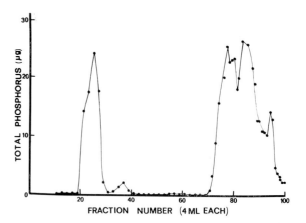

Figure 3. Chromatography on silicic acid column of the glycolipid (GL-DE). Column: BioSil A (200–325 mesh), 1.5 × 95 cm. Eluant: chloroform-methanol-water-ammonium hydroxide (100:50:8:4).

TLC by the same system, each fraction still contained the other components. Therefore, purification was done three times, and the isolated fractions were examined again. The chromatographic pattern is shown in figure 4 and indicates that the four components are still present in each of the four fractions.

Chemical analysis of the fractions. Although these five fractions had not been completely purified, some biological parameters and chemical analyses were done on them. The chemical analyses indicated that, whereas fractions *a, b, c,* and *d* contained KDO, hexosamine, phosphorus, and fatty acids, fraction *e* contained mainly phosphorus and fatty acids. The molar ratios of the fractions are shown in table 1. Fractions *a, b, c,* and *d* have quite similar molar ratios for KDO, hexosamine, fatty acids, and phosphorus that are in the range of reported data [8, 22] for GL from *S. minnesota* R595. The results indicate that *a, b, c,* and *d* are GL in nature, whereas fraction *e* con-

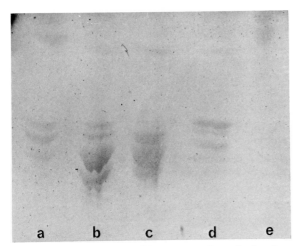

Figure 4. Thin-layer chromatography of the isolated fractions (*a, b, c, d,* and *e*) using the same system as in figure 2.

Table 1. Molar ratios of the glycolipid fractions.

Glycolipid fraction	Hexosamine	Fatty-acid ester	P	KDO*
GL-DE†	1	2.97	2.24	0.91
a	1	2.90	1.69	1.07
b	1	2.75	1.60	0.93
c	1	3.10	1.36	0.83
d	1	2.67	1.34	1.08

 * 2-keto-deoxy-octonate.
 † Glycolipid directly extracted with chloroform-methanol.

tains other phospholipids. The high phosphorus content of fraction *e* contributes to the higher phosphorus content in GL-DE.

GLC revealed that fractions *a, b, c,* and *d* contained the same major fatty acids, i.e., lauric, myristic, palmitic, and β-hydroxymyristic acids (figure 5). The relative ratios of these acids in the four fractions are similar. In comparison with these fractions, GL-DE contains more palmitic and stearic acids, which are probably the major fatty acids in fraction *e.*

Biological activities of the fractions. The biological activities of the GL fractions were determined and compared with the original GL material. The Schwartzman skin reactivity of GL-DE and the fractions is shown in figure 6. Compared with an endotoxin prepared from a smooth strain of *S. marcescens,* GL-DE and fractions *a, b, c,* and *d* were at least as active as the smooth-strain endotoxin, whereas fraction *e* had much less activity.

GL, GL-DE, and the fractions from TLC were examined for their ability to coagulate *Limulus* lysate. This method [10] can detect as little as 10^{-9}–10^{-10} g/ml of endotoxin. The results shown in table 2 indicate that both GL and GL-DE have activity comparable to that of endotoxin. It is noted that, with the exception of fraction *e,* all fractions showed similar activity.

The assay of chick-embryo lethality also showed that toxicity of both GL and GL-DE was similar to that of endotoxin. The activity of TLC fractions, expressed as LD_{50} (table 2), reveals again that fractions *a, b, c,* and *d* had similar activity, but fraction *e* was not active in this assay.

The data from the above three biological assays indicate that fraction *e* is most different from the GL and that the activity of fractions *a, b, c,* and *d* is almost identical and is in the range of the parent material.

The lethality of GL was investigated in both normal and tumor-bearing mice. The results shown in table 3 confirm the findings of Havas et al. [23] that tumor-bearing mice are more sensitive to endotoxin lethality. It is noteworthy that the C57B1/10 strain is more sensitive to the lethal effects of GL than any other strain of mice tested. This extreme sensitivity is unexplained and deserves further study. In the case of the ICR strain, the results indicate that although GL was toxic, it was not as active as endotoxin in this assay, thus confirming the findings of Tripodi

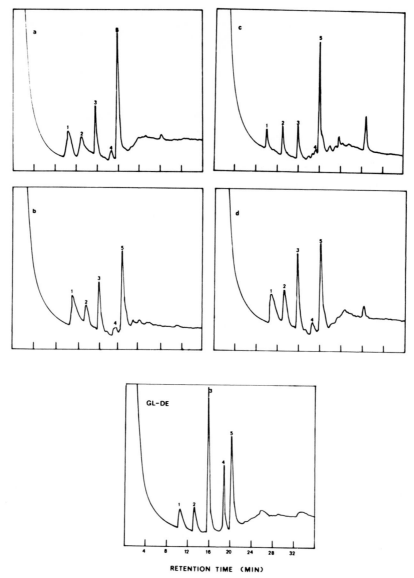

Figure 5. Gas-liquid chromatograms of fatty acids present in glycolipid fractions. Column: 3% ECNSS-M on 100/120 gas chrom Q. 12 ft. Temperature programmed from 50 C–185 C, 6.4 C/min. GL-DE = directly extracted glycolipid; *a, b, c, and d* = fractions obtained from glycolipid by thin-layer chromatography. 1 = lauric acid, 2 = myristic acid, 3 = palmitic acid, 4 = stearic acid, 5 = β-hydroxy-myristic acid.

and Nowotny [24]. Examination of the lethality of GL-DE in ICR mice showed activity similar to that of GL.

The tumor hemorrhagic activity of glycolipid (GL and GL-DE) was compared with that of endotoxin (table 4). The data clearly indicate that both materials are active in tumor-hemorrhage assay and, when compared on the basis of the 50% end point dose (THD_{50}), the endotoxic glycolipid appears to be more active than endotoxin.

The ability of GL-DE to act as an adjuvant was examined (table 5). When oRBC was used as an antigen, 100 µg of GL increased plaque-forming cells (PFC) to 44 PFC/10^6 from 3 PFC/10^6 in the control. With bovine serum albumin as the antigen, 10 µg of GL gave an antibody titer of 2,560, as compared with 40 in the control. The results show the strong adjuvant effect of GL. Both soluble and particulate antigens seem to be affected. There is a dose-response relationship to the amount of GL. When the humoral response is determined

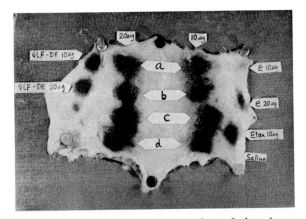

Figure 6. The Shwartzman reaction of the glycolipid fractions obtained from thin-layer chromatography. Etox = *Serratia marcescens* endotoxin; GLF-DE = directly extracted glycolipid; *a, b, c, d,* and *e* = fractions obtained from glycolipid by thin-layer chromatography.

Table 2. *Limulus* lysate activity and chick-embryo lethality of the glycolipid and the fractions.

Glycolipid fractions	Minimum (g/ml) to gel in *Limulus* assay	LD_{50} (µg) in chick-embryo lethality
Endotoxin	10^{-9}	0.007
GL*	10^{-8}	0.01
GL-DE†	10^{-9}	0.005
a	10^{-8}	0.03
b	10^{-9}	0.09
c	10^{-8}	0.05
d	10^{-8}	0.06
e	10^{-7}	>10

  * Glycolipid extracted by phenol-water method.

  † Glycolipid directly extracted with chloroform-methanol.

Table 3. Lethality (LD_{50} in µg) of endotoxic materials in normal and tumor-bearing mice.

Mouse strain and tumor	Endotoxin	Glycolipid
ICR normal	400	1,105
ICR with Sarcoma-37	80	90
C57B1/10 normal	230	70
C57B1/10 with Ehrlich ascites tumor	100	35

Table 4. Tumor hemorrhagic activity of endotoxic materials.

Endotoxic material	No. of mice injected	Lowest dose (µg) causing hemorrhage in 100% of animals	THD_{50}* (µg)
Saline control	12	. . .†	. . .
Glycolipid	30	20	5
Endotoxin	42	25	10

  * Fifty percent end point dose.

  † No hemorrhage induced.

Table 5. Adjuvant effect of glycolipid (directly extracted with chloroform-methanol) in the mouse.

Glycolipid (µg)	Antibody titer for BGG*	PFC/10^6 for oRBC†
0	40	3
0.1	160	2
1	1,280	10
10	2,560	24
100	640	44

  * Mice were bled eight days after injection; 500 µg of BGG (bovine gamma globulin) was used as antigen.

  † Spleens were removed four days after injection; 10^7 oRBC (ovine red blood cells) were used as antigen.

parative TLC, fractions *a, b, c,* and *d* have activities similar to those of the parent GL when examined in tests of the Shwartzman skin reaction, *Limulus* lysate clotting ability, and chick-embryo lethality. Fraction *e,* on the other hand, is much less active. The chemical analyses indicate that fractions *a, b, c,* and *d* have similar molar ratios and are GL in nature, whereas fraction *e* probably consists of other phospholipids that do not contain much KDO and hexosamine. The similarity in biological activities as well as chemistry suggests that fractions *a, b, c,* and *d* either are only slightly different in chemistry (e.g., in the substitution of fatty acids on glucosamine molecules) or differ chromatographically due to polymerism. Studies are continuing in our laboratories on this structure-function relationship.

Different components have not yet been purified by TLC. Chromatography on the silicic acid column appears to be a good step for removal of nonglycolipid material, i.e., fraction *e* and others. The repeated appearance of fractions *a, b, c,* and *d* in the purified fraction cannot be explained thus far. Although contamination of the adjacent fractions during purification cannot be totally ruled out, their contribution should be

by the hemagglutination method, an excess of GL appears to diminish the adjuvant effect.

It has been reported previously [8] that TLC of GL did not show a great degree of heterogeneity, but TLC with use of improved systems has demonstrated extensive heterogeneity of the product. Of the five fractions obtained by pre-

minor after purification has been repeated three times.

Discussion

The results reported here regarding the distribution of GL when extracted from *S. minnesota* R595 cells with 45% aqueous phenol are in agreement with those found earlier [8]; i.e., most of the endotoxic material is located in the middle, insoluble phase. This result is contrary to that reported by Lüderitz et al. [5], who found, using the same strain, that the GL could be located in the aqueous phase. Small differences in conditions of growth as well as in procedures of extraction may influence the different distribution of the lipophilic material.

The direct-extraction method from lyophilized R595 cells with a chloroform-methanol mixture indicates that GL is easily extracted. The yield of GL material (GL-DE) obtained by this method is four times higher than that obtained by the phenol-water method. The TLC patterns in several systems and the purity of GL-DE are very similar to those of GL. The biological activity of GL-DE is at least comparable to or even higher than that of GL. It is interesting that when the water-soluble GL isolated according to the procedure of Galanos et al. [25] is examined by our TLC method, a similar heterogeneity is observed (authors' unpublished observation). The similar molar ratios reported here in these four fractions support the possibility that they are different bands formed by polymerism of one molecular species. Preliminary experiments to dissociate these components have not been successful. GL in the presence of sodium dodecylsulfate or EDTA gave the same pattern of separation on the plates. The average molecular weight of the unpurified GL is approximately 4,000, as determined by the vapor pressure-depression method (authors' unpublished observation). Separation by size and other methods of purification will be done, and we hope that they will shed some light on this phenomenon.

The chromatographic behavior of phospholipids has long been known to depend on the type of metal ion combined with the lipid [26]. Kasai et al. recently found that divalent cations have profound influence on the physicochemical and biological properties of the glycolipid. The repeated appearance of the four bands in the purified fraction in our case also indicates the possible involvement of metal ions in this phenomenon.

The biological properties of the GL reported here have demonstrated that the activity of GL is comparable to that of a smooth-strain endotoxin in Shwartzman skin reaction, *Limulus* lysate clotting ability, chick-embryo lethality, mouse lethality, tumor-hemorrhage activity, and adjuvant effect. On the basis of these results, it would seem that some of the active sites of endotoxin are located in the lipid region.

The results demonstrate that the GL is heterogeneous and that there are at least four fractions that are chemically and biologically GL in nature. The object of the work in these laboratories will be to purify these components further and to study the structure-function relationship.

References

1. Nowotny, A. Chemical structure of a phosphomucolipid and its occurrence in some strains of *Salmonella*. J. Am. Chem. Soc. 83:501–503, 1961.
2. Nowotny, A. Relation of structure to function in bacterial O antigens. II. Fractionation of lipids present in Boivin-type endotoxin of *Serratia marcescens*. J. Bacteriol. 85:427–435, 1963.
3. Kasai, N., Yamano, A. Studies on the lipids of endotoxins. Thin-layer chromatography of lipid fractions. Jap. J. Exp. Med. 34:329–344, 1964.
4. Sweeney, M. J. Biological and chemical studies of purified "lipid A" fractions from *Serratia marcescens* endotoxin. Doctoral (Ph.D.) thesis, Temple University, 1971, p. 260.
5. Lüderitz, O., Galanos, C., Risse, H. J., Ruschmann, E., Schlecht, S., Schmidt, G., Schulte-Holthausen, H., Wheat, R., Westphal, O., Schlosshardt, J. Structural relationship of *Salmonella* O and R antigens. Ann. N.Y. Acad. Sci. 133:349–374, 1966.
6. Nowotny, A. Heterogeneity of endotoxic bacterial lipopolysaccharides revealed by ion-exchange column chromatography. Nature (Lond.) 210: 278–280, 1966.
7. Nowotny, A. Chemical and biological heterogeneity of endotoxins. *In* G. Weinbaum, S. Kadis, and S. J. Ajl [ed.]. Microbial toxins. vol. IV. Academic Press, New York, 1971, p. 309–329.
8. Kasai, N., Nowotny, A. Endotoxic glycolipid from a heptoseless mutant of *Salmonella minnesota*. J. Bacteriol. 94:1824–1836, 1967.
9. Nowotny, A. Basic exercises in immunochemistry. Springer Verlag, New York, 1969, p. 197.
10. Levin, J., Poore, T. E., Zauber, N. P., Oser, R. S. Detection of endotoxin in the blood of patients

with sepsis due to gram-negative bacteria. N. Engl. J. Med. 283:1313–1316, 1970.

11. Shear, M. J., Perreault, A., Adams, J. R., Jr. Chemical treatment of tumors. VI. Method employed in determining the potency of hemorrhage-producing bacterial preparations. J. Natl. Cancer Inst. 4:99–105, 1943.

12. Smith, R. T., Thomas, L. The lethal effect of endotoxins on the chick embryo. J. Exp. Med. 104: 217–231, 1956.

13. Finkelstein, R. A. Observations on mode of action of endotoxin in chick embryos. Proc. Soc. Exp. Biol. Med. 115:702–707, 1964.

14. Jerne, N. K., Nordin, A. A. Plaque formation in agar by single antibody-producing cells. Science 140:405, 1963.

15. Chen, P. S., Toribara, T. Y., Warner, H. Microdetermination of phosphorus. Anal. Chem. 28: 1756–1758, 1956.

16. Rondle, C. J. M., Morgan, W. T. J. The determination of glucosamine and galactosamine. Biochem. J. 61:586–589, 1955.

17. Snyder, F., Stephens, N. A simplified spectrophotometric determination of ester groups in lipids. Biochem. Biophys. Acta 34:244–245, 1959.

18. Weissbach, A., Hurwitz, J. The formation of 2-keto-3-deoxyheptonic acid in extracts of *Escherichia coli* B. J. Biol. Chem. 234:705–709, 1959.

19. Osborn, M. J. Studies on the gram-negative cell wall. I. Evidence for the role of 2-keto-3-deoxyoctonate in the lipopolysaccharide of *Salmonella typhimurium*. Proc. Natl. Acad. Sci. U.S.A. 50: 499–506, 1963.

20. Dittmer, J. C., Lester, R. L. A simple, specific spray for the detection of phospholipids on thin-layer chromatograms. J. Lipid Res. 5:126–127, 1964.

21. Gmeiner, J., Lüderitz, O., Westphal, O. Biochemical studies on lipopolysaccharides of *Salmonella* R mutants. VI. Investigations on the structure of the lipid A component. Eur. J. Biochem. 7:370–379, 1969.

22. Galanos, C., Rietschel, E. T., Lüderitz, O., Westphal, O. Interaction of lipopolysaccharides and lipid A with complement. Eur. J. Biochem. 19: 143–152, 1971.

23. Havas, H. F., Donnelly, A. J., Levine, S. I. Mixed bacterial toxins in the treatment of tumors. III. Effect of tumor removal on the toxicity and mortality rates in mice. Cancer Res. 20:393–396, 1960.

24. Tripodi, D., Nowotny, A. Relation of structure to function in bacterial O-antigens. V. Nature of active sites in endotoxic lipopolysaccharides of *Serratia marcescens*. Ann. N.Y. Acad. Sci. 133: 604–621, 1966.

25. Galanos, C., Lüderitz, O., Westphal, O. A new method for the extraction of R lipopolysaccharides. Eur. J. Biochem. 9:245–249, 1969.

26. Carter, H. E., Weber, E. J. Preparation and properties of various salt forms of plant phosphatidyl inositols. Lipids 1:16–20, 1966.

Summary of Discussion

Edward C. Heath

The comments and discussion during this session focused on three general aspects of bacterial cell-wall lipopolysaccharides (LPS).

(*1*) The remaining questions concerning the biochemistry of these polymers pertain primarily to their mode of biosynthesis and assembly in the complex membrane of the cell envelope. Of particular concern are the genetic and metabolic regulation of biosynthesis and the mechanisms for organization of the macromolecular constituents in the outer membrane.

(*2*) The physiologic role of LPS and other macromolecular constituents and lipids of the cell envelope in phage infection, processes of active transport, and cell integrity were discussed extensively.

(*3*) Significant progress has been made in defining in more concrete terms the molecular bases for the complex array of toxic effects of LPS. It is essential, however, to elucidate firmly the specific mechanisms of toxic and immunologic responses of mammalian cells to the complex LPS of the bacterial-cell envelope.

SESSION II: IMMUNOLOGY

Immunogenicity, Tolerogenicity, and Mitogenicity of Lipopolysaccharides

Göran Möller, Olof Sjöberg, and Jan Andersson

From the Division of Immunobiology, Karolinska Institute, Stockholm, Sweden, and the Basel Institute for Immunology, Basel, Switzerland

Lipopolysaccharide (LPS) is an immunogen that often induces synthesis of IgM only, and these antibodies exhibit regular cyclical fluctuations due to an antibody-mediated feed-back regulation. LPS can induce high-dose immunologic tolerance; however, tolerant animals have an increased number of antigen-binding cells, even though the number of antibody-producing cells is depressed. Tolerance to LPS is rapidly broken by incubation of lymphocytes in tissue culture for 24 hr or by adoptive transfer of tolerant cells to irradiated hosts, suggesting the existence of cells tolerant to this antigen. LPS is also a mitogen capable of activating DNA synthesis in bone-marrow (B) cells but not in thymus (T) cells. LPS-activated B cells secrete IgM antibodies against a variety of antigens. It is most likely that LPS-activated B lymphocytes differentiate into cells that produce antibody at a high rate and that express their genetically determined antibody specificity. When LPS is coated onto T-cell-dependent, heterologous red cells, the latter become T-cell-independent, presumably because of the mitogenic properties of LPS. LPS can substitute for both T cells and macrophages in the induction of a primary immune response in vitro, suggesting that LPS acts directly on B cells and that at least one function of the helper is immunologically nonspecific.

Lipopolysaccharides (LPS) from gram-negative bacteria have certain immunologic characteristics that distinguish them from most other antigens. These unique features of LPS concern its immunogenicity as well as its tolerogenicity. In addition, it has recently been found that LPS acts as a mitogen capable of stimulating DNA synthesis in subpopulations of lymphocytes from various species. This property of LPS has made it possible to study processes leading to activation of immunocytes and initiation of immunoglobulin synthesis. We shall consider in turn the various aspects of LPS that distinguish it from other immunogenic molecules.

Immunogenicity

LPS is an antigen and gives rise to synthesis of specific antibody when introduced into animals.

Please address requests for reprints to Dr. Göran Möller, Division of Immunobiology, Karolinska Institute, Wallenberglaboratory, Lilla Freskati, 104 05 Stockholm 50, Sweden.

However, the immune response to LPS is usually characterized by prolonged synthesis of IgM, often in the total absence of IgG. It has also been found that LPS causes a cyclical appearance of IgM antibodies [1]. The cyclicity is due to a feed-back mechanism, in which antibodies suppress their own synthesis by combining with the antigen molecule [1]. This event leads to a dramatic fall in the number of IgM molecules formed; thereafter, the antigen, which is immunogenic for prolonged periods in the host animal [2], becomes free to initiate a second cycle of antibody synthesis.

The mechanism responsible for the preferential stimulation of IgM synthesis is not known. In many cases, IgG synthesis against LPS appears, but it is not fully understood what properties of the LPS molecule determine the induction of IgG versus IgM synthesis.

Most protein antigens and cellular antigens require cooperation between thymus-derived (T) and bone-marrow-derived (B) lymphocytes for induction of antibody synthesis [3]. The T lymphocytes themselves do not produce antibodies but are necessary for activation of the B lymphocytes to division and immunoglobulin synthesis. The mechanism of cooperation of T and B cells has been the subject of various hypotheses, but it seems to have been established that very close physical contact is needed between the two cells. The most commonly accepted hypothesis is that antigen bridges the T and B cells. However, LPS does not require this cellular cooperation and is fully competent to initiate a primary immune response in the total absence of T cells [4]. This property is shared with certain other antigens, such as polyvinylpyrrolidone, pneumococcal polysaccharide, and polymerized flagellin [5–7]. The finding of T-independent immune responses is of great theoretical interest, because it demonstrates that cooperation of T and B cells is not an obligatory event in triggering of immunocytes but rather represents a helper mechanism. The basis for the T-cell-independent immune responses is unknown, but various hypotheses have been suggested [8–10]. It seems plausible that the repeated antigenic determinants characteristic of LPS molecules may be responsible for this effect in that they cause multiple bonding to the reactive B-cell receptors.

Although LPS is a potent immunogen in most strains of mice, it has been shown that at least one strain (C3H/HeJ) responds poorly [11]. The ability to respond is under genetic control, and several genes appear to be involved [11]. It has also been found that lymphocytes from the nonresponder strain failed to become activated by the mitogenic property of LPS. The existence of this genetic model opens the possibility of studying the mechanism of immunocyte activation by LPS.

Tolerogenicity

LPS is a conventional antigen in the sense that it induces immunologic tolerance, which is as specific as the immune response. Tolerance to certain protein antigens has been shown to exist in two zones of dosages [12]. However, LPS induces only high-dose tolerance. At present, it is generally accepted that high- and low-zone tolerance can be mapped onto the well-known T- and B-cell cooperation phenomena. Thus, low-zone tolerance is presumed to affect only T cells, whereas high-zone tolerance affects both T and B cells [13]. Since LPS is a T-cell-independent antigen, it is not surprising that tolerance exists only in a high-dose range.

Although tolerance to LPS is expressed as the absence of antibody-producing cells, it has nevertheless been found that animals tolerant to LPS have an increased number of antigen-binding cells [14]. Such antigen-binding cells increase in the immune response to all antigens studied. They are detected by their ability to fix antigen to their surface. If the antigen is present on a particle, such lymphocytes form rosettes. Rosette-forming cells (RFC) to ovine red-blood cells and haptens are of both T- and B-cell origin, whereas it was found that the rosette-forming cells against LPS were exclusively of B origin [15]. Thus, T cells appear to be incapable of recognizing LPS. In tolerant animals, it was found that the number of RFC was considerably increased over background but did not reach levels as high as those in specifically immune animals. Analogous findings have been made with certain haptens [16].

Various alternative suggestions have been made in explanation of the high numbers of antigen-binding cells in tolerant animals [17]. Although these alternatives will not be discussed here, it seems likely that the antigen-binding cells are probably of immunologic significance. This was shown by experiments in which spleen cells from tolerant animals either were incubated in tissue culture for 24 hr and thereafter adoptively transferred to irradiated animals or were directly transferred from the tolerant mouse to an irradiated host [17]. It was found that both procedures resulted in a rapid loss of tolerance, whereas tolerance persisted for the normal period when the cells remained in the original host. These findings suggest that tolerant cells do, in fact, exist but are blocked from expressing immunologic reactivity. Since the blockade is immunologically specific, it is tempting to suggest that it is due to antigen or antibody or both. Anyhow, these findings, which are analogous to those made with pneumococcal polysaccharide [18], do indicate the existence of tolerant cells. Furthermore, they suggest that the antigen-binding cells that increase in numbers in tolerant animals are in fact specifically competent cells prevented from carrying out their immunologic function.

46

Mitogenicity

It was originally observed by Peavy et al. [19] that LPS is a mitogen. It has subsequently been clearly shown that LPS affects DNA synthesis in B cells but not in T cells (figure 1). In contrast, other mitogens such as concanavalin A are exclusively T-cell mitogens. It is of importance to study the immunologic properties of B cells that have been nonspecifically activated by LPS. It was demonstrated initially that such cells showed an increased synthesis of IgM immunoglobulin (figure 2) as compared to untreated cells, whereas there was little or no synthesis of IgG [20]. Furthermore, it was demonstrated that the number of cells producing antibody to a variety of cellular antigens and haptens increased in parallel with induction of DNA synthesis (figure 3); this finding suggested strongly that LPS-induced mitoses of B cells resulted in division and differentiation into cells that secreted antibody at a high rate [20]. As shown in several experiments, LPS-activated lymphocytes express the genetically determined ability to synthesize specific antibodies [21].

The B-cell mitogenic property of LPS is presumably the mechanism responsible for the interesting observation that LPS-coated, heterologous red cells become independent of T cells [22]. Normally, ovine red cells require T and B cells

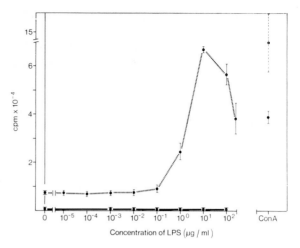

Figure 1. Uptake of ³H-thymidine on third day of culture of bone-marrow (●----●) and thymus (●——●) lymphocytes after exposure to various concentrations of lipopolysaccharide (LPS) (ConA = concanavalin A).

for the development of antibody-producing cells. However, if LPS is coated onto ovine red cells, and if these cells are subsequently injected into animals deprived of T cells, there is a normal immune response to the red cells (figure 4) [22]. LPS also converted red cells into T-independent antigens in LPS-tolerant animals, suggesting that an immune response to LPS was not responsible for this phenomenon. Actually, passive transfer

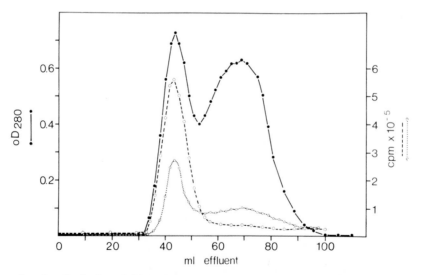

Figure 2. Separation by Sephadex G-200 of 50% ammonium sulfate-precipitated supernatants from untreated spleen-cell cultures (○ · · · ·○) and cultures treated with 10 μg/ml of lipopolysaccharide (○----○) after labeling with 10 μCi/ml of ³H-leucine for two days before harvest. The optical density at 280 nm of normal murine serum used as carrier in the precipitation (●——●) is also indicated.

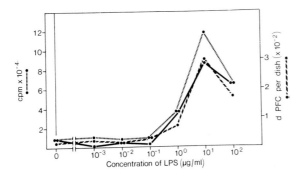

Figure 3. Number of antibody-forming cells (αPFC) to ovine red cells (●——●) and equine red cells (●————●) in normal spleen-cell cultures treated with various doses of lipopolysaccharide. Incorporation of ³H-thymidine into the cells from the same cultures is also indicated (●····●).

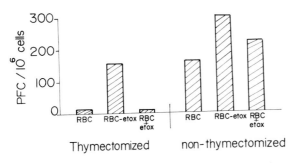

Figure 4. Plaque-forming-cell response against equine red cells in thymectomized or nonthymectomized, lethally irradiated, and bone-marrow-repopulated mice injected with the indicated cells. **RBC—etox** indicates red cells coated with lipopolysaccharide (LPS) from *Escherichia coli* O127. **RBC+etox** indicates that red cells and LPS were injected separately by different routes.

of antibody to LPS suppressed its ability to convert red cells into thymus-independent antigens. Presumably, the specific B cells, which recognized and bound ovine cells to their surface, were at the same time confronted with the mitogenic property of LPS, which made them respond to the ovine cells. This, together with a variety of other findings [21], suggests that at least one function of the T cell is nonspecific, because it can be replaced by a nonspecific mitogen such as LPS.

Further evidence for this theory comes from studies showing that LPS can be substituted for various helper cells in the induction of a primary immune response in vitro. Both T cells and macrophages are necessary for a primary immune re-

sponse to haptens and red cells in vitro. Thus, depletion of T cells by treatment of the suspension with anti-theta serum abolishes its ability to produce antibodies to ovine red cells. If adherent cells are removed by treatment with iron, followed by a removal of the cells with a magnet, there is also a drastic reduction in immunocompetence against ovine red cells [23]. However, that LPS can substitute both for T cells and adherent cells strongly indicates (*1*) that LPS acts directly on B cells and not on any type of helper cell, such as T cells and macrophages, and (*2*) that the function of these two helper cells is likely to be immunologically nonspecific, because it can be replaced by LPS.

References

1. Britton, S., Möller, G. Regulation of antibody synthesis against *Escherichia coli* endotoxin. I. Suppressive effect of endogenously produced and passively transferred antibodies. J. Immunol. 100: 1326–1334, 1968.
2. Britton, S., Wepsic, T., Möller, G. Persistence of immunogenicity of two complex antigens retained in vivo. Immunology 14:491–501, 1968.
3. Möller, G. [ed.] Antigen-sensitive cells: their source and differentiation. Transpl. Rev. 1:3–110, 1969.
4. Möller, G., Michael, G. Frequency of antigen-sensitive cells to thymus-independent antigens. Cell. Immunol. 2:309–316, 1971.
5. Andersson, B., Blomgren, H. Evidence for thymus-independent humoral antibody production in mice against polyvinylpyrrolidone and *E. coli* lipopolysaccharide. Cell. Immunol. 2:411–424, 1971.
6. Howard, J. G., Christie, G. H., Courtenay, B. M., Leuchars, E., Davies, A. J. S. Studies on immunological paralysis. VI. Thymic-independence of tolerance and immunity to type III pneumococcal polysaccharide. Cell. Immunol. 2:614–626, 1971.
7. Feldmann, M., Basten, A. The relationship between antigenic structure and the requirement for thymus-derived cells in the immune response. J. Exp. Med. 134:103–119, 1971.
8. Taylor, R. B., Iverson, G. M. Hapten competition and the nature of cell-cooperation in the antibody response. Proc. R. Soc. Lond. [Biol.] 176:393–418, 1971.
9. Mitchison, N. A. Cell cooperation in the immune response: the hypothesis of an antigen presentation mechanism. Immunopathology 6:52–60, 1971.
10. Möller, G. Editorial: immunocyte triggering. Cell. Immunol. 1:573–582, 1970.
11. Sultzer, B. M. Genetic control of host responses to endotoxin. Infec. Immun. 5:107–113, 1972.
12. Mitchison, N. A. Induction of immunological paralysis in two zones of dosage. Proc. R. Soc. Lond. [Biol.] 161:275–292, 1964.

13. Möller, G. [ed.]. Immunological tolerance. Effect of antigen on different cell populations. Transpl. Rev. 8:3–136, 1972.

14. Sjöberg, O. Antigen-binding cells in mice immune or tolerant to *Escherichia coli* lipopolysaccharide. J. Exp. Med. 133:1015–1025, 1971.

15. Möller, E., Sjöberg, O. Antigen-binding cells in immune and tolerant animals. Transpl. Rev. 8:26–49, 1972.

16. Möller, E., Sjöberg, O., Mäkelä, O. Immunological unresponsiveness against the 4-hydroxy-3,5-dinitro-phenacetyl (NNP) hapten in different lymphoid cell populations. Eur. J. Immunol. 218–220, 1971.

17. Sjöberg, O. Rapid breaking of tolerance against *Escherichia coli* lipopolysaccharide in vivo and in vitro. J. Exp. Med. 135:850–859, 1972.

18. Howard, J. G. Cellular events in the induction and loss of tolerance to pneumococcal polysaccharides. Transpl. Rev. 8:50–75, 1972.

19. Peavy, D. L., Adler, W. H., Smith, R. T. The mitogenic effects of endotoxin and staphylococcal enterotoxin B on mouse spleen cells and human peripheral lymphocytes. J. Immunol. 105:1453–1458, 1970.

20. Andersson, J., Sjöberg, O., Möller, G. Induction of immunoglobulin and antibody synthesis in vitro by lipopolysaccharides. Eur. J. Immunol. 2:349–353, 1972.

21. Andersson, J., Sjöberg, O., Möller, G. Mitogens as probes for immunocyte activation and cellular cooperation. Transpl. Rev. 11: 1972 (in press).

22. Möller, G., Andersson, J., Sjöberg, O. Lipopolysaccharides can convert heterologous red cells into thymus-independent antigens. Cell. Immunol. 4: 416–424, 1972.

23. Sjöberg, O., Andersson, J., Möller, G. Requirement for adherent cells in the primary and secondary immune response in vitro. Eur. J. Immunol. 2: 123–126, 1972.

Immunogenicity and Antigenicity of Endotoxic Lipopolysaccharides: Reversible Effects of Temperature on Immunogenicity

Erwin Neter, Hi Yun Whang, and Hubert Mayer

From the Departments of Microbiology and Pediatrics, State University of New York at Buffalo and the Laboratory of Bacteriology, Children's Hospital of Buffalo, Buffalo, New York, and the Max-Planck-Institut für Immunbiologie, Freiburg, Germany

The effect of heat on the immunogenicity of endotoxic lipopolysaccharide (LPS) preparations was determined in rabbits, and the antibody response to O antigens was measured by passive hemagglutination. HAI tests were used for quantitation of the antibody-combining capacity of these preparations. For all LPS tested, heating at 100 C for 1 hr reduced immunogenicity by more than 95%. Freezing at —20 C and thawing of heated preparations substantially restored immunogenicity. Rabbits immunized with the heated preparations, which engendered antibodies only to a minimal extent or not at all, nonetheless were immunologically primed and produced the corresponding O antibodies rapidly and to a significantly greater extent than nonimmunized controls. Unheated, heated, and heated-frozen LPS preparations neutralized antibodies to the same extent, indicating that heating and freezing affect immunogenicity but not antigenicity of LPS. To account for these observations, we suggest that heat may cause deaggregation and/or a structural change in configuration of the carrier molecule and that freezing causes a reversal. Heating also reduced the immunogenicity of enterobacterial suspensions, including typhoid vaccine for human use, while freezing and storage at 4 C restored significant immunogenicity.

This investigation was supported by research grant no. AI00658 from the National Institute of Allergy and Infectious Diseases, and by the Dr. Henry C. Buswell and Bertha H. Buswell Research Fellowship.

Please address requests for reprints to Dr. Erwin Neter, Children's Hospital of Buffalo, 219 Bryant Street, Buffalo, New York 14222.

Studies of the common enterobacterial antigen, first described by Kunin et al. [1, 2], have revealed that heat alters its immunogenicity but not its antigenicity [3]. The heat-induced decrease in immunogenicity was reversible, since repeated freezing and thawing restored immunogenicity to a substantial extent. Preliminary observations have revealed that at least one kind of preparation of lipopolysaccharide (LPS) (namely, endotoxins obtained by the aqueous-ether method) was similarly affected by heating and freezing. We undertook this investigation to extend these studies to LPS prepared by other procedures and to bacterial vaccines. It will be shown that heat strikingly reduces the O immunogenicity, but not the antigenicity, of various LPS preparations and of bacterial cultures of enteric bacteria, thus opening the way for identification of those components of the macromolecule, in addition to the antigenic determinants, that are responsible for immunogenicity.

Materials and Methods

Antigens. Purified LPS, obtained by the phenol-water extraction method, was kindly supplied by Dr. Jon A. Rudbach, University of Montana, Missoula, Mont., and Drs. Otto Westphal and Otto Lüderitz of the Max-Planck-Institut für Immunbiologie, Freiburg, Germany. The antigens were dissolved in phosphate-hemagglutination buffer at a concentration of 1,000 µg/ml and stored at 4 C until used. Additional LPS preparations were purchased from Difco, Detroit, Mich. Bacterial suspensions were prepared by growth of microorganisms on brain-veal agar in Kolle flasks for 18 hr at 37 C; growth was suspended in 25 ml of phosphate-hemagglutination buffer per flask.

Immunization. Albino rabbits weighing 2–3 kg were immunized iv with unheated, heated, or heated-frozen LPS daily for five days, with a total dose of 3.7 µg per rabbit. Each rabbit received 0.33, 0.33, 0.67, 0.67, and 1.7 µg in 2 ml. Heated antigens were kept in a waterbath at 100 C for 1 hr before each injection. Aliquots of heated LPS were frozen at −20 C and thawed repeatedly (as indicated below) or stored at 4 C before immunization. The rabbits were bled on days 0, 6, 8, and 11. Bacterial suspensions, heated (100 C for 1 hr) or unheated, were given in doses of 1 ml in various dilutions (see below) on days 0, 3, and 7. The rabbits were bled on days 0, 7, 10, and 14.

Rabbits that failed to respond with a significant antibody response were given a booster injection of immunogenic antigen on day 11 (0.7 µg of LPS in 2 ml, iv), and specimens of serum were obtained on days 14, 15, and 18 (i.e., three, four, and seven days after the booster injection). For control purposes, nonimmunized rabbits were given the same dose of antigen as was used for booster purposes and were bled on the corresponding days after this single injection.

Antibody titration. Antibodies were quantitated by passive hemagglutination [3]. Briefly, rabbit erythrocytes were modified with the corresponding O antigen and tested with serum in serial twofold dilutions ranging from 1:10 to 1:5,120. The mean titers of antibody in the sera of three rabbits per group were recorded. HAI tests were done as described previously [3] for determination of the antibody-combining capacity of unheated, heated, and heated-frozen antigen preparations.

Results

In the first series of experiments, the effect of heat (100 C, 1 hr) and of repeated freezing (−20 C) and thawing on the immunogenicity of purified LPS of *Escherichia coli* O113, obtained by the phenol-water method, was investigated. The LPS preparation was made available by Dr. Rudbach, who had previously shown its high immunogenicity in the mouse [4]. In addition, the antibody-combining capacity of these preparations was determined by the HAI test.

The results are summarized in table 1. It is clear that heating of the antigen caused a striking loss of immunogenicity, and the resulting titers of hemagglutinin were reduced by more than 95%. Immunogenicity was substantially or completely restored by storage of the preparation at 4 C or by repeated freezing (−20 C) and thawing. Of particular interest is the observation that animals immunized with the heated, poorly immunogenic preparation were immunologically primed. This is evident from the antibody response, which was early and significant when compared with that of nonprimed animals injected with the same dose of antigen as that used for booster purposes. In spite of these striking differences in immunogenicity that were caused in a reversible fashion by heat, the antibody-combining capacity

Table 1. Effect of temperature on immunogenicity of *Escherichia coli* O113 lipopolysaccharide.

Groups	Materials used for immunization (*Escherichia coli* O113 LPS*) Primary	Booster (unheated antigen)	Reciprocal mean O-hemagglutinin titers (days) 0 (A)†	6 (B)	8 (C)	11 (D)	14 (E)	15 (F)	18 (G)	Minimal inhibitory amounts of antigens (μg)
1	Unheated	...	<10	427	640	640	...	...	...	1.7
2	1 hr, 100 C	0.7 μg/rabbit	<10	17	20	13	107	160	320	1.7
3	1 hr, 100 C/4 C	...	<10	213	320	427	...	...	...	1.7
4	1 hr, 100 C/−20 C (2×)	...	<10	427	533	480	...	...	...	1.7
5	1 hr, 100 C/−20 C (4×)	...	<10	853	853	533	...	...	...	1.7
6	1 hr, 100 C/−20 C (6×)	...	<10	960	640	640	...	...	...	1.7
7	None	0.7 μg/rabbit	<10	...	...	...	<10	17	27	...

* LPS = purified lipopolysaccharide. Primary immunization: 3.7 μg/rabbit. Booster dose: 0.7 μg/rabbit.

† A = before immunization; B–D = before booster injection; E–G = after complete immunization.

of the antigens remained unaltered (table 1). Similar results were obtained with another highly immunogenic LPS preparation obtained by the phenol-water method from *E. coli* O111 and kindly supplied by Dr. Rudbach. Injection of 3.7 μg of the unheated antigen per rabbit, according to the above schedule, yielded a mean antibody titer of 1:960 on day 11. In contrast, the corresponding titer resulting from immunization with the heated preparation was only 1:13. Repeated freezing and thawing restored immunogenicity to a moderate degree; the maximal antibody titer reached was 1:170. It was also found that a booster dose resulted in a 10-fold higher antibody titer in the primed animals than in nonimmunized controls (1:427 vs. 1:43). It may be concluded, therefore, that heat causes a reversible loss of immunogenicity but not of antigenicity of these endotoxic LPS.

It has been recognized for years that many phenol-water-extracted, endotoxic LPS are rather poor immunogens. Therefore, additional experiments were done with such preparations obtained from the Max-Planck-Institut für Immunbiologie, Freiburg, Germany, through the courtesy of Drs. Westphal and Lüderitz. Also, since commercially available endotoxins are widely used in fundamental studies, such preparations, obtained by extraction with trichloroacetic acid (TCA) or with phenol and water, were purchased from Difco and were included in the experiments. The results on the effects of heat on the immunogenicity of four endotoxins obtained from *E. coli* O111 are summarized in table 2. It can be seen that one of the phenol-water preparations was only slightly immunogenic, and that heating for 1 hr at 100 C reduced the immunogenicity of all preparations. Freezing at −20 C and thawing of the heated preparations restored immunogenicity to a significant degree. Finally, it is evident that all heated

Table 2. Effect of heat on immunogenicity of *Escherichia coli* O111 endotoxins obtained by various procedures.

Treatment of endotoxins	*Escherichia coli* O111 endotoxins Phenol-water (NIAID)* a†	Phenol-water (NIAID)* b	TCA (Difco)* a	TCA (Difco)* b	Phenol-water (Difco) a	Phenol-water (Difco) b	Phenol-water (Max-Planck)* a	Phenol-water (Max-Planck)* b
Unheated	930	...	960	...	533	...	60	...
Heated (1 hr, 100 C)	13	100	13	320	<10	160	<10	240
Heated (1 hr, 100 C) and frozen (−20 C)	427	...	373	...	320	...	47	...
None	...	43	...	43	...	43	...	43

* NIAID = National Institute of Allergy and Infectious Diseases; Difco = Difco Laboratories, Detroit, Michigan; Max-Planck = Max-Planck-Institut für Immunbiologie, Freiburg, Germany.

† a = antibody titers after primary immunization (day 11). b = antibody titers after booster injection (day 18). Titers given are reciprocal mean O-hemagglutinin titers.

Table 3. Effect of heat on immunogenicity of endotoxins obtained by various procedures from group D *Salmonella.*

	Salmonella group D endotoxins							
	Phenol-water (NIAID)*		TCA (Difco)*		Phenol-water (Difco)		Phenol-water (Max-Planck)*	
Treatment of endotoxins	a†	b	a	b	a	b	a	b
Unheated	1200	...	960	...	187	...	67	...
Heated (1 hr, 100 C)	30	320	13	747	<10	347	<10	240
Heated (1 hr, 100 C) and frozen (−20 C)	220	...	747	...	187	...	50	...
None	...	20	...	23	...	23	...	23

* NIAID = National Institute of Allergy and Infectious Diseases; Difco = Difco Laboratories, Detroit, Michigan; Max-Planck = Max-Planck-Institut für Immunbiologie, Freiburg, Germany.

† a = antibody titers after primary immunization (day 11). b = antibody titers after booster injection (day 18). Titers given are reciprocal mean O-hemagglutinin titers.

preparations induced immunologic priming. Similar results were obtained with endotoxins of group D *Salmonella,* as shown in table 3.

It was of interest to determine whether heating also reduces the immunogenicity of enterobacterial suspensions. The results of a representative experiment are summarized in table 4. Heating of the *E. coli* O113 suspension at 100 C for 1 hr reduced the O-antibody response of the rabbits, and this reduction became more evident as the amount of antigen used for immunization was reduced.

Preliminary observations indicated that freezing at −20 C or storage at 4 C reversed significantly the loss of immunogenicity caused by heat. In one such experiment with *E. coli* O113, the mean titer of antibody elicited by the unheated suspension was 1:6,827; that by heated suspension 1:560; that by the heated and frozen antigen 1:2,027; and that by the heated and refrigerated material 1:1,493. Similar findings were obtained with *Salmonella typhi* vaccine for human use (Eli Lilly, Indianapolis, Ind.). In a representative experi-

ment, the antibody response to the vaccine in a dilution of 1:1,000 (3×10^6 bacteria/rabbit) resulted in a mean O-antibody titer of 1:1,173 on day 14; the titer of antibody to the vaccine that was heated at 100 C for 1 hr just before the injections was less than 1:10; that of the heated and six-times frozen (−20 C) and thawed vaccine was 1:800; and that of the heated and stored (4 C) vaccine was 1:320. Thus, these results parallel similar findings with LPS.

Discussion

This investigation has revealed that the immunogenicity of endotoxic LPS can be altered in vitro without a parallel change of antigenicity. Immunogenicity, in this connection, is defined as the capacity of LPS, when injected iv into rabbits, to induce the formation of O-specific circulating antibodies. The term antigenicity refers to the capacity of the antigens to neutralize the corresponding O antibodies under standardized

Table 4. Effect of heat on immunogenicity of *Escherichia coli* O113 suspension.

		Reciprocal mean O-hemagglutinin titers (days)			
Groups	Materials used for immunization (*Escherichia coli* O113 suspension)	0 (A)*	7 (B)	10 (C)	14 (D)
1	Unheated (1:100)	<10	68,000	205,000	123,000
	1 hr, 100 C (1:100)	<10	29,000	42,000	42,000
2	Unheated (1:10,000)	<10	1,707	2,560	2,560
	1 hr, 100 C (1:10,000)	<10	137	273	560
3	Unheated (1:1,000,000)	<10	1,280	2,560	2,560
	1 hr, 100 C (1:1,000,000)	<10	93	120	80

* A = before immunization, B–D = after immunization.

conditions in vitro. All tested LPS, extracted by the phenol and water, aqueous ether, or TCA method, lost immunogenicity when heated at 100 C for 1 hr just before immunization. This loss was reversible upon subsequent storage at 4 C or by repeated freezing (−20 C) and thawing. All preparations thus treated neutralized the same amounts of O antibodies, thus indicating that the antigenicity was not altered. These results represent additional examples indicating that various biologically important properties of LPS can be altered independently [5]; in addition, these observations focus on components of the macromolecules other than the O-specific side chains that account for immunogenicity, such as the configuration of the carrier molecule. It is also evident that various preparations, seemingly prepared by the same method (i.e., phenol-water extraction), may differ strikingly in immunogenicity when identical amounts are used for immunization. The basis for these differences requires further study.

The present investigation revealed that temperature similarly affects enterobacterial vaccines, including typhoid vaccine. In all likelihood, immunogenic antigen in these heated suspensions is present both in solution and on the surface of the bacterial cells. It remains to be determined whether heat reversibility affects the antigen in both states by an identical mechanism, such as de- and reaggregation and/or changes in configuration of the carrier of the antigenic determinants.

Of particular interest is the observation that immunization with heated preparations, which lead only to a slight or minimal O-antibody response, nonetheless primes the animals immunologically to a booster injection of the unheated antigens in subeffective amounts. The antibody response thus elicited, when compared to that of nonprimed animals, is more rapid and intense and, therefore, characteristic of a secondary response. Additional experiments are needed for determination of the duration of this priming effect and of whether it is due to the proliferation of "memory" cells. In this connection it can be pointed out that, in the mouse, LPS are among the antigens that do not require the participation of thymus-derived (T) cells for an effective immune response, as reported by Möller (see Session 2), Manning et al. [6], and Rudbach (see session 2). This conclusion is based, in part, on the study of the antibody response to E. coli LPS and pneumococcal polysaccharide of congenitally thymusless (nude) mice. It is noteworthy that endotoxins, as shown by Jones and Kind [7] apparently are capable of replacing T-cell requirement for an effective immune response of the mouse to ovine erythrocytes. Obviously, additional studies are needed to determine whether cells other than B cells are required in the immune response of the rabbit to unheated and heated LPS, whether immunologic memory is established by heated preparations, and, if so, which cells are responsible for this memory. In addition, it will be important to determine (1) whether the unheated, heated, or heated-frozen LPS are processed in different ways to account for decrease in immunogenicity by heat and restoration of immunogenicity by freezing or cold storage; and (2) whether, as was shown with the common enterobacterial antigen, the effect of heat and freezing is related to deaggregation and reaggregation [3] or whether it is related to altered configuration of the carrier molecule.

References

1. Kunin, C. M., Beard, M. V., Halmagyi, N. E. Evidence for a common hapten associated with endotoxin fractions of E. coli and other Enterobacteriaceae. Proc. Soc. Exp. Biol. Med. 111:160–166, 1962.
2. Kunin, C. M. Separation, characterization, and biological significance of a common antigen in Enterobacteriaceae. J. Exp. Med. 118:565–586, 1963.
3. Whang, H. Y., Mayer, H., Neter, E. Differential effects on immunogenicity and antigenicity of heat, freezing and alkali treatment of bacterial antigens. J. Immunol. 106:1552–1558, 1971.
4. Rudbach, J. A. Molecular immunogenicity of bacterial lipopolysaccharide antigens: establishing a quantitative system. J. Immunol. 106:993–1001, 1971.
5. Neter, E., Westphal, O., Lüderitz, O., Gorzynski, E. A., Eichenberger, E. Studies of enterobacterial lipopolysaccharides. Effects of heat and chemicals on erythrocyte-modifying, antigenic, toxic and pyrogenic properties. J. Immunol. 76:377–385, 1956.
6. Manning, J. K., Reed, N. D., Jutila, J. W. Antibody response to Escherichia coli lipopolysaccharide and type III pneumococcal polysaccharide by congenitally thymusless (nude) mice. J. Immunol. 108:1470–1472, 1972.
7. Jones, J. M., Kind, P. D. Enhancing effect of bacterial endotoxins on bone marrow cells in the immune response to SRBC. J. Immunol. 108:1453–1455, 1972.

Cellular and Molecular Aspects of the Immune Response to a Bacterial Somatic Antigen

Herman Friedman

*From the Departments of Microbiology, Albert Einstein
Medical Center and Temple University School of
Medicine, Philadelphia, Pennsylvania*

The immune response to the lipopolysaccharide (LPS) somatic antigen of *Shigella paradysenteriae* was studied at the levels of humoral antibody and individual antibody-forming cells. Mice immunized with various doses of either LPS or whole-cell vaccine rapidly developed serum agglutinins to the bacteria. Antibody activity was due to IgM globulins throughout the primary and secondary responses. LPS-sensitized ovine erythrocytes permitted assessment of the serum antibody response and were useful for detection and enumeration of specific antibody plaque-forming cells in spleens of immunized animals. Antibody-forming cells rapidly appeared after immunization and peaked on days 5 and 6. Only 19S IgM antibody-forming cells were stimulated by the shigella antigen. Immunologic tolerance to shigella LPS was induced by injection of a relatively large inoculum into neonatal mice and persisted for at least six to eight weeks, with markedly depressed cellular and humoral antibody responses after challenge. Extracts rich in ribonucleic acid (RNA) from spleens of normal immune and tolerant donor mice transferred antibody-forming activity to normal or shigella-primed mice. Immunogenic activity of these RNA extracts was abolished by in-vitro treatment with RNAase or specific antibody to *Shigella*; thus both RNA and a persisting antigen fragment appear to contribute to biologic activity. The significance of immunogenic RNA extracts during immunogenesis and tolerogenesis was studied.

Much current work concerning cellular and molecular aspects of antibody formation is based on model systems using natural or synthetic antigens such as serum proteins, chemical haptens, polypeptides, etc. [1–3]. In cases in which the immune response has been studied in regard to complex antigens, emphasis has been placed on histocompatibility and tumor antigens. Bacteria, on the other hand, have generally fallen into disfavor as immunologic tools for study of broad aspects of the immune response, mainly because of their marked chemical complexity and the knowledge that many individuals have previously been sensitized or exposed to similar or cross-reacting antigens in their environment. Nevertheless, recent studies with relatively purified protein antigens derived from salmonella flagella and with lipopolysaccharide (LPS) antigens from gram-negative bacilli or pneumococcal polysaccharides have revealed important information concerning many aspects of cellular and molecular events during formation of antibody [4–9]. Knowledge obtained with microbial antigens is of value in the understanding of the complex series of events that occurs between contact of an individual with an immunogen and the appearance of specific immunocytes. Several model systems with bacteria have been used in this laboratory for study of antibody formation and immunologic tolerance at the level of individual antibody-forming cells; direct and indirect bacteriolytic plaque assays in agar gel have been used [10–14]. This paper reports detailed studies of the immune response to LPS antigen derived from *Shigella paradysenteriae* in terms of cellular formation of antibody during immunity and tolerance. In addition, the role of immunogenic RNA in the immune response is examined.

This study was supported in part by research grants from the United States National Science Foundation and the National Institute of Allergy and Infectious Diseases.

I thank Mrs. Leony Mills, Mrs. Joyse Jenderowski, Mr. Leonard Silverman, and Mr. Jerry Rosenberg for their capable technical assistance during various portions of this study.

Please address requests for reprints to Dr. Herman Friedman, Department of Microbiology, Albert Einstein Medical Center, Philadelphia, Pennsylvania 19141.

Materials and Methods

All experiments were performed with male NIH albino A mice bred and maintained by a local dealer. LPS antigen was derived from *S. paradysenteriae* by trypsin digestion of overnight cultures of bacteria grown on brain-heart-infusion agar for 18 hr at 37 C [10]. When the organisms had been killed by treatment with trypsin for 3 hr, a soluble extract was obtained. For induction of both tolerance and immunity, LPS was suspended in saline and injected ip. Tolerance was induced by treatment of mice on day of birth with 20–40 µg of antigen [10, 15]. Immunity was induced by injection of similar doses of antigen when mice were six to eight weeks old.

The immune response was followed at both the cellular and the humoral levels by appropriate assays. For determinations of serum antibody, direct bacterial-agglutination and indirect-hemagglutination tests were done with microtiter plates and 0.025-ml volumes of saline. The antigen for serologic titration consisted of either alcohol-killed shigella or a suspension of ovine erythrocytes (oRBCs) first sensitized with NaOH-treated LPS. All sera were tested before and after treatment with 2-mercaptoethanol (2ME) for estimation of activity of 19S and 7S antibodies. The localized, indirect, hemolytic-plaque assay in agar gel was used for enumeration of individual antibody plaque-forming cells (PFC) to the shigella antigen [10]. For this assay, suspensions of lymphoid cells from tested and control mice were incubated in agar plates containing LPS-sensitized oRBCs. After appropriate incubation with complement, distinct zones of hemolysis appeared and were considered to be due to 19S IgM PFCs. The number of PFCs per million viable spleen cells plated was determined, and the number per whole spleen was calculated. Unsensitized RBCs were used as controls. RBCs sensitized with an unrelated LPS, such as one derived from *Escherichia coli,* were used as a further control. The antibody nature of each plaque was determined in control experiments by demonstration that specific LPS could inhibit all PFCs from shigella-immunized mice. Other LPS extracts had no inhibitory effect. Furthermore, specific rabbit antiserum to murine globulin also inhibited development of PFCs, indicating the γ-globulin nature of the plaques.

Extracts rich in RNA were prepared from pooled spleens of immune, normal, or tolerant mice by a standard cold phenol-extraction procedure [16, 17]. RNA was standardized spectrophotometrically, and 50–100 µg was injected ip into either normal or antigen-primed recipient mice. For in-vitro studies, spleens were obtained aseptically from individual mice at autopsy, and dispersed cell suspensions were made by "teasing" in sterile Hanks' balanced salt solution or medium 199 fortified with 20% isologous murine serum or agammaglobulinemic fetal-calf serum. Washed cell suspensions (at a concentration of approximately 5×10^6 or 5×10^7 nucleated cells per ml) were incubated in tissue-culture tubes in roller drums in an atmosphere of 5% CO_2 and 95% air.

Results

Humoral immune response to shigella antigen. Agglutinating antibody to *Shigella* was readily induced in young adult mice by injection of either shigella vaccine or LPS. As is evident from table 1, mice given graded doses of either whole-vaccine or LPS antigen developed specific agglutinins to the intact bacteria, as detected by direct serologic assay with alcohol-killed organisms. Most mice showed little, if any, antibody before immunization, with titers ≤1:2–1:4. Mice given 50–100 µg of whole-cell vaccine generally responded with titers ≥1:1,024–1:2,048 by seven to 10 days after immunization. There was a gradual decline thereafter; by 20–30 days after immunization, most mice had titers <1:64. Similar results were obtained with LPS as antigen, except that, in general, peak responses were often somewhat lower than those obtained with intact bacteria as the immunogen. Higher or lower doses of antigen in either form resulted in relatively lower antibody responses. All antibody appeared to be due to 19S IgM globulin, since agglutinating activity was completely abolished by incubation with 2ME. Indirect-hemagglutination or hemolytic assays with LPS-sensitized oRBCs also permitted detection of serum antibody to the bacteria. Serologic tests with LPS-sensitized RBCs revealed a generally similar antibody response. Moreover, most serum titers were higher by this test, including those from mice immunized with whole-cell vaccine rather than LPS.

Secondary immune response to shigella LPS.

Table 1. Effect of dose and form of antigen on humoral antibody response of mice to shigella antigen.

Antigen injected* (μg/mouse)	Bacterial agglutinin titer on day†					
	3	5	7	10	15	20
Whole-cell vaccine						
250	1:12	1:192	1:640	1:1,086	1:631	1:58
100	1:22	1:236	1:2,176	1:2,050	1:1,120	1:93
50	1:18	1:128	1:2,360	1:1,975	1:730	1:47
25	1:10	1:38	1:785	1:932	1:520	1:110
10	1:8	1:32	1:286	1:370	1:120	1:28
Lipopolysaccharide						
250	1:8	1:73	1:528	1:764	1:315	1:60
100	1:18	1:69	1:1,265	1:1,052	1:473	1:82
50	1:12	1:51	1:760	1:758	1:283	1:61
25	1:8	1:40	1:315	1:400	1:128	1:40
10	1:8	1:31	1:238	1:205	1:32	1:18

* Groups of young adult mice were injected ip with indicated form and dose.

† Average serum agglutination titer of five to 10 mice on indicated day after immunization.

Groups of mice primed with shigella antigen showed an increased antibody response after a second injection of LPS. The time and dose of antigen used for priming markedly affected the magnitude of the secondary response. Mice given either whole bacteria or LPS approximately eight weeks before secondary immunization showed maximal responses. The response was much lower when the primary injection was given one to two weeks before secondary immunization. Responses were also lower when the time between the first and second injections was 10–12 weeks or longer. It is noteworthy that, regardless of the magnitude of the secondary response, all antibody activity still appeared to be due to 19S IgM globulins, since treatment of sera with 2ME abolished all serologic activity. Furthermore, sucrose-gradient centrifugation of selected specimens of serum confirmed the 19S globulin nature of the antibody.

Antibody response at the cellular level. The LPS-sensitized oRBCs used for indirect-hemagglutination and hemolytic assays were also useful for detection of individual cells forming antibody to the bacteria. Enumeration of individual PFCs by the plaque assay showed that all mice tested had a background of several hundred or more PFCs to *Shigella* in their spleens and lymph nodes before immunization. After immunization with 50–100 μg of LPS, there was a rapid rise in the number of PFCs, with a peak of 20,000–40,000 by days 5 and 6 (table 2). The kinetics of the PFC response was generally similar to that observed in agglutination tests, except that the peak occurred three or four days earlier. The number of PFCs declined relatively rapidly to a low of several hundred to several thousand PFCs by days 10–15 after immunization. Mice given larger doses of antigen often showed a second increase in number of PFCs, usually about 20–25 days

Table 2. Antibody plaque-forming cell (PFC) response in spleens of young adult mice (normal, shigella-primed, or tolerant) challenged with shigella lipopolysaccharide (LPS).

Group* and challenge dose (μg)	PFC/spleen†	Serum titer†
Normal controls		
100	26,100 ± 1,200	1:2,048
50	32,500 ± 1,730	1:2,048
25	30,300 ± 1,830	1:2,048
10	15,120 ± 876	1:1,536
LPS-primed‡		
50	68,500 ± 2,870	1:4,096
10	53,200 ± 4,300	1:4,096
LPS-tolerant§		
100	1,230 ± 180	1:11
50	385 ± 63	1:20
10	410 ± 39	1:8

* Groups of mice were injected ip with indicated dose of LPS and tested five or six days later for PFC and serum antibody responses.

† Antibody response was detected at cellular and serum levels with ovine erythrocytes sensitized with shigella LPS.

‡ Mice were given 20 μg of LPS six weeks earlier.

§ Mice were given 20 μg of LPS on day of birth.

after the first peak. All PFCs appeared to be due to 19S IgM antibody, since facilitation antiserum (rabbit antiserum to murine globulin) capable of detecting 7S PFCs to oRBCs and protein antigens failed to increase the number of detectable PFCs in agar plates containing spleen cells from mice late in the primary response or even after secondary immunization. For example, a rabbit antiserum to murine globulin, at dilutions of 1:100–1:500, did not increase the number of PFCs to *Shigella* when spleen cells were used 10, 15, or 20 days after primary immunization or 10 days after secondary immunization.

The antibody nature of the PFCs to *Shigella* was readily demonstrable by inhibition experiments with specific LPS. When 50–100 μg of shigella LPS was added to the agar plates, together with splenocytes from immunized mice, there was a 90%–95% reduction in the number of PFCs. Similar quantities of LPS from unrelated bacteria, such as *E. coli,* had no effect on development of PFCs to *Shigella.* In addition, relatively low dilutions of rabbit antiserum to murine globulin (1:10–1:20) also inhibited PFCs when added to the agar plates; there was a reduction of 95% or more in number of PFCs, indicating the globulin nature of the plaques.

Immunologic tolerance to shigella antigen in neonatal mice. Immunologic unresponsiveness was readily induced in neonatal mice by injection of a relatively large amount of LPS antigen. In contrast to the rapid appearance of serum antibody to *Shigella* in normal immunized mice, very little antibody appeared in sera of mice given a single tolerance-inducing injection of LPS at birth and challenged later in life with shigella antigen (table 2). Maximal suppression lasted until about eight weeks of age (table 3). A 90%–95% (or greater) reduction in number of PFCs was demonstrable in neonatally treated mice after challenge, as compared with responses of control, nontolerant animals. However, the tolerance-inducing injection of LPS sometimes stimulated a small but consistent rise in number of PFCs in neonatal mice. This occurred only when the dose of LPS was rather large (50–100 μg). Mice given 20 μg at birth developed tolerance but did not form significantly more PFCs during the first weeks of life, as compared with noninjected control mice.

In additional studies, it was found that suspensions of spleen cells from shigella-primed mice,

Table 3. Duration of immunologic tolerance to *Shigella* in neonatally treated mice.

Age at challenge (weeks)*	PFC/spleen†		Serum titer after challenge‡
	Before challenge	After challenge	
2–3	< 100	182 ± 176	< 1:4
4–5	138 ± 19	432 ± 123	1:18
6–7	165 ± 38	310 ± 87	1:31
8–9	218 ± 45	495 ± 120	1:49
10–14	209 ± 37	11,700 ± 2,260	1:258
16 or more	197 ± 52	32,500 ± 1,970	1:763

* Mice were given 20 μg of lipopolysaccharide (LPS) at birth and were challenged at age indicated with 50 μg of LPS.

† Average response of five or six mice, either on day before or five days after challenging immunization at age indicated. PFC = plaque-forming cells.

‡ Mean titers of mice five days after challenge.

cultured in vitro, were readily stimulated by LPS antigen. Normal mice failed to develop more than background numbers of PFCs when cultured in vitro for five to six days after in-vitro treatment with LPS. Spleen cells from tolerant mice also failed to respond to stimulation by in-vitro LPS. However, lack of response was observed only when spleen cells were obtained from tolerant mice less than six to eight weeks old. Older mice (10–12 weeks) given the tolerance-inducing injection of LPS at birth developed some PFCs to the shigella antigen when challenged in vitro.

Transfer of antibody formation with RNA extracts from donors immune to Shigella. Spleen cells from mice immunized by ip injection of LPS readily transferred antibody-forming capacity to syngeneic recipient mice, either normal or X-irradiated 24 hr earlier. RNA-rich extracts prepared from the spleens of such donors were also capable of transferring immunity to *Shigella* to similar recipients (table 4). For example, normal nonimmune mice given 50–100 μg of RNA extract from donors immune to *Shigella* showed a five- to 10-fold or greater increase in number of PFCs within three or four days. Similar preparations of RNA from nonimmune donors or from mice immunized with an unrelated antigen (such as oRBCs or *E. coli*) failed to stimulate formation of specific PFCs to *Shigella.* The biologic activity of "immune" RNA extracts was readily abolished by treatment with RNAase in vitro. Incubation of RNA extracts from immune donors

Table 4. Effects of RNA extracts from immune and tolerant donor mice on antibody response to *Shigella* in normal, antigen-primed, and tolerant recipient mice.

RNA donors*	Day after injection of LPS into donor	PFC/spleen†		
		Normal	Shigella-primed‡	Shigella-tolerant§
Normal adult	1–2	348	876	138
	3–5	958	2,300	123
	7–10	1,042	2,930	185
	15–20	541	1,895	176
	30–40	282	1,150	150
Neonatal tolerant	3–5	1,330	3,850	165
	7–10	1,950	4,260	138
	15–20	2,100	3,150	196
	30–40	1,620	2,080	173
Normal control	None	< 100	362	146

  * Mice were injected either as adults (six to eight weeks old) or as neonates (tolerant) with 20 µg of lipopolysaccharide (LPS) and were used as donors on day indicated.

  † Average plaque-forming cell (PFC) response in spleens of three to six recipient mice per group six to eight days after injection of 50–100 µg of RNA extract from indicated donor group.

  ‡ Mice were primed four to six weeks earlier with 20 µg of LPS.

  § Mice were treated on day of birth with LPS to induce tolerance and were used as recipients at seven or eight weeks of age.

at a concentration of 100–500 µg/ml with 5–10 units of pancreatic RNAase at 25 C completely abolished all stimulatory activity. Furthermore, incubation of the RNA extracts with rabbit or murine immune antiserum to *Shigella* also abolished biologic activity. The most active RNA extracts were obtained from donor mice two or three days after immunization. RNA extracts obtained from mice within one or two days after immunization were much less active (table 4). Furthermore, extracts obtained 10–15 days or longer after immunization were also much less active.

Effect of immunogenic RNA on secondary antibody responses. RNA extracts from shigella donor mice capable of converting normal recipients to active formation of antibody were injected into mice previously primed with LPS (table 4). The PFC response to *Shigella* was markedly higher in these mice and was similar to the response induced by a second injection of antigen. For example, whereas RNA from immune donors induced approximately 30–60 PFCs per million spleen cells in normal mice, the same extracts stimulated several hundred PFCs per million spleen cells in primed recipients. RNA-rich extracts from control donors had no effect.

Absence of immunogenic effect in tolerant recipients. We were interested in determining

whether mice evincing a specific impairment of their immune response to shigella antigen (i.e., tolerant) after neonatal injection of LPS could be induced to form specific antibody by RNA extracts from LPS-immune donors. Mice made tolerant at birth could not respond normally to a challenge injection of shigella LPS for a period of at least six to eight weeks. Groups of such tolerant mice, four to six weeks old, were given 50–100 µg of RNA from immune donors. There was no significant increase in number of PFCs in spleens of these mice (table 4). It was shown in control experiments that unresponsiveness after injection of RNA was specific, since injection either of oRBCs alone or of RNA from oRBC-immune donors stimulated appearance of PFCs to the RBCs in the spleens of shigella-tolerant mice, as with normal control recipients.

Immunogenicity of RNA from shigella-tolerant donor mice. Although shigella-tolerant mice remained unresponsive to LPS after being given RNA from immune donors, it was of interest to examine the reverse situation; i.e, would injection of LPS into tolerant mice stimulate appearance of an immunogenic RNA, even though no antibody could be detected? For this purpose prospective donor mice were made tolerant to shigella antigen by neonatal treatment with LPS. RNA-rich extracts prepared from spleens of such mice (which ranged in age from one to six weeks) were injected into nontolerant recipient mice, either normal or antigen-primed. The PFC responses in these recipients were essentially similar to those of mice given RNA from normal immune donors. Maximal responses were noted in mice given RNA extracts prepared from spleens of one- to two-week-old tolerant mice (table 4). However, RNA extracts obtained as long as six weeks after neonatal induction of tolerance were still immunogenic, especially in mice that had been primed with *Shigella*. RNA extracts derived from tolerant donors were essentially indistinguishable from RNA extracts from immune donors in terms of biologic activity and sensitivity to RNAase and antiserum to *Shigella*.

Involvement of bone-marrow-derived (B) and thymus-derived (T) cells in the immune response to shigella antigen. It is now widely accepted that both B and T lymphocytes are involved in the immune response to certain antigens, such as oRBCs, serum proteins, and polypeptides. The immune response to bacterial somatic antigens is thought to involve only B lymphocytes, since T cells do not appear to be necessary for formation of antibody to strongly immunogenic gram-negative microorganisms. For example, Möller and associates have indicated that only B cells are stimulated by *E. coli* [9]. Thus, it was of interest to determine which of these cell types (if not both) were involved in the specific response to shigella antigen. For these experiments, X-irradiated recipient mice (650 rads) were reconstituted by intravenous (iv) injection of spleen, T, or B, cells or with a mixture of the latter two (table 5).

The mice were then immunized by ip injection of shigella LPS (20 μg), and the number of specific PFCs to *Shigella* was determined eight to 10 days later. Irradiated control mice given no cells showed essentially no response. In contrast, irradiated mice injected iv with viable spleen cells from normal donor mice formed relatively large numbers of PFCs to the shigella antigen eight to 10 days later. Mice given B cells alone responded with about one-third fewer PFCs than those given spleen cells only. Furthermore, mice given only T cells developed very few, if any, PFCs in their spleens after challenge. Mice injected with a mixture of B and T cells pooled from normal donors showed a slightly better response than those given B cells only, but not nearly as good a response as those given spleen cells. On day 10 (the peak day) after challenge, nearly the same results were obtained in mice given both B and T cells. When two or three times more B cells were given to irradiated mice, there was a moderately higher PFC response to LPS. An increased challenge dose of LPS (two- and fourfold) also resulted in a higher response. An even larger dose of LPS generally stimulated even more PFCs in spleens of mice given either spleen or B cells but not T cells. However, mortality due to toxic LPS was extremely high.

Discussion

Many cellular and molecular events leading to formation of antibody are still not understood. Recent studies from numerous laboratories have provided important information concerning the

Table 5. Effect of transfer of spleen, thymus, or bone-marrow cells or a combination of these types of cells on antibody response to shigella lipopolysaccharide in X-irradiated mice.

Lymphoid cells transferred*	PFC/spleen on day†					Serum titer
	2	5	7	10	15	
None	< 100	< 100	122	116	< 100	< 1:2
Spleen	< 100	530	2,160	4,350	186	1:138
Thymus	< 100	< 100	< 100	138	< 100	1:4
Bone-marrow‡	< 100	460	875	1,520	643	1:78
Thymus plus bone-marrow‡	< 100	638	1,472	1,675	1,240	1:98

* Indicated lymphoid-cell suspension from donor mice was injected iv into groups of 25–30 mice X-irradiated 24 hr earlier; 4–8 × 10⁷ nucleated cells were transferred per recipient.

† Average plaque-forming cell (PFC) response of three or four recipients per group on day indicated after transfer of cells and challenging immunization with 50 μg of lipopolysaccharide.

‡ Statistically insignificant ($P < 5$) on day 10 (both groups of mice showed about one-third of the response of mice given spleen cells); statistically significant ($P < .02$) on days 7 and 15 (mice given thymus and bone-marrow cells, as compared with bone-marrow cells only).

participation of more than one type of cell in the immune response [1–3]. There is strong evidence for the participation of at least three types of cells in the immune response to many antigens. These include T cells, B cells (or bursa equivalent), and phagocytic cells (such as macrophages). The complex interaction among these types is a matter of active investigation in basic immunology. In this regard, strongly immunogenic microbial antigens provide a tool for analysis of cellular and molecular events during the immune response. Although the chemical complexity and ubiquity of bacteria in nature are drawbacks, it is apparent that gram-negative microorganisms are highly immunogenic and that relatively small amounts, without adjuvant, rapidly stimulate a marked immune response readily assayed at the levels of serum antibody and specific immunocytes [4–6].

Cells forming antibody to many bacterial antigens can be assessed by indirect-hemolytic or direct-bacteriolytic plaque methods in agar gel or by related single-cell assays. Furthermore, LPS antigens preferentially stimulate IgM antibody responses. Such antibody readily fixes complement and presumably is involved in the protective mechanism against pathogenic microorganisms. Similar complement-fixing cytotoxic antibodies are involved in many other in-vivo immunologic phenomena, including the humoral aspects of graft rejection, as well as in tumor immunity and autoimmunity. Thus, gram-negative bacteria are useful in studies of various parameters of the immune response in regard to natural immunologic events. Immunologic manipulations are often needed for induction of detectable immune responses to haptens, polypeptides, etc., which are generally effective immunogens only when coupled with a carrier or administered with an adjuvant such as oil emulsion containing mycobacteria.

In the studies described here, the LPS-rich somatic antigen derived from *Shigella* was used both as an immunogen and for induction of specific immunologic tolerance. This antigen, like other gram-negative bacterial cell-wall extracts, is strongly immunogenic and apparently triggers only B lymphocytes to form antibody, without involvement of T cells. However, some synergism between T and B cells was suggested, since additional PFCs occurred in X-irradiated mice given both types of cells, as compared with those given B cells alone. This was most evident on day 7 but

could be due to the use of suboptimal numbers of B cells in the earlier transfer experiments. Increased numbers of B cells resulted in a moderate increase in formation of PFCs, even without transfer of T cells. It is noteworthy that an increased dose of LPS also resulted in a heightened PFC response. This could be due either to direct immunogenicity of the LPS or to its marked adjuvant activity. It is widely recognized that bacterial endotoxins may serve as adjuvants, possibly because of their toxic activity, which results in release of endogenous nucleic acids [18]. Synthetic nucleotides can affect antibody formation and may replace T cells when B cells plus oRBCs are injected into X-irradiated recipient mice [19]. Thus, shigella LPS could preferentially stimulate B cells only, and the endotoxin moiety of the antigen could augment this response without the assistance of T cells.

The role of subcellular factors in formation of antibody is still controversial. Studies from many laboratories appear to confirm that RNA-rich extracts from immunized donor animals or from normal lymphoid cells exposed to antigen in vitro may convert nonimmune individuals or lymphoid cells to active formation of antibody [20]. In the present study, the shigella system offered a highly sensitive model for detailed study of the role of RNA-rich extracts in the immune response. Relatively small amounts of LPS induced a rapid antibody response in normal mice and in mice specifically primed with shigella antigen. RNA-rich extracts obtained from spleens of such immune animals were highly immunogenic in recipient mice, either normal or antigen-primed.

The kinetics of the response was similar to that of the response in mice given antigen alone, although the magnitude of the response was lower. Results of such studies suggested that both an RNA moiety and a persisting antigenic determinant or fragment, complexed with the RNA, were necessary for biologic activity of the extracts. This contrasts with reports from other laboratories, in which an informational RNA seems to be induced after in-vivo or in-vitro immunization with antigens such as oRBCs or bacteriophage [20, 21]. Informational RNA is thought to be free of antigen. Allotype-marker experiments have shown that the donor allotype, rather than the recipient's, is expressed on the molecules of antibody stimulated by immunogenic RNA. How-

60

ever, it should be noted that, in most of those experiments, RNA was added in vitro to cultures of lymphoid cells or fragments. Furthermore, the RNA was usually extracted from lymphoid cells exposed in vitro to antigen. In the experiments described here, donor animals were actively immunized with shigella antigen and were already producing some antibody at the time when RNA extracts were prepared. The actual cell source of the RNA moiety is unknown but presumably could be splenic macrophages. However, mere persistence of antigen in spleen-cell suspensions or extracts could not explain the immunogenic effects of the RNA extracts in recipient animals. For example, immunogenicity was maximal when donors had been immunized two to four days earlier. RNA extracts from donor mice given antigen one or two days earlier were less active, as were those from spleens of mice immunized 10–15 days earlier. Thus, it seems unlikely that the mere persistence of antigen in the spleen would account for immunogenic activity of the RNA extracts.

Inactivation experiments with RNAase indicated that the RNA moiety was essential for biologic activity. However, the biochemical nature and specificity of the RNA is not clear. It is certainly possible that the RNA may serve merely as a nonspecific stimulator with an associated "superantigen." Thus, the experiments with RNA extracts from shigella-tolerant donor mice are pertinent to the question of biologic specificity. Although immunogenic RNA extracts could be readily obtained from mice evincing a state of immunologic tolerance, these mice did not themselves form normal levels of antibody to shigella antigen, as assessed at serum and cellular levels. Despite lack of detectable formation of antibody in the donors, the RNA-rich extracts from spleens of these mice were as immunogenic as extracts from nontolerant mice. Tolerant mice, when used as recipients, did not respond to immunogenic RNA from other mice, indicating that cells capable of responding to shigella antigen, even if associated with RNA, are absent or nonresponsive during tolerance. Furthermore, it appears to be evident that lack of antibody formation in mice tolerant to a strongly immunogenic antigen such as shigella LPS is not due to absence of antigen "processing" into an RNA-antigen complex.

Studies such as these on the cellular and molec-

ular characteristics of the immune response to strongly immunogenic bacterial antigens, at both cellular and humoral levels, should be of value in further understanding of various parameters of the immune response from the viewpoint of immunogenesis and tolerogenesis. Evaluation of the nature and significance of persistence of such bacterial antigens with RNA may also provide valuable information on the regulation of formation of antibody to microbial antigens at the subcellular level.

References

1. Mitchison, N. A. Antigen recognition responsible for the induction in vitro of the secondary response. Cold Spring Harbor Symp. Quant. Biol. 32:431–439, 1967.
2. Steiner, L. A., Eisen, H. N. Variation in the immune response to a simple determinant. Bacteriol. Rev. 30:383–396, 1966.
3. Siskund, G. W., Benacerraf, B. Cell selection by antigen in the immune response. Adv. Immunol. 10:1–50, 1969.
4. Ada, G. L. Antigen binding cells in tolerance and immunity. Transpl. Rev. 5:105–129, 1970.
5. Diener, E., Armstrong, W. D. Immunologic tolerance in vitro. J. Exp. Med. 129:591–603, 1969.
6. Makela, O. Analogies between lymphocyte receptors and the resulting humoral antibodies. Transpl. Rev. 5:3–18, 1970.
7. Britton, S. Regulation of antibody synthesis against E. coli endotoxin. J. Exp. Med. 129:469–482, 1969.
8. Baker, P. J., Bernstein, M., Pasanen, V., Landy, M. Detection and enumeration of antibody producing cells by specific adherence of antigen-coated bentinite particles. J. Immunol. 97:767–777, 1966.
9. Möller, G., Sjoberg, O., Möller, E. Cellular aspects of immunologic tolerance to E. coli lipopolysaccharide. Ann. N.Y. Acad. Sci. 181:134–142, 1971.
10. Friedman, H. Prevention of immunologic unresponsiveness to shigella antigens in neonatal mice by homologous spleen cell transplants. J. Immunol. 92:201–207, 1964.
11. Friedman, H., Allen, J., Landy, M. Assessment of a bacterial adherence colony (B.A.C.) method for enumeration of antibody forming cells to E. coli somatic antigen. J. Immunol. 103:204–214, 1969.
12. Field, C., Allen, J. L., Friedman, H. The immune response of mice to Serratia marcescens LPS and intact bacteria. J. Immunol. 105:193–203, 1970.
13. McAlack, R. F., Cerny, J., Allen, J. L., Friedman, H. Vibriolytic antibody-forming cells: a new application of the Pfeiffer phenomenon. Science 168:141–142, 1970.
14. Friedman, H., Landy, M. Separate spleen cell populations synthesizing bacteriolytic versus agglutinating antibody in mice immunized with E. coli somatic antigen. Cell. Immunol. 2:153–163, 1971.

15. Friedman, H. Immunologic tolerance to microbial antigens, II. Suppressed antibody plaque formation to shigella antigen by spleen cells from tolerant mice. J. Bacteriol. 92:820–827, 1966.

16. Friedman, H. Antibody plaque formation by normal spleen cell cultures exposed in vitro to RNA from immune mice. Science 146:934–936, 1964.

17. Friedman, H. The nature of immunogenic RNA-antigen complexes in nucleic acids in immunology. Rutgers University Press, New Brunswick, N.J., p. 505–526.

18. Winchurch, R., Braun, J. W. Antibody formation: premature initiation of antibody formation by endotoxin or synthetic polynucleotides in newborn mice. Nature (Lond.) 223:843–844, 1969.

19. Johnson, A. G., Cone, R. E., Friedman, H. M., Han, I. H., Johnson, H. G., Schmidtke, J. R., Stout, R. D. Stimulation of the immune system by homopolyribonucleotides. *In* R. F. Beers and W. Braun [ed.] Biological effects of polynucleotides. Springer-Verlag, New York, 1971, p. 157–179.

20. Fishman, M., Adler, F. L. The role of macrophage-RNA in the immune response. Cold Spring Harbor Symposium Quant. Biol. 32:343–348, 1967.

21. Bell, C., Dray, S. Conversion of non-immune spleen cells by ribonucleic acid of lymphoid cells from an immunized rabbit to produce 8-M antibody of foreign light chain allotype. J. Immunol. 103:1196–1211, 1969.

Immunologic Responses of Mice to Lipopolysaccharide from *Escherichia coli*

Norman D. Reed, Judith K. Manning,
and Jon A. Rudbach

*From the Immunobiology Unit, Department of Botany
and Microbiology, Montana State University, Bozeman,
Montana and the Department of Microbiology,
University of Montana, Missoula, Montana*

The immune response of athymic and normal mice to lipopolysaccharide (LPS) from *Escherichia coli* 0113 was studied. Normal mice could be primed by sub-nanogram amounts of LPS for a secondary response to a dose of the same LPS; the magnitude of the antibody response was proportional to the size of the primary dose, as was the specific plaque-forming cell (PFC) response to a similar regimen of LPS. The amount of LPS required for priming was 10^6-fold less than that required to initiate a primary PFC response. The thymus-independent nature of the immune response of mice to LPS was confirmed and expanded. Congenitally athymic (nude) mice responded in both primary and secondary fashions to LPS in a manner similar to normal littermates that had intact thymuses. This phenomenon could be demonstrated at various dosage levels. Furthermore, the response to LPS in nude mice was demonstrable concomitantly with the demonstration of a severely impaired response to equine erythrocytes; the latter response is thymus-dependent.

This paper presents the results of the successful coalescence of research projects from two separate laboratories. In the first laboratory, bacterial lipopolysaccharides (LPS) had been studied for several years with the hope of elucidating their essential biophysical structure and the manner(s) in which they interact with and are detoxified by host tissues [1, 2]. More recently, the immunologic reactions of LPS, in particular the readily demonstrable capacity of minute doses of LPS antigens to stimulate primary responses or to sensitize for secondary responses, were studied [3].

In the second laboratory, the immune responses of congenitally athymic (nude) mice [4] had been under investigation. These athymic mice failed to reject homografts and heterografts of rat, rabbit, and hamster skin (C. F. Shaffer, N. D. Reed, and J. W. Jutila, unpublished observation); they failed to make normal immune responses to heterologous erythrocytes [5] and plasma-protein antigens (N. D. Reed, unpublished observation). However, nude mice produced humoral antibodies to *Diplococcus pneumoniae* polysaccharide and *Escherichia coli* lipopolysaccharide [6].

The data presented below are the result of a comprehensive study of immunologic responses to *E. coli* LPS of three types of mouse: (*1*) normal Rocky Mountain Laboratory stock albino, (*2*) athymic (nude), and (*3*) the phenotypically normal littermates of the nudes. The results characterize primary and secondary responses to LPS antigen and establish the thymus-independent nature of these responses.

Materials and Methods

Antigens. The LPS endotoxin was extracted from cell walls of *E. coli* 0113 (Braude) by the phenol-water procedure [7]. The details of cultivation, fractionation, and extraction have been described previously [8]. Whole cells of *E. coli* 0113, boiled for 150 min, were used as a bacterin. Equine and ovine erythrocytes were purchased from Colorado Serum Co. The LPS and the bacterin were injected either into the lateral tail vein or ip; equine erythrocytes were injected ip.

Mice. Normal laboratory mice were obtained from the albino stock maintained at the Rocky

This work was supported in part by grants no. AI-10384 and AI-09876 from the National Institute of Allergy and Infectious Diseases, National Institutes of Health. We thank Dr. John W. Jutila for his cooperation.

Please address requests for reprints to Dr. Norman D. Reed, Department of Botany and Microbiology, Montana State University, Bozeman, Montana 59715.

Mountain Laboratory, Hamilton, Montana. Athymic (nude) mice and their phenotypically normal littermates were raised at Montana State University, as described elsewhere [5].

Immunoassays. Numbers of plaque-forming cells (PFC) in the spleens and titers of humoral antibodies were determined by standard methods [3, 6], using ovine erythrocytes coated with LPS as indicator cells [9].

Results

The LPS isolated from *E. coli* 0113 is a very potent immunogen in normal laboratory mice. For example, only a few hundred molecules are necessary to prime mice for a secondary antibody response. In an experiment designed to demonstrate this fact (figure 1), the immune response was quantitated from the humoral antibody titers. Mice receiving a primary dose of as little as 10^{-9}–10^{-11} µg of LPS reproducibly showed a heightened immune response to a second dose of 1 µg of the same LPS given 21 days after the primary dose. On this log-log plot, there is a linear relationship between the primary dose and the titers of antibody after the constant secondary dose.

Similar results could be obtained if the same type of experiment was set up with the immune response expressed as the number of specific PFC (figure 2). Mice receiving as little as 10^{-9} µg of LPS were immunologically primed to respond in a heightened fashion to a secondary dose of LPS. However, the lowest single dose of the *E. coli*

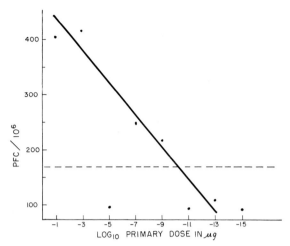

Figure 2. Specific plaque-forming cell (PFC) response of Rocky Mountain Laboratory mice receiving various primary doses of lipopolysaccharide (LPS) from *Escherichia coli* 0113 and a constant secondary dose (1 µg on day 21). The dashed line parallel to the abscissa is the response of normal mice to 1 µg of LPS. The responses were assayed on day 25 and are represented as PFC per 10^6 nucleated spleen cells.

LPS that gave rise to reproducible PFC above background was 10^{-3} µg (figure 3); below this level no overt immune response to a single dose of LPS was mustered. Therefore, the minimal amount of LPS necessary to prime mice for a secondary response (figure 2) was about 10^6-fold less than that necessary to elicit a primary response (figure 3).

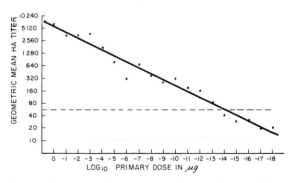

Figure 1. Peak hemagglutinating-antibody (HA) response of Rocky Mountain Laboratory mice receiving various primary doses of lipopolysaccharide (LPS) from *Escherichia coli* 0113 and a constant secondary dose (1 µg on day 21). Mice were bled on days 24 and 26. The dashed line parallel to the abscissa is the response of normal mice to 1 µg of LPS.

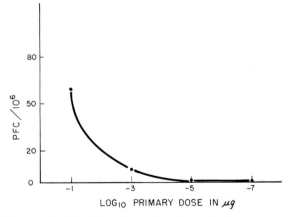

Figure 3. Specific plaque-forming cell (PFC) response of Rocky Mountain Laboratory mice receiving single doses of lipopolysaccharide from *Escherichia coli* 0113. The responses were assayed on day 4 and are represented as PFC per 10^6 nucleated spleen cells.

Reports from other laboratories have shown that mice that are thymus-deprived (by artificial means) generate strong immune responses to LPS [10, 11]. Thus, LPS has been classified as a thymus-independent antigen. However, there is some controversy about the validity of the concept of a thymus-independent immune response [12], because this concept is based on data obtained from neonatally thymectomized or adult thymectomized, irradiated, and reconstituted animals. The availability of the congenitally athymic (nude) mouse allowed the thymus-dependency of the immune response to LPS to be tested directly.

Nude mice responded as well as their normal littermates to a single dose of *E. coli* LPS (table 1). Because the thymus-dependency of the immune response to certain antigens can be altered with the quantity of antigen [13], the responses of nudes and normal littermates were compared at various doses of LPS. The primary response to LPS appeared to be thymus-independent over the 1,000-fold dose range studied (table 1). Furthermore, a primary dose of LPS sensitized both nudes and normal littermates for a heightened immune response to a second dose of the same LPS. As shown in table 2, the response of mice receiving both primary and secondary doses of LPS was markedly stronger than the response of mice receiving only one dose.

It has been suggested that the physical state of an antigen is one factor that determines the thymus-dependency of the immune response to that antigen [14]. In the experiments described above, LPS was injected in the form of a soluble extract. Accordingly, nudes and normal littermates were immunized with 10^9 heat-killed *E. coli* whole cells. The response of the bacteria-immunized nudes (17,768 PFC per spleen) was comparable to the response of their normal littermates (20,576 PFC

Table 1. Primary immune response of nude mice and normal littermates to various doses of lipopolysaccharide (LPS) from *Escherichia coli* 0113.

| Phenotype | LPS dose (µg) | | | |
	0.1	1	10	100
Nude	880 (3)*	3,495 (4)	5,753 (14)	4,100 (2)
Normal	1,413 (3)	2,050 (4)	5,848 (14)	5,300 (1)

* Mean number of plaque-forming cells per spleen four days after immunization (number of mice studied).

Table 2. Secondary immune response of nude mice and normal littermates to lipopolysaccharide (LPS) from *Escherichia coli* 0113.

Phenotype	LPS dose	Mean PFC/ spleen*
Nude (8)†	1 µg day 0 and	54,515
Normal (6)	10 µg day 21	23,287
Nude (1)	10 µg day 21	5,080
Normal (1)		3,200
Nude (2)	1 µg day 0	400
Normal (2)		500

* Mean number of plaque-forming cells per spleen on day 25.
† Number of mice studied.

per spleen); the number of PFC still was assessed with LPS-coated ovine erythrocytes as the indicator system. In a preliminary experiment it was also observed that nudes and littermates responded similarly to LPS, even though the LPS in the immunizing dose was attached to the thymus-dependent antigen, equine erythrocytes.

To establish that athymic mice could respond to LPS antigen as well as could normal littermates while the former maintained an impaired response to a thymus-dependent antigen, we did the double-immunization experiment outlined in table 3. These data show that the same nudes that failed to respond normally to the thymus-dependent antigen (equine erythrocytes) responded to LPS as well as did their normal littermates.

Finally, the possibility was explored that there could have been a maternal or littermate thymic influence on the nudes in utero. This influence could have altered the subsequent response of the nudes to LPS. The "nude" gene (nu) is an autosomal recessive gene [15], and because nude mice show low fertility, phenotypically normal heterozy-

Table 3. Immune response of nude mice and normal littermates to equine erythrocytes (EE) and lipopolysaccharide (LPS) from *Escherichia coli* 0113.

| Phenotype | Mean PFC*/spleen specific for | |
	EE	LPS
Nude (4)†	2,530	4,710
Normal (6)	105,333	2,023

NOTE. All mice were given 10^8 EE on day 0 and 10 µg of LPS on day 1. The number of plaque-forming cells specific for EE and LPS was determined on day 5.
* Plaque-forming cells.
† Number of mice studied.

gous (+/nu) males and heterozygous females are mated to produce nudes (nu/nu). Thus, nudes may be influenced in utero by the thymuses of their mother and/or their normal littermates. Several nude litters were obtained by mating homozygous (nu/nu) males and homozygous females. Nude mice produced by this special mating procedure, which excluded any in utero thymic influence, responded as well as normal mice to LPS.

Discussion

Certain factors associated with LPS make it admirably suited as a model antigen in studies such as those presented above. In many cases the chemical structure of the immunodeterminant groups is known [16]; biologic systems in vitro and in vivo appear to be capable of dissociating the quaternary structure of LPS only to 10,000–20,000 particle-weight subunits [2]; LPS is an extremely potent antigen, in that only a few thousand molecules can stimulate a primary response in rabbits [17] or sensitize mice for a secondary response [3]. Furthermore, LPS appears to be a truly thymus-independent antigen [6]. This theory was confirmed in our study by demonstration of the fact that the congenitally athymic (nude) mouse would respond immunologically to LPS in a manner quantitatively similar to littermate mice with functional thymuses. The thymus-independent nature of the immune response could be demonstrated over a wide dose range of LPS. Presentation of LPS attached to cells or in the soluble form did not alter the thymus-independency of the immune response of the nude mouse to LPS. Furthermore, both primary and secondary immune responses to LPS were thymus-independent. The observation that a secondary response could be elicited in nude mice indicates that the cells carrying the immunologic memory were not thymus-derived cells.

Recently, it has been reported that the immune response of mice to a quaternary complex of polymerized flagellin was thymus-independent, whereas the response to monomeric forms of flagellin was thymus-dependent [14]. On the basis of these studies, the suggestion was put forth that thymus-dependency was not a property of the nature of the immunodeterminant groups on an antigen; rather, it depended on the physical form and mode of presentation of the antigen. This hypothesis

can be tested further with LPS antigens and the nude mouse. The physical form of LPS can be manipulated readily with surfactants [1], and it should be possible to test the immunogenicity of LPS subunits in the athymic mouse. Perhaps the LPS-LPS subunit system will be analogous to the polymerized flagellin-monomeric flagellin system. This is only one example of the many experimental questions that can be asked of this model.

References

1. Ribi, E., Anacker, R. L., Brown, R., Haskins, W. T., Malmgren, B., Milner, K. C., Rudbach, J. A. Reaction of endotoxin and surfactants. I. Physical and biological properties of endotoxin treated with sodium deoxycholate. J. Bacteriol. 92:1493–1509, 1966.
2. Rudbach, J. A., Anacker, R. L., Haskins, W. T., Johnson, A. G., Milner, K. C., Ribi, E. Physical aspects of reversible inactivation of endotoxin. Ann. N.Y. Acad. Sci. 133:629–643, 1966.
3. Rudbach, J. A. Molecular immunogenicity of bacterial lipopolysaccharide antigens: establishing a quantitative system. J. Immunol. 106:993–1001, 1971.
4. Pantelouris, E. M. Absence of thymus in a mouse mutant. Nature (Lond.) 217:370–371, 1968.
5. Reed, N. D., Jutila, J. W. Immune response of congenitally thymusless mice to heterologous erythrocytes. Proc. Soc. Exp. Biol. Med. 139:1234–1237, 1972.
6. Manning, J. K., Reed, N. D., Jutila, J. W. Antibody response to Escherichia coli lipopolysaccharide and type III pneumococcal polysaccharide by congenitally thymusless (nude) mice. J. Immunol. 108:1470–1472, 1972.
7. Westphal, O., Lüderitz, O., Bister, F. Über die Extraktion von Bakterien mit Phenol/Wasser. Z. Naturforsch. [B] 7:148–155, 1952.
8. Fukushi, K., Anacker, R. L., Haskins, W. T., Landy, M., Milner, K. C., Ribi, E. Extraction and purification of endotoxin from Enterobacteriaceae: a comparison of selected methods and sources. J. Bacteriol. 87:391–400, 1964.
9. Neter, E., Westphal, O., Lüderitz, O., Gorzynski, E. A., Eichenberger, E. Studies of enterobacterial lipopolysaccharides. Effects of heat and chemicals on erythrocyte-modifying, antigenic, toxic and pyrogenic properties. J. Immunol. 76:377–385, 1956.
10. Möller, G. Triggering mechanisms. In R. T. Smith and M. Landy [eds.] Immune surveillance. Academic Press, New York, 1970, p. 87–116.
11. Andersson, B., Blomgren, H. Evidence for thymus-independent humoral antibody production in mice against polyvinylpyrrolidone and E. coli lipopolysaccharide. Cell. Immunol. 2:411–424, 1971.
12. Mitchell, G. F., Mishell, R. I., Herzenberg, L. A.

66

Studies on the influence of T cells in antibody production. *In* B. Amos [ed.] Progress in immunology. Academic Press, New York, 1971, p. 323–335.

13. Taylor, R. B., Wortis, H. H. Thymus dependence of antibody response: variation with dose of antigen and class of antibody. Nature (Lond.) 220:927–928, 1968.

14. Feldmann, M., Basten, A. The relationship between antigenic structure and the requirement for thymus-derived cells in the immune response. J. Exp. Med. 134:103–119, 1971.

15. Flanagan, S. P. "Nude," a new hairless gene with pleiotropic effects in the mouse. Genet. Res. 8:295–309, 1966.

16. Lüderitz, O., Staub, A. M., Westphal, O. Immunochemistry of O and R antigens of *Salmonella* and related Enterobacteriaceae. Bacteriol Rev. 30:192–255, 1966.

17. Landy, M., Baker, P. J. Cytodynamics of the distinctive immune response produced in regional lymph nodes by salmonella somatic polysaccharide. J. Immunol. 97:670–679, 1966.

Fate of Endotoxin in Macrophages: Biological and Ultrastructural Aspects

C. A. Bona

From the Centre de Physiologie et d'Immunologie Cellulaires, Hôpital Saint-Antoine, Paris, France

Endotoxin is taken up by macrophages by pinocytosis. Various properties (i.e., the ability to induce the Shwartzman phenomenon and the immunogenicity of endotoxin taken up by macrophages) are not altered for 48 hr. Endotoxin taken up by macrophages is transferred to autologous lymphocytes that adhere to macrophages, forming lymphocyte-macrophage islets.

Endotoxins exert their multiple biological and immunologic effects only after liberation from dead bacteria. Such bacterial and immunologic effects become evident after adsorption of endotoxin onto the cellular membrane or penetration of endotoxin into the cells.

In this paper, we present data concerning the mechanism of penetration of endotoxin into macrophages and the fate of endotoxin taken up by these cells.

For this purpose, we have investigated three topics: (*1*) pinocytosis of endotoxin by macrophages; (*2*) alteration of some immunogenic and toxic properties of endotoxin taken up by macrophages; and (*3*) transfer of endotoxin from macrophages to lymphocytes.

Please address requests for reprints to Dr. C. A. Bona, Département Chimiothérapie Expérimentale, Institut Pasteur, 28, rue du Dr. Roux, 75015 Paris, France.

Materials and Methods

Macrophages. Peritoneal exudate of guinea pigs was obtained by injection of 15 ml of mineral oil into the animals. Three days later, cells from the peritoneal exudate were harvested by washing of the peritoneum with heparinized isotonic saline solution. The cells (10^7/ml) were cultivated for 2 hr in Falcon flasks in medium 199 (Pasteur Institute) supplemented with 5% heat-inactivated calf serum (Pasteur) in an atmosphere of 5% CO_2. After 2 hr, the supernatant containing the nonadherent cells was decanted. The monolayer of adherent cells was 98% macrophages as shown by Wright staining.

Pinocytosis. The macrophages (10^7/ml) were incubated for 15 min at 37 C with 150 mg of *Salmonella typhimurium* endotoxin labeled with fluorescein isothiocyanate, with uranyl acetate-labeled *Salmonella enteritidis* endotoxin, or with

S. enteritidis endotoxin biosynthetically labeled with [14]C.

The labeling of endotoxins with fluorescein and uranyl acetate was done by a previously described technique [1]; the biosynthetic labeling with ^{14}C was done by the method of Ribi et al. [2].

After incubation for 15 min, the monolayers of macrophages were washed three times and incubated again in the same medium; samples were harvested at 15 min, 45 min, 24 hr, and 48 hr and were examined by fluorescent microscopy or prepared either for electron microscopy or for high-resolution autoradiography by a technique previously described [1].

Alteration of ability to elicit Shwartzman phenomenon. Macrophages (10^7) incubated for 30 min with 150 µg of endotoxin were cultivated in the same medium, harvested after 3, 6, 9, 24, and 48 hr, and homogenized in a Potter-Elvejem apparatus. This homogenate was used as the preparatory injection, and 5 µg of homologous endotoxin was used as the injection for challenge of rabbits. After 4 hr, the skin of rabbits was excised for histopathologic examination.

Immunogenicity of lysosomal fractions prepared from macrophages with ingested endotoxin. Macrophages incubated in the same conditions as those described above were harvested after 3, 9, and 24 hr. Lysosomal granular fractions were prepared from macrophages according to the method of Clarke [3]. The standard lysosomal suspension obtained from 10^7 macrophages/ml showed 0.6 mg of nitrogen per ml and an acid-phosphatase activity (at *p*H 5.5) corresponding to 24 µmoles of phenol released/30 min per mg of nitrogen.

Protein nitrogen was determined by the method of Kjeldahl [4]. The method of Pal Yu et al. [5] was used to estimate the acid-phosphatase activity.

The lysosomal suspension was injected into male New Zealand rabbits by a technique previously described [6], and the antibody assays were performed either according to the method of Jerne as modified by Landy et al. [7] or by passive hemagglutination [8].

Transfer of endotoxin. The macrophagic monocellular layer (10^7 cells) was incubated for 1 hr at 37 C with 150 µg of ^{14}C-labeled endotoxin in 10 ml of medium 199 supplemented with 5% heat-inactivated calf serum. After incubation,

the cell layer was washed five times in the same medium and incubated with 5×10^8 autologous lymph-node lymphocytes. After standing for 1 hr at 37 C, the supernatant was carefully decanted, and the cell layer was fixed for 1 hr at 4 C with 1.5% buffered glutaraldehyde. This layer was then collected and embedded for electron microscopy. Autoradiography on thick and thin sections was done by the method of Rogers [9].

The following controls were included. (*1*) Transfer was done as described above but was incubated at 4 C. (*2*) Macrophages were incubated simultaneously with the same labeled antigen but in glutaraldehyde instead of medium 199; the cells were washed five times and later incubated again with lymphocytes. (*3*) The supernatant fraction from macrophages incubated for 1 hr with ingested antigen in absence of lymphocytes was subsequently added to the lymphocyte population for 1 hr at 37 C. We included this control to determine whether or not the antigen excreted from macrophages could account for the silver grains contained by the lymphocytes. (*4*) Fibroblast "L" cells with ingested endotoxin were incubated with lymphocytes.

Results

Pinocytosis of endotoxin by macrophages. Using endotoxins labeled by three different methods (i.e., fluorescein, uranyl acetate, and ^{14}C), we observed by three techniques (fluorescence microscopy, electron microscopy, and high-resolution autoradiography) the uptake of endotoxin by macrophages through pinocytosis.

The first 15 min were characterized by strong adsorption of endotoxin onto the membrane, and small invaginations containing the labeled endotoxin were observed. After 45 min, the endotoxin was always found on the membrane but was located within heterophagosomes. At 24 hr, the endotoxin was present either within heterophagosomes or free in the cytoplasm and, at 48 hr, only in the residual bodies. It is noteworthy that, after 24 hr, we did not observe endotoxin bound to the membrane of macrophages (figures 1 and 2).

Alteration of ability to induce Shwartzman phenomenon. A local Shwartzman reaction could be obtained after a preparatory injection

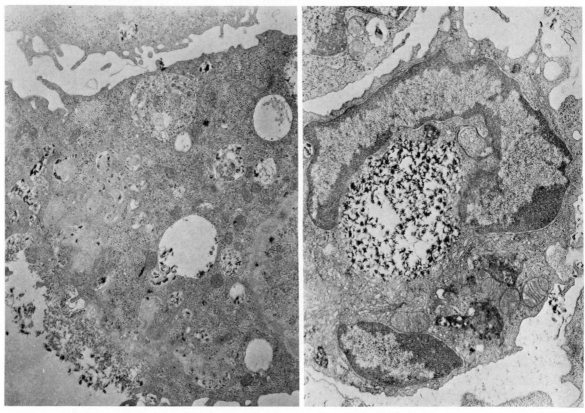

Figure 1. Pinocytosis of uranyl acetate-labeled endotoxin by guinea-pig macrophages. **Left:** Thirty-min pulse uranyl acetate-labeled endotoxin; macrophages were cultivated for 45 min. Note strong adsorption of endotoxin onto the membrane (×8,500). **Right:** Thirty-min pulse uranyl acetate-labeled endotoxin; macrophages were cultivated for 24 hr. Endotoxin is observed either free in the cytoplasm or within heterophagosomes (×8,000).

of the homogenate of macrophages containing endotoxin.

The challenging injection of endotoxin was followed by the progressive development of a local inflammatory reaction at the site of injection. The area of local inflammation showed little change for 2–3 hr, but then petechia and purpuric spots appeared and rapidly coalesced until the entire area of inflammation underwent severe hemorrhagic necrosis with destruction of all elements of the epidermis and the dermis. The microscopic aspects of these zones were characterized by an accumulation of granuloctyes around the adventitia of small veins, which produced perivascular cuffing, thrombosis, fragmentation of collagen, myolysis, and edema (table 1).

Immunogenicity of lysosomal fractions. When lysosomal fractions prepared from macrophages with ingested endotoxin were injected into rabbits, they were able to induce a specific immune response estimated either by the plaque technique or by the passive hemagglutination procedure (table 2).

Transfer of endotoxin. Autologous lymphocytes adhered in vitro to macrophages, forming "lymphocyte-macrophage islets" (LMI). In our experimental system, 1.8% of the lymphocytes were recovered in LMI, and 0.2% showed silver grains in two types of cells (positive LMI); this indicated that there was a transfer of ^{14}C-labeled endotoxin in only 10% of the lymphocytes recovered in LMI. The controls were negative (table 3).

High-resolution autoradiography showed that one-third of the silver grains were found on the surface of the macrophages or on their process, while two-thirds were found in the cytoplasm of macrophages that were not recovered in the LMI.

At the level of LMI, the silver grains from the surface of macrophages disappeared, and in the

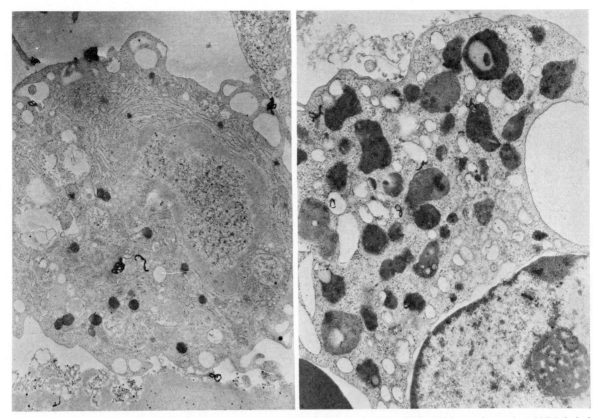

Figure 2. Pinocytosis of endotoxin labeled biosynthetically with ¹⁴C. **Left:** Thirty-min pulse ¹⁴C-labeled endotoxin; macrophages were cultivated for 45 min. Note that 40% of silver grains were bound to macrophage membranes (×6,000). **Right:** Thirty-min pulse ¹⁴C-labeled endotoxin; macrophages were cultivated for 24 hr. Silver grains were observed only in heterophagosomes (×8,000).

lymphocytes 80% of radioactivity was found in the nuclei and 20% either free within cytoplasm or bound to the lymphocyte surface. The statistics on the distribution of silver grains within lymphocyte nuclei [corrected by eradication of the resolution error of Ilford L4 emulsion, geometric error of strong, energetic ¹⁴C particles (155 MeV)] established that 12.6% of the silver

Table 1. Shwartzman phenomenon induced by homogenate prepared from macrophages that had ingested endotoxin.

Homogenate used as preparatory injection	Culture time of macrophages (hr)	Percentage of dead cells after culture	No. of experiments	Results			
				Positive		Negative	
				Macro-scopic	Micro-scopic	Macro-scopic	Micro-scopic
Macrophages	0	8	5	5	5	0	0
that had	3	20	5	5	5	0	0
ingested	6	22	5	5	5	0	0
endotoxin	9	28	5	5	5	0	0
30 min before	24	30	5	5	5	0	0
	48	33	5	5	5	0	0
Macrophages without endotoxin	None	10	5	0	0	5	5
Endotoxin (10 μg)	None	None	7	7	7	0	0

Table 2. Immunogenicity of lysosomal fraction from macrophages that had ingested endotoxin.

| Lysosome prepared from macrophages | Culture time of macrophages (hr) | No. of experiments | Hemagglutination titer | | Mean no. of PFC*/10^6 cells |
			Before injection of fraction	Six days after injection of fraction	
30 min after ingestion of endotoxin	0	4	0	425.6	1.8
	1	4	0	160.0	3.8
	3	4	0	259.1	14.5
	9	3	0	320.0	10.0
	24	3	0	226.0	3.0
No endotoxin ingested	None	1	0	0	0

* Plaque-forming cells.

grains found in the lymphocytes were localized within the nuclei [10].

Twenty-four and 48 hr later, the silver grains were observed rarely in macrophages and were more numerous within lymphocytes (figure 3).

Discussion

Our pictures from the fluorescent electron microscope and high-resolution autoradiography showed clearly that endotoxin is taken up by macrophages through pinocytosis, as is the case for various protein bacterial toxins [11].

At an early stage, the endotoxin is adsorbed onto the surface, and after 15 min it penetrates into the cytoplasm through invaginations or small channels. At 45 min, the endotoxin is observed in the heterophagosomes. In contrast with hemocyanin, which remains bound to the macrophage membrane long after contact [12, 13], the endotoxin is completely absorbed after 24 hr.

At 24 hr, the endotoxin is observed either within heterophagosomes or free in the cytoplasm. We previously reported that free endotoxin is associated at this interval with rich ^{3}H-uridine-labeled ribosome zones [14].

Table 3. Transfer of endotoxin from macrophages to lymphocytes.

| Cellular interaction | Percentage of lymphocytes in LMI* | Percentage of labeled lymphocytes in LMI | Percentage of labeled cells in LMI | | Percentage of free, labeled lymphocytes | No. of scanned lymphocytes |
			Macrophages	Lymphocytes		
Macrophages with ingested ^{14}C-labeled endotoxin plus lymphocytes	1.8	0.2	100	10.1	0.2	13,664
Macrophages incubated simultaneously with glutaraldehyde and ^{14}C-labeled endotoxin plus lymphocytes (control)	0	0	0	0	0	12,000
"L" fibroblasts with ingested ^{14}C-labeled endotoxin plus lymphocytes (control)	2.1	0	100	0	0.04	1,700
Macrophages with ingested ^{14}C-labeled endotoxin plus lymphocytes incubated at 4 C	0.5	0	100	0	0.01	27,000
Culture medium of macrophages that had ingested ^{14}C-labeled endotoxin 1 hr before plus lymphocytes	0	0	0	0	0.009	18,000

* Lymphocyte-macrophage islets. See "Results" for definition.

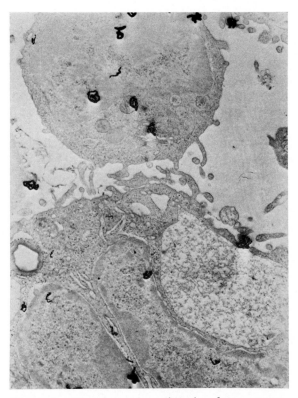

Figure 3. Transfer of endotoxin from macrophages to lymphocytes. Lymphocytes were incubated for 24 hr with macrophages that had ingested ^{14}C-labeled endotoxin 1 hr before. At the level of the lymphocyte-macrophage islet, one lymphocyte shows numerous silver grains ($\times 8,500$).

The presence of endotoxin within residual bodies after 48 hr suggests that this substance, which is not broken down by lysosomal hydrolases, is stored in these organelles; apparently these bodies can load not only the nondegraded material [15] but also the biologically active substances [14].

Evidence that endotoxin contained in the heterophagosomes or stored in the residual bodies had not been broken down was provided by the Shwartzman experiments and by data concerning the immunogenicity of the lysosomal fraction. Indeed, the lysosomal fraction or homogenate from macrophages with ingested endotoxin elicited a specific immune response in normal rabbits; it also prepared the rabbits for the Shwartzman phenomenon. In contrast with neutrophils, which alter the ability of endotoxin to prepare rabbits for the Shwartzman phenomenon within 1–3 hr after ingestion [5], the macrophages were unable to destroy this capacity.

These results are comparable with those obtained by other authors [16–22], who have showed that various antigens bound to the macrophagic lysosomal fraction retain their immunogenicity.

A more striking observation concerns the transfer of endotoxin from macrophages to lymphocytes at the level of LMI.

LMI could represent the cytologic and functional unit of cooperation between macrophages and lymphocytes during the immune response. These islets have been observed on smears or sections from normal or immunized lymphoid organs [23–27], in the living state by microcinematography [28], and in the in-vitro system [29–34]. Lymphocytes containing silver grains are observed at the level of LMI.

The disappearance of radioactivity from LMI-located macrophage membranes (where the lymphocytes contain silver grains) and the regular inhibition of antigen transfer after pronase treatment of macrophages containing ingested antigen [35] strongly suggest that antigen bound to the macrophage membrane was transferred to lymphocytes. The free, labeled lymphocytes, found in a proportion of 0.2%, could be explained by the fact that lymphocytes can spontaneously detach themselves from an islet, as shown by microcinematography [28].

It is impossible at present to establish the exact significance of the radioactivity within the lymphocytes, but the facts that endotoxin was biosynthetically labeled and that its immunogenicity was not altered with the lysosomal fraction suggest that silver grains could correspond to whole antigen or to antigen fragments that have retained their immunogenicity.

It seems satisfactory to associate the presence of antigen in the lymphocytes with synthesis of antibody. It was observed that the number of endotoxin-binding lymphocytes harvested at various intervals after contact with macrophages that had ingested antigen increased regularly between 24 and 72 hr. Furthermore, the majority of endotoxin-binding lymphocytes contained radioactivity, as shown by autoradiographs on thick and thin sections [36].

References

1. Bona, C. A. La pinocytose de toxines bactériennes par les macrophages et les implications immunologiques. Arch. Biol. (Liège) 82:323–392, 1971.

72

2. Ribi, E., Haskins, W. T., Landy, M., Milner, K. C. Preparation and host-reactive properties of endotoxin with low content of nitrogen and lipid. J. Exp. Med. 114:647–663, 1961.

3. Clarke, A. E. Hydrolytic enzymes of human polymorphonuclear leucocytes and rat monocytes. Aust. J. Exp. Biol. Med. Sci. 43:201–212, 1965.

4. Bertrand, D. Techniques de dosage. In F. Loisleur [ed.] Technique de laboratoire. Vol. 1. Masson, Paris, 1963, p. 980–983.

5. Pal Yu, B., Kummerow, F. A., Nishida, T. Acid phosphatases of rat polymorphonuclear leucocytes. Proc. Soc. Exp. Biol. Med. 122:1045–1048, 1966.

6. Mesrobeanu, L., Mesrobeanu, I., Bona, C., Vranialici, D. Le destin des endotoxines thermostables pinocytées par les leucocytes. In L. Chedid [ed.]. La structure et les effets biologiques des produits bactériens provenant de germes gram-négatifs. Colloque International No. 174, Paris, 1969, p. 429–445.

7. Landy, M., Sanderson, R. P., Jackson, A. L. Humoral and cellular aspects of the immune response to the somatic antigen of Salmonella enteritidis. J. Exp. Med. 122:483–504, 1965.

8. Landy, M., Trapani, R.-J., Clark, W. R. Studies of the O antigen of Salmonella typhosa. III. Activity of the isolated antigen in the haemagglutination procedure. Am. J. Hyg. 62:54–65, 1955.

9. Rogers, A. W. Techniques of autoradiography. Elsevier, Amsterdam, London-New York, 1967, p. 335.

10. Bona, C. La résolution de l'autoradiographie en microscopie électronique utilisant les émetteurs β— avec une haute énergie maximale. Ph.D. thesis, University of Paris, 1972, p. 5–14.

11. Mesrobeanu, I., Mesrobeanu, L., Bona, C. Uptake of bacterial protein toxins by cells. In S. J. Ajl, S. Kadis, and T. C. Montie [ed.]. Microbial toxins. Vol. 1. Academic Press, New York, 1970, p. 445–469.

12. Unanue, E. R., Cerottini, J.-C., Bedford, M. The persistence of antigen on the surface of macrophages. Nature (Lond.) 222:1193–1195, 1969.

13. Unanue, E. R., Cerottini, J.-C. The immunogenicity of antigen bound to the plasma membrane of macrophages. J. Exp. Med. 131:711–725, 1970.

14. Anteunis, A., Robineaux, R., Bona, C. The intracellular association of antigen and newly synthesized RNA in "in vitro" cultured macrophages. An autoradiographic, cytochemical and ultrastructural study. R.E.S. Journal 1973 (in press).

15. De Duve, C., Wattiaux, R. Functions of lysosomes. Ann. Rev. Physiol. 28:435–492, 1966.

16. Franzl, R. E. Immunogenic sub-cellular particles obtained from spleens of antigen-injected mice, Nature (Lond.) 195:457–458, 1962.

17. Uhr, J. W., Weissman, G. Intracellular distribution and degradation of bacteriophage in mammalian tissues. J. Immunol. 94:544–550, 1965.

18. Ada, G. L., Williams, J. M. Antigen in tissues. I. State of bacterial flagella in lymph nodes of rats injected with isotopically-labelled flagella. Immunology 10:417–429, 1966.

19. Mitchison, N. A. The immunogenic capacity of antigen taken up by peritoneal exudate cells. Immunology 16:1–14, 1969.

20. Kölsch, E., Mitchison, N. A. The subcellular distribution of antigen in macrophages. J. Exp. Med. 128:1059–1079, 1968.

21. Cruchaud, A., Unanue, E. R. Fate and immunogenicity of antigen endocytosed by macrophages: a study using foreign red cells and immunoglobulin G. J. Immunol. 107:1329–1340, 1971.

22. Schmidtke, J. R., Unanue, E. R. Macrophage-antigen interaction: uptake, metabolism and immunogenicity of foreign albumin. J. Immunol. 107:331–338, 1971.

23. Nossal, G. J. V., Abbot, A., Mitchell, J., Lummus, Z. Antigens in immunity. XV. Ultrastructural features of antigen capture in primary and secondary lymphoid follicles. J. Exp. Med. 127:277–290, 1968.

24. Robineaux, R., Anteunis, A., Bona, C. Ultrastructure des macrophages de cobaye. Ann. Inst. Pasteur (Paris) 120:329–336, 1971.

25. Schoenberg, M. D., Mumaw, V. R., Moore, R. D., Weisberger, A. S. Cytoplasmic interaction between macrophages and lymphocytic cells in antibody synthesis. Science 143:964–965, 1964.

26. Thiery, J. P. Ultrastructure et fonctions de cellules impliquées dans la réaction immunitaire. Bull. Soc. Chim. Biol. 50:1077–1100, 1968.

27. Sharp, J. A., Burvell, R. G. Interaction (periplolesis) of macrophages and lymphocytes after skin homografting or challenge with soluble antigens. Nature (Lond.) 188:474–475, 1960.

28. Robineaux, R., Pinet, J., Kourilsky, R. Etude microcinématographique de la rate en culture sous membrane de dialyse. Nouv. Rev. Franç. Hématol. 2:797–811, 1962.

29. Bona, C., Anteunis, A., Robineaux, R., Astesano, A. Etude radioautographique ultrastructurale du transfert in vitro de l'ARN immunogène produit par les macrophages qui ont capté l'endotoxine de Salmonella typhimurium "S". C.R. Acad. Sci. [D] (Paris) 269:1145–1147, 1969.

30. Cline, M. J., Swett, V. C. The interaction of human monocytes and lymphocytes. J. Exp. Med. 128:1309–1324, 1968.

31. Fishman, M., Hammerstrom, R. A., Bond, V. P. In vitro transfer of macrophage RNA to lymph node cells. Nature (Lond.) 198:549–551, 1963.

32. Hanifin, J. M., Cline, M. J. Human monocytes and macrophages. Interaction with antigen and lymphocytes. J. Cell Biol. 46:97–105, 1970.

33. Hersh, E. M., Harris, J. E. Macrophage-lymphocyte interaction in the antigen-induced blastogenic response of human peripheral blood leukocytes. J. Immunol. 100:1184–1194, 1968.

34. Siegel, I. Natural and antibody-induced adherence

of guinea pig phagocytic cells to autologous and heterologous thymocytes. J. Immunol. 105:879–885, 1970.

35. Bona, C., Anteunis, A., Robineaux, R., Astesano, A. Transfer of antigenic macromolecules from macrophages to lymphocytes. I. Autoradiographic and quantitative study of ^{14}C endotoxin and ^{125}I hemocyanin transfer. Immunology 23:799–816, 1972.

36. Bona, C., Robineaux, R., Anteunis, A., Heuclin, C., Astesano, A. Transfer of antigen from macrophages to lymphocytes. II. Immunological significance of the transfer of lipopolysaccharide. Immunology 1973 (in press).

Role of O-antigen (Lipopolysaccharide) Factors in the Virulence of *Salmonella*

P. Helena Mäkelä, V. V. Valtonen, and M. Valtonen

From the Central Public Health Laboratory (State Serum Institute) and the Department of Serology and Bacteriology, University of Helsinki, Helsinki, Finland

The lipopolysaccharide of *Salmonella* is an important virulence factor; loss of its O-specific side chains results in loss of virulence. We have examined the effect of qualitative alterations of the O side chains by determining the LD_{50} values of *Salmonella typhimurium* derivatives after intraperitoneal infection of the mouse. Isogenic transductants were most virulent if the O antigen was 1,4,12, less virulent when the O antigen was 1,9,12, and nearly avirulent when the O antigen was 6,7. Modifications of the 4,12 type (antigen factors 1, 5, or 12_2) did not affect virulence [1]. The 4,12 and 9,12 lipopolysaccharides of the transductants, although quantitatively similar [2], differ qualitatively: 4,12 has abequose in the position where 9,12 has another dideoxyhexose, tyvelose. How then does the mouse discriminate between these two very similar structures in a way that leads to the observed difference in virulence? So far we have obtained no evidence that either the immune system or the phagocytic defenses account for this difference. Suppression of immune responses did not alter the discrimination [3], and the 4,12 and 9,12 types survived to the same extent inside peritoneal macrophages of mice.

The O antigen (lipopolysaccharide or LPS) has long been known to be an important virulence factor in *Salmonella*. Rough (R) forms of the bacteria that have lost the O-specific side chains of the LPS are avirulent. Semirough forms, in which the length of these normally polymeric side chains is reduced, have reduced virulence in the mouse [4, 5].

We have studied the effect on virulence of more subtle variations of the O side chains. Our experimental system is infection of the mouse with *Salmonella typhimurium*. Intraperitoneal injection of as few as 100 organisms of virulent *S. typhimurium* results in generalized infection and death in five to 10 days. We took advantage of the recently accumulated knowledge of the genetic determination of various parts of LPS [6] in altering separate O-antigen factors one at a time. We thus compared in the test for virulence nearly isogenic pairs of *S. typhimurium* derivatives with different O antigens to establish the role of the quality of the LPS in virulence. Such pairs were prepared by transduction (or in some cases by lysogenic conversion or conjugation), in which phage P22 was used as vector of the genetic material to be exchanged. The size of the phage restricted the amount of foreign gene material in the transductants to not more than 1% of the

This study was supported by a grant from the Sigrid Jusélius Foundation.

Please address requests for reprints to Dr. P. Helena Mäkelä, Central Public Health Laboratory (State Serum Institute), Helsinki, Finland.

bacterial genome. The procedure of selection further limited the genetic difference between the sister transductants that were compared.

Materials and Methods

We determined the LD_{50} 10 days after injection of a series of 10-fold dilutions of overnight broth cultures into groups of 10 mice each [7]. Randomly bred strain C and Swiss Albino Webster mice weighing 22–25 g were used with essentially the same results. The method and its statistical evaluation have been presented and examined in detail by Valtonen [8].

In most experiments derivatives of the *S. typhimurium* line LT2 strain SL1027 (from B. A. D. Stocker, Stanford University, Palo Alto, California) were used. A recent fecal isolate from a human case of diarrhea, line IH2, was used as a more virulent line. New mutations were selected in each line after treatment with diethylsulfate [8], and each strain was checked before use for unaltered virulence in mice.

The transducing phage P22 was obtained from *S. typhimurium* derivatives in which the original O antigen (4,5,12 or 4,12) had first been altered by conjugation with either a *Salmonella enteritidis* (O antigens 9,12) or a *Salmonella montevideo* (O antigens 6,7) donor. Thus they had the *rfb* genes that determine the O-antigen types (9,12 or 6,7) of these parents, but yielded phage that plated with normal efficiency on *S. typhimurium*. P22 was grown in the normal manner on the O–9,12 derivative SH 1036 [8]. O–6,7 derivatives are not sensitive to P22 [9], but a transducing phage preparation was obtained after induction from the P22-lysogenic O–6,7 strain SH 1011 [8]. The O-antigen characteristics of the strains were determined serologically and with smooth- and rough-specific bacteriophages [8, 10]. Details of the bacteriologic methods have been presented by Valtonen [8].

Peritoneal macrophages were obtained from untreated C-strain mice and used without washing; 5×10^6 cells were suspended in minimal essential medium [11] with 10% fetal-calf serum. Log-phase bacteria were added, bringing the ratio of bacteria to cells to 1:1 in a total volume of 3 ml. The mixture was incubated at 37 C; 0.1-ml samples were drawn at 0, 1, and 3 hr, mixed with 0.9 ml of 1% Triton X-100, and shaken for 2 min for disruption of the phagocytic cells [12]. The bacteria were then counted after serial dilution for determination of the number of surviving bacteria.

Results

The *S. typhimurium* LT2 line that has been extensively used in genetic work is of rather low virulence; the LD_{50} is approximately 10^5 [8]. Its LPS is of the smooth, complete type, and its O-antigenic formula is 4,5,12. Presence or absence of factors 1, 5, or 12_2 does not affect its virulence [1]. All three of these factors are nonessential components of the LPS, and their presence or absence does not disturb the synthesis of the complete, smooth LPS. Instead they represent modifications, O-acetyl or glucosyl groups, added to the completed O side chains; even the genes directing their synthesis are far removed from the genes directing the synthesis of the basic structure of the O side chains [6, 10].

The basic structure of the O side chains is determined by genes in a cluster called *rfb* and situated close to the *his* genes (which determine the biosynthesis of histidine). When we used P22 grown on an O–9,12 strain for transduction and selected transductants that had received the *his*+ genes carried over from the O–9,12 strain by the phage, some of them had also received the *rfb* cluster of the O–9,12 strain and produced an O antigen with the specificity 9,12. Other *his*+ transductants still had the original *rfb* cluster of *S. typhimurium* and consequently produced O antigen of the 4,5,12 type (factor 5 is O acetylation of abequose, which is not present in the 9,12 LPS). Both types of transductants also had the antigen factor 1 determined by P22, which had been established as a prophage.

When the transductants were used in the mouse-infection assay (table 1), those with O–1,9,12 had a LD_{50} 10-fold greater than those with O–1,4,(5,)12. The same was true when the parent strain had been a LT2 derivative, with a LD_{50} of 10^5, as well as when the parent had been of the more virulent IH2 line.

A similar experiment was done as a comparison of sister transductants with either O–6,7 or O–1,4,5,12 antigens; the O–6,7 transductants were of very low virulence (table 1) [8].

Observed differences in virulence have always followed the O-antigen type in the manner indi-

Table 1. LD$_{50}$ values 10 days after intraperitoneal injection of a series of 10-fold dilutions of overnight broth cultures into mice. In each case sister *his*+ transductants were compared that had either the original O–4,(5,)12 or donorlike O–9,12 or O–6,7 antigens.

Phage P22 grown on strains with	*his*+ transductants	LD$_{50}$ of *Salmonella typhimurium* transductants	
		LT2 [8]	Line IH2
O–9,12	O–1,4,(5,)12	10^5	4 × 10^2
	O–1,9,12	10^6	4 × 10^3
O–6,7	O–1,4,5,12	10^5	
	O–6,7	5 × 10^7	

For O–9,12: $P < 0.001$; For O–6,7: $P < 0.001$

NOTE. The LD$_{50}$ was the same for closely related strains with and without antigens 1 or 5 [1].

cated above. A total of 12 transductants (or conjugational recombinants) of the O–4,12 type, nine of the O–9,12 type, and five of the 6,7 type have been tested, and the range of virulence has always been the same: O–4,12 more virulent than O–9,12, with O–6,7 nearly avirulent. We have shown in other experiments that neither P22 lysogenization nor antigen factors 1 or 5 affect the LD$_{50}$ values in this system [1], and thus the basis of the observed differences in virulence seems to be correlated with the basic O-antigenic types. No other differences in the phenotypes have been detected when the O–4,12 and O–9,12 transductants are compared; they grow in broth at equal rates even in competitive conditions, and they both have the enterobacterial common antigen ([13] and H. Mayer, personal communication). The O–6,7 transductants have pleiotropic changes in their phenotype, i.e., a slightly reduced growth rate and absence of the common antigen (H. Mayer, personal communication).

Therefore, we shall restrict the present paper to a comparison of the O–4,12 and O–9,12 sister strains.

The O–4,12 and the O–9,12 transductants differ qualitatively in their LPS structure (see below). In addition, there could be quantitative differences; e.g., if the O–9,12 LPS was less complete than the O–4,12, this could easily account for the lower virulence of the former. Therefore, the LPS of both sorts of transductants (two strains of each) was extracted by the phenol-water method at 65 C and analyzed by gas chromatography and mass spectrometry after methylation [2]. The relevant conclusions are collected in figure 1 and table 2; the length of the O side chains (determined from the ratio of terminal mannose to internal mannose) was the same in all strains, as was the degree of substitution of core by O side chains. The amount of LPS extracted was also the same in all the strains; in other words, no indication of even partial roughness was found by this very accurate method.

The qualitative difference between the O–4,12 and O–9,12 LPS types previously shown in natural salmonella B and D group strains was, as expected, present in the transductants. The main difference is in the immunodominant dideoxyhexose "branch" of each O-specific unit (figure

Table 2. Quantitative aspects of the structure of lipopolysaccharides (LPS) of *Salmonella typhimurium* transductants.

Proportions	LPS of transductants	
	1,4,(5,)12	1,9,12
Terminal Man/internal Man	1/3	1/3
Terminal Man/core Gal$_1$	1/1	1/1
Core Glc$_2$ substituted at carbon 4	All	All

NOTE. Man = mannose; Gal = D-galactose; Glc = D-glucose.

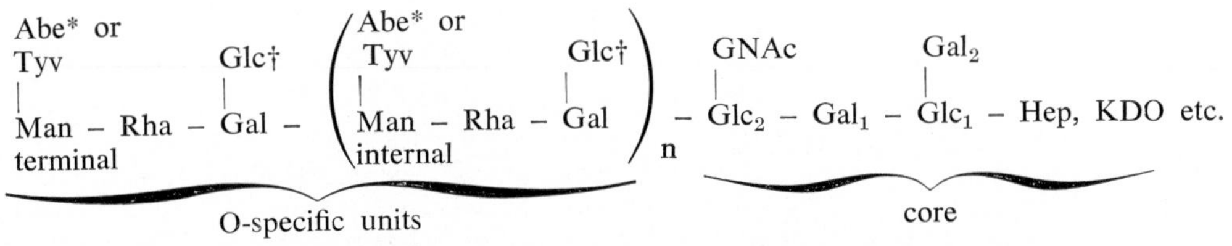

Figure 1. Structure of the lipopolysaccharide of transductants with O antigens 1,4,(5,)12 or 1,9,12 [2]. (Abe = abequose; Tyv = tyvelose; Man = D-mannose; Rha = L-rhamnose; Gal = D-galactose; GNAc = 2-acetamido-2-deoxy-D-glucosamine; Glc = D-glucose; Hep = heptose; KDO = 2-keto-3-deoxyoctonate.) *In strains with antigen 5, 2-o-acetylabequose partially replaces abequose (2, 15). †In different transductants 25% to 53% of the O-specific unit D-galactoses were substituted by D-glucose.

1, table 2): abequose in O–4,12 and tyvelose in O–9,12 LPS.

The qualitative difference in the two sorts of LPS seems small, but it is the only difference that we could find. The possibility that the two types of transductants differ in some other genes (unrelated to LPS) cannot be excluded but does not look probable. For example, such genes would need to be very close to the *rfb* cluster to account for the fact that the virulence type of the transductants always followed the LPS type, and such close association by chance is fairly unlikely.

If we then accept the conclusion that the 4,12-type LPS makes otherwise similar salmonella strains more virulent for mice than does 9,12-type LPS, we are led to inquire about the means by which the mouse discriminates between these two rather similar types of LPS. Immunologic mechanisms could readily account for such specificity; specific antibodies to factors 4 and 9 are easily produced. On the other hand, nonimmunologic host-defense mechanisms, most notably phagocytosis, play an important role, especially in rapidly progressing infections, but might not be expected to be equally specific.

We have tested the immunologic-discrimination hypothesis in the following ways. A test for natural antibodies to the two sorts of smooth transductants was negative; neither bactericidal antibodies nor those active in a hemagglutination assay with LPS-coated murine red cells were found in several individual sera of our mice. We then repeated the virulence assay in immunosuppressed (thymectomized and X-irradiated) mice (table 3) [3]. Immunosuppression may not have been complete, but it was sufficient to increase the sensitivity of the mice to infection 1,000-fold. However, the difference in virulence between the

O–4,12 and the O–9,12 transductants was retained. Thus, immunologic discrimination is probably not of prime (if any) importance in the observed difference in virulence.

We then hoped to obtain positive evidence for a discrimination between the O–4,12 and O–9,12 types by the phagocytic system of the mouse. We collected peritoneal macrophages and let them ingest the two sorts of bacteria. As controls, we used a rough mutant (with the complete LPS core) of LT2 and the avirulent O–6,7 transductant. All of these bacteria seemed to be ingested with equal efficiency, but the measurements were not very exact. When intracellular survival rates of the bacteria were compared (figure 2), the O–4,12 and O–9,12 transductants were equal and survived (actually multiplied) better than the rough strain and the O–6,7 transductant. Thus, we obtained no evidence for a discrimination between the O–4,12 and O–9,12 strains, but the result is not conclusive since the sensitivity of the test is rather poor. While the difference in virulence between the O–4,12 strain and the rough mutant is 1,000-fold, the difference in their survival was only 30-fold; thus the 10-fold difference in virulence between O–4,12 and O–9,12 might be too small to be reflected at all in the test for phagocytosis.

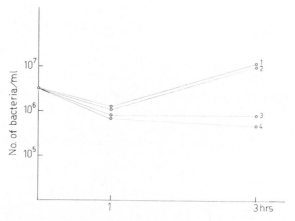

Figure 2. Survival of *Salmonella typhimurium* derivatives in peritoneal macrophages of mice. Approximately 5×10^6 bacteria and the same number of cells were mixed in 3 ml of medium at 37 C; bacteria were counted after disruption of the cells with 1% Triton X-100 at 0, 1, and 3 hr. (1) *his+* O–1,9,12; (2) *his+* O–1,4,(5,)12; (3) *his+* O–6,7 transductants; (4) a rough (*rfb*) mutant SL 748 with the complete lipopolysaccharide core.

Table 3. LD$_{50}$ of sister transductants in mice with and without immunosuppression.

| Transductants | LD$_{50}$ in mice | |
	Untreated	Immuno-suppressed
1,4,(5,)12	3×10^5	50
1,9,12	2×10^6	400
6,7	5×10^7	5×10^5

NOTE. Mice were thymectomized at age seven to eight weeks; 10 days later they were X-irradiated (600 rad). Fifteen to 18 hr after irradiation, they were infected ip with transductants [3].

Discussion

We believe that we have demonstrated that a small qualitative difference in the structure of LPS can affect the virulence of salmonellae. In two different lines of *S. typhimurium,* LPS of the O–4,12 type induced higher virulence in mice than LPS of the O–9,12 type. A similar experiment was performed with *S. enteritidis,* originally a mouse-virulent O–9,12 strain [14]. When its O antigen was changed to O–4,12, its ability to grow in tissues of mice improved. In both of these experiments, nearly isogenic strains were compared, an approach that makes it possible to examine the effect of single bacterial factors on virulence.

Why O–4,12 would be better for pathogenicity than O–9,12 is not easy to see. Both sorts of LPS are compatible with pathogenicity; in the mouse both *S. typhimurium* (O–4,12) and *S. enteritidis* (O–9,12) are pathogenic, and in humans both *S. typhimurium* or *S. paratyphi* B (O–4,12), and *S. typhi* (O–9,12) are pathogenic.

The chemical difference between O–4,12 and O–9,12 LPS seems to be small; it does not, for example, affect the lipophilic or hydrophilic properties of the LPS. The mechanism by which the murine host discriminates between the two types must be rather specific, but so far we know nothing about this mechanism. The approaches that we have tried have not been very sensitive, and by improving them we still hope to learn more about the host factors involved in this system.

References

1. Valtonen, V. V., Mäkelä, P. H. The effect of lipopolysaccharide modifications—antigenic factors 1, 5, 12_2 and 27—on the virulence of salmonella strains for mice. J. Gen. Microbiol. 69:107–115, 1971.
2. Nurminen, M., Hellerqvist, C. G., Valtonen, V. V., Mäkelä, P. H. The smooth lipopolysaccharide character of 1,4,(5),12 and 1,9,12 transductants formed as hybrids between groups B and D of *Salmonella.* Eur. J. Biochem. 22:500–505, 1971.
3. Valtonen, V. V., Aird, J., Valtonen, M., Mäkelä, O., Mäkelä, P. H. Mouse virulence of *Salmonella:* antigen-dependent differences are demonstrable also after immunosuppression. Acta Pathol. Microbiol. Scand. [B] 79:715–718, 1971.
4. Nakano, M., Saito, K. Chemical components in the cell wall of *Salmonella typhimurium* affecting its virulence and immunogenicity in mice. Nature (Lond.) 222:1085–1086, 1969.
5. Valtonen, V. Virulence of salmonella strains with a reduced amount of O-antigen. J. Gen. Microbiol. 57:28–29, 1969.
6. Mäkelä, P. H., Stocker, B. A. D. Genetics of polysaccharide biosynthesis. Annu. Rev. Genet. 3:291–322, 1969.
7. Reed, L. J., Muench, H. A simple method of estimating fifty per cent endpoints. Am. J. Hyg. 27:493–497, 1938.
8. Valtonen, V. V. Mouse virulence of salmonella strains: the effect of different smooth-type O side-chains. J. Gen. Microbiol. 64:255–268, 1970.
9. Zinder, N. D., Lederberg, J. Genetic exchange in *Salmonella.* J. Bacteriol. 64:679–699, 1952.
10. Stocker, B. A. D., Mäkelä, P. H. Genetic aspects of biosynthesis and structure of salmonella lipopolysaccharide. *In* G. Weinbaum, S. Kadis, and S. J. Ajl [ed.] Microbial toxins. vol. 4. Academic Press, New York, 1971, p. 369–438.
11. Eagle, H. Amino acid metabolism in mammalian cell cultures. Science 130:432–437, 1959.
12. Ruutu, T. Effect of phenothiazines and related compounds on phagocytosis and bacterial killing by human neutrophilic leukocytes. Ann. Med. Exp. Biol. Fenn. 50:24–36, 1972.
13. Kunin, C. M., Beard, M. V., Halmagyi, N. E. Evidence for a common hapten associated with endotoxin fractions of *E. coli* and other Enterobacteriaceae. Proc. Soc. Exp. Biol. Med. 111:160–166, 1962.
14. Eisenstein, T. K., Ornellas, E. P., Roantree, R. J., Steward, J. P. Importance of 0:4 and 0:9 antigenic determinants in vaccines of heat-killed *Salmonella* (abstract). Bacteriol. Proc. 70:94, 1970.
15. Hellerqvist, C. G., Lindberg, B., Svensson, S., Holme, T., Lindberg, A. A. Structural studies on the O-specific side-chains of the cell wall lipopolysaccharide from *Salmonella typhimurium* 395 MS. Carbohydrate Res. 8:43–55, 1968.

Activation of Complement by Endotoxin

Stephan E. Mergenhagen, Ralph Snyderman,*
and Jean K. Phillips

*From the Laboratory of Microbiology and Immunology,
National Institute of Dental Research, National
Institutes of Health, Bethesda, Maryland*

Ovine erythrocytes (E) coated with endotoxin (LPS) from *Salmonella typhosa*
0901 have been used for characterization of the nature of a pathway by which
LPS activates complement (C). E-LPS consumed C in normal guinea-pig serum
in a manner identical to that in which C is consumed by LPS, that is, with
marked consumption of C3–C9 and little depletion of C1, C4, and C2. A "natural" γ2-globulin, isolated from serum by ion-exchange chromatography, was
required for C-mediated lysis of E-LPS in dilute serum. When it was reacted with
guinea-pig γ2-globulin, C1, C4, and C2, followed by C-EDTA, E-LPS was lysed.
Deletion of γ2-globulin, C1, C4, or C2 from the reaction prevented lysis. The small
amounts of the early-acting components of C used indicated that it would be
difficult to measure their loss in whole serum after incubation with LPS. This
hypothesis could explain the earlier findings of a relative sparing of C1, C4, and
C2 during consumption of C3–C9 by LPS in normal serum. Our present concepts
of C activation by LPS must consider two pathways: a conventional pathway involving "natural" antibody (γ2-globulin) and the formation of the C3 convertase
(C$\overline{42}$) and an alternate pathway involving the formation of a C3 activator, which
proceeds without obvious participation of C1, C4, or C2. The significance of
either or both of these pathways to the induction of tissue injury by endotoxin
awaits further investigation.

There is now considerable evidence to suggest a
role for the complement (C) system in mediating
certain of the host responses to endotoxic lipopolysaccharide (LPS). Several of the in-vitro biological activities attributed to components of C or
products of their cleavage are significant in the
pathophysiology of the endotoxin reaction (table 1)
[1–3]. This conclusion is supported by recent
work in vivo, which implicates C in the early
phase of the endotoxin-shock syndrome in dogs
[4], in the local and generalized Shwartzman reaction [5], and in the early accumulations of polymorphonuclear leukocytes in inflammatory exudates produced with LPS [6].

It has been recognized for some time that LPS,
unlike preformed antigen-antibody complexes, can
preferentially consume the terminal components
of C (C3–C9) in normal mammalian serum with
little detectable inactivation of C1, C4, or C2 [7].
One question, therefore, that continually pervades
interpretation of the available data on the interaction of LPS with C is whether LPS, like certain
other antigens, requires an antibody-C$\overline{142}$ pathway
to activate terminal components of C or whether the
early components are by-passed or used in a qualitatively different fashion by LPS. Indeed, recent
data demonstrate this latter mode of activation of
C by LPS as well as by certain polysaccharides
and aggregated immunoglobulins [8–12]. When
serum is treated with plant or bacterial polysaccharides, a noncomplement protein in plasma (C3
proactivator) is cleaved into two fragments. It is
the larger fragment (mol wt, 60,000) that has
been designated the C3 activator and that has the
ability to cleave C3 into C3a and C3b subunits
[8, 9].

Since LPS is partially soluble in serum, it has
been difficult to explore the role of ubiquitous
"natural" antibody and other serum-derived factors in the C-activation pathway by the building
of intermediates on LPS in a manner similar to the

Please address requests for reprints to Dr. Stephan E.
Mergenhagen, Laboratory of Microbiology and Immu-
nology, National Institute of Dental Research, National
Institutes of Health, Bethesda, Md. 20014.
* Presently a Howard Hughes Medical Investigator,
Department of Medicine, Duke University, Durham, N.C.

Table 1. Biological activities produced in vitro as a consequence of interactions between endotoxin and complement.

Biological activity	Complement components involved

Anaphylatoxin
  Contraction of smooth muscle
  Increased capillary permeability
	C3, C5

Release of histamine and heparin from mast cells	C3, C5
Adherence and degranulation of platelets	C3
Coagulation of blood	C6

Chemotactic factors
  Polymorphonuclear leukocytes
  Mononuclear leukocytes
	C5

building of C intermediates on the surface of ovine erythrocytes (E) sensitized with antibody (A). Recently, however, a method was devised that allows the sequential addition of serum-derived components to E coated with LPS (E-LPS) [13]. By measurement of C-dependent lysis of E-LPS, the nature of a pathway by which LPS can activate C3–C9 has been elucidated.

Table 2 shows that E coated with LPS from *Salmonella typhosa* 0901 can be lysed by pooled normal guinea-pig serum. Similar results can be obtained with other LPS preparations and with sera of individually bled, unimmunized guinea pigs [13]. However, although hemolytic C activity is sufficient for lysis of EA at high dilutions of serum,

there appears to be a factor(s) other than C that limits the lysis of E-LPS at such high serum dilutions. On the other hand, the addition of heated (56 C, 30 min) guinea-pig serum to the reaction of E-LPS and dilutions of serum enhanced the hemolytic titer of guinea-pig serum for E-LPS to that observed with EA. This indicated that a heat-stable component of serum other than "classical" C, required for hemolysis of E-LPS, was a limiting factor in diluted serum. In parallel experiments on determination of the C-consumption profile in normal guinea-pig serum by E-LPS and LPS alone, it was also found that both reagents consumed considerable quantities of terminal C components (C-EDTA)[1] but spared C1, C4, and C2 (table 3). Therefore, these data suggested that E-LPS would be a useful reagent for study of the nature of the heat-stable factor required for the lysis of E-LPS by normal serum, as well as the nature of the pathway involved in the activation of C by LPS.

For characterization of the heat-stable factor in normal guinea-pig serum that limited lysis of E-LPS, the serum was fractionated by ion-exchange chromatography on DEAE-cellulose in a manner similar to that previously described for the isolation of guinea-pig immunoglobulins [14]. Fractions from the column were tested with E-LPS; a 1:100 dilution of guinea-pig serum, which would supply ample hemolytic C activity but which in itself would not lyse the coated cells, was used. As can be seen in figure 1, a sharp peak of

Table 2. Enhanced lysis of endotoxin-coated erythrocytes by addition of heated serum to various dilutions of normal guinea-pig serum.

Type of cell*	Material tested	Hemolytic titer†
E-LPS	GPS	12
E-LPS	GPS + heated GPS	384
EA	GPS	384
E	GPS + heated GPS	0

NOTE. Data are adapted from [13] with permission of The Williams and Wilkins Co., Baltimore, Md.

* Endotoxin (LPS) from *Salmonella typhosa* 0901 was heated at 100 C for 30 min to increase its affinity for the erythrocyte. Ovine erythrocytes (E) or E sensitized with antibody (EA) or endotoxin (E-LPS) were standardized to contain 1×10^8 cells/ml. Twenty-five thousandths milliliter of respective cell type was added to 0.025 ml of serially diluted, normal guinea-pig serum (GPS), and 0.025 ml of heated GPS (56 C, 30 min) or buffer was then added to each reaction mixture.

† Reciprocal of dilution of GPS required for 50% lysis.

Table 3. Consumption of components of complement (C) in normal guinea-pig serum by endotoxin-coated erythrocytes.

Material tested*	Total C	C1	C4	C2	C-EDTA
E-LPS	64	<16	<10	<10	46
LPS alone	86	<16	<10	<10	84
E	2	<10	<10	<10	<10

NOTE. Data are adapted from [13] with permission of The Williams and Wilkins Co., Baltimore, Md.

* One-half milliliter of buffer containing 1×10^9 ovine erythrocytes (E), E sensitized with endotoxin from *Salmonella typhosa* (E-LPS) or endotoxin (LPS) alone were incubated with 0.5 ml of guinea-pig serum at 37 C for 1 hr. Residual hemolytic activities of C were assayed, and the percentage of the available C consumed was calculated.

[1] Serum incubated with 0.02 M EDTA to block out C1, C4, and C2 to provide a source of C3–C9.

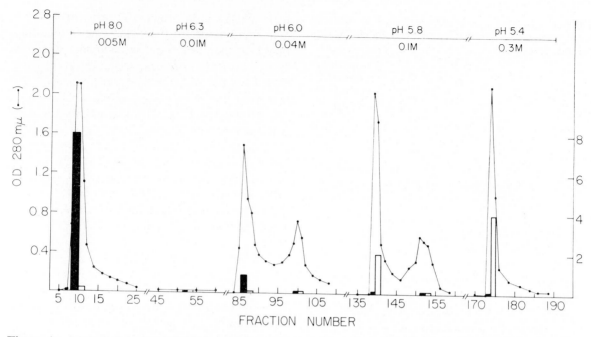

Figure 1. Elution from DEAE-cellulose of heat-stable factors that are functionally absent in diluted serum for lysis of endotoxin-coated erythrocytes. Fractions tested for titer of hemolysis-limiting factor (■) were serially diluted, and a 1:100 dilution of guinea-pig serum was added to each dilution. This was followed by the addition of endotoxin-coated erythrocytes. Also shown is the titer of agglutinating activity (□) eluted from the column, which was determined in the presence of a 1:100 heated guinea-pig serum. Data are adapted from [13] with permission of The Williams and Wilkins Co., Baltimore, Md.

a heat-stable factor that restored the hemolytic activity of diluted guinea-pig serum for E-LPS to the full C titer appeared in the first protein peak. This factor was termed hemolysis-limiting factor (HLF), since it was functionally absent in diluted guinea-pig serum. The peak of protein containing HLF consisted almost entirely of γ2 migrating globulins. This conclusion was substantiated by the observation that the HLF, isolated from the DEAE-cellulose column, produced a single precipitin band, corresponding to the mobility of γ2-globulin when it was electrophoresed and developed with a rabbit antibody to highly purified guinea-pig γ2-globulin.

The foregoing experiments indicated that a "natural" γ2-globulin was required for the C-mediated lysis of E-LPS in normal, diluted guinea-pig serum. We then performed experiments to determine whether or not this γ2-globulin could produce lysis of E-LPS in the presence of early-acting C components and C-EDTA (C3–C9) (table 4). E-LPS reacted with γ2-globulin and purified C1, C4, and C2, followed by C-EDTA, resulted in lysis. Deletion of any one of these components

from the reaction mixture prevented lysis. It should be stressed that the quantities of functionally pure C4 and C2 used in these experiments were well below those amounts the loss of which could be detected by conventional hemolytic assay of serum activated with LPS. This could explain earlier findings of a relative sparing of early-acting C components when LPS is incubated in normal guinea-pig serum [7].

Table 4. Lysis of endotoxin-coated erythrocytes (E-LPS) by "natural" guinea-pig γ2-globulin and components of complement (C).

Type of cell	γ2-globulin	Reagents added				Lysis (%)
		C1	C4	C2	C-EDTA	
E-LPS	+	+	+	+	+	57
E-LPS	−	+	+	+	+	4
E-LPS	+	−	+	+	+	3
E-LPS	+	+	−	+	+	5
E-LPS	+	+	+	−	+	8
E	+	+	+	+	+	14
EA*	−	+	+	+	+	99

NOTE. Data are adapted from [13] with permission of The Williams and Wilkins Co., Baltimore, Md.

* Ovine erythrocytes sensitized with antibody.

Since various preparations of LPS are capable of converting the C3 proactivator into an enzymatically active fragment (C3 activator) that can cleave C3 into C3a and C3b, it is highly probable that LPS can initiate consumption of C3–C9 by either of two pathways. Our present concept of C activation by LPS is summarized in figure 2, which depicts both a conventional pathway proceeding via γ2-globulin with the formation of the C3 convertase (C$\overline{42}$) and the alternate or "shunt" pathway involving the C3-activator system. It is tempting to speculate that the alternate C3-activator system may be especially important in the generation of factors that initiate the inflammatory response and coagulation of the blood. This notion is based on the observation that LPS and immune complexes can induce decided consumption of C3–C9 with the production of phlogistic polypeptides by the C3 shunt pathway in normal serum and in the sera of guinea pigs genetically deficient in C4 [11, 15, 16]. In addition, it has recently been shown that activation of C through the C3-activator system results in the initiation of blood coagulation [3]. On the other hand, the antibody-mediated classical pathway may be more efficient in damaging the cell membrane. This latter conclusion is substantiated by the present findings and also by a recent report suggesting that antibody-dependent cell damage proceeds inefficiently via the alternate C pathway [17]. Since endotoxin has a high affinity for cell surfaces such as those on erythrocytes, the antibody-mediated C-activation pathway, such as that described in the present communication, may be of considerable importance in the action of endotoxin on granulocytes, macrophages, platelets, and mast cells. In any event, the precise role and significance of either or both of these C-activation pathways in the induction of injury to tissues by endotoxins will have to await further investigation.

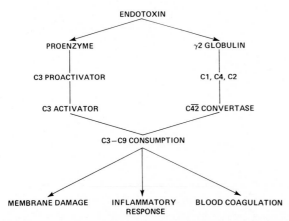

Figure 2. Two pathways of complement activation by endotoxin. The classical mechanism involves the γ2-globulin and subsequent formation of the C$\overline{42}$ convertase. The alternate pathway bypasses C1, C4, and C2 by using the C3 activator. Both pathways lead to consumption of C3–C9, which induces damage to membranes, the inflammatory response, and coagulation of the blood.

References

1. Mergenhagen, S. E., Snyderman, R., Gewurz, H., Shin, H. S. Significance of complement to the mechanism of action of endotoxin. Curr. Topics Microbiol. Immunol. 50:37–77, 1969.
2. Gewurz, H., Snyderman, R., Mergenhagen, S. E., Shin, H. S. Effects of endotoxic lipopolysaccharides on the complement system. *In* Microbial toxins. Vol. 5. Academic Press, New York, 1971, p. 127–149.
3. Zimmerman, T. S., Müller-Eberhard, H. J. Blood coagulation initiation by a complement-mediated pathway. J. Exp. Med. 134:1601–1607, 1971.
4. From, A. H. L., Gewurz, H., Gruninger, R. P., Pickering, R. J., Spink, W. W. Complement in endotoxin shock: effect of complement depletion on the early hypotensive phase. Infec. Immun. 2:38–41, 1970.
5. Fong, J. S. C., Good, R. A. Prevention of the localized and generalized Shwartzman reactions by an anticomplementary agent, cobra venom factor. J. Exp. Med. 134:642–655, 1971.
6. Snyderman, R., Phillips, J. K., Mergenhagen, S. E. Biological activity of complement *in vivo*: role of C5 in the accumulation of polymorphonuclear leukocytes in inflammatory exudates. J. Exp. Med. 134:1131–1143, 1971.
7. Gewurz, H., Shin, H. S., Mergenhagen, S. E. Interactions of the complement system with endotoxic lipopolysaccharide: consumption of each of the six terminal complement components. J. Exp. Med. 128:1049–1057, 1968.
8. Götze, O., Müller-Eberhard, H. J. The C3-activator system: an alternate pathway of complement activation. J. Exp. Med. 134(Suppl.):90s–108s, 1971.
9. Müller-Eberhard, H. J., Götze, O. C3 proactivator convertase and its mode of action. J. Exp. Med. 135:1003–1008, 1972.
10. Sandberg, A. L., Osler, A. G. Dual pathways of complement interaction with guinea pig immunoglobulins. J. Immunol. 107:1268–1273, 1971.
11. Frank, M. M., May, J., Gaither, T., Ellman, L. *In vitro* studies of complement function in sera of

82

C4-deficient guinea pigs. J. Exp. Med. 134:176–187, 1971.

12. Müller-Eberhard, H. J., Fjellström, K. E. Isolation of the anticomplementary protein from cobra venom and its mode of action on C3. J. Immunol. 107:1666–1672, 1971.

13. Phillips, J. K., Snyderman, R., Mergenhagen, S. E. Activation of complement by endotoxin: a role for γ2-globulin, C1, C4 and C2 in the consumption of terminal complement components by endotoxin-coated erythrocytes. J. Immunol. 109:334–341, 1972.

14. Oliveira, B., Osler, A. G., Siraganian, R. P., Sandberg, A. L. The biologic activities of guinea pig antibodies I. Separation of gamma 1 and gamma 2 immunoglobulins and their participation in allergic reactions of the immediate type. J. Immunol. 104:320–328, 1970.

15. Clark, R. A., Frank, M. M., Kimball, H. R. Generation of chemotactic factors in normal and C4 deficient guinea pig serum by activation with immune complexes and endotoxin [abstract]. Fed. Proc. 31:788, 1972.

16. Sandberg, A. L., Snyderman, R., Frank, M. M., Osler, A. G. Production of chemotactic activity by guinea pig immunoglobulins following activation of the C3 complement shunt pathway. J. Immunol. 108:1227–1231, 1972.

17. May, J. E., Green, I., Frank, M. M. The alternate complement pathway in cell damage: antibody mediated cytolysis of erythrocytes and nucleated cells [abstract]. Fed. Proc. 31:788, 1972.

Selective Effects of Bacterial Endotoxins on Various Subpopulations of Lymphoreticular Cells

Duane L. Peavy, Joseph W. Shands,
William H. Adler, and Richard T. Smith

*From the Tumor Biology Unit, Departments of
Pathology and Immunology and Medical Microbiology,
University of Florida College of Medicine,
Gainesville, Florida*

Evidence has been presented that the in-vitro mitogenic activity of bacterial endotoxins on lymphoreticular cells of mice is a property of the lipid A fraction and that this mitogenic activity affects approximately one-half of the spleen cells of murine strains studied. This mitogenic effect was shown to be thymus-independent and a property of a low-density cell subpopulation. These transformed cells have unique characteristics (including a highly developed endoplasmic reticulum) that distinguish them from cells transformed by thymus-dependent mitogens such as phytohemagglutinin. Evidence was reviewed that the initial mitogen-cell interaction involves both membrane binding and internalization and that mitogenic activity depends more on the nature and rate of the internalization process that results from binding than on the actual degree of binding. It is proposed that bacterial lipopolysaccharide (LPS) has a bifunctional nature, acting as both antigen and mitogen, and that this bifunctional nature makes LPS thymus-independent as an antigen. No evidence presently exists showing that the antigen- and mitogen-responsive subpopulations are identical.

Despite extensive investigation of the biological effects of endotoxin over the past 30 years, no adequate explanation for its toxicity is known. However, evidence that has accumulated suggests that there is an immunologic basis for the biological activities of endotoxin [1, 2]. Stetson postulated that endotoxins do not possess toxic moieties but rather are antigenic macromolecules that will react with antibody or immunologically committed lymphocytes in the host [3].

It was reported previously that an endotoxin derived from *Salmonella typhimurium* was mitogenic for spleen cells of mice; mitogenicity was measured by blast transformation and ^{3}H-thymidine uptake [4]. With use of the in-vitro technique of Adler et al. [5, 6], which allows examination of murine lymphoreticular cell-population responses to various general mitogens, antigens, and alloantigens, the direct effects of bacterial endotoxins on lymphoid cells have been further examined. Evidence has emerged that (*1*) the mitogenic response induced in vitro by bacterial lipopolysaccharides (LPS) is a property of lipid A and not of antigenic polysaccharide, (*2*) the mitogenic effect is selective for a thymus-independent cell subpopulation, (*3*) the target population responding to LPS has morphologic features that distinguish it from cell populations that respond to thymus-dependent mitogenic substances, and (*4*) contact between LPS and susceptible cells involves membrane binding and internalization similar to the process associated with the mitogenic activity of thymus-dependent mitogens such as phytohemagglutinin (PHA).

This paper was supported in part by Frances K. Braden Memorial Grant no. ACS-ET-6B for Cancer Research from the American Cancer Society, by grant no. F-7-OUF from the Florida Division of the American Cancer Society, by grant no. HD-00384 from the National Institute of Child Health and Human Development, by institutional grant no. IN-62-G from the American Cancer Society, and by research grant no. AI-17257 and training grant no. AI-00401 in cellular immunobiology from the National Institute of Allergy and Infectious Diseases.

Dr. Peavy is National Institute of Allergy and Infectious Diseases Trainee no. 5TI-AI-0128.

Please address requests for reprints to Dr. Richard T. Smith, Department of Pathology, University of Florida College of Medicine, Gainesville, Florida 32601.

84

Methods and Results

General characteristics of the mitogenic behavior of LPS on murine lymphoreticular-cell populations. The culture techniques involved in these studies have been described elsewhere in detail [5, 6]. Briefly, 2×10^6 cells from various lymphoid organs in 0.5 ml of complete tissue-culture medium (RPMI-1640 containing 5% heat-inactivated human serum and antibiotics) were incubated for two to three days at 37 C in a 5% CO_2 environment and assayed during the final 24 hr of culture for the incorporation of tritiated thymidine into acid-precipitable material.

It was initially shown that when an endotoxin derived from *S. typhimurium* was added to murine spleen cells, blast transformation (as measured by morphology and incorporation of tritiated thymidine [4]) occurred in approximately 50% of the cells cultured. Endotoxins derived from the genera *Salmonella, Escherichia, Shigella, Serratia, Brucella,* and *Neisseria* are mitogenic [7]. It appears that mitogenic activity is a general property of LPS, since each of 30 preparations of endotoxin possessing in-vivo toxicity was active in this assay. The in-vitro response to all mitogenic substances studied thus far is characterized by an almost linear relationship between concentration of mitogen and incorporation of thymidine up to an optimal value. Beyond this point, increasing concentration inhibits incorporation of thymidine by the responding cell population. This relationship was also found for stimulation by LPS (figure 1). However, the range of effective dosage appeared broader for LPS preparations than for PHA and pokeweed mitogen (PWM).

Preparations of endotoxin derived from rough mutants lacking polysaccharide side chains were also highly active. In addition, Rc and Re mutant endotoxins lacking core polysaccharides were mitogenic. However, methods of detoxification that removed ester- and amide-bound fatty acids markedly decreased mitogenicity.[1] This finding suggested strongly that in-vitro mitogenicity was probably related to the lipid components rather than to 0- and R-polysaccharides. This hypothesis was

[1] D. L. Peavy, J. W. Shands, W. H. Adler, and R. T. Smith, "Mitogenicity of Bacterial Endotoxins: Effects of Immunization, Detoxification, and Characterization of the Mitogenic Principle," manuscript in preparation.

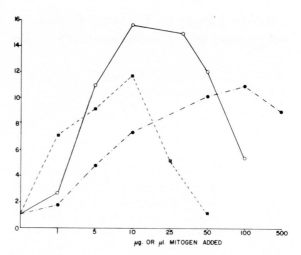

Figure 1. Dose-response relationship of spleen cells of A/J mice reacting to phytohemagglutinin (■– – –■), endotoxin (●– - –●), and pokeweed mitogen (○———○). Fifteen $\times 10^6$ cells per tube in 3 ml of culture medium were incubated with the indicated substances for 72 hr. Incorporation of ^{3}H-thymidine was measured during the last 24 hr of culture. The ratio of incorporation was calculated from triplicate stimulated and control cultures.

confirmed when it was shown that lipid A prepared by Westphal's method [8] was fully stimulatory in this system. On the other hand, the lipid-free antigenic polysaccharide was inactive. The dose-response relationship for lipid A was the same as that for other LPS preparations (figure 2).

Preparations of cells from various lymphoid tissues of mice reacted differently to LPS preparations (table 1). The most responsive population was found in the spleen; less responsiveness was noted in lymph-node cells. Bone-marrow cells from C57BL and AKR mice, on the other hand, were quite responsive to LPS. Variations in responsiveness of cells from mice of various strains and ages were significant, as exemplified by CBA and A strains. A-strain spleen cells were highly responsive to PHA and minimally responsive to LPS, while CBA spleen cells, on a cell-for-cell basis, were more responsive to LPS and less responsive to PHA. Variability by age in in-vitro responses to endotoxins was also striking (figure 3). The capacity of spleen cells to respond increases rapidly early in life, peaks at seven to eight weeks, remains elevated for up to 15–20 weeks, and rapidly falls off thereafter to minimal responsiveness at 30–50 weeks.

Response of various subpopulations of lympho-

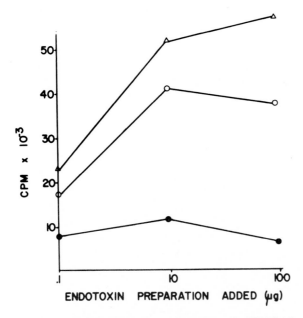

Figure 2. Dose-response relationship of C57BL/6 spleen cells reacting to degraded polysaccharide (●——●), lipid A (△——△), and untreated endotoxin derived from *Salmonella typhimurium* (○——○). Two × 10⁶ cells in 0.5 ml of culture medium were incubated with the indicated amounts of each preparation for 48 hr. Incorporation of ³H-thymidine was measured during the last 24 hr of culture. Results are expressed as the mean counts per min (CPM) of triplicate tubes.

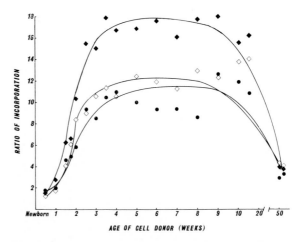

Figure 3. Effects of age on mitogenic stimulation. Spleen cells of A/J mice of various ages were reacted to phytohemagglutinin (◇——◇), endotoxin (●——●), and staphylococcal enterotoxin B (◆——◆). Data are given as the mean ratio of incorporation of six to eight experiments from animals of identical age.

reticular cells to LPS preparations. Andersson and Blomgren [9] and Möller and Michael [10] found that the antigenic properties of LPS are thymus-independent. Because LPS, in contrast to

other mitogens, does not significantly stimulate thymus cells in most strains, it was of interest to see if the mitogenic effect of LPS was thymus-independent. Direct tests of populations depleted of thymus lymphocytes (T cells) were made on neonatally thymectomized and adult thymecto-mized, irradiated, and bone-marrow–reconstituted animals. The data, summarized in figure 4, illustrate the thymus-independence of stimulation by LPS in this model system, in contrast to the response to PHA. Figure 4C shows that, in contrast to the findings of Janossy and Greaves [11] and Stockman et al. [12], PWM is partially thymus-dependent in this system. The reasons for this discrepancy may be found in the greater sensitivity of the assay used in this system, or they may reside in differences in interpretation of the data.

Thus, the data on LPS responses in vitro suggest that there are two thymus-independent modes of action for this class of substances: (*1*) that of a general mitogen with the property of stimulating a high proportion of the heterogeneous spleen-cell populations, as indicated by blast transformation at 24 hr, and (*2*) that of a specific antigen with the property of eliciting nearly pure gamma-M antibody responses in vitro [13]. Möller and Michael [10] have shown that the antigenic component of an LPS preparation from *Escherichia coli* involves a reactive subpopulation approximat-

Table 1. Comparison of in-vitro responses to various mitogens.

Mitogen*	Mouse strain	Spleen	Lymph nodes	Thymus
PHA	C57BL/6	++	++	+
	CBA	+++	ND†	+
	A/J	+++–++++	ND	++
PWM	C57BL/6	+++	++	++
LPS	C57BL/6	++++	++	0–+
	CBA	+++	ND	0
	A/J	++–+++	ND	0

NOTE. Relative values are estimates based upon comparisons between animals four to eight weeks old from many experiments and taken at optimal dose-response levels. Results are expressed as + to ++++, according to level of response.

* PHA = phytohemagglutinin; PWM = pokeweed mitogen; LPS = bacterial endotoxin.

† Not determined.

86

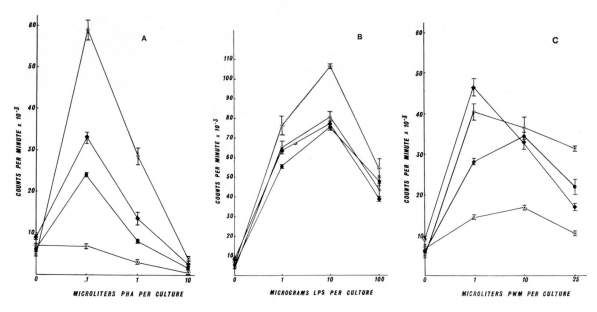

Figure 4. Effects of thymectomy and reconstitution on mitogenic stimulation. Four-week-old C57BL/6 mice were thymectomized, lethally irradiated, and reconstituted with syngeneic bone-marrow and/or thymus cells. Spleen cells were harvested four weeks later and cultured in vitro in the presence of (**A**) phytohemag-glutinin (PHA), (**B**) endotoxin (LPS), and (**C**) pokeweed mitogen (PWM). Two × 10^6 cells in 0.5 ml of culture medium were cultured for 48 hr. Incorporation of ^{3}H-thymidine was measured during the last 24 hr. Results are expressed as the counts per min (CPM) of three to five replicate tubes ± standard error. ⊙ = normal mice; ● = lethally irradiated mice reconstituted with bone-marrow cells; ⊘ = thymectomized mice lethally irradiated and reconstituted with bone-marrow cells; ◆ = thymectomized mice lethally irradiated and reconstituted with bone-marrow and thymus cells.

ing one in 10^5 spleen cells. As will be described, the mitogenic activity involves approximately 50% of spleen cells.

Characteristics of subpopulations responding to the mitogenic activity of LPS. It was evident early in these studies that the cells responding to LPS were lymphoblasts with the appearance of transformed cells, similar to those described in the morphologic studies of Douglas et al. [14]. For most strains, however, the size of the originally responding spleen-cell populations was somewhat greater with LPS than with PHA. Attempts have been made to characterize further these subpopulations as to average density, size of the originally responding cell populations, and ultrastructural characteristics, comparing in every case the cells of the thymus-dependent, PHA-responding populations with those of the thymus-independent, LPS-responding ones.

The average density of cells responding to the mitogenic effects of LPS was low in bovine serum-albumin density-gradient centrifugation [7]. By the discontinuous technique used in this laboratory

[15], thymus-dependent functions have been located in both low- and high-density bands, but antibody-forming cells and certain other functions are concentrated in the low-density fractions. It has been concluded that T cells can be either low- or high-density, depending on their physiologic state, but that bone-marrow lymphocytes (B cells) are generally of lower density as they occur in the spleen. Mitogenic activity of LPS was greatest in the populations of lower density found in the A and B fractions, whereas PHA activity in the same studies was concentrated in the higher-density fractions [7]. The dissociation, however, was not clean in this semiquantitative separation technique, and considerable overlap was observed.

The size of the originally responding cell population was estimated by a count of the number of blast cells after various intervals in culture and determination of the proportion of cells incorporating tritiated thymidine with higher-than-background grain counts in autoradiography. The estimated size of the LPS-responding cell population was approximately 50% of the spleen cells in

Table 2. Autoradiography of the mitogenic response of spleen cells to bacterial lipopolysaccharides (LPS).

Cell type	Response at 24 hr	
	Control	LPS
Blasts	1/1*	45/49
Small lymphocytes	0/86	0/36
Large lymphocytes	0/13	2/13
Total percentage labeled	1	47

* Percentage labeled/percentage total.

the individual cultures (table 2). In contrast, the PHA-responding population was somewhat smaller (between 20% and 40%) in the same strain [4].

Electron microscopy of the cell populations comparing PHA-, LPS-, and PWM-responding subpopulations was done by adequate sampling techniques; the results are shown in figures 5 and 6. The data clearly showed that the LPS-responding cell possesses a highly developed, dilated, rough endoplasmic reticulum. Cells responding to PHA and those responding to PWM have less rough endoplasmic reticula than those responding to LPS [16]. Both possessed diffuse polyribosomes in their cytoplasm, however. Because this type of study is subject to selective viewing by electron microscopy, attempts have been made to quantitate the differences observed by comparison of randomly cut cells with sufficient sample size to permit quantitative comparison; these are shown in table 3 and clearly indicate that the characteristic of a highly developed endoplasmic reticulum is that of the cell population responding to LPS.

These morphologic studies add further evidence to support the conclusion that the LPS-responding cell population is morphologically as well as biologically distinct from those cells responding to thymus-dependent mitogens. One may conclude that the population is not thymus-derived and is independent of thymus influence. It is tempting to conclude further that these are B cells, but critical tests for recognized B-cell properties have yet to be made; such properties include the presence of

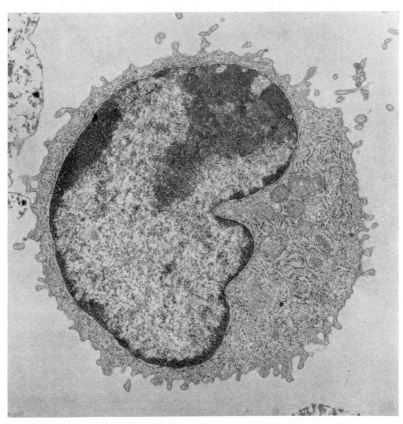

Figure 5. Blast from 48-hr, endotoxin-stimulated culture. Note development of cytoplasm with numerous mitochondria and channels of rough endoplasmic reticulum (× 8,000).

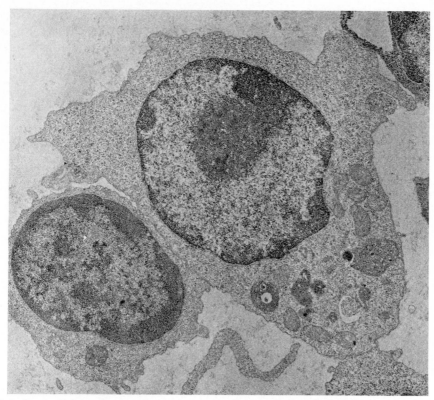

Figure 6. Blast from 48-hr culture stimulated by phytohemagglutinin. The cell possesses a prominent nucleolus and many ribosomal aggregates but no rough endoplasmic reticulum ($\times$ 8,000).

μ chains in large quantity in the surface membrane and complement receptors and the capability for secreting antibody. Tests of LPS-stimulated cells by these criteria are currently under way in this laboratory.

Mechanisms involved in interaction between LPS and lymphoreticular cells. A variety of evidence exists that the mitogenic response to most

Table 3. Response of murine spleen cells to phytohemagglutinin (PHA), pokeweed mitogen (PWM), and endotoxin (LPS).

	Percentage of "blastoid cells" with	
Mitogen	Dilated, rough ER*	Poorly developed or no rough ER*
Controls	ND†	ND
PHA	8	92
PWM	15	85
LPS	41	59

 * Determined by electron microscopy; ER = endoplasmic reticulum.

 † Not determined because too few "blastoid cells" were found.

materials is first apparent shortly after the cells and mitogen come into contact with one another; changes in the membrane, both biochemical and morphologic, are soon obvious. Examination of some of these early events of mitogen-cell interaction enabled us to trace the fate of the mitogen and to examine factors that alter this fate [17].

The model chosen for most of these studies depended on the use of fluorescent antibody to PHA or to LPS for detection of initial binding of mitogens to the membrane; a variation of the techniques of Osunkoya et al. [18] was used. The same reagents were used for examination of the internalization process on fixed cells after various manipulations. Thus, by fluorescence microscopy it could be determined what proportion of cells bear membrane-bound PHA or LPS and, at various intervals, at what rate cells become positive for cytoplasmic (presumably internalized) PHA or LPS.

The effects of temperature, duration of culture, state of the cell cycle, and treatment with neuraminidase have been examined in this system. The initial binding of PHA was not detectable at 37 C, because a high rate of internalization prevents even transient observation of fluorescent PHA on the surface membrane. These findings are summarized and compared with those concerning LPS in table 4. LPS was bound and internalized more slowly than PHA, presumably by a different population of cells. However, the effect of temperature on the prevention of internalization was the same. Treatment with neuraminidase increased the degree of membrane binding detectable by this method but essentially prevented internalization, inasmuch as no cytoplasmic LPS or PHA could be found in cells so treated. The effect of neur-

Table 4. Localization of phytohemagglutinin (PHA) and *Salmonella* endotoxin (LPS) in murine spleen cells by fluorescent-antibody technique.

Experimental group	Relative degree of fluorescent staining	
	Membrane-associated mitogen	Cytoplasmic mitogen
Untreated cells incubated at 37 C with PHA	0	++++
Untreated cells incubated at 4 C with PHA	++++	+
Cells cultured 24 hr alone, then PHA added*	++	+++
Cells cultured 48 hr alone, then PHA added*	++++	+
Cells cultured 48 hr with PHA†	++++	++++
Neuraminidase-treated cells incubated at 37 C with PHA‡	++++	+
Untreated cells incubated at 37 C with LPS	++	++
Neuraminidase-treated cells incubated at 37 C with LPS‡	++++	+

NOTE. In these experiments murine spleen cells were suspended in medium with PHA for 20 min or with LPS for 1 hr, then washed, and either examined directly in the living state after appropriate incubations with antibody to mitogen and fluorescent antibody for determination of membrane-bound mitogen or spread on slides, fixed, and subjected to the same fluorescent technique in order to determine cytoplasmic localization. Results are expressed as + to ++++, according to the degree of staining and number of stained cells observed. Data are taken from [15].

* Cells were cultured in regular medium with PHA for the designated period, then washed and incubated with PHA (20 min, 37 C) and examined as above.

† Cells were cultured 48 hr with PHA, then washed and examined as above with no additional PHA.

‡ Neuraminidase-treated cells were incubated with the enzyme (75 μ) for 40 min at 37 C, then washed and suspended in medium with the mitogen.

aminidase was temporary, and cells could recover from this deficiency after about 12 hr of continued incubation.

The question of whether the binding of mitogen or the internalization process itself was more significant in bringing about a mitogenic effect was raised by these data; accordingly, we designed studies to examine the rate of transport of mitogenic activity from the surrounding media into the cells. Such studies confirmed that it is the internalization process that is critical to mitogenesis.

Data such as these have led to the concept that mitogenic effects on lymphoid cells have two components. The first constitutes binding of the mitogenic material by a specific receptor on the cell membrane. After binding, internalization by some version of micropinocytosis or endocytosis is initiated, and the rate and sustained character of this internalization process determine the need of the cell to develop cytoplasmic machinery required to replace membrane used in the internalization process. While the lymphocyte is in a G_o state with little membrane turnover, minimal synthetic machinery is required to maintain membrane turnover. Such a cell can sustain a high rate of internalization only by developing the cytoplasmic machinery necessary to support this process, and, ultimately, it must enter mitosis. This concept, which relates stimulated transformation and mitosis to the need to preserve membrane integrity, also appears relevant to the morphologic antecedents of mitogenic stimulation of lymphocytes by antigens and antibodies to immunoglobulins, as recently described by Taylor et al. [19].

Discussion

These data lead to the conclusion that the target for LPS activity is a large subpopulation of lymphoreticular cells that does not require thymic influence. While doubtless of bone-marrow origin, this subpopulation has not yet been rigorously characterized as B cells. The mitogenic effect of LPS on this large subpopulation, comprising nearly 50% of spleen cells, is elicited by the lipid-A component. Although intrinsically antigenic [8], the effect of lipid A is separable from the O-antigen–induced immune response described by Möller and Michael [10]. LPS elicits formation of specific antibody in a much smaller thymus-independent subpopulation, estimated to be one in 10^5 cells.

It seems possible that cells responding to the antigenic activity of LPS could be a part of the larger thymus-independent population stimulated by mitogenic activity, but there are no data to support or deny this possibility. Specific stimulation of an exclusively B-cell population by LPS as an antigen would result in a population with different kinetics in vitro from those of the cell population driven by the more inclusive mitogenic effect. Thus, the clonal expansion of such rare specific-recognition cells would probably not be easily detected in the three-day culture period used here. It could not, therefore, contribute substantially to the observed in-vitro mitogenic effect.

These considerations raise the possibility that LPS might be a thymus-dependent antigen by virtue of a bifunctional nature; i.e., it functions both as an antigen and as a mitogen for B cells. The mitogenic effect on B cells may substitute for all of the local influences thought to be effected by T cells activated by specific antigens [20–22]. This hypothesis predicts that there would be thymus cells with recognition receptors for LPS polysaccharide but that their incidence is low, as is that of specific receptors for other antigens, and that they would not be detected in a short-term mitogenic assay. It should be possible to educate thymus cells either in vitro or in vivo with the oligosaccharide preparation and, thereby, to demonstrate that specific receptors do, indeed, exist on thymus cells. Such experiments are currently under way.

References

1. Farr, R. S., Campbell, D. H., Clark, S. L., Jr., Proffitt, J. E. The febrile response of sensitized rabbits to the intravenous injection of antigen [abstract]. Anat. Rec. 118:385–386, 1954.
2. Stetson, C. A., Jr. Similarities in the mechanisms determining the Arthus and Shwartzman phenomena. J. Exp. Med. 94:347–358, 1951.
3. Stetson, C. A. Role of hypersensitivity in reactions to endotoxins. In M. Landy and W. Braun [ed.]. Bacterial endotoxins. Rutgers University, New Brunswick, N.J., 1964, p. 658–662.
4. Peavy, D. L., Adler, W. H., Smith, R. T. The mitogenic effects of endotoxin and staphylococcal enterotoxin B on mouse spleen cells and human peripheral lymphocytes. J. Immunol. 105:1453–1458, 1970.
5. Adler, W. H., Takiguchi, T., Marsh, B., Smith, R. T. Cellular recognition by mouse lymphocytes in vitro. I. Definition of a new technique and results of stimulation by phytohemagglutinin and specific antigens. J. Exp. Med. 131:1049–1078, 1970.
6. Adler, W. H., Takiguchi, T., Marsh, B., Smith, R. T. Cellular recognition by mouse lymphocytes in vitro. II. Specific stimulation by histocompatibility antigens in mixed cell culture. J. Immunol. 105:984–1000, 1970.
7. Peavy, D. L., Adler, W. H., Shands, J. W., Smith, R. T. Mitogenic effects of endotoxins on mouse lymphoid cells: thymus independence of LPS. Cell. Immunol. 1973 (in press).
8. Galanos, C., Lüderitz, O., Westphal, O. Preparation and properties of antisera against the lipid-A component of bacterial lipopolysaccharides. Eur. J. Biochem. 24:116–122, 1971.
9. Andersson, B., Blomgren, H. Evidence for thymus-independent humoral antibody production in mice against polyvinylpyrrolidone and E. coli lipopolysaccharide. Cell. Immunol. 2:411–424, 1971.
10. Möller, G., Michael, G. Frequency of antigen-sensitive cells to thymus-independent antigens. Cell. Immunol. 2:309–316, 1971.
11. Janossy, G., Greaves, M. F. Lymphocyte activation. I. Response of T and B lymphocytes to phytomitogens. Clin. Exp. Immunol. 9:483–498, 1971.
12. Stockman, G. D., Gallagher, M. T., Heim, L. T., South, M. A., Trentin, J. J. Differential stimulation of mouse lymphoid cells by phytohemagglutinin and pokeweed mitogen. Proc. Soc. Exp. Biol. Med. 136:980–982, 1971.
13. Britton, S., Möller, G. Regulation of antibody synthesis against Escherichia coli endotoxin. I. Suppressive effect of endogenously produced and passively transferred antibodies. J. Immunol. 100:1326–1334, 1968.
14. Douglas, S. D., Hoffman, P. F., Borjeson, J., Chessin, L. N. Studies on human peripheral blood lymphocytes in vitro. III. Fine structural features of lymphocyte transformation by pokeweed mitogen. J. Immunol. 98:17–30, 1967.
15. Adler, W. H., Peavy, D., Smith, R. T. The effect of PHA, PPD, allogeneic cells, and sheep erythrocytes on albumin gradient-fractionated mouse spleen cell populations. Cell. Immunol. 1:78–91, 1970.
16. Shands, J. W., Peavy, D. L., Smith, R. T. Differentiated morphology of mouse spleen cells stimulated in vitro by endotoxin, phytohemagglutinin, pokeweed mitogen, and staphylococcal enterotoxin B. Am. J. Pathol. 1973 (in press).
17. Adler, W. H., Osunkoya, B. O., Takiguchi, T., Smith, R. T. The interactions of mitogens with lymphoid cells and the effect of neuraminidase on the cells' responsiveness to stimulation. Cell. Immunol. 3:590–605, 1972.
18. Osunkoya, B. O., Williams, A. I. O., Adler, W. H., Smith, R. T. Studies on the interaction of phyto-

mitogens with lymphoid cells. I. Binding of phyto-hemagglutinin to cell membrane receptors of cultured Burkitt's lymphoma and infectious mononucleosis cells. Afr. J. Med. Sci. 1:3–16, 1970.

19. Taylor, R. B., Duffus, W. P. H., Raff, M. C., de Petris, S. Redistribution and pinocytosis of lymphocyte surface immunoglobulin molecules induced by anti-immunoglobulin antibody. Nature [New Biol.] 233:225–229, 1971.

20. Dutton, R. W., Falkoff, R., Hirst, J. A., Hoffmann, M., Kappler, J. W., Kettman, J. R., Lesley, J. F., Vann, D. Is there evidence for a non-antigen specific diffusable chemical mediator from the thymus-derived cell in the initiation of the immune response? *In* B. Amos [ed.]. Progress in immunology. Academic Press, New York, 1971, p. 355–368.

21. Britton, S. When allogeneic mouse spleen cells are mixed *in vitro* the T-cells secrete a product which guides the maturation of B-cells. Scand. J. Immunol. 1:89–98, 1972.

22. Feldmann, M., Basten, A. Specific collaboration between T and B lymphocytes across a cell impenetrable membrane. Nature [New Biol.] 237:13–15, 1972.

Etiology of the Wasting Diseases

John W. Jutila

From the Department of Botany and Microbiology, Montana State University, Bozeman, Montana

The various neonatally induced and adult forms of murine wasting disease appear to include common pathologic events related to the activities or products of the intestinal flora. Therefore, the role of endotoxins and gram-negative flora in the development of wasting was assessed in neonatal, germfree, conventionally reared mice and in congenitally athymic mice. The results indicate that, although they exhibited an impaired response to ovine erythrocytes, germfree mice treated with endotoxin failed to develop a fatal wasting disease readily produced in conventionally reared mice. Adult athymic mice developed a wasting condition that appeared to be related to an impaired secretory immune response and a coliform-overgrowth syndrome. This led to a profound increase in susceptibility to endotoxins. These data suggest that in wasting disease, microorganisms of the gut condition the immunosuppressed animal to the lethal effects of endotoxin or act synergistically with endotoxins to yield fatal infections.

Wasting syndromes can be produced by treatment of newborn mice with allogeneic lymphoid cells (runt disease) [1], cortisol acetate [2], estradiol [3], multiple doses of bacterins [4], or certain infectious agents [5]. A wasting syndrome can be produced in adults by neonatal thymectomy [6], by injection of parental lymphoid cells into F_1 hybrid mice [7], or by injection of allogeneic bone marrow into lethally irradiated mice (secondary disease) [8]. Pathologic findings commonly observed in these wasting syndromes include failure to gain weight, ruffled fur, diarrhea, atrophy of the lymphoid organs, diminution of immunologic competence, and, in many cases, death.

In 1965 we proposed that these syndromes may have an important pathogenic event in common, i.e., an infectious or toxemic process engendered by an impaired immune response of the host [9]. This concept was later supported by the observation that most symptoms of the postneonatal-thymectomy syndrome [10] and the steroid [11, 12] and bacterin-induced wasting diseases [13] and secondary disease [14] fail to develop in germfree mice. The exception cited is runt disease, which develops normally in germfree mice [10]. Significantly, the course of the disease in nearly

These studies were supported by grant no. RO1 AI06552-07 from the U.S. Public Health Service.

I thank Norman D. Reed and John Barnett for their cooperation in these studies.

Please address requests for reprints to Dr. John W. Jutila, Department of Botany and Microbiology, Montana State University, Bozeman, Montana 59715.

all forms of wasting (including runt disease) is mitigated by treatment with antibiotics [4, 11, 15].

Although infections have been implicated in many types of wasting disease, they are rarely overwhelming, are often difficult to detect, and may be absent [12, 15, 16]. Yet wasting may occur in all these situations, suggesting the complicity of another factor(s) present in conventionally reared but not in germfree mice. That a component of the gut is involved is strongly suggested by the observation that monocontamination of germfree mice with *Escherichia coli* renders them susceptible to wasting by neonatal administration of cortisol acetate [9]. Thus, we designed these studies to explore the role of endotoxin, a product of the gram-negative intestinal flora in the development of symptoms of wasting in neonatal and adult mice. The rationale for these studies predicted that if endotoxin was a factor in wasting of immunodepressed mice, then injection of endotoxin into immunologically null neonates or immunodepressed adult mice would produce or accentuate symptoms of wasting. To distinguish between the true pathologic effects of endotoxin and the potentiating or synergistic effects of endogenous flora and their endotoxins, we used germfree mice in the neonatal studies as recipients of comparable doses of endotoxin.

Materials and Methods

Mice. Germfree (GF) and conventionally reared (CR) Balb/c mice and specific-pathogen-free (SPF), congenitally athymic mice were used. The GF mice were obtained from the National Institutes of Health in 1966 and maintained by gnotobiotic methodology developed in our laboratory in 1964. Homozygous athymic mice (*Nu* +/*Nu* +), hereafter referred to as nudes, and littermate control mice (+ +/+ + or *Nu* +/+ +) were the offspring of heterozygous mice obtained by crossing Rex (+ *Re*/*Nu* +) males [17] with females from our SPF Balb/c colony. All SPF mice were given sterilized Purina 5010C and acidified-chlorinated water.

Endotoxin. Endotoxin was prepared by the aqueous-ether and dioxane methods of Ribi [18] or by the phenol-water method of Westphal [19]. Ten-hour broth cultures of *Salmonella enteritidis* and a strain of *Escherichia coli* isolated from the normal flora of the athymic mouse

served as sources of endotoxin. An endotoxin prepared by the phenol-water method from *E. coli* strain 0113 was obtained from Dr. Jon Rudbach (University of Montana, Missoula, Mont.). The endotoxins for neonatal studies were autoclaved for 10 min and frozen at -25 C until used.

Immunologic procedures. Immunoglobulin was quantitated in serum and fecal samples, according to the procedure of Masseyeff and Zisswiller [20]. The assay for plaque-forming cells (PFC) described by Mishell and Dutton [21] was used for estimation of the direct PFC response of the spleen after immunization with 1×10^8 or 4×10^8 ovine erythrocytes (ORBC).

Bacteriologic procedures. Fresh fecal samples or the contents of the large bowel were weighed, diluted in 0.01 M phosphate-buffered saline (pH 7.2), and cultured anaerobically for 48 hr on Tergitol-7 (Difco, Detroit, Mich.), MacConkey agar, and blood agar.

Results

Studies in neonatal mice. To determine the ability of endotoxin to produce symptoms of wasting disease in neonatal mice, we injected 35–75 μg of *S. enteritidis* endotoxin ip into GF or CR mice within 24 hr of birth; a dose of 150 μg/g of body weight was given every third day thereafter until day 18. *E. coli* 0113 endotoxin, which is somewhat more toxic to neonates than is *S. enteritidis* endotoxin, was administered initially in a dose of 2–10 μg on day of birth and then in gradually increasing doses every other day until day 18. Total doses of endotoxin administered during the 18-day period varied from 1.7 to 3.6 mg.

The effects of treatment of neonatal mice with endotoxin prepared by the dioxane method are presented in figure 1. The data characterize the course of the wasting syndrome in CR mice as an increasing disparity in weight (runting index or RI) between treated and control mice within a litter during 30 days [22]. Each curve represents the mean of one litter, and the variation in severity of wasting among litters and even within litters treated with the same dose of endotoxin is demonstrated. Thus, litter A exhibited an RI of 4.8, whereas litter D had an RI of 9.2 on day 28. Significantly, the RI of GF mice treated with comparable doses of endotoxin was 0, indicating that

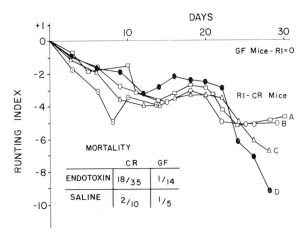

Figure 1. Runting index (RI) and mortality of conventionally reared (CR) and germfree (GF) Balb/c mice treated with 150 μg of *Salmonella enteritidis* endotoxin per g of body weight every third day until day 18 after birth. The RI is the difference between the mean weight of tested and normal animals within the same litter. Curves A-D represent litters A-D. Mortality data are presented with the numerator designating the number of mice dying of treatment with endotoxin and the denominator designating the total number of mice injected.

GF mice were resistant to wasting by endotoxin. In addition, mortality among nine litters of CR and four litters of GF mice treated with endotoxin was 51% (18 of 35) and 7% (1 of 14), respectively. The mortality rate among both CR and GF control mice injected with saline was 20% (2 of 10 and 1 of 5, respectively).

Hematologic findings in CR mice were represented by a pronounced leukocytosis (largely a neutrophilia) as early as day 10. No differences in hematocrit were found. Splenomegaly, slight to moderate thymic atrophy, and, at times, gray necrotic plaques in the liver characterized the gross pathology of wasting in CR mice. A mild atrophy of spleen and liver and a slight hypertrophy of the thymus were observed in GF mice.

The PFC response of CR mice treated with a total dose of 2.1 mg of *E. coli* 0113 endotoxin was markedly impaired when assayed as late as 30 days after birth. Immunosuppression appeared to be related to the severity of wasting. Thus, the spleen PFC response of mice with an RI of 7.4 was 3.6% of the PFC response in intralitter control mice, whereas the response of mice with an RI of 3.90 was 45% of control values. Although they showed no symptoms of wasting, GF mice

given a total dose of 2.3 mg of *E. coli* 0113 endotoxin demonstrated a 49% depression of spleen PFC when compared at 30 days of age with intralitter control mice.

Studies in adult mice. More recent studies on the role of endotoxin in adult forms of wasting have been done in congenitally athymic mice, which developed a wasting syndrome similar to that described in neonatally thymectomized mice. Indeed, an infectious etiology for wasting in the congenitally athymic mouse has been ascertained by our studies, which have also established a relationship between a coliform-overgrowth syndrome and an impaired secretory immune response in nudes.[1] Further studies have sought to relate the massive proliferation of gut coliforms to a dramatic increase in susceptibility to endotoxin.

In these studies, the weight gain of nudes, maintained under SPF conditions with their normal-appearing littermates, was subnormal throughout the experimental period of six months. A disparity, for example, of 3.8, 5.2, and 10.8 g between nudes and littermate controls was observed on days 20, 50, and 130, respectively. Maximal growth was achieved at about two months of age, and the animals usually remained healthy for an additional two months before a wasting process, distinguished by a gradual loss of weight, began to erode their health. Loss of weight was often precipitous in the last two weeks of life. The mean survival time of nudes reared under SPF conditions was 132.86 days.

To assess the effect of the athymic condition on levels of intestinal flora and the secretory (IgA) immune response, we simultaneously assayed nudes and littermate controls for their fecal coliform and IgA content from one week to six months of age. In brief, these studies (figure 2) showed that nudes and littermates developed an early coliform flora that peaked at two weeks of age at 350,000 organisms/mg of feces. Thereafter, the organisms rapidly declined in number until many fewer could be detected at three weeks of age. In adult control mice, coliforms increased to and fluctuated within the range of 500–3,800 organisms/mg of feces. On the other hand, counts of fecal coliform in nudes increased rapidly dur-

[1] J. W. Jutila and N. D. Reed, "The Etiology of Wasting Disease in Congenitally Athymic Mice," manuscript in preparation.

94

ing the same interval to numbers exceeding 300,000/mg. This overgrowth phenomenon persisted for the remaining months of the animals' lives.

IgA, presumably derived from the dam, rose and fell normally in nude mice during the first three weeks of life (figure 2). In young adult littermate controls, IgA again rose to levels within the range of 1.4–3.3 µg/mg of feces, presumably through de-novo synthesis in the gut. Nudes, on the other hand, slowly synthesized fecal IgA, reaching a peak of 1.3 µg/mg of feces (40.6% of littermate values) on days 65–67. Thereafter, IgA levels declined to approximately 25% of those of littermate controls. Levels of IgA in sera of one-month-old nudes were 11% of control values but increased to 61% of control values in two-month-old nudes. These data suggest that a diminished ability to synthesize IgA may contribute, at least in part, to an increased number of coliforms in the gut; i.e., IgA may have a regulating effect on the normal flora of the gut.

Since a marked proliferation of *E. coli* was detected in nudes, we designed studies to establish whether overgrowth with coliforms contributed to an increased sensitivity to their endotoxins. To this end, the LD50 of an endotoxin extracted from an endogenous strain of *E. coli* was estimated in 28-, 42-, 60-, and 78-day-old nudes and their littermate controls. Older nudes showing symptoms of wasting were included in a fifth group. The LD50, expressed as µg of endotoxin,

is shown for each interval (arrows) in figure 2. The data show that the nude mouse developed an increased sensitivity to endotoxin as it aged. Nudes experiencing symptoms of wasting were found to be profoundly sensitive to endotoxin; doses of 20 µg and (more recently) 10 µg produced rapidly fatal endotoxin shock.

Discussion

These data indicate that endotoxin produces in neonatal mice symptoms of and mortality from wasting disease similar to those phenomena produced by neonatal administration of cortisol acetate or of estradiol or by multiple doses of bacterin. These observations largely confirm the findings of Keast [15, 22]. The observations that mortality rates from wasting were low and that few or no symptoms of wasting developed in germfree mice treated with endotoxin suggest that microorganisms or their products act synergistically with or capitalize on the pathologic effects of endotoxin in CR neonates. Thus, the severe wasting of CR mice may be explained by a predisposition to an infectious process; this predisposition would result from the immunosuppressive effects of endotoxin. Significantly, the development of a "natural" coliform overgrowth (figure 2) in the guts of young mice corresponds to the course of the wasting disease induced with endotoxin in CR mice. That endotoxin exerts a primary immunodepressive effect is shown by the failure of GF mice treated with large doses of endotoxin to respond normally to immunization with ORBC. These findings are supported by the observations of Schaedler and Dubos [23], who showed that treatment of mice with 1.0 µg or less of endotoxin helped establish a rapidly fatal septicemia with *Staphylococcus aureus*.

Studies on adult wasting in congenitally athymic mice (the "experiment-of-nature" analogue of the postneonatal-thymectomy syndrome) also implicate the participation of gut flora and their endotoxins in the development of wasting in adult mice. These data show that athymic mice develop an overgrowth syndrome with coliform, presumably due to a diminished or impaired secretory IgA response. In turn, the increased exposure of the animal to coliforms and their endotoxins probably conditions or sensitizes the animal to later challenges with endotoxin [23]. Thus, in the

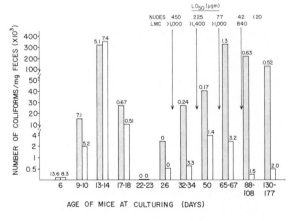

Figure 2. Number of coliform organisms (per mg of feces) and associated fecal IgA (µg/mg of feces) estimated in nude (▨) and littermate control (☐) mice of various ages. The LD50 of endotoxin were estimated at days 28, 42, 60, and 78 (arrows).

hyperreactive animal, bursts of minute amounts of endotoxin from the gut into the vascular system initiate pathologic events, evidenced morphologically as symptoms of wasting and explicable etiologically as protracted endotoxin toxicity.

The precise mode of action of endotoxin that yields a fatal outcome to wasting disease is unknown. However, it is questionable whether hypersensitivity phenomena contribute to toxicity of endotoxin, since athymic mice fail to develop any form of cell-associated immune response or immediate type of hypersensitivity to protein antigens (author's unpublished observations).

It is proposed that adult wasting, which follows or accompanies graft-vs.-host reactions in secondary disease or F_1 hybrid wasting, and the post-neonatal-thymectomy syndrome may have a similar etiology.

References

1. Billingham, R. E., Brent, L. A simple method for inducing tolerance to skin homografts in mice. Transplant. Bull. 4:67–71, 1957.

2. Schlesinger, M., Mark, R. Wasting disease induced in young mice by administration of cortisol acetate. Science 143:965–966, 1964.

3. Reilly, R. W., Thompson, J. S., Bielski, R. K., Severson, C. D. Estradiol-induced wasting syndrome in neonatal mice. J. Immunol. 98:321–330, 1967.

4. Ekstedt, R. D., Hayes, L. L. Runt disease induced by non-living bacterial antigens. J. Immunol. 98:110–118, 1967.

5. Brooke, M. S. Experimental runt disease in mice caused by *Salmonella typhimurium* var. *Copenhagen*. J. Exp. Med. 120:375–387, 1964.

6. Miller, J. F. Effect of neonatal thymectomy on the immunological responsiveness of the mouse. Proc. Roy. Soc. [Biol.] 156:415–428, 1962.

7. Kaplan, H. S., Rosston, B. H. Studies on a wasting disease induced in F_1 hybrid mice injected with parental strain lymphoid cells. Stanford Med. Bull. 17:77–92, 1959.

8. Russell, P. S., Monaco, A. P. The biology of tissue transplantation. Little, Brown, Boston, 1965. 207 p.

9. Reed, N. D., Jutila, J. W. Wasting disease induced with cortisol acetate: studies in germ-free mice. Science 150:356–357, 1965.

10. McIntire, K. R., Sell, S., Miller, J. F. Pathogenesis of the post-neonatal thymectomy wasting syndrome. Nature (Lond.) 204:151–155, 1964.

11. Reed, N. D., Jutila, J. W. Wasting disease induced with cortisol acetate. I. Studies in germ-free and conventionally reared mice. J. Immunol. 99:238–245, 1967.

12. Hatch, G. G., Reed, N. D. Estradiol-induced wasting syndrome in conventionally reared and germfree mice. J. Bacteriol. 99:902–903, 1969.

13. Ekstedt, R. D., Nishimura, E. T. Runt disease induced in neonatal mice by sterile bacterial vaccines. J. Exp. Med. 120:793–804, 1964.

14. Connell, M. S., Wilson, R. The treatment of X-irradiated germfree CFW and C3H mice with isologous and homologous bone marrow. Life Sci. 4:721–729, 1965.

15. Keast, D., Walters, M. N. The pathology of murine runting and its modification by neomycin sulphate gavages. Immunology 15:247–262, 1968.

16. Jutila, J. W., Reed, N. D. Wasting disease induced with cortisol acetate. II. Bacteriologic studies. J. Immunol. 100:675–681, 1968.

17. Rygaard, J. Immunobiology of the mouse mutant "nude." Acta Pathol. Microbiol. Scand. 77:761–762, 1969.

18. Ribi, E., Milner, K. C., Perrine, T. D. Endotoxic and antigenic fractions from the cell wall of *Salmonella enteritidis*. Methods for separation and some biologic activities. J. Immunol. 82:75–84, 1959.

19. Westphal, O., Lüderitz, O., Bister, F. Uber die Extraktion von Bacterien mit Phenol/Wasser. Z. Naturforsch. [B] 7:148–155, 1952.

20. Masseyeff, R. F., Zisswiller, M.-C. A versatile method of radial immunodiffusion assay employing microquantities of antiserum. Anal. Biochem. 30:180–189, 1969.

21. Mishell, R. I., Dutton, R. W. Immunization of dissociated spleen cell cultures from normal mice. J. Exp. Med. 126:423–442, 1967.

22. Keast, D. A simple index for the measurement of the runting syndrome and its use in the study of the influence of the gut flora in its production. Immunology 15:237–245, 1968.

23. Schaedler, R. W., Dubos, R. J. The susceptibility of mice to bacterial endotoxins. J. Exp. Med. 113:559–570, 1961.

Role of Bacterial Endotoxin in the Graft-vs.-Host Syndrome

D. Keast

From the Department of Microbiology, School of Medicine, University of Western Australia, Perth, Western Australia

A comparison of the histopathology and the general macroscopic features of the graft-vs.-host reaction (GVHR) with those produced by bacterial endotoxin suggests an important role of bacterial endotoxicity in the GVHR and related experimental models for immune-deficiency states. The protective effect of antibiotic therapy and the lack of development of a lethal syndrome in germfree animals undergoing the GVHR or after thymectomy further supports the importance of endotoxemia in these syndromes. It is proposed that bacterial endotoxicity may influence immune senescence and lead to the increasing incidence of disease and cancer that occurs with age; this phenomenon may also explain some of the complications presently associated with immune suppression and lymphoid transplantation in man.

The graft-vs.-host reaction (GVHR), or homologous disease, was first developed as a laboratory model for study of immunologic mechanisms associated with the grafting of skin and organs and the acquisition of tolerance to histo-incompatible antigens [1–5].

It was soon established that lymphoid cells played a decisive role [1, 6] in the GVHR, and Gowans [7] showed that the small lymphocyte was involved as the key member of the lymphoid cells. Nisbet and Heslop [8, 9] reported in detail on the GVHR, including the histopathology of the disease. Even at this early stage in the development of the GVHR as an experimental model, variability in pathology was becoming apparent. During subsequent years extension of the system to many species has led to an accumulation of information on the variability of the histopathology [10]. During these years other systems have been developed that produce runted animals during their course, and there has been a great tendency to compare the pathology seen in these animals with that of the GVHR.

In 1968 I reviewed the literature in terms of runting syndromes, autoimmunity, and neoplasia [11] and found that research workers often compared the pathology of their models with that of the GVHR. If the two were similar, this similarity was considered to be sufficient evidence for ascribing to immune mechanisms contributory roles in the production of the systems being studied. While in some cases these grounds were sufficient, it was apparent that early work on the beneficial effects of antibiotic treatment of the GVHR and other immunoablative models [12, 13] was being largely ignored [11]. It had been clearly shown that by treatment of animals with antibiotics, the GVHR syndrome could be substantially reduced in intensity, and larger numbers of animals would survive. However, several groups of workers ascribed the deaths of their animals to infectious processes, and while only one group produced evidence for bacterial invasion of the animals, there are many instances in which photomicrographs show no evidence of microbial involvement [11]. The finding by Howard [14], indicating that animals undergoing the GVHR are more resistant to bacterial invasion due to a transient increase in reticuloendothelial activity, may account for lack of bacterial invasion of the animals. During the GVHR, however, animals are more sensitive to the lethal effects of bacterial endotoxin [14]. While inconclusive evidence was available for the increased survival of germfree mice undergoing the GVHR (N. F. Stanley, unpublished observations, 1964), germfree mice survived thymectomy without the wasting induced in the conventional state [15]. Since we had no access to germfree animals, we

This work was made possible by a grant from the National Health and Medical Research Council of Australia and the West Australian Cancer Council.

Please address requests for reprints to Dr. D. Keast, Department of Microbiology, School of Medicine, University of Western Australia, Perth, Western Australia.

decided to reappraise in one strain of mice several of the neonatal systems producing runted animals, and we also repeated earlier experiments, using antibiotic therapy as one experimental system. We added to these groups neonatal animals treated with small amounts of crude preparations of bacterial endotoxin from the coliforms of their own intestinal flora [16–19].

We first developed a simple "runting index" (which did not involve killing of animals) and showed that it was applicable to results already published by other workers [16] and to our experimental systems.

We extended our studies to show the influence of neomycin-sulfate gavages during the first 30 days of the experiment. Figure 1 illustrates (1) the reduced severity of macroscopic features of the syndrome and (2) the fact that the effects of endotoxin on growth of the animals could be titrated to low levels. Furthermore, it is clear that the number of animals capable of surviving was greatly increased in the cases of animals with the GVHR and animals treated with cortisone. While there was no effect of antibiotic therapy on survival of neonatally thymectomized animals, the animals died earlier if they were given bacterial endotoxin and antibiotic therapy [16].

In conjunction with these experiments, we examined histopathology in detail throughout the course of the disease [17]. Much of the histopathology was similar, irrespective of the initiating regimes, and was closely mimicked by the treatments with endotoxin. In none of our animals was there evidence that active infections were responsible for the pathology seen. However, when I repeated the treatments with cortisone on weanling mice, although the initial pathology was similar to that seen in the neonatal series, these animals invariably developed fatal bacteremia in which the only organisms isolated from blood cultures were *Escherichia coli* (author's unpublished observations). One of the most interesting observations of the neonatal series was the development of intestinal mucosal erosion in runted animals (figure 2). Again, there was no evidence of microbial invasion at these lesions.

In the GVHR and treatments with cortisone, all histopathology peripheral to that of the lymphoid organs was reduced markedly if antibiotic therapy was used [17]. Finally, we examined by electron microscopy the livers of mice runted by the GVHR and bacterial endotoxin [20]. Once again, there were strong similarities between the two treatments. There was no evidence of microbial infection; platelet aggregation, leukocyte accumulation, and vesicle morphology suggested that effects of endotoxin predominated. In both cases stores of glycogen were markedly reduced [20]. It had already been established that glycogen stores are depleted in hepatocytes by endotoxin [21], probably because endotoxin induces an insulinlike glycolysis [22, 23].

We also found that chronic forms of the GVHR, cortisone-induced runting, infection with reovirus 3, runting due to bacterial vaccines and runting due to endotoxin led to a continuing pathology and, in some cases (especially after infection with reovirus 3 and treatments with endotoxin), our strain of mice finally presented with significant numbers of tumors: thymic lymphomas, spindle-cell sarcomas, and cystic developments of the frontal lobes of the skull [18, 19, 24–27].

In many of the endotoxin-treated animals and in some of the reovirus 3-infected animals, almost complete thymic ablation was seen [18, 23]. While we have no formal evidence as yet, it is conceivable that the presently proposed immune-surveillance system for tumor immunity [26, 28] was damaged; this would have allowed the murine oncogenic viruses to express themselves. Another possibility is that these viruses favor an anaerobic type of metabolism, again induced by endotoxin, to initiate their transforming cycle.

In recent attempts to determine what type of lymphocyte was stimulated by bacterial endotoxin, we have shown that, as was reported by Peavy, Adler, and Smith [29], nanogram amounts of endotoxin will stimulate murine lymphocytes in culture, and that this ability is reduced after thymectomy (figure 3). The system is proving difficult to reproduce, apparently because of antibodies (present in some of the human sera used for maintenance of murine lymphocytes in culture) that are directed against our endotoxin preparations. However, if endotoxin-stimulated lymphocytes subsequently die or cannot be committed to other antigens, the immune-surveillance system will be weakened. Should this prove to be correct in the long run, we would have an indication of how the natural microenvironment may be able to induce immune senescence, which in turn would lead to the increased incidence of disease and establish-

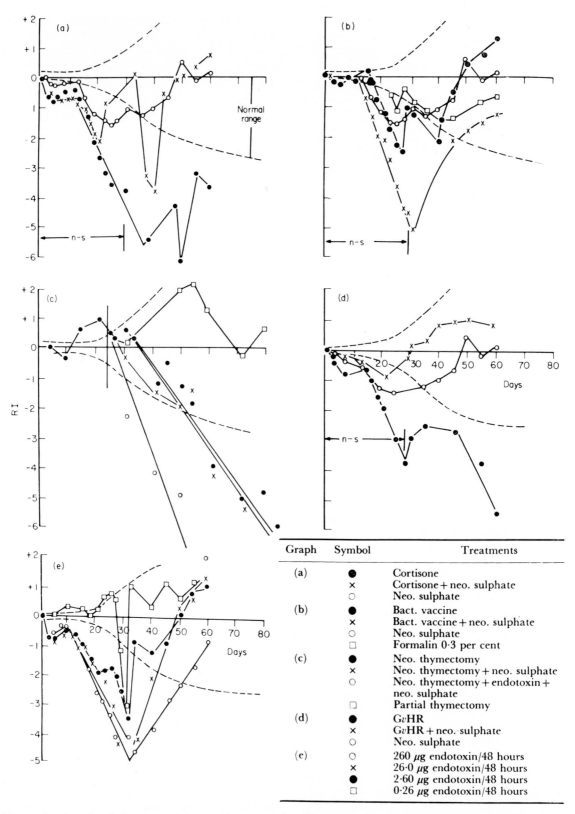

Figure 1. Runting index for runting syndrome produced by several methods (see key, lower right) in mice of the PH strain. Modifications in index after neomycin-sulphate gavaging of animals are also presented. (a) Cortisone runts; (b) bacterial-vaccine runts; (c) thymectomy runts; (d) graft-vs.-host syndrome (GVHR) runts; (e) endotoxin runts. (Reprinted with permission of Blackwell Scientific Publications Ltd.)

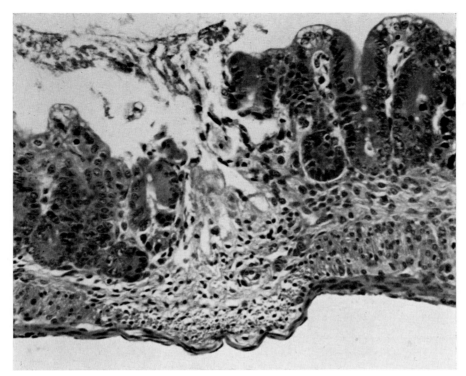

Figure 2. Development of intestinal mucosal erosion in runted animals. (Reprinted with permission of Blackwell Scientific Publications Ltd.)

ment of tumors that occur with age [26, 28]. Landy et al. [30] have shown that small amounts of bacterial endotoxin given to adult rabbits causes almost complete emptying of thymocytes from their thymus glands and that these glands take at least three weeks to recover their normal architecture. Therefore, unremitting low-grade endotoxemia may well influence the long-term function of the thymus.

Recently, it has been shown that the GVHR, like neonatal thymectomy [15], is not lethal to germfree animals [31], and that while typical lesions of the GVHR develop in the lymphoid organs, much of the peripheral pathology fails to develop. However, when the animals are conventionalized, even after 300 days of presumably reactive GVHR in the germfree state, they die quickly. Therefore, the real point at issue in these experiments may be the relative importance of microbial infection or bacterial endotoxins in producing death.

Schaedler and Dubos [32] have shown that pathogenfree mice that had no *E. coli* and few gram-negative bacteria of any sort in a predominantly lactobacillary intestinal flora were almost wholly nonsusceptible to endotoxic shock. However, this susceptibility could be restored if the animals were given heat-killed gram-negative bacteria or suckled onto mice of conventional stock. This suggests that humoral factors of some kind are important and may be IgM antibody [33]. It is known that bacterial endotoxins are rapidly lethal in several species of animals; however, it is also established that bacterial endotoxins function immunologically in many ways, depending on concentration [33]. They appear to be able to function in tolerance, as adjuvants, as antigens, and as toxins; therefore, the absolute role of bacterial endotoxins in the GVHR and associated runting syndromes may prove to be difficult to establish.

With the increasing occurrence of immunosuppression in man as an adjunct to organ transplantation and the recognition of immune-deficiency diseases and their rectification by such regimes as bone-marrow transplantation, it is becoming increasingly obvious that the microbial component of the host often plays a major role in survival of the patient. It seems, therefore, that the neglected role of bacterial endotoxin in the mediation of pathology of the GVHR in experimental systems

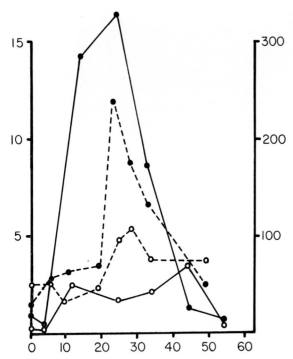

Figure 3. Incorporation of radioactivity (^{3}H-6T-thymidine) into DNA of murine lymphocytes stimulated by crude bacterial endotoxin that was prepared from the gut coliforms of the mice after culture for 48 hr. Lymphocytes were obtained from either peripheral blood (●) or spleen (○) of normal intact mice (solid line) or thymectomized mice (broken line). Abscissa: ng of endotoxin per culture well (0.5 × 10^4 mononuclear cells). Left ordinate: dpm × 10^{-3} (intact mice) in DNA. Right ordinate: dpm (thymectomized mice) in DNA. The micromethod of lymphocyte culture and processing has been reported [34]. Results represent the mean of triplicate determinations at each concentration.

may be a clue to some of the unexpected deaths and complications in patients under immunosuppression and lymphoid transplantation. Man is known to be extremely sensitive to the effects of bacterial endotoxemia [33].

References

1. Billingham, R. E., Brent, L., Medawar, P. B. Quantitative studies on tissue transplantation immunity (III) Actively acquired tolerance. Phil. Trans. Roy. Soc. London [B] 239:357–414, 1956.
2. Billingham, R. E., Brent, L. A simple method for inducing tolerance of skin homografts in mice. Transp. Bull. 4:67–71, 1957.
3. Siskind, G. W., Thomas, L. Studies on the runting syndrome in newborn mice. Bull. Soc. Int. Chir. 18:208–218, 1959.
4. Billingham, R. E. Reactions of grafts against their hosts. Science 130:947–953, 1959.
5. Simonsen, M. Graft-versus-host reactions. Their natural history, and applicability as tools of research. Progr. Allerg. 6:349–367, 1962.
6. Billingham, R. E., Brown, J. B., Defendi, V., Silvers, W. K., Steinmuller, D. Quantitative studies on the induction of tolerance of homologous tissues and on runt disease in the rat. Ann. N.Y. Acad. Sci. 87:457–471, 1960.
7. Gowans, J. L. The role of lymphocytes in the destruction of homografts. Br. Med. Bull. 21:106–110, 1965.
8. Nisbet, N. W., Heslop, B. F. Runt disease. I. Br. Med. J. 1:129–135, 1962.
9. Nisbet, N. W., Heslop, B. F. Runt disease. II. Br. Med. J. 1:203–213, 1962.
10. McBride, R. A. Graft-versus-host reaction in lymphoid proliferation. Cancer Res. 26:1135–1151, 1966.
11. Keast, D. Runting syndromes, autoimmunity and neoplasia. Adv. Cancer Res. 11:43–71, 1968.
12. van Bekkum, D. W., van Putten, L. M., de Vries, M. J. Anti-host reactivity and tolerance of the graft in relation to secondary disease in radiation chimeras. Ann. N.Y. Acad. Sci. 99:550–563, 1962.
13. Duhig, J. T. Beneficial effect of oxytetracycline in cortisone-induced wasting disease. Nature (Lond.) 207:651–652, 1965.
14. Howard, J. G. Increased sensitivity to bacterial endotoxin of F$_1$ hybrid mice undergoing graft-versus-host reaction. Nature (Lond.) 190:1122, 1961.
15. Wilson, R., Sjodin, K., Bealmar, M. The absence of wasting in thymectomized germfree (axenic) mice. Proc. Soc. Exp. Biol. Med. 117:237–239, 1964.
16. Keast, D. A simple index for the measurement of the runting syndrome and its use in the study of the influence of the gut flora in its production. Immunology 15:237–245, 1968.
17. Keast, D., Walters, M. N-I. The pathology of murine runting and its modification by neomycin sulphate gavages. Immunology 15:247–262, 1968.
18. Keast, D. The murine runting syndrome and neoplasia. Immunology 16:693–697, 1969.
19. Keast, D., Stanley, N. F. The induction of murine neoplasia. Pathology 1:19, 1969.
20. Papadimitriou, J. M., Keast, D. The electron microscopy of livers of mice runted as a result of the graft versus host reaction and bacterial endotoxin. Br. J. Exp. Pathol. 50:574–577, 1969.
21. Boler, R. K., Bibighaus, A. J., III. Ultrastructural alterations of dog livers during endotoxin shock. Lab. Invest. 17:537–561, 1967.
22. Berry, L. J., Smythe, D. S., Young, L. G. Effects of bacterial endotoxin on metabolism. (I) Carbohydrate depletion and the protective role of cortisone. J. Exp. Med. 110:389–405, 1959.
23. Woods, M. W., Landy, M., Whitby, J. L., Burk, D. Symposium on bacterial endotoxins. (III) Meta-

bolic effects of endotoxins on mammalian cells. Bacteriol. Rev. 25:447–456, 1961.

24. Stanley, N. F., Keast, D. Murine infection with reovirus 3 as a model for the virus induction of autoimmune disease and neoplasia. *In* M. Pollard [ed.] Perspectives in virology. Vol. V. Academic Press, New York, 1967, p. 281–289.

25. Keast, D., Stanley, N. F., Phillips, P. A. The association of murine lymphoma with reovirus type 3 infection: the development of neoplasia in an animal suffering from chronic reovirus 3 disease. Proc. Soc. Exp. Biol. Med. 128:1033–1038, 1968.

26. Keast, D. *In* R. T. Smith and M. Landy [ed.] Immune surveillance. Academic Press, New York, 1970, p. 324.

27. Phillips, P. A., Keast, D., Walters, M. N-I., Stanley, N. F. Murine lymphoma induced by reovirus 3. Pathology 3:133–138, 1971.

28. Keast, D. Immunosurveillance and cancer. Lancet 2:710–712, 1970.

29. Peavy, D. L., Adler, W. H., Smith, R. T. The mitogenic effects of endotoxin and staphylococcal enterotoxin B on mouse spleen cells and human peripheral lymphocytes. J. Immunol. 105:1453–1458, 1970.

30. Landy, M., Sanderson, R. P., Bernstein, M. T., Lerner, E. M., II. Involvement of thymus in immune response of rabbits to somatic polysaccharides of gram-negative bacteria. Science 147:1591–1592, 1965.

31. Jones, J. M., Wilson, R., Bealmar, P. M. Mortality and gross pathology of secondary disease in germfree mouse radiation chimeras. Radiat. Res. 45:577–588, 1971.

32. Schaedler, R. W., Dubos, R. Relationship of intestinal flora to resistance. *In* M. Landy and W. Braun [ed.] Bacterial endotoxins. Rutgers University Press, New Brunswick, N.J., 1964, p. 390–396.

33. Lee, L., Stetson, C. A., Jr. The local and generalized Shwartzman phenomena. *In* B. W. Zweifach, L. Grant, and R. T. McCluskey [ed.] The inflammation process. Academic Press, New York, 1965, p. 791–817.

34. Keast, D., Bartholomaeus, W. N. A micromethod for stimulation of lymphocytes by phytohaemagglutinin. Aust. J. Biol. Med. Sci. 50:603–609, 1972.

Summary of Discussion

Jon A. Rudbach

Dr. Barnet M. Sultzer opened the discussion by expanding on aspects of his work referred to by Dr. Göran Möller. He noted that there is a variant strain of C_3H mice whose spleen cells are low responders to the mitogenic activity of lipopolysaccharides (LPS). Preliminary work has shown that these mice have a deficiency in B-cell lymphocytes that is specific for endotoxin. An exchange between Dr. Werner Braun and Dr. Möller established that most of the endotoxins used by Dr. Möller were alkali-treated for reduction of toxicity.

In reply to a question by Dr. S. A. Broitman, Dr. John W. Jutila stated that he had not assessed the small-bowel motility in his nude and littermate control mice. Dr. Edward H. Kass suggested that bacteria other than coliforms could be influencing Dr. Jutila's observations. Dr. Jutila replied that enterococci were elevated in the overgrowth syndrome, but that he had not examined the non-coliforms very closely.

The role of cell-mediated mechanisms, as opposed to humoral systems, in accounting for the more acute effects of endotoxin was questioned by Dr. Peter Ward. Dr. Richard T. Smith replied that no T-cell system has yet been detected to respond to endotoxin. However, we do not know enough at this point to evaluate the complex effects of endotoxin in the whole animal. Dr. Stephan E. Mergenhagen and Dr. Ward concurred that any discussion of humoral or cellular reactions had to be undertaken within the context of a specific activity of endotoxin. Dr. Louis Chedid stated that, in his experience, mice were very sensitive to an injection of endotoxin within the first 24 hr after birth. Dr. Jutila replied that he had made similar observations and had excluded these early deaths from his data. Dr. Gerard Renoux commented on a reduction in weight of newborn and adult mice after an injection of brucella endotoxin. Inasmuch as no brucella are found in the fecal flora of mice, this might be a pure endotoxic effect without an allergic component. Dr. Sheldon M. Wolff commented that hypogammaglobulinemic humans

(and animals treated with 6-MP) gave normal biologic responses to endotoxin. Furthermore, an immune fever induced with protein antigen-antibody complexes differed markedly from an endotoxin-induced fever. The logical assumption was that classical humoral antibody was not playing a primary role in these endotoxic reactions.

Dr. Herman Friedman, in reply to a question from Dr. Loretta Leive, stated that he could not induce his RNA complex by adding antigen in vitro. He felt that the complex represented an in vivo processing step. Dr. Möller pointed out that both he and Dr. Smith found that LPS induced mitogenic effects in about 50% of murine spleen cells. However, specific precursor antibody-forming cells to LPS occurred at a frequency of only 1 in 100,000. When questioned by Dr. Georg Springer, Dr. Smith replied that neuraminidase enhanced the binding of LPS by cells, but that it blocked the internalization of LPS. The neuraminidase-treated cells returned to normal in 24 hr. Dr. Abraham Braude mentioned that intradermal injection of endotoxin induced a histologically classic Arthus reaction in humans; this supports some of Dr. Mergenhagen's ideas on the activation of complement by endotoxin. In reply to a question by Dr. Bernhard Urbaschek, Dr. Mergenhagen stated that consumption of complement was associated with toxicity and not with antigenicity of endotoxin. Dr. Smith commented that he and others have been unable to show any transformation or mitogenic stimulation of human peripheral-blood lymphocytes, despite the demonstration of considerable numbers of B cells in the peripheral blood of humans. Dr. Möller suggested that the lack of activation of peripheral human B cells may be only a quantitative matter, and that the experiment should be performed with spleen cells from humans. He felt, also, that sufficient evidence was at hand to say that B cells are being transformed. Dr. Wolff mentioned that Dr. J. J. Oppenheim induced 6%–8% transformation of human peripheral leukocytes with purified endotoxin.

In reply to a question from Dr. P. Helena Mäkelä, Dr. Mergenhagen suggested that his "globulin" probably is an immunoglobulin and an antibody. Dr. Mergenhagen went on to comment that there was a material in dental plaque of patients with peridontal disease that would transform their leukocytes; subsequent work showed that this transforming material was not LPS. Dr. Masaya Kawakami asked Dr. Mäkelä if she measured the amount of LPS in her strains of *Salmonella.* She replied that the bacterial strains contained the same amounts of LPS, within the limits of the assay.

The general problem of the role of antibody in mediating the several biological activities of endotoxin was discussed. Dr. Dennis W. Watson was called upon to state his position.

DR. WATSON: During the endotoxin conference at Rutgers in 1963, Dr. Yoon Berm Kim and I presented a scheme in which we postulated the presence of a specific antigenic determinant within lipid A. Antibodies directed against this determinant in collaboration with a normal reticuloendothelial system (RES) could account for immunity against the lethal and pyrogenic activities of endotoxins. Later we were able to transfer this immunity to normal animals with fractions containing 19S immunoglobulins. At the time, I remember that a distinguished endotoxicologist commented that it was "boloney," however we sliced it! As I listened to Dr. Christos Galanos today, I wished that this same gentleman could have been present, because Dr. Galanos and his colleagues have, by a different approach, established the existence of a specific antigenic determinant in lipid A, and in addition, have proved that primary toxicity is associated with this molecule, a prediction we had made based on more indirect evidence.

It was at the Rutgers meeting that Dr. Chandler A. Stetson suggested that there may be no such thing as an endotoxin, because all of the reactions attributed to endotoxins could be simulated with antigen-antibody reactions. The implication was made that the primary toxicity could be attributed to an antigenic determinant within the so-called endotoxin reacting with "normal" antibodies present in the sera of most adult animals. Dr. Kim and I hurried home and immediately injected endotoxin into our "immunologically virgin" piglets; these we knew had no antibodies in their sera. The piglets died, and we concluded that there must be a primary toxic unit in endotoxin, because we were certain there was no antigen-antibody reac-

tion in these animals. This observation, however, does not preclude the possibility, as Dr. Mergenhagen pointed out today, that complement and antigen-antibody reactions play a role. We have always stressed the secondary role of hypersensitivity reactions. Thus, there is a primary toxicity associated with lipid A, and a secondary interdependent role can be attributed to complement action. Delayed hypersensitivity could also play a secondary role, as we have continually stressed. The point remains, however, that the bovine serum albumin-lipid A complexes described by Dr. Galanos today give us something with which we can work, and I am confident that these should help to resolve the mechanism of primary toxicity to endotoxin and the role of antibodies in resistance. Whether or not the antibody Dr. Galanos described today is the same as we have described is not known. We are working hard at it, and perhaps at the next endotoxin conference, there will be more definitive answers.

Dr. Watson's statement was followed by extensive discussion of the possibility that antibody was present before the initial injection of endotoxic antigen, rather than being made in response to injection. It was pointed out that there were early observations that suggested that it might be possible to reduce the toxicity of endotoxin and yet have the less toxic materials produce tolerance. In fact, these less toxic materials were used in treatment of febrile states such as typhoid fever, and it is often forgotten that bacterial extracts as well as whole bacterial vaccines were given intravenously in the treatment of typhoid fever and related infections. However, phenomena similar to the Shwartzman phenomenon occurred. Some patients developed severe hypotension, others multiple hemorrhagic reactions and other similar responses; thus the approach to treatment of typhoid fever was discontinued. Further discussion raised the suggestion that Dr. Möller might be working with a secondary response in animals that had earlier experienced contact with the antigen. The offer was made that Dr. Möller come to Minneapolis to examine newborn piglets, delivered by hysterectomy in a germfree environment; it was predicted that there would be no antibodies in the piglets, although these would be demonstrable in the sow. However, in rebuttal, it was indicated that the clonal-selection hypothesis, stating that antibodies are formed before any contact with antigen, is now widely accepted. It becomes of the utmost importance to discover reasons for which the piglets do not react before the clonal-selection hypothesis is discarded. When an LPS antigen is slightly altered by insertion or transposition of a sugar, a new antibody is demonstrated; such a demonstration would be expected on the basis of selection of a new antibody population in accordance with the clonal hypothesis. In brief, there seems to be no clear resolution of the problem at present, and it may be that the validity of the clonal-selection hypothesis may hinge on the possibility that piglets may not react for reasons other than that they lack preformed antibody capacity.

SESSION III: BIOLOGICAL RESPONSES

Possible Role of Endotoxemia during Immunologic Imbalance

Louis Chedid

From the Institut Pasteur and Centre National de la Recherche Scientifique, Paris, France

Transplantation of immunocytes (the graft-versus-host reaction, or GVHR) and, in many cases, immunosuppression (X-irradiation, cortisone overdosage, actinomycin D) or immunostimulation (Bacille Calmette Guerin, *Corynebacterium parvum,* etc.) decrease the host's resistance to endotoxins. In the present period of immunologic engineering, this situation may constitute a serious hazard. The data reported here demonstrate that (*1*) it is possible to obtain mycobacterial preparations that have lost their capacity to render mice susceptible to endotoxins, although they are still potent adjuvants or immunostimulants, and (*2*) the GVHR can be inhibited by in-vivo or in-vitro pretreatment with endotoxins.

It has long been taken for granted that resistance or susceptibility to endotoxins was closely related to stimulation or blockade of the reticuloendothelial system (RES). However, since Beeson's original work [1], certain paradoxical situations have been recognized, such as the hyperreactivity of mice treated with Bacille Calmette Guerin; these mice nevertheless clear carbon and phagocytose gram-negative organisms. It is now well established that, in many cases, immunologic imbalance produces susceptibility to endotoxins, and more recently Jutila and Keast have insisted that similarities exist between the wasting syndrome and chronic endotoxicity. These authors have hypothesized that such a situation could be the end result of an identical mechanism that, in the case of runting, is related to a greater permeability of the host's gut to products of his intestinal flora [2–4].

A rapid survey of the literature reveals that the host becomes hyperreactive to endotoxins after various immunosuppressive treatments, such as irradiation, neonatal thymectomy, cortisone overdosage, or actinomycin D. Little information exists concerning antiserum to lymphocytes, which apparently does not influence the host's reactivity to endotoxins [5]. Resistance also decreases after immunostimulation by BCG, *Hemophilus pertussis, Corynebacterium parvum,* polyriboinosinic-polyribocytidylic acid (poly I:C), and, in certain

Please address requests for reprints to Dr. Louis Chedid, Institut Pasteur, Immunothérapie Expérimentale, 28, rue du Dr. Roux, 75015 Paris, France.

physiologic or pathologic situations such as pregnancy, the graft-versus-host reaction (GVHR), or parabiosis.

All of the situations enumerated above have in common their ability to influence the immune response in one way or another. Whether or not all of these effects have a common mechanism, proper attention should be given to the end result. Indeed, we are now in a period of immunologic engineering in which immunosuppressive or immunostimulating treatments are readily undertaken in therapy of cancer, transplantation of organs, or transfer of bone marrow. Since these procedures can sensitize the host to endotoxins, efforts should be made to take advantage of the abundant information on the many biological manifestations of lipopolysaccharide and the various experimental methods proposed to control endotoxemia.

This effort would be especially rewarding, since all of this information has not been very fruitful as a means of improving the treatment of gram-negative infections.

Before presenting some of our data concerning nontoxic, active mycobacterial preparations and the influence of tolerance to endotoxins on the GVHR, let us review the situations mentioned above.

Immunosuppression

X-irradiation. Three days after irradiation with dosages capable of suppressing antibody formation and delayed sensitivity, mice become 100 times more susceptible to endotoxins [6]. This susceptibility, according to the authors, is related to damage to the intestinal tract, since, if the abdomen is shielded, the LD_{50} is not modified.

Neonatal thymectomy. The lethal effects of endotoxin are increased three and one-half- to eightfold four weeks after neonatal thymectomy [7, 8].

Cortisone overdosage. After adrenalectomy, mice are more susceptible to endotoxins and to histamine or anaphylactic shock and can be protected by cortisone. However, when this same hormone is given repeatedly, resistance to endotoxins decreases [9]. Thus, 1 µg of endotoxin is the LD_{50} for mice that have previously received eight injections of cortisone, whereas the LD_{50} for controls is 200 µg [10]. Likewise, a local or gen-

eral Shwartzman reaction can be obtained with a single injection of endotoxin if the rabbit has previously been treated with glucocorticoids [11].

Prolonged treatment with corticoid abolishes formation of antibody in different animal species [12–14], but the effect of corticoid on hypersensitivity is not clearly understood. The hormone, in addition to its inhibition of antibody synthesis, acts at several different levels, such as histamine induction or reduction of the inflammatory reactions of the target cells [15]. Nevertheless, cortisone has been shown to depress delayed-hypersensitivity responses without influencing immediate hypersensitivity. Thus, cortisone suppresses dermal reactivity and systemic shock produced by injections of old tuberculin in rabbits and guinea pigs sensitized by BCG, whereas it does not prevent anaphylactic shock or the appearance of the Arthus reaction in animals sensitized with normal equine serum [16].

Actinomycin D. Simultaneous injection of this antimetabolite produces such an important decrease of resistance that it is one of the most sensitive tests for endotoxin dosage [17–19].

Immunostimulation

H. pertussis and C. parvum. Like mycobacteria, these organisms are potent immunoadjuvants and render the host susceptible to endotoxins [20]. However, unlike mycobacteria, they can also sensitize mice to histamine and to passive anaphylaxis [21, 22].

Poly I:C. Huang and Landay have observed that poly I:C or Newcastle Disease Virus renders mice more susceptible to endotoxins and have assumed that this susceptibility was mediated by the release of interferon [23]. However, we have recently observed that a potent preparation of interferon extracted from brains of mice does not sensitize to endotoxins (I. Gresser, M. Parant, and L. Chedid, unpublished observations). Whatever the mechanism may be, poly I:C, which is also an immunoadjuvant, sensitizes to endotoxins.

Parabiosis and pregnancy. After parabiosis, mice and rats become 10 times more susceptible to endotoxins [24]. They also become 10 times more susceptible to endotoxins during pregnancy [25]. These two situations are reminiscent of the GVHR. The two other procedures that increase susceptibility of mice to endotoxins (treatment with BCG and the GVHR) will be treated in greater detail. We shall see in the first case that the "package deal" obtained by using whole mycobacteria produces toxic effects and that preparations obtained by chemical fractionation can fully induce or even reinforce immunostimulation without eliciting many of the pathologic effects. Finally, data will be reported that show that pretreatment with endotoxin may inhibit the GVHR.

BCG. Hyperreactivity to endotoxins establishes itself rapidly in BCG-treated mice and lasts for several weeks [26–28]. Other strains of mycobacteria produce the same effect, i.e., both an increased susceptibility to endotoxins and (paradoxically) a stimulation of the RES and an increased resistance to infection [29].

The fact that infection with BCG (or other agents of the same type) can simultaneously induce both stimulation of the RES (leading to an increased resistance to gram-negative bacterial infections) and increased susceptibility to endotoxin is rather baffling. These paradoxical effects may, however, be related to different active factors in BCG that can be separated by chemical treatment. According to Suter, hyperreactivity to endotoxins of BCG-treated mice is unrelated to delayed hypersensitivity and is mediated by a component of the cell wall, cord factor [30].

In the field of tumor immunity, mycobacteria have been used either as adjuvants or as stimulants of nonspecific immunity [31–33]. As an operational definition of these two approaches, one can say that an immunoadjuvant increases the immune response to a given antigen with which it is administered simultaneously; in opposition, a nonspecific immunostimulant must not be administered in the same site or at the same time as the tumor cells (it is usually more active if injected several days before) and gives a response that is unrelated to a specific antigen. Whether the nonspecificity resides only in the first step (stimulation) and, thereafter, immunologic responses are turned on, or whether these substances induce nonspecific responses in the second phase can be debated [33]. In any case, our aim is to show that preparations can be obtained that have retained their capacity to stimulate immunity, although they no longer exert the toxic effects observed after administration of whole mycobacterial cells.

In opposition to most adjuvants, including

Freund's incomplete adjuvant, Freund's complete adjuvant not only increases humoral antibodies but can also induce delayed hypersensitivity and autoimmune diseases [34]. Furthermore, it contains whole mycobacterial cells that render mice susceptible to endotoxins. It has been recently shown that the active component of mycobacteria in Freund's complete adjuvant is, contrary to what had been previously observed, a water-soluble oligomer of the cell wall, containing few, if any, fatty acids [35–37]. This cell-wall component, hereafter referred to as WSA (water-soluble adjuvant), which is capable of increasing the level of precipitins and of inducing delayed hypersensitivity towards a given antigen, does not induce allergic polyarthritis. Our experiments also clearly established that WSA did not sensitize mice to histamine nor induce hypertrophy of the spleen and liver, hyperreactivity to endotoxins, or sensitivity to tuberculin [38, 39]. Therefore, this substance seems to be a good candidate for experimentation of adjuvant activity directed against tumors. However, we were unable to demonstrate any nonspecific stimulation by injection of WSA before inoculation of the tumor cells, as opposed to the protection observed if phenol-killed BCG or whole bacterial cells are injected under the same conditions. Such an effect can also be obtained, however, with delipidated whole cells or even with delipidated and enzyme-treated cell walls [40].

These preparations, which had lost their capacity to induce hyperreactivity to endotoxins, were still able to increase the host's resistance to Ehrlich ascitic carcinoma or to a syngeneic lymphoid leukemia as effectively as or more effectively than BCG. Moreover, one preparation that had been obtained by digestion of purified cell walls with pronase (treatment that greatly inhibits tuberculin sensitization) was equally active. Thus, tuberculin sensitivity and hyperreactivity to endotoxins are not prerequisites of antitumoral activity, since a strong response can be demonstrated in their absence. Therefore, immunostimulants of bacterial origin should be checked, and those that do not affect resistance to endotoxins should be systemically preferred.

Tolerance to endotoxin and GVHR. Liacopoulos and Merchant have reported that if adult $C_{57}Bl$ mice were rendered tolerant to endotoxins by repeated injections of this antigen, the transfer of their splenocytes into adult irradiated C_3H recipients did not induce a GVHR. This result was interpreted by the authors as being the consequence of an immunologic commitment [41].

We repeated these experiments, using as donors adult $C_{57}Bl$ mice that had previously received 30 μg of endotoxin daily for seven days and, as recipients, either newborn Swiss or newborn F_1 (Cba T_6T_6/Akr) hybrids [42]. In all cases, 10^7 viable cells were transferred. As can be seen in table 1, when the splenocytes had been recovered from tolerant adults, the mortality rate was reduced, to nearly 0%, whereas 76.5%–82% of the controls had died at day 40 (no deaths were observed later) (table 1). Besides these differences in survival rate, the controls had a typical stunted appearance, whereas the animals that had received splenocytes from tolerant donors had a normal morphology.

In the course of these experiments, it was also observed that Swiss newborns were very susceptible to endotoxins, especially if the toxic antigen was injected a few hours after birth. However, simultaneous administration of splenocytes with endotoxins increased their resistance (table 2). Forty-eight hours after injection of either 1.5 μg or 3 μg of endotoxin alone, all animals were dead; after the same interval, only four of 15 animals

that received 3 μg of endotoxin plus splenocytes were dead.

Inhibition of the GVHR could also be obtained by incubation of the donors' splenocytes with endotoxin (table 3). In these experiments, 2×10^8 cells of $C_{57}Bl$ adult mice were incubated with 30 μg of radioactive endotoxin in a volume of 1 ml for 1 hr at 37 C. The splenocytes injected into the controls were previously incubated under the same conditions but without endotoxin. All samples were then centrifuged and resuspended in 3 ml of medium and then centrifuged a second time; the cells were suspended again in the initial volume of 1 ml. The viability of the incubated cells was measured by the exclusion of trypan blue; in all cases, cell mortality was about 10%. The amount of radioactivity attached to the pellet was measured. Care was taken to make sure that the radioactivity recovered in the supernatant after incubation was not dialyzable and was related to the presence of antigen, as measured by inhibition of passive hemagglutination.

It was found that a maximal amount of 0.07 μg of endotoxin (if any) was injected into the newborn Swiss recipients with 10^7 splenocytes.

Table 1. Graft-versus-host reaction induced by transfer of splenocytes from normal or endotoxin-tolerant donors into newborn mice.

Recipients	Donors	Mortality	
		Day 15	Day 40
F_1 hybrids	Normal	8/17*	13/17 (76.5)†
	Tolerant	0/21	0/21 (0)
Swiss	Normal	4/28	23/28 (82)
	Tolerant	2/24	2/24 (8.5)

* Dead/total.

† Number in parentheses is percentage mortality at day 40.

Table 3. Inhibition of graft-versus-host reaction by prior incubation of splenocytes with endotoxin.

Treatment of donor cells	Mortality	
	Day 15	Day 30
Normal cells*	38/86 (44.2)†	55/86 (64)
Normal cells plus injection of 0.5 μg endotoxin	26/46 (56.5)	44/46 (95.6)
Cells incubated with endotoxin	8/55 (14.5)‡	15/55 (27.3)‡

* In all cases, 10^7 splenocytes from $C_{57}Bl$ mice were injected into newborn Swiss mice.

† Dead/total (percentage).

‡ $P < .01$.

Table 2. Toxicity of endotoxin in newborn mice.

	Swiss				F_1 hybrids	
Age	No.	Weight (g)	LD_{50} (μg)	LD_{50}/kg (mg)	No.	LD_{50} (μg)
< 24 hr	101	1.2	0.74	0.6	90	4.8
1–2 days	40	1.6	3	1.9	...	...
10–12 days	36	7	>25*	>3.6	...	...

* The highest dose used.

As can be seen in table 3, under these conditions preincubation of the splenocytes with endotoxin greatly reduced the GVHR; mortality was 27.3%, while it was 64% among controls injected with cells only. As can also be seen in table 3, simultaneous administration of 0.5 µg of endotoxin and 10^7 splenocytes (but without inoculation) increased the severity of the GVHR in a second group of controls. Therefore, resistance to the GVHR could in no case be attributed to the transfer of antigen.

Although the difference observed is statistically very highly significant, these experiments should be repeated with inbred neonates. Whether the donor cells persist and proliferate and, if so, what has become of their immunologic responsiveness are topics under investigation.

Conclusion

It is known that in many cases of immunostimulation or immunosuppression the host's sensitivity is increased. Whether or not there exists a causal relationship between those facts, this problem deserves serious consideration because of its theoretical and clinical implications. Three practical approaches can be followed: (1) control of endogenous source of endotoxins, i.e., intestinal flora, by appropriate antibiotics; (2) whenever possible, the choice of an "immunomodificator" that does not influence resistance to endotoxins; and (3) pharmacologic and serologic treatments increasing resistance to endotoxins.

References

1. Beeson, P. B. Tolerance to bacterial pyrogens. J. Exp. Med. 86:29–38, 1947.
2. Keast, D. A simple index for the measurement of the runting syndrome and its use in the study of the influence of the gut flora in its production. Immunology 15:237–245, 1968.
3. Jutila, J. W. Wasting disease induced with cortisol acetate. 3. Immunologic studies. J. Immunol. 102:963–969, 1969.
4. Jutila, J. W. The etiology of wasting diseases. J. Infect. Dis. 128(Suppl.):S99–S103, 1973.
5. Abdelnoor, A., Chang, C. M., Nowotny, A. Effect of antilymphocyte serum on endotoxin reactions. Bacteriol. Proc., p. 92, 1972.
6. Smith, W. W., Alderman, I. M., Schneider, C., Cornfield, J. Sensitivity of irradiated mice to bacterial endotoxin. Proc. Soc. Exp. Biol. Med. 113:778–781, 1963.
7. Berry, L. J. Discussion of part IV. In M. Landy and W. Braun [ed.]. Bacterial endotoxins. Rutgers University Press, New Brunswick, N.J., 1964, p. 353.
8. Salvin, S. B., Peterson, R. D. A., Good, R. A. The role of the thymus in resistance to infection and endotoxin toxicity. J. Lab. Clin. Med. 65:1004–1022, 1965.
9. Chedid, L., Boyer, F. Action de la cortisone sur la sensibilité de la souris à une endotoxine bactérienne. C. R. Soc. Biol. (Paris) 146:239–241, 1952.
10. Chedid, L., Parant, M., Boyer, F., Skarnes, R. C. Non-specific host responses in tolerance to the lethal effect of endotoxins. In M. Landy and W. Braun [ed.]. Bacterial endotoxins. Rutgers University Press, New Brunswick, N.J., 1964, p. 500–516.
11. Thomas, L., Good, R. A. The effect of cortisone on the Shwartzman reaction. J. Exp. Med. 95:409–428, 1952.
12. Germuth, F. G., Jr. The role of adrenocortical steroids in infection, immunity and hypersensitivity. Pharmacol. Rev. 8:1–24, 1956.
13. Ward, P. A., Johnson, A. G. Studies on the adjuvant action of bacterial endotoxins on antibody formation. II. Antibody formation in cortisone-treated rabbits. J. Immunol. 82:428–434, 1959.
14. Blumer, H., Richter, M., Cua-Lim, F., Rose, B. Precipitating and nonprecipitating antibodies in the primary and secondary immune responses. J. Immunol. 88:669–678, 1962.
15. Raffel, S. Immunity. 2nd ed. Appleton, New York, 1961. 646 p.
16. Harris, S., Harris, T. N. Effect of cortisone on some reactions of hypersensitivity in laboratory animals. Proc. Soc. Exp. Biol. Med. 74:186–189, 1950.
17. Pieroni, R. E., Broderick, E. J., Bundeally, A., Levine, L. A simple method for the quantitation of submicrogram amounts of bacterial endotoxin. Proc. Soc. Exp. Biol. Med. 133:790–794, 1970.
18. Dowling, J. N., Feldman, H. A. Quantitative biological assay of bacterial endotoxins. Proc. Soc. Exp. Biol. Med. 134:861–864, 1970.
19. Parant, M., Chedid, L. Sensibilisation de la souris aux endotoxines par la surrénalectomie ou par l'administration d'actinomycine D. C. R. Acad. Sci. [D] (Paris) 272:1308–1311, 1971.
20. Howard, J. G. Mechanisms concerned with endotoxin sensitivity during graft-versus-host reaction. In L. Chedid [ed.]. La structure et les effets biologiques des produits bactériens provenant de germes gram-négatifs. Colloque International CNRS no. 174 (Paris), 1969, p. 331–340.
21. Parfentjev, I. A., Goodline, M. A. Histamine shock in mice sensitized with Hemophilus pertussis vaccine. J. Pharmacol. Exp. Ther. 92:411–413, 1948.
22. Kind, L. S. The altered reactivity of mice after immunization with Hemophilus pertussis vaccine. J. Immunol. 70:411–420, 1953.
23. Huang, K. Y., Landay, M. E. Enhancement of the

lethal effects of endotoxins by interferon inducers. J. Bacteriol. 100:1110–1111, 1969.

24. Chedid, L., Boyer, F., Pophillat, F., Parant, M. Etude de la toxicité d'une endotoxine radioactive (^{51}Cr) injectée à des parabiontes normaux et hypophysectomisés. Ann. Inst. Pasteur (Paris) 104:197–207, 1963.

25. Chedid, L., Boyer, F., Saviard, M. Disparition de différents effets de la cortisone chez la femelle gestante du rat. C. R. Acad. Sci. (Paris) 238:156–158, 1954.

26. Halpern, B. N., Biozzi, G., Howard, J., Stiffel, C., Mouton, D. Exaltation du pouvoir toxique d'*Eberthella typhosa* tuée chez la souris inoculée avec le BCG vivant. C. R. Soc. Biol. (Paris) 152:899–902, 1958.

27. Suter, E., Ullman, G. E., Hoffman, R. G. Sensitivity of mice to endotoxin after vaccination with BCG. Proc. Soc. Exp. Biol. Med. 99:167–169, 1958.

28. Suter, E. Hyperreactivity to endotoxin after infection with BCG. J. Immunol. 92:49–54, 1964.

29. Biozzi, G., Stiffel, C., Halpern, B. N., Mouton, D. Recherches sur le mécanisme de l'immunité non spécifique produite par les mycobactéries. Rev. Franç. Etudes Clin. Biol. 5:876–890, 1960.

30. Suter, E., Kirsanow, E. M. Hyperreactivity to endotoxin in mice infected with mycobacteria. Induction and elicitation of the reactions. Immunology 4:354–365, 1961.

31. Zbar, B., Bernstein, I., Tanaka, T., Rapp, H. J. Tumor immunity produced by the intradermal inoculation of living tumor cells and living *Mycobacterium bovis* (strain BCG). Science 170:1217–1218, 1970.

32. Mathé, G. Active immunotherapy. Adv. Cancer Res. 14:1–36, 1971.

33. Yashphe, D. J. Immunological factors in non-specific stimulation of host resistance to syngeneic tumors. A review. *In* D. W. Weiss [ed.]. Immunological parameters of host-tumor relationships. Academic Press, New York, 1971, p. 90–107.

34. Asherson, G. L., Allwood, G. G. Immunological adjuvants. *In* E. E. Bittar and N. Bittar [ed.]. The biological basis of medicine. Vol. 4. Academic Press, London, 1969, p. 327–355.

35. Adam, A., Ciorbaru, R., Petit, J.-F., Lederer, E. Isolation and properties of a macromolecular water-soluble immuno-adjuvant fraction from the cell wall of *Mycobacterium smegmatis*. Proc. Natl. Acad. Sci. U.S.A. 69:851–854, 1972.

36. Hiu, I. J. MAAF, a fully water-soluble lipid-free fraction from BCG with adjuvant and antitumour activity. *In* G. Mathé [ed.]. Investigation and stimulation of immunity in cancer patients. Colloque International CNRS (Paris) 1972 (in press).

37. Jollès, P., Migliore, D. Le pouvoir adjuvant des cires D des mycobactéries: relations structure/activité et importance de la partie hydrosoluble "native." *In* G. Mathé [ed.]. Investigation and stimulation of immunity in cancer patients. Colloque International CNRS (Paris) 1972 (in press).

38. Chedid, L., Parant, M., Parant, F., Gustafson, R. H., Berger, F. M. Biological study of a nontoxic, water-soluble immunoadjuvant from mycobacterial cell walls. Proc. Natl. Acad. Sci. USA. 69:855–858, 1972.

39. Parant, M., Chedid, L. Biological properties of nontoxic water-soluble immunoadjuvant from mycobacterial cells. *In* G. Mathé [ed.]. Investigation and stimulation of immunity in cancer patients. Colloque International CNRS (Paris) 1972 (in press).

40. Chedid, L., Lamensans, A. Experimental screening of systemic adjuvants extracted from mycobacteria. *In* G. Mathé [ed.]. Investigation and stimulation of immunity in cancer patients. Colloque International CNRS (Paris) 1972 (in press).

41. Liacopoulos, P., Merchant, B. Effet du traitement des donneurs avec des endotoxines sur la maladie homologue des receveurs adultes irradiés. *In* L. Chedid [ed.]. La structure et les effets biologiques des produits bactériens provenant de germes gram-négatifs. Colloque International CNRS no. 174 (Paris), 1969, p. 341–356.

42. Damais, C., Lamensans, A., Chedid, L. Endotoxines bactériennes et maladie homologue du nouveau-né. C. R. Acad. Sci. (Paris) 274:1113–1116, 1972.

Effect of Endotoxin on Induced Liver Enzymes

L. Joe Berry and Delfin F. Rippe

From the Department of Microbiology, University of Texas, Austin, Texas

Phosphoenolpyruvate carboxykinase (PEPCK) was purified from murine liver and used as an antigen in rabbits for stimulation of antibody to enzymes. Experiments with the antibody showed that endotoxin inhibits the synthesis of PEPCK rather than inactivating the enzyme present in liver. Two techniques were used. Increasing amounts of immune globulin added to the high-speed supernatant of liver homogenates reduced PEPCK activity, so that the slope of the resulting graph was the same for normal and poisoned livers. Radial-immunodiffusion assays for the amount of enzyme in a liver homogenate revealed that closely similar results followed with time after an injection of the LD_{50} of either endotoxin or actinomycin D. Both poisons, regardless of mode of action, prevented the induction of PEPCK due to fasting.

There are three ways to sensitize mice and rats to endotoxin at least 1,000-fold: by adrenalectomy [1], by an injection of actinomycin D [2,3], or by an injection of lead acetate [4]. The reason for sensitization in each case is not fully understood, but a functioning adrenal cortex is known to be essential for an animal's normal response to stress [5,6]. The adrenocorticoids are known also to be inducers of a number of hepatic enzymes, some of which have half-lives of no more than 2 hr [7–10]. Actinomycin D inhibits DNA-dependent synthesis of RNA and, consequently, synthesis of protein [11], while lead, as a heavy metal, acts as a poison to enzymes. The common effect of the three sensitizers appears to be either impairment of enzyme synthesis or inhibition of enzyme action. The usual response of an animal to endotoxin must require, therefore, a metabolic adjustment dependent on increases in enzyme activity. Homeostasis cannot be maintained otherwise.

Previous work from the laboratory of Dr. Berry showed that endotoxin inhibited the induction by adrenocortocoids of tryptophan oxygenase in a manner similar to the inhibition seen with actinomycin D [12]. Even more dramatic were the indistinguishable effects on the enzyme when either poison was injected at different stages of hormonal induction of tryptophan oxygenase. The activity of the enzyme after injection was eventually identical in both cases (compare figures 7 and 9 in [12]). By contrast, another hormonally inducible hepatic enzyme, tyrosine aminotransferase, was not influenced by endotoxin, while actinomycin D inhibited its induction [12]. The lack of uniformity of action of endotoxin against the two enzymes, compared with that of actinomycin D, raised doubt as to whether the observations with tryptophan oxygenase were, in fact, the result of impaired synthesis of the enzyme. If endotoxin does block synthesis of this enzyme, then its action would have to be more selective than that of actinomycin D. This is especially true since Shtasel and Berry [13] found that endotoxin increases net synthesis of RNA and protein in murine liver. Actinomycin D inhibits both [14].

It was, however, of interest to find that endotoxin prevents the induction of phosphoenolpyruvate carboxykinase (PEPCK) that follows either an injection of glucocorticoid [15] or fasting [16]. The effects of endotoxin on this enzyme are similar to those obtained with actinomycin D [16].

The above observations pose the question of whether endotoxin inhibits the synthesis of PEPCK and tryptophan oxygenase or releases an inhibitor under in-vivo conditions that reduces the in-vitro activity of the enzymes. A mediator must be postulated, since endotoxin added directly to an assay mixture results in no change in measured activity [2].

To answer this question, we have purified

This study was supported in part by a grant from the National Institute of Allergy and Infectious Diseases.

Please address requests for reprints to Dr. L. Joe Berry, Department of Microbiology, University of Texas, Austin, Texas 78712.

PEPCK and tryptophan oxygenase from murine liver and used them to develop antibody to enzymes in rabbits. The immune globulin was used for assay of the enzymes as described below. The results presented are those obtained with PEPCK alone, since work has progressed more rapidly with this enzyme than with the other. The full details for purifying PEPCK have been presented elsewhere [17] along with details of the results summarized here [18].

The rabbit globulin immune to PEPCK was placed in the center well of an Ouchterlony diffusion plate. Surrounding it was (1) purified normal PEPCK, (2) partially purified PEPCK from endotoxin-poisoned liver, (3) the high-speed supernatant from a normal-liver homogenate, and (4) the high-speed supernatant from a poisoned-liver homogenate. A precipitin line without spur formation was obtained. This result argues against the presence of an enzyme-inhibitor complex. Were such a complex present, it would either have to alter totally antigenicity of the enzyme in an all-or-none manner, or it would have to block catalysis without interfering in any way with antigenicity. Either possibility seems unlikely.

Identical immunoelectrophoretic patterns were developed with supernatants from normal- and poisoned-liver homogenates. The similarity of mobility of PEPCK in the two samples is additional evidence in favor of identity of the enzyme. Were an inhibitor bound to the poisoned enzyme, its charge or its mobility would likely be altered.

Addition of the partially purified immune globulin to high-speed supernatants of liver homogenates from normal or poisoned mice before assays for activity of PEPCK gave the results shown in figure 1. Assays were done by the procedure of Phillips and Berry [19]. The lower initial activity in the poisoned liver is apparent, as is the lack of effect of normal rabbit-serum globulin on the enzyme. The addition of increasing amounts of anti-enzyme globulin resulted in the loss of increasing amounts of enzyme. The similarity of slopes of the two curves suggests that no inactive form of PEPCK was present in the livers of endotoxin-poisoned mice. Had there been inactive enzyme, a curve of different slope would have been obtained. There seems to be, therefore, less PEPCK in poisoned livers.

The use of radial immunodiffusion as an assay for amount of antigen was first developed by

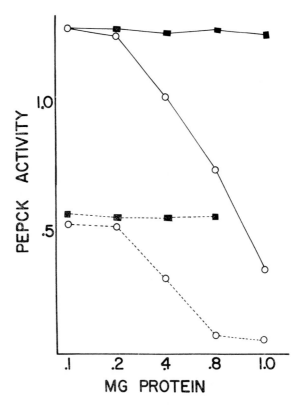

Figure 1. Precipitation of phosphoenolpyruvate carboxykinase (PEPCK) from poisoned and control liver supernatants by anti-enzyme IgG. Increasing amounts of rabbit anti-PEPCK IgG (O——O) or normal rabbit IgG (■– – –■), expressed as mg of protein, were added to high-speed supernatants of livers of control (——) and poisoned (– – – –) mice. The mixture was incubated and centrifuged. The supernatant was assayed for enzymatic activity. PEPCK activity is expressed as hundredths of μmoles of PEP formed per min per 0.2 ml of supernatant.

Mancini et al. (20). This technique was applied to livers of normal, endotoxin-poisoned, and actinomycin D-poisoned mice; the results are presented in figure 2. The upper row contained the high-speed supernatants of pooled homogenates of livers from three mice starting at time zero on the left. During a period of 20 hr, three mice were sacrificed at intervals of 4 hr. The animals were fasted throughout the period, and the increase in PEPCK that resulted was apparent from the larger precipitin rings that formed. The middle and lower rows were from mice given the LD$_{50}$ of endotoxin or actinomycin D, respectively. These animals were also fasted from the beginning of the experiment. The failure of the precipitin rings to

112

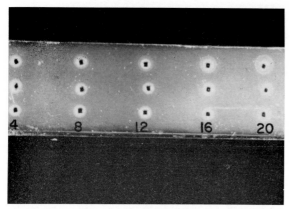

Figure 2. Radial immunodiffusion assay for phosphoenolpyruvate carboxykinase (PEPCK). High-speed supernatants of homogenates were introduced into wells of an agar-coated slide containing anti-PEPCK IgG. Injections were given at time zero, and animals were sacrificed at intervals of 4 hr for 20 hr. Top row: control fasted mice. Middle row: mice injected with the LD_{50} of endotoxin. Bottom row: mice injected with the LD_{50} of actinomycin D.

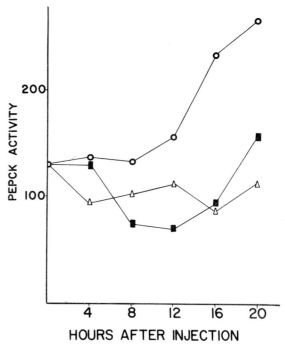

Figure 3. Activity of phosphoenolpyruvate carboxykinase (PEPCK) in the high-speed supernatants of liver homogenates of normal fasted mice (○—○), endotoxin-poisoned mice (■—■), and actinomycin D-poisoned mice (△—△). The same liver homogenates as those used in figure 2 were assayed for PEPCK activity, and the values were plotted against time after injection. Activity of PEPCK is expressed as μmoles of PEP formed per min per 0.2 ml of supernatant.

increase in diameter in the poisoned mice indicates that both poisons prevent induction of PEPCK.

A plot of the activity of PEPCK vs. time gives the three curves shown in figure 3. The similarity of the inhibitory effect of the two poisons is evident and in agreement with ring sizes seen in figure 2.

Other tests described elsewhere [18] established the validity of relating the size of radial precipitin rings to PEPCK activity as measured catalytically. Values obtained by each method agreed within 10%.

These results show that endotoxin interferes with the increase in synthesis of PEPCK that accompanies a period of fasting. At a time during fasting when gluconeogenesis is augmented in normal mice, no such increase can be detected in endotoxin-poisoned mice.[1] Thus a consistency exists between the observations made under in-vivo conditions and those based on in-vitro measurements of PEPCK. Endotoxin, through what is believed to be a mediated effect, interferes with the synthesis of selected inducible enzymes that are essential for the maintenance of homeostasis. It is at this level of metabolism that the animal's ability to cope with stress is probably compromised. Since a mediator

[1] R. E. McCallum and L. J. Berry, "Effect of Endotoxin on Glucose and Glycogen Synthesis in Mouse Liver," manuscript in preparation.

is postulated as necessary for this effect, events before those reported here must occur. The ultimate consequences of earlier changes might well be at the regulatory level of enzyme synthesis. This concept serves the valuable purpose of at least guiding future experiments.

References

1. Lewis, L. A., Page, I. H. Method of assaying steroids and adrenal extracts for protective action against toxic materials (typhoid vaccine). J. Lab. Clin. Med. 31:1325–1329, 1946.
2. Berry, L. J., Smythe, D. S. Effects of bacterial endotoxins on metabolism. VII. Enzyme induction and cortisone protection. J. Exp. Med. 120:721–732, 1964.
3. Pieroni, R. E., Broderick, E. J., Bundeally, A., Levine, L. A simple method for the quantitation of submicrogram amounts of bacterial endotoxin. Proc. Soc. Exp. Biol. Med. 133:790–794, 1970.
4. Selye, H., Tuchweber, B., Bertók, L. Effect of lead

acetate on the susceptibility of rats to bacterial endotoxins. J. Bacteriol. 91:884–890, 1966.

5. Lewis, J. T. Sensibilité des rats privés de surrénales envers les toxiques. C. R. Soc. Biol. (Paris) 84: 163–164, 1921.

6. Jaffe, H. L. On diminished resistance following suprarenalectomy in the rat and the protection afforded by autoplastic transplants. Am. J. Pathol. 2:421–430, 1926.

7. Knox, W. E., Auerbach, V. H. The hormonal control of tryptophan peroxidase in the rat. J. Biol. Chem. 214:307–313, 1955.

8. Kenney, F. T., Flora, R. M. Induction of tyrosine-α-ketoglutarate transaminase in rat liver. I. Hormonal nature. J. Biol. Chem. 236:2699–2702, 1961.

9. Shrago, E., Lardy, H. A., Nordlie, R. C., Foster, D. O. Metabolic and hormonal control of phosphoenolpyruvate carboxykinase and malic enzyme in rat liver. J. Biol. Chem. 238:3188–3192, 1963.

10. Weber, G. Study and evaluation of regulation of enzyme activity and synthesis in mammalian liver. Adv. Enzyme. Reg. 1:1–35, 1963.

11. Reich, E., Franklin, R. M., Shatkin, A. J., Tatum, E. L. Effect of actinomycin D on cellular nucleic acid synthesis and virus production. Science 134: 556–557, 1961.

12. Berry, L. J., Smythe, D. S., Colwell, L. S. Inhibition of inducible liver enzymes by endotoxin and actinomycin D. J. Bacteriol. 92:107–115, 1966.

13. Shtasel, T. F., Berry, L. J. Effect of endotoxin and cortisone on synthesis of ribonucleic acid and protein in livers of mice. J. Bacteriol. 97:1018–1025, 1969.

14. Feigelson, P., Feigelson, M., Greengard, O. Comparison of the mechanisms of hormonal and substrate induction of rat liver tryptophan pyrrolase. Rec. Progr. Horm. Res. 18:491–507, 1962.

15. Berry, L. J., Smythe, D. S., Colwell, L. S. Inhibition of hepatic enzyme induction as a sensitive assay for endotoxin. J. Bacteriol. 96:1191–1199, 1968.

16. Rippe, D. F., Berry, L. J. Effect of endotoxin on the activation of phosphoenolpyruvate carboxykinase by tryptophan. Infec. Immun. 6:97–98, 1972.

17. Rippe, D. F., Berry, L. J. Inhibition of induction of phosphoenolpyruvate carboxykinase by bacterial endotoxin. I. Purification of the enzyme. Infec. Immun. 1972 (in press).

18. Rippe, D. F., Berry, L. J. Inhibition of induction of phosphoenolpyruvate carboxykinase by endotoxin. II. Immunological quantitation of the enzyme. Infec. Immun. 1972 (in press).

19. Phillips, L. J., Berry, L. J. Circadian rhythm of mouse liver phosphoenolpyruvate carboxykinase. Am. J. Physiol. 218:1440–1444, 1970.

20. Mancini, G., Carbonara, A. O., Heremans, J. F. Immunochemical quantitation of antigens by single radial immunodiffusion. Int. J. Immunochem. 2: 235–254, 1965.

Host Resistance to Bacterial Endotoxemia: Mechanisms in Endotoxin-Tolerant Animals

Sylviane C. Moreau and Robert C. Skarnes

From the Worcester Foundation for Experimental Biology, Shrewsbury, Massachusetts

We designed this study to determine the contribution to host defense of the humoral detoxifying mechanism in endotoxin-tolerant animals. Compared with normal serum, sera from tolerant rabbits proved to be a much more potent medium for the detoxification of endotoxins. A fivefold increase in the activity of heat-stable, organophosphate-resistant esterase in tolerant serum was shown to be associated with the low-density lipoprotein fraction. The response of tolerant rabbits to challenge with endotoxin was immediate and was manifested by a sharp drop in ionized calcium in serum and a rapid formation of endotoxin-lipoprotein (esterase) complexes in the circulation. Tolerant rabbits were capable of effecting a rapid intravascular detoxification of endotoxin in spite of total occlusion of blood flow to the liver and spleen. These results clearly support the thesis that intravascular detoxification plays a dominant role in the increased resistance of tolerant animals to endotoxin.

Tolerance or increased resistance to the damaging effects of bacterial endotoxins ensues after exposure of experimental animals to single or multiple injections of such toxins [1]. Host mechanisms responsible for tolerance are not well understood, despite intensive studies of this phenomenon [1, 2]. We reported previously that organophosphate-resistant esterases in plasma are of major significance in defense against experimental endotoxemia in the normal, nontolerant animal [3, 4]. The present report deals with the contribution of this humoral detoxifying system to the endotoxin-tolerant state.

Materials and Methods

Endotoxins. Endotoxins were prepared from *Escherichia coli* 0-111B4 ($LD_{80} = 1$ mg/kg in rabbits) and from the Danysz strain of *Salmonella enteritidis,* according to Boivin and Mesrobeanu [5]. The latter preparation was sedimented in an ultracentrifuge to obtain a more homogeneous

This study was supported by grant no. GB29298 from the National Science Foundation and, in part, by grant no. AI 08374 from the National Institute of Allergy and Infectious Diseases.

Please address requests for reprints to Dr. Robert C. Skarnes, Worcester Foundation for Experimental Biology, Shrewsbury, Massachusetts 01545.

product [6, 7]. The *E. coli* preparation was used to render rabbits and mice tolerant to endotoxin and also to examine changes in ionized calcium after injections of endotoxin. Endotoxin from *S. enteritidis* was used in the in-vitro assay for detoxification and also for studies on the persistence of toxin in the circulation after iv injection into rabbits and mice. In certain experiments, a ^{51}Cr-labeled preparation of *S. enteritidis* [6] was also used. For immunodiffusion experiments, antiserum to *S. enteritidis* was raised in both the horse and the rabbit and was combined in appropriate amounts to obtain a high level of precipitating antibodies to the two major antigens, designated A and C [7, 8]. The equine antiserum is essential for augmentation of the concentration of precipitating antibodies to the fast-diffusing, detoxified endotoxin (A antigen), whereas rabbit antiserum reinforces the concentration of precipitins to the slow-diffusing, toxic C antigen. The toxicity of the *S. enteritidis* preparation was 0.5 mg/kg (LD_{25}) in rabbits and 20 mg/kg (LD_{50}) in mice.

Induction of tolerance. Common white rabbits (average weight, 2.5 kg) and mice (average weight, 35 g) were rendered tolerant to endotoxin in the following manner. Rabbits received daily or alternate-day iv injections of 100×2, 200×2, and 400×2 μg of *E. coli* endotoxin, with a total of six injections. Mice received a single ip injection of 50 μg of this toxin. Studies on tolerant animals were conducted 48 hr after the last injection of endotoxin.

Collection of normal, tolerant, and postendotoxin sera. Samples of blood were obtained from normal and tolerant rabbits by cardiac puncture and were immediately put into small glass tubes beneath a layer of mineral oil. The samples were held at room temperature (24 C) for 1 hr before clots were sedimented by light centrifugation. Serum was drawn off with a syringe and needle and added under mineral oil to tubes in an ice bath. (All procedures described in this report were carried out with freshly drawn serum.) Samples of blood from normal and tolerant mice were collected by cardiac puncture from groups of five animals and handled in the same fashion. Samples of serum obtained in the above manner were used for studies of detoxification and/or measurements of ionized calcium. Individual samples of serum were removed from beneath the oil layer with a 1-ml syringe and were immediately put into a serum flow-through electrode (model no. 99-20, Orion Research, Cambridge, Mass.) for measurement of ionized calcium under essentially anaerobic conditions.

For studies on the persistence of endotoxin in the circulation, blood samples were collected at various intervals after injection of endotoxin, placed immediately into tubes in an ice bath, and allowed to clot 2–3 hr before centrifugation in the cold for separation of the serum. In these experiments, rabbits received 500 µg/kg of *S. enteritidis* endotoxin, and mice received 1.4 mg/kg.

For preparation of the lipoprotein fractions in serum, groups of five or six normal rabbits were bled by cardiac puncture (35 ml each), and, after separation from clots, the sera were pooled for ultracentrifugal flotation. The same animals were subsequently rendered tolerant and were bled again to obtain a pool of tolerant serum. Rabbits were fasted 14–16 hr before each collection of blood. The lipoprotein fractions were prepared in the ultracentrifuge as described elsewhere [9]. Both cellulose acetate and acrylamide-disk electrophoresis [10] showed that the two lipoprotein fractions were free of contaminating proteins. Concentrations of lipoproteins were determined by the method of Lowry et al. [11] and were expressed in terms of protein content.

Immunodiffusion assay for detoxification. Detoxification was measured in vitro by an immuno-

diffusion method described in detail elsewhere [8, 12]. In brief, the assay is based on the transformation of the slow-diffusing, toxic molecules (C antigen) of endotoxin into fast-diffusing, nontoxic polysaccharides (A antigen). The disappearance of detectable precipitation of C antigen, which results in maximal precipitation of A antigen, is taken as the point of complete detoxification. The validity of this method as an assay for detoxification has been judged against several biologic assays [4, 7, 13, 14]. In all tests for detoxification, we took care to keep serum from air and thus to prevent critical changes in *p*H. A small volume (0.02 ml) of *S. enteritidis* endotoxin was added to the bottom of test tubes, after which the serum being tested was added; the mixture was immediately covered with mineral oil and incubated at 37 C. No buffers or cation-binding agents were used. The patterns of immunoprecipitation in the incubated mixtures were allowed to develop in a moist atmosphere at room temperature for 36 hr. The agar patterns were then washed for 24 hr in a constantly stirred bath of buffered saline, rapidly dried in a jet stream of air under filter paper, and stained with azocarmine.

Occlusion of hepatic blood flow. Laparotomies were performed on a group of four normal rabbits and on two groups of four each of endotoxin-tolerant rabbits. In the group of normal rabbits and in one group of tolerant animals, a single ligature was placed on the portal vein-hepatic artery. In the third group, a thread was placed in position but not tried. While the abdominal incision was being closed and while the rabbits were under deep ether sedation, 500 µg of ^{51}Cr-labeled endotoxin/kg was injected iv. Ten minutes later, samples of blood were taken from the heart and immediately put into tubes in an ice bath. All samples were centrifuged in the cold to obtain serum for immunodiffusion studies. All rabbits with occlusions died within 30–40 min, at which time the spleen and liver were removed and counted for radioactivity to ascertain the efficacy of the occlusions. The sham-operated rabbits were sacrificed 30–40 min after injection of endotoxin for removal of tissues. This experiment was repeated a second time with 12 rabbits.

In a subsequent experiment, endotoxin-tolerant rabbits were divided into two groups of five each. One group was heavily sedated with Halothane while receiving an iv injection of 500 µg of unlabeled endotoxin/kg; the second group received the same dose of endotoxin and served as non-anesthetized tolerant controls. Blood samples were collected at various intervals from both groups, and sera were prepared for immunodiffusion studies as described above.

Radial immunodiffusion. Equine antiserum to *S. enteritidis* was used for radial immunodiffusion; this favored the precipitation of the fast-diffusing, detoxified fraction of endotoxin (A antigen). A final dilution of antiserum of 1:20 was prepared in liquefied agar (1.25%) at 46 C. The mixture was poured onto level glass plates to a depth of 4 mm and allowed to gel. Samples of serum (0.07 ml) were taken after injection of endotoxin and were added to duplicate wells and diffused for 48 hr at room temperature. The plates were washed in a large volume of buffered saline (*p*H 7.4), dried, and stained with azocarmine. When auto-radiographs were desired, they were prepared before staining as previously described [6]. Vernier caliper measurements of diffusion diameters were made in duplicate and at right angles for each well, and the readings were averaged. Graded concentrations of a standard preparation of plasma-detoxified endotoxin were included in each diffusion plate for construction of the standard curve. The validity of this method for the quantitation of low concentrations of detoxified endotoxin was established in a series of experiments with known quantities of serum-detoxified endotoxin.

Results

Detoxifying capacity of endotoxin-tolerant serum. The inability to demonstrate that serum from tolerant animals is a more potent detoxifying medium than nontolerant serum has posed an argument against the functional role of a humoral agency in increased resistance to endotoxin. However, these experiments show that an important difference can be demonstrated between the detoxifying capacity of normal serum and that of tolerant serum.

When care is taken to test fresh undiluted serum under oil, a marked difference in endotoxin-detoxifying capacity is seen. The immunodiffusion pattern in figure 1 shows that normal rabbit serum does not fully detoxify endotoxin in 60 min (C antigen still in evidence), whereas tolerant serum

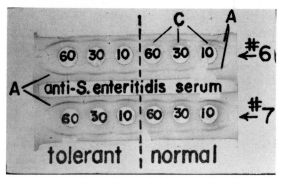

Figure 1. Immunodiffusion assay for detoxification: comparison of the detoxifying capacity of normal and tolerant sera from two rabbits. *Salmonella enteritidis* endotoxin (20 μg) was incubated under oil for 10, 30, or 60 min with 0.4 ml of each of four samples of serum. Each well in the pattern received 0.035 ml of the indicated incubation mixture containing 1.7 μg of endotoxin. Antiserum was put into elongated troughs on both sides of the circular wells in this and all subsequent double-diffusion patterns. **C** = slow-diffusing, toxic antigen; **A** = fast-diffusing, nontoxic antigen. The agar pattern was dried and stained with azocarmine [8].

from the same two animals detoxifies endotoxin within 10 min, as ascertained by the absence of slow-diffusing C antigen. A similar rate of detoxification in serum taken from normal rabbits 5-hr after injection of endotoxin was previously reported; this immunodiffusion assay was used in conjunction with a biologic assay [4]. We also observed a similar marked increase in the detoxifying rate of tolerant guinea-pig serum. Tolerant murine serum, on the other hand, detoxifies endotoxin at a rate only slightly faster than that of normal serum. However, the detoxifying capacity of normal murine serum is significantly greater than that of serum from normal rabbits or guinea pigs and may account, in part, for the greater resistance of this species to endotoxins.

Previous in-vivo and in-vitro studies [3, 4, 8] have shown a direct correlation between the rate and intensity of interaction of endotoxin with a serum lipoprotein (esterase) and the rate of detoxification. Therefore, the finding that tolerant serum detoxifies much faster than normal serum implies a more rapid and intense endotoxin-lipoprotein interaction. Lipoprotein staining of immunodiffusion patterns similar to that in figure 1 revealed that this was the case. This rapid, intense interaction of endotoxin with serum lipoprotein in tolerant rabbits was also demonstrable in vivo

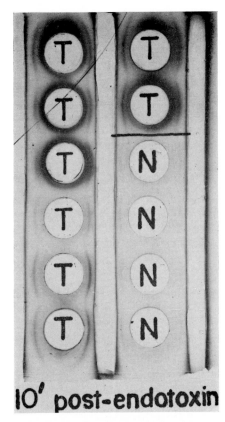

Figure 2. In-vivo complexing of endotoxin with lipoprotein in serum of eight tolerant (T) and four normal (N) rabbits. The samples were collected 10 min after injection of 500 μg of *Salmonella enteritidis* endotoxin/kg. Each well received 0.2 ml of serum. The washed and dried agar pattern was stained with sudan black [8] to reveal the presence of lipoprotein on the specifically precipitated endotoxin.

(figure 2). Lipoprotein staining was barely discernible on the precipitation arcs formed in sera from the four normal rabbits 10 min after endotoxin. In additional experiments, the lipoprotein-endotoxin interaction was readily demonstrated in sera from tolerant rabbits even 5 min after endotoxin, and esterase activity, measured as described elsewhere [8], could be demonstrated on the arcs of precipitated endotoxin.

According to earlier in-vitro studies [7, 8, 15], decreased ionized calcium in serum promotes an increased rate of detoxification of endotoxin. A gradual drop in serum Ca^{++} has been demonstrated in vivo after administration of endotoxin to normal rabbits: the change in Ca^{++} parallels the marked increase in detoxifying capacity of serum taken after administration of endotoxin [4].

It was thus of interest to determine the in-vivo changes in ionized calcium in endotoxin-tolerant rabbits and thereby to acquire further evidence for relating the humoral detoxifying mechanism to the established capacity of tolerant animals to rid their vascular compartment of toxicity shortly after an iv injection of endotoxin. The data in figure 3 demonstrate a rapid calcium response in sera taken from tolerant rabbits after they received endotoxin. These measurements showed a remarkable drop in ionized calcium that was evident within 10 min after endotoxin and that returned toward normal within 1 hr. The maximal Ca^{++} decrement, -1.4 mV, occurred 10 min after endotoxin and represents a drop of 12% in the ionized form of calcium [16]; Ca^{++} had returned to control values within 24 hr in both normal and tolerant animals.

In a previous study, it was shown that sera taken from nontolerant rabbits 5 hr after administration of endotoxin exhibited a threefold increase in activity of heat-stable, organophosphate-resistant esterase. This esterase was thought to be associated with lipoprotein, and its increase was thought to account, in part, for the potent detoxifying capacity of serum taken after endotoxin [4]. In the present experiments, tolerant sera, as compared with normal sera from the same rabbits,

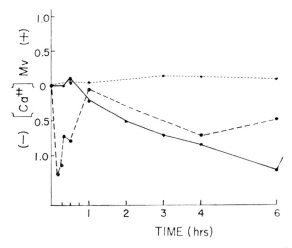

Figure 3. Average decrease in concentrations of ionized calcium in serum after injection of 100 μg of *Escherichia coli* endotoxin/kg into 13 normal (●——●) and nine endotoxin-tolerant (●–––●) rabbits. The control group of seven rabbits (●···●) received pyrogenfree saline. One-milliliter blood samples were collected under oil before administration of endotoxin and at several intervals after injection.

118

showed a four- to sixfold increase in activity of this esterase when tested in the same manner, i.e., after being heated at 60 C for 25 min in the presence of 2×10^{-4} M diisopropylfluorophosphate (DFP). To determine whether the increased esterase activity in tolerant serum was associated with lipoprotein, we isolated the two major lipoprotein fractions from pools of normal and tolerant sera by ultracentrifugal flotation. The results revealed no significant changes in either concentration or esterase activity of the high-density lipoprotein (HDL) fraction. By contrast, there was an average increase of 2.8-fold in the low-density lipoprotein (LDL) fraction in nine rabbits after they were made tolerant to endotoxin. In addition, the LDL fraction from tolerant rabbits exhibited an average increase of 1.77-fold in activity of organophosphate-resistant esterase (figure 4). Taken together, these data indicate a 4.95-fold increase in the activity of DFP-resistant esterase of tolerant serum. It should be noted that the esterase activity of the isolated LDL fraction was not stable at 60 C. However, when LDL was heated in the presence of an amount of purified rabbit γ-globulin equivalent to the protein concentration of whole serum, esterase activity was essentially preserved. The degradation of endotoxin after interaction with the purified LDL fraction from tolerant serum is shown in figure 5. Esterase activity could be demonstrated quite readily on the precipitated endotoxin-lipoprotein complexes.

Role of liver and spleen in uptake of endotoxin.

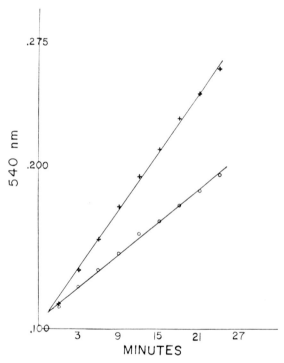

Figure 4. Esterase activity of purified low-density lipoprotein (LDL) fractions from sera of normal (O——O) and tolerant (X——X) rabbits. One-half milliliter of each of the two LDL preparations (0.4 mg) was added to standard cuvettes; this was followed by addition of 2.5 ml of a saturated, filtered solution of 1-naphthyl acetate in 0.1 M phosphate buffer (*p*H 7.4). Saturated diazo blue solution (0.1 ml) and diisopropylfluorophosphate at a final concentration of 2×10^{-4} M were present in the substrate solution. The release of 1-naphthol to form the blue-violet azo-dye complex was monitored spectrophotometrically at 540 nm.

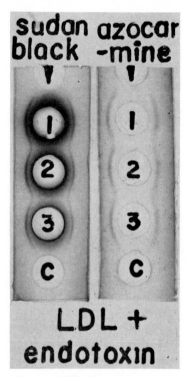

Figure 5. Immunodiffusion patterns demonstrating the interaction of purified low-density lipoprotein (LDL) from tolerant rabbit serum with *Salmonella enteritidis* endotoxin. The numbers in the wells denote different proportions of LDL incubated in vitro with 10 μg of endotoxin: 1 = 0.12 mg; 2 = 0.06 mg; 3 = 0.03 mg; c = saline and endotoxin control. After incubation for 60 min at 37 C, 0.04-ml aliquots containing 1.5 μg of endotoxin were added to each well and diffused against the specific antiserum to endotoxin. Precipitation of this concentration of undegraded endotoxin is barely discernible in the control well stained with azocarmine.

It has been established that the toxicity associated with parenterally administered endotoxin disappears rapidly from the circulation of tolerant animals. What has not been established, however, is whether this disappearance is due to removal of the endotoxin by the reticuloendothelial system or whether an intravascular detoxifying system is primarily responsible. The following experiments were designed for examination of these two possibilities.

Occlusion of the portal vein-hepatic artery in a group of four normal and four endotoxin-tolerant rabbits was followed by iv injection of labeled endotoxin while the animals were heavily sedated. Serum obtained from blood samples taken from each animal 10 min after injection of endotoxin was diffused against antiserum to *S. enteritidis* for precipitation of endotoxin in the circulation. In figure 6 it is seen that sera from normal, occluded rabbits (left) contained small amounts of detoxified endotoxin and substantial amounts of the slow-diffusing, radiopositive C antigen, i.e., the toxic moiety. Sera from the tolerant, occluded group (center) contained a substantial amount of detoxified, fast-diffusing antigen A, whereas the autoradiograph of this immunodiffusion pattern revealed a barely detectable amount of radiopositive material around the circular wells, indicating nearly complete absence of toxicity in the

circulation. No radioactivity was found in livers or spleens of animals with hepatic occlusion; 83% of the injected radioactivity was recovered in livers and spleens from sham-operated, tolerant rabbits. Sera from the four rabbits in the sham-operated, tolerant group contained neither precipitable endotoxin nor radiopositive material, indicating the total absence of endotoxin in circulating blood.

The above experiment was repeated on another group of 12 rabbits and yielded nearly identical results. In this series, lung tissue was also removed within 40 min after injection of endotoxin and counted. The average radioactivity in the lungs from the three groups of rabbits, expressed as percentage of injected dose, was 2.7% in occluded normal, 8.8% in occluded tolerant, and 7.3% in sham-operated rabbits.

Inability to demonstrate endotoxin in either toxic or detoxified form in the plasma from sham-operated tolerant rabbits indicated that the anesthetic had dramatically influenced in-vivo disposition of the toxin. This effect was examined again in nonoperated tolerant rabbits. The immunodiffusion patterns in figure 7 demonstrate that endotoxin disappeared from the circulation of anesthetized rabbits within 5 min of injection and did not reappear during the next 6 hr. This effect was seen in four of five animals in this group. Endotoxin was demonstrable in the serum of the control tolerant rabbit (figure 7) for 6 hr after injection, as was the case with four other tolerant controls. In other experiments, it was noted that medium-to-light stages of anesthesia were not sufficient to permit total clearance of endotoxin from blood of tolerant animals. It was also noted that nontolerant rabbits were unable to effect a complete clearance of endotoxin even under heavy sedation.

Quantitation of circulating endotoxin. Concentrations of detoxified endotoxin in the blood of normal and tolerant rabbits that had received endotoxin were determined by radial immunodifusion (figure 8). Precipitation of the nontoxic A antigen was seen in samples of both normal and tolerant serum at all intervals after endotoxin. The autoradiograph of the immunodiffusion pattern obtained from a normal rabbit demonstrates the continued presence of the slow-diffusing, radiopositive endotoxin from 10 min through 5 hr after endotoxin. No radiopositive material was detectable in tolerant serum at any interval, indicating

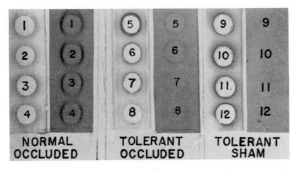

NORMAL OCCLUDED   **TOLERANT OCCLUDED**   **TOLERANT SHAM**

Figure 6. Immunodiffusion patterns with autoradiographs prepared from samples of serum taken from three groups of four rabbits each 10 min after an iv injection, under anesthesia, of 500 µg of *Salmonella enteritidis* endotoxin/kg. In the two occluded groups, hepatic blood flow was arrested before administration of endotoxin. Immunodiffusion patterns on the left in each group were stained with azocarmine to show detoxified A antigen. The autoradiographs on the right in each group were prepared from their respective dried agar patterns to reveal the presence of the slow-diffusing, toxic moiety (C antigen), which is radiopositive [6].

120

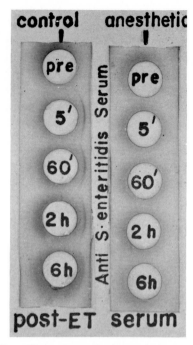

Figure 7. Effect of anesthesia (Halothane) on clearance of endotoxin (ET) from blood of tolerant rabbits receiving an injection of 500 µg of unlabeled *Salmonella enteritidis* preparation/kg. The tolerant control received no anesthetic. Blood samples were collected at the indicated intervals from 5 min to 6 hr after infection; each well contained 0.2 ml of serum. Immunodiffusion patterns were stained with azocarmine.

that the toxic form of endotoxin was absent within 10 min of injection. Radial immunodiffusion was not suitable for quantitation of the slow-diffusing endotoxic fraction, due to its antigenic heterogeneity [6, 7].

The concentrations of nontoxic A antigen persisting in the circulation of normal and tolerant rabbits are shown in figure 9. The average amount of detoxified endotoxin in serum during the 5 hr after injection was 60% of the injected dose in normal rabbits and 25% in tolerant rabbits. The slower elimination of radioactivity from the blood of normal rabbits correlates with the continued presence of radiopositive endotoxin witnessed in figure 8. The results on persistence of detoxified endotoxin in the normal and tolerant mouse are given in figure 10. Although the patterns of elimination of radioactivity were similar to those in the rabbit, the amount of detoxified endotoxin in the circulation of the tolerant mouse during the 5 hr after endotoxin was considerably higher than that in the tolerant rabbit.

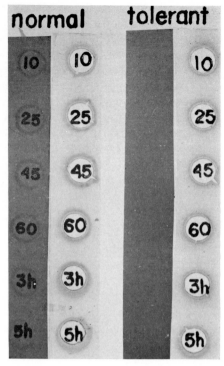

Figure 8. Radial immunodiffusion patterns from a normal and a tolerant rabbit. Blood samples were collected at the indicated intervals from 10 min to 5 hr after administration of 500 µg of ^{51}Cr-labeled endotoxin/kg. Radial immunodiffusion patterns (right) were stained with azocarmine to show detoxified A antigen. The autoradiographs (left of each pattern) demonstrate the presence or absence of slow-diffusing, radiopositive C antigen.

Discussion

The host mechanisms responsible for increased resistance or tolerance to bacterial endotoxins are not established. To date, interpretations of observations favoring a dominant role for the reticuloendothelial system (RES) and for specific antibody are based on indirect or inconclusive results, which by no means exclude other interpretations [17]. For example, the view that stimulation of the RES by repeated exposure to endotoxin is causally related to an increased resistance to endotoxin can be countered by the fact that other agents that stimulate RES capacity render such animals highly sensitive to endotoxins [1, 18, 19].

It is unclear whether or not specific antibody mediates tolerance, be it actively induced or passively transferred. In the former case, there is strong evidence against the view that an active

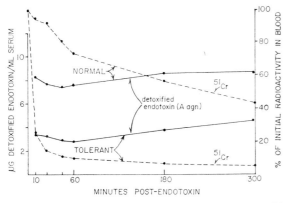

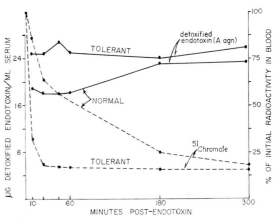

Figure 9. Persistence of detoxified endotoxin (A antigen) and clearance of radioactivity from blood of normal and tolerant rabbits after injection of 500 μg of ^{51}Cr-labeled endotoxin/kg. Concentrations of endotoxin were estimated by radial immunodiffusion and represent the averages of four normal and four tolerant rabbits. On the assumption that blood volume is 6% of body weight, the concentration of endotoxin at the time of injection was 13.8 μg/ml of circulating plasma.

Figure 10. Persistence of detoxified endotoxin (A antigen) and clearance of radioactivity from blood of normal and tolerant mice after injection of 50 μg of ^{51}Cr-labeled endotoxin/kg. Concentrations of endotoxin were estimated by radial immunodiffusion and represent values obtained from pooled sera of five mice at each interval. On the assumption that blood volume is 7% of body weight in this species, the concentration of endotoxin at the time of injection was 33 μg/ml of circulating plasma.

immune response is required for development of tolerance to endotoxin [20, 21]. In the latter case, partial tolerance to the toxic actions of endotoxins can be transferred to normal recipients via sera or serum fractions from tolerant [1, 22–24] or immune [25, 26] donors. In the latter reports, it was suggested that the mediation of passive tolerance is specific, although the presumed antibodies have never been identified.

On the other hand, Greisman et al. [27] reported that, after a single small dose of endotoxin, rabbits exhibited an early nonspecific phase of pyrogenic tolerance followed by a later, specific phase mediated by O antibodies. In view of most other reports [1, 2], it is doubtful that specific O antibodies have any protective effect against larger doses of endotoxin, such as those used in tests of survival of tolerant animals. Furthermore, other recent evidence appears to rule against an important role for specific O antibody, even in pyrogenic tolerance, if the tolerant state is induced in the usual manner, i.e., by repeated injections of endotoxin [28].

The fact that endotoxin persists for many hours in the circulation of normal and tolerant animals after parenteral administration [3, 4, 6, 7, 21] implies that the vascular compartment represents a major site of detoxification. The important role of plasma in detoxification of endotoxin in normal

animals was recently described [4]. It was shown that the response of nontolerant rabbits to modest doses of endotoxin resulted in marked activation of the humoral detoxifying system. The in-vivo response was characterized by a gradual reduction in ionized calcium in serum and a threefold increase in the level of heat-stable, organophosphate-resistant esterase within 5 hr after injection of endotoxin. The significance of these responses to host defense against endotoxins has been described [4, 17]. Briefly, the interaction of endotoxin with serum lipoprotein (esterase) is greatly intensified when Ca^{++} is reduced either in vitro [8] or in vivo [3, 4]. The resultant complex brings about a rapid disaggregation or degradation of the large endotoxin molecules and is requisite to eventual detoxification by a second serum esterase [8].

We designed the present experiments to determine the extent to which the humoral detoxifying system participates in resistance of tolerant animals to endotoxin. The finding that, under appropriate conditions, the detoxifying potency of sera from tolerant rabbits was markedly increased over that of sera from the same animals before development of tolerance provided the initial evidence that the humoral detoxifying system contributes to increased resistance to endotoxin. The increased level of heat-stable organophosphate-resistant es-

122

terase in tolerant serum could explain, in part, the increased detoxifying capacity.

An immediate in-vivo activation of the humoral detoxifying system in tolerant rabbits was indicated by the sharp drop in serum ionized calcium and by the rapid formation of endotoxin-lipoprotein (esterase) complexes after challenge with endotoxin. This immediate response was significantly different from the delayed activation observed in nontolerant rabbits [4]. In the latter instance, the drop in Ca^{++} and the strong interaction of endotoxin with lipoprotein (esterase) was not seen until 4–5 hr after exposure to endotoxin. The mechanism responsible for the initial rapid drop in serum ionized calcium in tolerant rabbits receiving endotoxin is unknown. However, since leukocyte mobilization and granulopoiesis occur in response to repeated injections of endotoxin [29, 30] and since rabbit neutrophilic leukocytes have been shown to release lactate under the influence of small concentrations of endotoxin [31], it is possible that granulocytes provide a surge of lactate that accounts for this transient drop in ionized calcium in tolerant serum.

The host mechanism responsible for the delayed decrement in Ca^{++} in sera taken after endotoxin injection from normal and tolerant rabbits could be due to Ca^{++} binding resulting from increases in plasma lactate, phosphate, and lipoproteins, all of which have been reported to increase within a few hours after exposure to endotoxic products [32–35].

Isolation of the two major lipoprotein fractions from normal and tolerant serum revealed that the concentration (and esterase activity) of the LDL fraction was significantly higher in tolerant serum. This increase, in terms of total organophosphate-resistant esterase per ml of tolerant serum, was five times that in normal sera from the same rabbits and fits in well with the observed increase in the level of heat-stable, DFP-resistant esterase in whole tolerant serum.

These findings thus substantiate the importance of the serum detoxifying system in tolerance to endotoxins. The tolerant rabbit responds quickly to a challenge dose of endotoxin. The rapid formation of endotoxin-lipoprotein complexes correlates well with both drop in ionized calcium and rapid rate of detoxification.

The experiments in tolerant rabbits with portal-vein occlusions led to two pertinent findings.

First, a marked difference was seen between detoxification of circulating endotoxin in normal and in tolerant groups of occluded rabbits. Whereas there was little detoxification in the normal group, detoxification was essentially complete within 10 min in tolerant rabbits, in which blood flow through the liver and spleen had been effectively blocked. These results demonstrate that the tolerant rabbit is capable of detoxifying a substantial dose of endotoxin in a short time, without participation of liver or spleen. Furthermore, the presence of substantial amounts of detoxified polysaccharide (A antigen) in the circulation of these rabbits provides a further indication that detoxification occurred in the vascular compartment.

The striking capacity of intact, tolerant animals to eliminate toxicity from their blood within a few minutes after injection of endotoxin (figure 8) is in agreement with the intravascular detoxifying rate observed in occluded, tolerant rabbits. Participation of the liver or spleen in removal of circulating endotoxins in the intact, tolerant animal is probably minimal in view of the impairment of RES function after injection of endotoxin [36]; this effect has been reported to last beyond the time required for elimination of toxicity from the blood.

The second pertinent observation that emerged from the occlusion experiments was the dramatic effect of anesthesia on the removal of endotoxin from the bloodstream and its deposition in the liver in nonoccluded (sham-operated), tolerant rabbits. This result, which was also confirmed in nonoperated tolerant rabbits under deep anesthesia, bears witness to the capacity of the RES of the liver to clear endotoxin rapidly. The opportunity to do so in this case is most likely provided by the action of the anesthetic to impede or block vasoconstriction, allowing reasonably normal passage of blood through the liver.

The failure to detect endotoxin, either toxic or nontoxic, in the circulation of anesthetized, tolerant rabbits throughout the 6 hr after endotoxin indicates that, once endotoxin has been taken up by the RES, it is not subsequently released into the vascular compartment in an immunologically identifiable form. This observation stands in sharp contrast to the persistence of detoxified fragments of endotoxin in the circulation of unanesthetized, tolerant rabbits during the 6 hr after endotoxin. This persistence was likewise observed in unanesthetized, tolerant mice during the 5 hr after administration of endotoxin. In fact, approximately equal amounts of detoxified endotoxin (40% of injected dose) were present in the circulation of both normal and tolerant mice 24 hr after injection of endotoxin.

Elimination of detoxified fragments of endotoxin from the circulation did not necessarily correlate with elimination of radioactivity (figures 9 and 10). In the nontolerant mouse, the level of detoxified endotoxin did not change significantly during the first hour after injection; this is interpreted to signify a lag period before full activation of the humoral detoxifying system. This lag period was previously observed in vivo [4]. At some point between 1 and 3 hr after injection, the intravascular detoxifying rate increases, as indicated by an increased concentration of circulating detoxified polysaccharides (A antigen).

Toxicity is rapidly eliminated from plasma of the tolerant mouse [7, 21]; the concentration of detoxified endotoxin is high initially and remains so for 5 hr after endotoxin. From 70% to 75% of the total dose of endotoxin was found in the circulation of both normal and tolerant mice. The remaining 25%–30% of the injected dose not accounted for in the vascular compartment is likely to have been distributed throughout the tissues and fluids. A portion of the detoxified fragments has been demonstrated in murine urine collections made between 0 and 6 hr after endotoxin [6]. These results provide firm evidence to support the thesis that the major site of detoxification is in the vascular compartment of normal and tolerant mice and that the humoral system is of primary importance in host defense against endotoxemia in this species.

In the nontolerant rabbit, persistence of a fairly constant amount of detoxified endotoxin (about 60% of the injected dose) was observed. This finding, together with previous results [3, 4], indicates a major role for the humoral detoxifying system in normal rabbits. In the tolerant rabbit, however, a much smaller amount of detoxified endotoxin (about 25% of the injected dose) was demonstrable in the circulation after challenge with endotoxin. The explanation for this is unclear but if, for reasons stated above, participation of the liver and spleen is excluded, then the possibility must be considered that intravascular granulocytes play an important role in the uptake of endotoxin in the tolerant rabbit.

Injections of endotoxin stimulate the mobilization of the marginating pool of leukocytes and cause marked hyperplasia of the bone-marrow neutrophil compartment [29, 30]. Polymorphonuclear leukocytes from the rabbit have been shown to take up endotoxin by micropinocytosis and to inactivate it [37]. Detoxification of endotoxin by extracts of rabbit neutrophils has been reported [38], and we have isolated a potent endotoxin-degrading component from these cells (authors' unpublished observations). The increased uptake of labeled endotoxin by the lungs of tolerant rabbits (observed in the present study and by others [39]) may be a reflection of a sequestration (after endotoxin) of neutrophils containing endotoxin. All of these reports considered together strongly suggest that granulocytes contribute significantly to defense against endotoxins, particularly in the tolerant host. In view of the rapid interaction of endotoxin with lipoprotein (esterase) in the circulating plasma of tolerant rabbits, the contribution of granulocytes could, in fact, be enhanced if the lipoprotein were capable of functioning as an opsonin. This possibility is currently being investigated.

Participation of granulocytes in the normal or tolerant mouse is probably inconsequential in the present experiments. When the total number of granulocytes available in the mouse is compared with the number of endotoxin molecules injected, it is doubtful that these cells could have any measurable effect on host defense. This doubt is supported by the data showing persistence of high levels of detoxified endotoxin in the circulation of mice.

Thus a dominant role for the humoral detoxifying mechanism in defense against endotoxemia has been demonstrated in normal rabbits and in normal and tolerant mice. While the present data provide clear evidence for an important role of the humoral system in tolerant rabbits, a cellular mechanism involving the intravascular granulocyte may be equally important in this experimental animal.

References

1. Atkins, E. Pathogenesis of fever. Physiol. Rev. 40: 580–646, 1960.

2. Nowotny, A. Molecular aspects of endotoxic reactions. Bacteriol. Rev. 33:72–98, 1969.

3. Skarnes, R. C. In vivo interaction of endotoxin with a plasma lipoprotein having esterase activity. J. Bacteriol. 95:2031–2034, 1968.

4. Skarnes, R. C. Host defense against bacterial endotoxemia: mechanism in normal animals. J. Exp. Med. 132:300–316, 1970.

5. Boivin, A., Mesrobeanu, L. Recherches sur les antigènes somatiques et sur les endotoxines des bactéries. I. Considérations générales et exposé des techniques utilisées. Rev. Immunol. (Paris) 1:553–569, 1935.

6. Chedid, L., Skarnes, R. C., Parant, M. Characterization of a ^{51}Cr-labeled endotoxin and its identification in plasma and urine after parenteral administration. J. Exp. Med. 117:561–571, 1963.

7. Skarnes, R. C., Chedid, L. Biological degradation and inactivation of endotoxin. In M. Landy and W. Braun [ed.] Bacterial endotoxins. Rutgers University Press, New Brunswick, N.J., 1964, p. 575–587.

8. Skarnes, R. C. The inactivation of endotoxin after interaction with certain proteins of normal serum. Ann. N.Y. Acad. Sci. 133:644–662, 1966.

9. Havel, R. J., Eder, H. A., Bragdon, J. H. The distribution and chemical composition of ultracentrifugally separated lipoproteins in human serum. J. Clin. Invest. 34:1345–1353, 1955.

10. Ornstein, L., Davis, B. J. Acrylamide electrophoresis. In L. P. Cawley. Electrophoresis and immunoelectrophoresis. Little, Brown, Boston, 1969, p. 308–341.

11. Lowry, O. H., Rosebrough, N. J., Farr, A. L., Randall, R. J. Protein measurement with the Folin phenol reagent. J. Biol. Chem. 193:265–275, 1951.

12. Skarnes, R., Rutenburg, S., Fine, J. Fractionation of an esterase from calf spleen implicated in the detoxification of bacterial endotoxin. Proc. Soc. Exp. Biol. Med. 128:75–80, 1968.

13. Rojas-Corona, R. R., Skarnes, R., Tamakuma, S., Fine, J. The Limulus coagulation test for endotoxin. A comparison with other assay methods. Proc. Soc. Exp. Biol. Med. 132:599–601, 1969.

14. Rutenburg, S. H., Skarnes, R. C., Palmerio, C., Fine, J. Detoxification of endotoxin by perfusion of liver and spleen. Proc. Soc. Exp. Biol. Med. 125:455–459, 1967.

15. Skarnes, R. C., Rosen, F. S., Shear, M. J., Landy, M. Inactivation of endotoxin by a humoral component. II. Interaction of endotoxin with serum and plasma. J. Exp. Med. 108:685–699, 1958.

16. Moore, E. W. Studies with ion-exchange calcium electrodes in biological fluids: some applications in biomedical research and clinical medicine. In Symposium on ion-selective electrodes. National Bureau of Standards, no. 314. U.S. Government Printing Office, Washington, D.C., 1969, p. 215–285.

17. Skarnes, R. C., Rosen, F. S. Host-dependent detoxification of bacterial endotoxin. In S. Kadis, G. Weinbaum, and S. J. Ajl [ed.] Microbial toxins. Vol. 5. Academic Press, New York, 1971, p. 151–164.

18. Cooper, G. N., Stuart, A. E. Sensitivity of mice to bacterial lipopolysaccharide following alteration of activity of the reticulo-endothelial system. Nature (Lond.) 191:294–295, 1961.

19. Suter, E. Hyperreactivity to endotoxin after infection with BCG. J. Immunol. 92:49–54, 1964.

20. Wolff, S. M., Mulholland, J. H., Ward, S. B., Rubenstein, M., Mott, P. D. Effect of 6-mercaptopurine on endotoxin tolerance. J. Clin. Invest. 44:1402–1409, 1965.

21. Chedid, L., Parant, M., Boyer, F., Skarnes, R. C. Nonspecific host responses in tolerance to the lethal effect of endotoxins. In M. Landy and W. Braun [ed.] Bacterial endotoxins. Rutgers University Press, New Brunswick, N.J., 1964, p. 500–516.

22. Freedman, H. H. Passive transfer of tolerance to pyrogenicity of bacterial endotoxin. J. Exp. Med. 111:453–463, 1960.

23. Shulman, J. A., Petersdorf, R. G. Relationship of endogenous pyrogen and serum augmenting factor to endotoxin tolerance. Proc. Soc. Exp. Biol. Med. 119:218–221, 1965.

24. Kim, Y. B., Watson, D. W. Modification of host responses to bacterial endotoxins. II. Passive transfer of immunity to bacterial endotoxin with fractions containing 19S antibodies. J. Exp. Med. 121:751–759, 1965.

25. Davis, C. E., Brown, K. R., Douglas, H., Tate, W. J., Braude, A. I. Prevention of death from endotoxin with antisera: I. The risk of fatal anaphylaxis to endotoxin. J. Immunol. 102:563–572, 1969.

26. Creech, H. J., Hankwitz, R. F., Jr., Wharton, D. R. A. Further studies of the immunological properties of polysaccharides from Serratia marcescens (Bacillus prodigiosus). I. The effects of passive and active immunization on the lethal activity of the polysaccharides. Canc. Res. 9:150–157, 1949.

27. Greisman, S. E., Young, E. J., Carozza, F. A., Jr. Mechanisms of endotoxin tolerance. V. Specificity of the early and late phases of pyrogenic tolerance. J. Immunol. 103:1223–1236, 1969.

28. Mulholland, J. H., Wolff, S. M., Jackson, A. L., Landy, M. Quantitative studies of febrile tolerance and levels of specific antibody evoked by bacterial endotoxin. J. Clin. Invest. 44:920–928, 1965.

29. Boggs, D. R., Athens, J. W., Cartwright, G. E., Wintrobe, M. M. Leukokinetic studies. IX. Experimental evaluation of a model of granulopoiesis. J. Clin. Invest. 44:643–656, 1965.

30. Chervenick, P. A., Boggs, D. R. Granulocytic hyperplasia and induction of tolerance in response to chronic endotoxin administration. J. Reticuloendothel. Soc. 9:288–297, 1971.

31. Cohn, Z. A., Morse, S. I. Functional and metabolic properties of polymorphonuclear leucocytes. II.

The influence of a lipopolysaccharide endotoxin. J. Exp. Med. 111:689–704, 1960.

32. Delafield, M. E. A comparison of the changes in blood sugar and blood phosphorus in rabbits following the injection of suspensions of different dead bacteria. J. Pathol. Bacteriol. 35:53–68, 1932.

33. Kun, E., Miller, C. P. Effect of bacterial endotoxins on carbohydrate metabolism of rabbits. Proc. Soc. Exp. Biol. Med. 67:221–225, 1948.

34. LeQuire, V. S., Hutcherson, J. D., Hamilton, R. L., Gray, M. E. The effects of bacterial endotoxin on lipide metabolism. I. The responses of the serum lipides of rabbits to single and repeated injections of Shear's polysaccharide. J. Exp. Med. 110:293–309, 1959.

35. Gallin, J. I., Kaye, D., O'Leary, W. M. Serum lipids in infection. N. Engl. J. Med. 281:1081–1086, 1969.

36. Moses, J. M., MacIntyre, W. J. Effect of endotoxin on simultaneously determined cardiac output and hepatic blood flow in rabbits. J. Lab. Clin. Med. 61:483–493, 1963.

37. Mesrobeanu, L., Mesrobeanu, I., Bonna, C., Vranialici, D. *In* L. Chedid [ed.] La structure et les effets biologiques des produits bactériens provenant de germes gram-négatifs. C.N.R.S., no. 174, Paris, 1969, p. 429–446.

38. Chedid, L., Lamensans, A., Prixova, J. Comparison of the effects of an antibacterial leukocytic extract and the serum endotoxin-detoxifying component of lipopolysaccharides extracted from rough and smooth salmonellae. J. Infect. Dis. 121:634–639, 1970.

39. Carey, F. J., Braude, A. I., Zalesky, M. Studies with radioactive endotoxin. III. The effect of tolerance on the distribution of radioactivity after intravenous injection of *Escherichia coli* endotoxin labelled with Cr[51]. J. Clin. Invest. 37:441–457, 1958.

Evidence for Participation of Granulocytes in the Pathogenesis of the Generalized Shwartzman Reaction: A Review

Robert G. Horn

From the Department of Pathology, Vanderbilt University, Nashville, Tennessee

Evidence is reviewed that indicates that granulocytes are involved in the syndrome of intravascular clotting produced by sequential injections of endotoxin, i.e., the generalized Shwartzman reaction (GSR). Animals are not susceptible to the GSR during the phase of granulocytopenia induced by treatment with nitrogen mustard. Transfusion of granulocytes, infusion of polymorphonuclear-leukocyte granules, or infusion of a granule-supernatant fraction induces an immediate state of preparation for the GSR; in this state, one injection of endotoxin will trigger massive clotting. There is substantial evidence suggesting that acid mucosubstance and/or lysosomal cationic proteins from granulocytes may interact with endotoxin-induced soluble fibrin-monomer complexes or related fibrinogen derivatives, leading to formation of the fibrinous or fibrinlike coagulum that characterizes the GSR.

The generalized Shwartzman reaction (GSR) is a pathologic entity characterized primarily by the development of glomerular capillary thrombosis in response to the second of two sequential injections of endotoxin, given about 24 hr apart [1]. Glomerular thrombosis and secondary renal cortical necrosis can be produced by a variety of mechanisms, perhaps the most straightforward of which is the slow infusion of thrombin into renal arteries [2]. Other techniques for stimulating the GSR involve use of thromboplastic materials [3], fibrinolysis inhibitors [4], so-called reticuloendothelial-blockading agents [5], antigens (in sensitized animals) [6], granulocyte products [7–10], and antisera to platelets [11] and granulocytes [12]. Dietary manipulations and pregnancy have also been extensively studied as elements involved in disseminated intravascular coagulation [13]. It is important to recognize that, while each of these systems may be of significance, both basic and clinical, it is almost certain that a single pathogenetic sequence will not be found to be common to all of these entities. Therefore, in attempts to analyze the effects of endotoxin that elicit glomerular thrombosis, it is important to avoid conclusions that factors leading to disseminated intravascular coagulation in one circumstance are necessarily operative in any or all other conditions that resemble the GSR. In this review we shall attempt to analyze some of the factors involved in the pathogenesis of the classical GSR as produced by sequential injections of endotoxin.

It is likely that low-grade intravascular clotting occurs to some degree at all times in the normal animal, but that there are efficient mechanisms for removal and degradation of clots [14]. A clot-promoting factor could exert its effect by causing increased production of clots, impaired removal of clots, or some combination of these effects.

Many of the multitudinal sequelae of an injection of endotoxin, such as those discussed in other papers in this volume, may be completely unrelated to enhanced formation of clots. However, many physiologic effects of endotoxin are known to bear upon the mechanisms of coagulation [15]. Possible primary effects of endotoxin on initiation of clotting include aggregation of platelets [16], activation of factor XII (Hageman factor) [15], and activation of the complement system [17, 18]. There is no consensus concerning the relative importance of these factors [19, 20]. Endotoxin may produce injury to endothelial cells and thereby promote coagulation of the blood [21]. Impaired fibrinolytic capacity after injection of endotoxin has been proposed as a clot-promoting factor [22].

This work was supported in part by grant no. HE-10048 and Research Career Development Award no. GM-28110 from the U.S. Public Health Service.

The previous collaborations of Sam Spicer, Jacek Hawiger, and Robert D. Collins in various facets of this work are gratefully acknowledged.

Please address requests for reprints to Dr. Robert G. Horn, Department of Pathology, Vanderbilt University, Nashville, Tennessee 37232.

Impaired ability to remove particulate carbon from the circulation has been regarded as evidence of reticuloendothelial blockade by endotoxin, and it has been proposed that impaired phagocytic capacity of reticuloendothelial cells to clear the blood of fibrin and clotting factors accounts for the marked clotting effects of a second injection of endotoxin [4, 13, 23].

With so many clot-producing effects, the casual reader may ask why there is a problem in understanding the pathogenesis of the GSR. There are several aspects to the answer of this important question. In the first place, some of the postulated clot-promoting effects of endotoxin are not well documented or are so mild that questions are raised as to whether or not they can account for the massive clotting of the GSR. More significantly, since so many effects are produced by endotoxin, it has been difficult to determine which factors account for enhanced clotting after the second injection of endotoxin. The first injection of endotoxin into the normal animal produces a variety of changes, some of which lead to transient and low-grade intravascular coagulation, but no significant thrombosis occurs. However, if another identical injection of endotoxin is given during the following 24 hr, the animal becomes exquisitely sensitive to massive intravascular clotting. The two doses of endotoxin are not additive. Rather, the first injection induces a profound alteration in the host. A crucial question requiring an answer, if the pathogenesis of the GSR is to be explained, is this: What are the changes, induced in the host by a first injection of endotoxin, that cause this enhanced susceptibility to the clot-promoting effects of a second injection?

Possible "Preparing" Effects of the First Injection of Endotoxin

During the past 20 years, three major hypotheses have been offered to account for the preparatory role of the first injection of endotoxin. It has been postulated that the crucial preparatory effect of the first injection is due to (1) reticuloendothelial blockade [4, 13, 23], (2) impaired fibrinolysis [22], and (3) a granulocyte-mediated effect on clotting [7, 24].

When comparing the relative merits of these three hypotheses, several considerations are worthwhile. First, the effect of the first injection, which leads to "preparation," should coincide in time with the period of enhanced susceptibility to the GSR. Second, if an effect is measurable not before, but only after, the provoking endotoxin, it is likely that the effect is a consequence, rather than a cause, of intravascular clotting. Third, these hypotheses are not necessarily mutually exclusive, but may merely be different or related aspects of the same effect. For example, it is possible that a granulocyte-mediated effect might cause reticuloendothelial blockade and inhibited fibrinolytic activity, although, I hasten to add, no such interrelationship has been demonstrated.

Role of reticuloendothelial blockade. It has long been thought that iv-injected endotoxin is cleared from the circulation by reticuloendothelial cells of the liver, spleen, and other reticuloendothelial organs [25]. Substances regarded as reticuloendothelial-blockading agents potentiate some effects of endotoxin; this blockading effect is thought to be due to impaired removal of endotoxin, and, thus, to a prolonged exposure to endotoxin [5].

Lee took cognizance of the fact that reticuloendothelial cells also participate in the removal of fibrin from the blood and postulated that the massive clotting produced by the second injection of endotoxin was a consequence of the reticuloendothelial blockade produced by the first injection, with resulting impaired removal of fibrin [4, 23]. This hypothesis and variations of it have been accepted by a number of other students of the generalized Shwartzman reaction [13, 26]. However, there are a number of bothersome observations that create reservations about this concept.

McKay postulated that it was actually the fibrin aggregates themselves, produced by the first injection, that led to blockade of the reticuloendothelial system (RES) and impaired the ability of the RES to clear fibrin and procoagulant factors formed after a second injection [13]. This particular variation of the reticuloendothelial-blockade hypothesis is not tenable if one accepts the observations of Good and Thomas in 1953 [27] that aggressive treatment with heparin at the time of the first injection of endotoxin fails to prevent the induction of preparation.

Lee recognized that clotting per se was not a prerequisite for preparation caused by a first injection and postulated that the endotoxin somehow "sensitized" the RES in an unspecified manner [4, 23].

When attempts have been made to measure reticuloendothelial blockade after injection of endotoxin, it has been found that ability to remove particulate carbon from the circulation is impaired for the first few hours after endotoxin, but by 24 hr (the time of maximal susceptibility to clotting with a second injection), reticuloendothelial capacity often returns to normal levels [28]. Recognizing the lack of a direct correlation of GSR susceptibility to the reticuloendothelial-blockading effect of a first injection of endotoxin, Lee postulated that the reticuloendothelial cells had been "sensitized" to a second endotoxin, because phagocytic capacity of reticuloendothelial cells was even more markedly impaired after a second dose [4, 23]. This reticuloendothelial impairment after the second injection was thought not to be a consequence of clotting, since treatment with heparin did not prevent reticuloendothelial injury as a result of the second injection [23]. Thus, if one accepts that reticuloendothelial blockade is the essence of "preparation," one is left to speculate as to the pathogenesis of this "sensitization" of the RES.

A second area of reservation about the importance of reticuloendothelial blockade in the GSR is suggested by certain other effects of thorotrast, the reticuloendothelial-blockading agent that has been most widely studied in relation to the generalized Shwartzman phenomenon. Although thorotrast has commonly been considered to exert its preparatory effect only by blockading the RES, it is pertinent to note (1) that thorotrast also causes substantial granulocytosis [5], and (2) that thorotrast is ineffective in eliciting preparation in a nitrogen mustard-treated granulocytopenic animal [5]. These observations seem particularly significant in view of the observations suggesting that granulocytes are necessarily involved in the pathogenesis of the GSR [7, 24].

Recent studies of reticuloendothelial blockade have cast some doubt on this whole concept, if it is regarded in a mechanistic way as the phagocytic capacity of a set of cells lining certain sinusoidal spaces. Koenig et al. [29] have shown, in a carefully studied system, that measurable reticuloendothelial blockade was specific for the particle under study, and that the apparent blockade was not due to saturation of hepatic removal mechanisms but due to high levels of the blockading agent per se. At the least, this study indicates that methods for measuring reticuloendothelial blockade are likely not actually measuring phagocytic capacity in any general way. With these reservations in mind, it seems premature to conclude that the first injection of endotoxin produces a state of susceptibility to disseminated intravascular clotting by virtue of reticuloendothelial blockade.

Role of impaired fibrinolysis. Because of the obvious importance of the plasminogen-plasmin system in removal of clots, the possibility has been investigated that the first dose of endotoxin might prepare hosts for the GSR by causing inhibition of fibrinolysis. Thus, in the face of impaired fibrinolysis resulting from a preceding dose, the clot-promoting effects of a second injection of endotoxin might be adequate to account for the observed increase in accumulation of clots. Lee investigated this possibility and, finding no evidence of an active fibrinolytic system in rabbits, concluded that it was unlikely that inhibition of an ineffectual system could account for the difference in the amount of clotting after two injections of endotoxin [4].

More recently, evidence has been adduced that there is significant fibrinolytic activity in rabbits [22, 26], but we are aware of no studies coupling a phase of inhibited fibrinolysis with the phase of enhanced susceptibility to the GSR. It has been proposed that the particular susceptibility of the rabbit to the GSR, in comparison with that of other species, is due to a poor fibrinolysis system [23].

For the moment, the available evidence would seem to suggest that, while impaired fibrinolysis may be a factor in the GSR, it is likely that other preparatory effects of the first injection of endotoxin are of primary importance.

Role of Granulocytes in the GSR

It is common knowledge that an injection of endotoxin has pronounced effects on blood granulocytes [25]. Endotoxin produces an abrupt granulocytopenia, followed quickly by a marked granulocytosis that is maximal in 12–24 hr [1]. This granulocytosis is associated with a marked hyperplasia of the granulocyte precursors in the bone marrow; cells released to the circulating blood from the marrow are immature and contain a high proportion of azurophile granules [30].

The possible significance of components of these granules is elaborated below.

Inhibition of the reaction with nitrogen mustard. The initial, and, indeed, some of the most powerful evidence that granulocytes are involved in the pathogenesis of the GSR came from the experiments of Thomas and associates in the early 1950s. Early in these studies, which defined many of the characteristics of the GSR, the crucial observation was made that, in animals treated with nitrogen mustard, the GSR was suppressed [1]. Several supporting lines of evidence indicated that this suppressive effect of nitrogen mustard was, in fact, attributable to the granulocytopenia that was induced. First, the period of suppression by nitrogen mustard corresponded to the period of granulocytopenia. Secondly, it was shown by experiments involving protection of part of the marrow from the effects of the nitrogen mustard that the suppression by nitrogen-mustard is almost certainly exerted through hematopoietic tissue of the bone marrow and is not attributable to some effect on the spleen or kidney or lymphoid cells of the body generally [1, 7]. Although Thomas and his associates expressed confidence that the suppressive effect of nitrogen mustard was attributable to its effect on granulocytes [1], much doubt was expressed in the literature of subsequent years that the effect of mustard was actually attributable to granulocytopenia [31], and it was thought by some that an effect on platelets probably accounted for this suppressive effect of nitrogen mustard [32]. We again explored this aspect of the suppressive effect of nitrogen mustard several years ago and found that doses of nitrogen mustard that were entirely adequate for complete inhibition of the GSR produced no substantial qualitative effect on platelets [33]. Transfusions of platelets failed to restore susceptibility to the GSR after treatment with nitrogen mustard. Furthermore, we could find no evidence that tests of blood-clotting function in general were altered by such treatment.

In a rather brief study, Wendt et al. found incomplete suppression of the GSR by nitrogen mustard and little correlation between granulocyte levels and susceptibility to the GSR [34]. This report has been cited by Margaretten and McKay as evidence that granulocytes are not involved in the GSR [12]. Most of Wendt's animals that developed the GSR after treatment with nitrogen mustard were treated with endotoxin four days after the mustard, at a time when marrow recovery is often substantial [35] and when release of granulocytes from the recovered marrow could be expected to occur under the stimulus of endotoxin. Wendt's data do not allow for this likelihood. In view of the extensive evidence [1, 7, 33] that treatment with nitrogen mustard suppresses the GSR via an effect on marrow, it seems likely that Wendt's anomalous findings do not justify the conclusions based thereon.

In summary, a number of experiments concerning the effect of nitrogen mustard on suppression or inhibition of the GSR can be readily explained by the assumption that granulocytes are somehow necessary to the pathogenesis of the GSR. However, none of these experiments provided positive or direct evidence of granulocytic involvement.

Induction of "preparation" with granulocytes and products of granulocytes. McKay et al. studied the possibility that leukocytes trigger clotting in the GSR by infusing leukocyte lysates into animals "prepared" in various ways for the GSR [32]. Since no clotting was produced in vivo, and no substantial thromboplastic activity was demonstrated in vitro in the leukocyte lysates, the authors concluded that leukocytes ". . . are not concerned directly in the production of the glomerular capillary thrombi." It is significant that, in these experiments, McKay et al. substituted the leukocyte lysates for the provoking rather than the preparing injection of endotoxin.

Because of evidence implicating granulocytes in the pathogenesis of the GSR, we devised experiments in an attempt to see whether granulocytes or products of granulocytes could be substituted for the first injection of endotoxin [7]. In these experiments we were able to show that, if otherwise normal rabbits were given a transfusion (i.e., an iv injection) of granulocytes simultaneous with one injection of endotoxin, massive intravascular coagulation, glomerular thrombosis, and renal cortical necrosis would be produced. For understanding of these experiments, it is important to bear in mind that the endotoxin was given at the same time as the granulocyte transfusion. When polymorphonuclear leukocytes (PMNs) alone were injected into normal rabbits, no renal lesions were found. However, when PMNs were injected along with an injection of endotoxin, striking lesions of renal cortical necrosis and glomerular thrombosis were produced regularly. This regimen of PMNs, along with a single injection of endotoxin, was also effective in animals made severely granulocytopenic with nitrogen mustard. In other control animals given serum with endotoxin or endotoxin alone, no lesions were found. In these experiments it appeared that the PMN transfusion had, in effect, substituted for the effect of the first injection of endotoxin.

Thus, in the animal in which granulocytosis is produced by transfusion of PMNs, a single injection of endotoxin acts like a second injection of endotoxin in the GSR. These experiments offered the most compelling evidence that granulocytes are involved in the pathogenesis of the GSR. Particularly when considered with the studies with nitrogen mustard [1, 33], these experiments [7] indicated that the granulocytosis produced by endotoxin was a sufficient explanation for its ability to elicit preparation for the GSR.

Forman et al. [8] have shown that a transfusion of granulocytes coupled with one injection of endotoxin leads to a glomerular thrombosis in thorotrast-treated granulocytopenic rabbits, affording additional evidence for the granulocyte requirement in the GSR.

It has been shown that isolated leukocyte granules, when given with a single injection of endotoxin, also produce massive disseminated intravascular coagulation [7]. When the granules were given alone (i.e., without endotoxin) to normal animals, no clots were produced, and the animals tolerated the injection perfectly well. If the granules were disrupted and the debris sedimented at 8,000 g, material with in-vivo clot-promoting activity remained in the supernatant.

The rapidity of the induction of clottting in the endotoxin-treated animal is illustrated by the fact that, when the granules were given iv, the clots occurred predominantly in the lungs, and animals treated in this fashion rapidly became dyspneic and agitated and often died during the course of the iv infusion. In contrast, when the granules were given via the aorta to endotoxin-treated animals, the animals tolerated the injection without developing symptoms of agitation or anxiety, but massive clotting was found in the kidney. These experiments [7] indicate the remarkable rapidity of the effect by which these isolated granules potentiated clotting, since the clots apparently were

induced almost instantly before any substantial recirculation.

To summarize the indirect evidence indicating that the first injection of endotoxin induces a state of preparation for the GSR through its ability to elicit granulocytosis, we should consider the following points. (1) Granulocytosis is a characteristic effect of endotoxin [1, 25]. (2) The phase of granulocytosis is correlated with the phase of enhanced susceptibility to clotting by a second injection of endotoxin [1]. (3) Nitrogen mustard inhibits the GSR, almost certainly by its ability to produce injury to the granulocyte precursors in the marrow, and, thereby, to prevent the usual granulocyte response to endotoxin [1, 7, 33]. (4) Transfusion of granulocytes into the systemic circulation induces immediately a state of preparation for the GSR [7]. This technique of preparation is equally effective in normal and in nitrogen mustard-treated, severely granulocytopenic rabbits [7, 8]. (5) Isolated granulocyte granules from peritoneal-exudate leukocytes or a soluble product from such granules that are disrupted induce clotting when infused immediately after an injection of endotoxin (but not in normal animals) [7]. (6) Other substances that induce granulocytosis also cause a state of preparation for the GSR (e.g., thorotrast [5], pneumococcal infection [36]). (7) Disseminated intravascular clotting and hypofibrinogenemia are significant clinical correlates of certain other diseases in which granulocytosis is a prominent feature (e.g., promyelocytic leukemia [37, 38]).

Possible Specific Roles for Granulocytes in the GSR

The role of granulocytes in the GSR could be accounted for by assuming (1) that they contribute a procoagulant or "thromboplastic" factor, or (2) that they supply a conditioning or predisposing factor that enhances other clot-promoting effects of endotoxin. The possibility that leukocytes might contribute highly charged, anionic macromolecules active in promoting blood coagulation, first suggested by Thomas et al. in 1955 [24], has received additional support from various sources over the years. This hypothesis, which was originally highly speculative, has been rendered more credible by the observation that neutrophil granules do, in fact, contain highly charged macromolecules, both anionic and cationic, and by the emergence of the concept of paracoagulation.

Soluble fibrin-monomer complexes and paracoagulation. Fibrin monomer is produced by the specific action of the proteolytic enzyme, thrombin, on fibrinogen [14]. Under usual circumstances of coagulation, fibrin monomer spontaneously polymerizes into the highly ordered fibrillar protein called fibrin. However, there are other possible fates for fibrin monomer. It may interact with unaltered fibrinogen to form an unstable intermediate, which has been referred to as cryoprofibrin or cryofibrinogen [39]. In addition, after plasmin has acted on fibrinogen or fibrin to produce fibrin-degradation products (FDP) these plasmin-induced FDP may react with fibrin monomer to produce another type of complex [40]. Such complexes of fibrin monomer and possibly other similar complexes have been referred to as soluble fibrin-monomer complexes (SFMC). The formation of such SFMC may be particularly significant in states of low-grade intravascular coagulation, such as that produced by a single injection of bacterial endotoxin [22].

Such SFMC may be coagulated by the addition of any of a variety of highly charged macromolecules, either anionic or cationic [40, 41]. The formation of fibrin or a fibrinlike material in such a reaction has been called paracoagulation. Paracoagulation of SFMC has been produced by protamine [40], a synthetic acid polymer called "Liquoid" [41], platelet factor IV [42], and a lysosomal cationic protein (LCP) derived from PMNs [41]. The susceptibility of SFMC to paracoagulation by a granulocyte product affords a possible specific mechanism for granulocytic involvement in the pathogenesis of the GSR.

Acidic mucosubstance and LCP in neutrophils. In histochemical studies with Samuel Spicer during the early 1960s, we were able to show that the azurophile granules, a normal component of immature granulocytes, are azurophilic (stain with basic azure dyes) because they contain polysaccharide and ester sulfate, not because of ribonucleoprotein, as had been previously supposed [43]. Subsequently, Fedorko and Morse [44] have isolated a chondroitin sulfatelike moiety from PMN granules, confirming the earlier histochemical findings.

In addition to acidic polymers, histochemical techniques also identified the presence of arginine-rich cationic proteins in PMN granules [43]. These LCP have been extensively characterized and studied for antibacterial and other properties [45, 46].

These granules containing acidic mucosubstance (AMS) and LCP are, by histochemical techniques, inconspicuous in circulating granulocytes of normal rabbits, but during an endotoxin-induced granulocytosis, the number of these granules appears to be greatly increased [30]. Thus, these histochemical studies indicate that neutrophil granulocytes do, in fact, contain highly charged macromolecules, and that, after an injection of endotoxin, there is apparently an increased delivery of these AMS- and LCP-containing granules to the peripheral blood. These observations tend to support the prediction by Thomas et al. that granulocytes might contain macromolecular substances similar to the synthetic acid polymers that they had explored in relation to endotoxin-triggered intravascular coagulation [24].

Evidence that such components of PMNs are active in the GSR may be summarized as follows. (*1*) Certain PMN granules contain AMS and LCP [43]. (*2*) These PMN granules are increased in circulating PMNs during endotoxin-induced granulocytosis [30]. (Fragmentation of granulocytes intravascularly after injection of endotoxin affords a potential mechanism for release of these components from granulocytes [47], but we are not aware of a direct demonstration that these PMN components appear free in the blood.) (*3*) Macromolecular substance of the general class of the LCP and AMS of granulocytes cause in-vitro paracoagulation of SFMC [40, 41]. Injection of any of several synthetic acid polymers [24] or of protamine sulfate (R. D. Collins, J. Hawiger, and R. G. Horn, unpublished observations) along with one injection of endotoxin will cause a syndrome of intravascular coagulation. (*4*) LCP derived specifically from rabbit PMN granules causes in-vitro paracoagulation of SFMC [41].

Thus, a variety of experimental observations are consistent with the hypothesis that the first injection of endotoxin induces a state of preparation for the GSR by causing (*1*) granulocytosis, (*2*) release of numerous and relatively immature granulocytes from the marrow, and (*3*) release from

PMNs of LCP and AMS, which are capable of interacting with SFMC or related fibrinogen derivatives, thus producing an insoluble fibrinous or fibrinlike coagulum.

The precise nature of the interaction between the LCP and AMS of PMN origin and the SFMC is still speculative. Presumably, the highly charged polymers might force the SFMC to dissociate, releasing fibrin monomer to polymerize spontaneously into fibrin. Alternatively, some of these highly charged macromolecules might actually complex with, precipitate with, and be incorporated as an element of the coagulum. There is histochemical evidence to indicate that the fibrinoid thrombi of the GSR are composed in part of a sulfated mucosubstance [48]. It is of interest that a fibrillar fibrinlike protein with regular ultrastructural periodicity may be formed by the paracoagulation reaction between protamine and SFMC in the absence of thrombin [49].

Procoagulant factors of PMN origin. The hypothesis implicit in the foregoing remarks assumes that the role of PMNs in the GSR is to produce a conditioning factor that enhances net formation of clots, but that an effect of endotoxin on platelets, factor XII, or the complement system is required to trigger coagulation. In other words, it does not seem necessary to postulate that granulocytes are directly involved in the generation of thrombin, as, for example, by release of a tissue thromboplastin from their cytoplasm. In fact, most of the experiments on the GSR suggest that endotoxin is required for initiation of clotting, and, in general, granulocyte products given without endotoxin have not caused clotting [7]. McKay found little thromboplastic activity in leukocyte lysates [32]. However, Niemetz has recently reported significant levels of procoagulant activity in leukocytes from endotoxin-treated animals and has produced clotting by infusing products of leukocytes from endotoxin-treated rabbits [9, 10]. In these experiments the recipient animals were not endotoxin-treated, but the animals that donated the leukocytes had received endotoxin both iv and ip. Although it is suggested that the procoagulant factor from PMNs was active in a normal (i.e., not endotoxin-treated) animal, it seems likely that endotoxin carried over from the exudate leukocytes to the recipient animal could produce an endotoxin effect in the tested animal. If this were the case, Niemetz would in effect have repeated

our experiments of some years ago [7]. Nevertheless, the experiments of Niemetz indicate the possibility that leukocytes of endotoxin-treated animals may be the source of a significant procoagulant factor, as well as of other factors promoting clot formation.

Concluding Remarks

Figure 1 outlines a hypothetical sequence to account for the role of granulocytes in the pathogenesis of clotting in the GSR.

Two considerations are of particular significance in the evaluation of this hypothesis. First, it is suggested that the primary effects of each of the two injections of endotoxin are the same, and it is useful to consider that the second injection produces clotting because of a change in the host, which has ensued in response to the first injection. This is a subtle but obvious point. Second, it is proposed, on the basis of the evidence cited in this review, that changes in the host that render him "hypercoagulable" 24 hr after a first injection can be reasonably attributed to the effects of that first injection on the host's system of granulocytes.

It is emphasized that all of the effects of endotoxin that bear on the pathogenesis of the GSR are not spelled out in this outline. (In an excellent and recent review, Müller-Berghaus has outlined a lucid theory of pluricausality, which incorporates a number of additional significant elements [26].) For example, effects of endotoxin on vasomotor reactivity, fibrinolysis, and reticuloendothelial function may be of significance in the GSR. However, there is no clear evidence that alteration of these factors is primarily responsible for the increased susceptibility to clotting 24 hr after a first injection of endotoxin.

Thus, it is agreed that there are a multitude of factors, sine qua non, that are essential to endotoxin-induced clotting. However, present evidence indicates that the factors determining that the phase of increased susceptibility to clotting is maximal at about 24 hr after a first injection of endotoxin are probably primarily dependent on a granulocyte response to endotoxin. There is much indirect evidence to indicate that highly charged anionic and/or cationic macromolecules of PMN origin may interact with SFMC or related fibrinogen derivatives to produce the fibrinlike coagulum that characterizes the GSR.

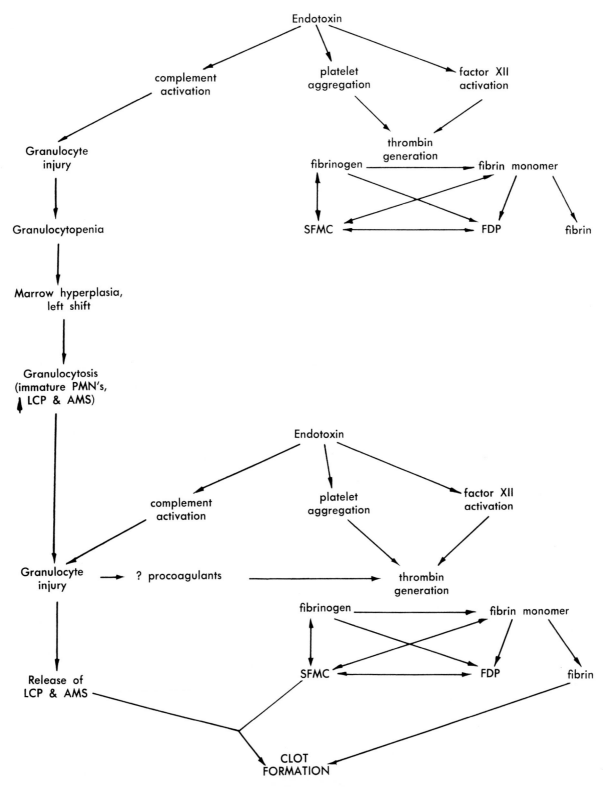

Figure 1. Scheme to indicate the possible role of granulocytes in the pathogenesis of clotting after two sequential injections of endotoxin. AMS = acid mucosubstance; LCP = lysosomal cationic proteins; SFMC = soluble fibrin-monomer complexes; FDP = fibrin-degradation products.

134

Further studies are required for clarification of the nature of that effect of granulocytes on blood clotting and the significance of that effect in clinical syndromes of disseminated intravascular clotting.

References

1. Thomas, L., Good, R. A. Studies on the generalized Shwartzman reaction. I. General observations concerning the phenomenon. J. Exp. Med. 96:605–624, 1952.
2. Robbins, B., Collins, R. D. Studies on the Shwartzman reaction. Production of the renal lesions by intra aortic infusion of thrombin (abstract). Fed. Proc. 20:261, 1961.
3. Rodriguez-Erdmann, F. Intravascular activation of the clotting system with phospholipids. Production of the generalized Shwartzman reaction with platelet factor 3. Blood 26:541–553, 1965.
4. Lee, L. Reticuloendothelial clearance of circulating fibrin in the pathogenesis of the generalized Shwartzman reaction. J. Exp. Med. 115:1065–1082, 1962.
5. Good, R. A., Thomas, L. Studies on the generalized Shwartzman reaction. II. The production of bilateral cortical necrosis of the kidneys by a single injection of bacterial toxin in rabbits previously treated with thorotrast or trypan blue. J. Exp. Med. 96:625–641, 1952.
6. Lee, L. Antigen-antibody reaction in the pathogenesis of bilateral renal cortical necrosis. J. Exp. Med. 117:365–376, 1963.
7. Horn, R. G., Collins, R. D. Studies on the pathogenesis of the generalized Shwartzman reaction. The role of granulocytes. Lab. Invest. 18:101–107, 1968.
8. Forman, E. N., Abildgaard, C. F., Bolger, J. F., Johnson, C. A., Schulman, I. Generalized Shwartzman reaction: role of the granulocyte in intravascular coagulation and renal cortical necrosis. Br. J. Haematol. 16:507–515, 1969.
9. Niemetz, J., Fani, K. Role of leukocytes in blood coagulation and the generalized Shwartzman reaction. Nature (New Biol.) 232:247–248, 1971.
10. Niemetz, J. Coagulant activity of leukocytes. Tissue factor activity. J. Clin. Invest. 51:307–313, 1972.
11. Levin, J., Cluff, L. E. Platelets and the Shwartzman phenomenon. J. Exp. Med. 121:235–246, 1965.
12. Margaretten, W., McKay, D. G. The effect of leukocyte antiserum on the generalized Shwartzman reaction. Am. J. Pathol. 57:299–305, 1969.
13. McKay, D. G. A partial synthesis of the generalized Shwartzman reaction. Fed. Proc. 22:1373–1379, 1963.
14. Seegers, W. H. Use and regulation of the blood clotting mechanisms. In W. H. Seegers [ed.] Blood clotting enzymology. Academic Press, New York, 1967, p. 1–21.
15. Ratnoff, O. D. Some relationships among hemostasis, fibrinolytic phenomena, immunity, and the inflammatory response. Adv. Immunol. 10:145–227, 1969.
16. Des Prez, R. M., Horowitz, H. I., Hook, E. W. Effects of bacterial endotoxin on rabbit platelets. I. Platelet aggregation and release of platelet factors in vitro. J. Exp. Med. 114:857–874, 1961.
17. Mergenhagen, S. E., Snyderman, R., Gewurz, H., Shin, H. S. Significance of complement to the mechanism of action of endotoxin. Curr. Top. Microbiol. Immunol. 50:37–77, 1969.
18. Fong, J. S. C., Good, R. A. Prevention of the localized and generalized Shwartzman reactions by an anticomplementary agent, cobra venom factor. J. Exp. Med. 134:642–655, 1971.
19. Müller-Berghaus, G., Goldfinger, D., Margaretten, W., McKay, D. G. Platelet factor 3 and the generalized Shwartzman reaction. Thromb. Diath. Haemorrh. 18:726–735, 1967.
20. Müller-Berghaus, G., Schneberger, R. Hageman factor activation in the generalized Shwartzman reaction induced by endotoxin. Br. J. Haematol. 21:513–527, 1971.
21. Still, W. J. S., Scott, G. B. D. An electron microscopic study of the endothelial changes and the nature of the "fibrinoid" produced in the generalized Shwartzman reaction. Exp. Molec. Pathol. 5:118–124, 1966.
22. Lipinski, B., Jeljaszewicz, J. A hypothesis for the pathogenesis of the generalized Shwartzman reaction. J. Infect. Dis. 120:160–168, 1969.
23. Lee, L. Mechanisms involved in the production of the generalized Shwartzman reaction. In M. Landy and W. Braun [ed.] Bacterial endotoxins. Rutgers University Press, New Brunswick, N.J., 1964, p. 648–657.
24. Thomas, L., Brunson, J., Smith, R. T. Studies on the generalized Shwartzman reaction. VI. Production of the reaction by the synergistic action of endotoxin with three synthetic acid polymers. J. Exp. Med. 102:249–261, 1955.
25. Bennett, I. L., Jr., Cluff, L. E. Bacterial pyrogens. Pharmacol. Rev. 9:427–475, 1957.
26. Müller-Berghaus, G. Pathophysiology of disseminated intravascular coagulation. Thromb. Diath. Haemorrh. 36(Suppl.):45–61, 1969.
27. Good, R. A., Thomas, L. Studies on the generalized Shwartzman reaction. IV. Prevention of the local and generalized Shwartzman reactions with heparin. J. Exp. Med. 97:871–888, 1953.
28. Benacerraf, B., Sebestyen, M. M. Effect of bacterial endotoxins on the reticuloendothelial system. Fed. Proc. 16:860–867, 1957.
29. Koenig, M. G., Heyssel, R. M., Melly, M. A., Rogers, D. E. The dynamics of reticuloendothelial blockade. J. Exp. Med. 122:117–142, 1965.
30. Horn, R. G., Spicer, S. S. Sulfated mucopolysaccharide and basic protein in certain granules of circulating heterophiles of rabbits during endo-

toxin-induced leukocytes. Am. J. Pathol. 44:905–919, 1964.

31. Lee, L., Stetson, C. A., Jr. The local and generalized Shwartzman phenomena. *In* B. W. Zweifach, L. Grant, and R. T. McCluskey [ed.] The inflammatory process. Academic Press, New York, 1965, p. 791–817.

32. McKay, D. G., Margaretten, W., Phillips, L. L. The role of the leukocyte in the generalized Shwartzman reaction. Lab. Invest. 16:511–515, 1967.

33. Smith, S. P., Nuckolls, J. W., Horn, R. G., Collins, R. D. Pathogenesis of the generalized Shwartzman reaction. Evaluation of platelets as a mediator of the suppressive effect of a nitrogen mustard. Arch. Pathol. 85:459–462, 1968.

34. Wendt, F., Kappler, C., Burckhardt, K., Bohle, A. Production of generalized Shwartzman reaction in rabbits with nitrogen mustard-induced granulocytopenia. Proc. Soc. Exp. Biol. Med. 125:486–488, 1967.

35. Herion, J. C., Walker, R. I., Herring, W. B., Palmer, J. G. Effects of endotoxin and nitrogen mustard on leukocyte kinetics. Blood 25:522–540, 1965.

36. Collins, R. D., Goolsby, J. P., Jr., Horn, R. G. Studies on preparation for the generalized Shwartzman reaction (GSR) by bacterial septicemia (abstract). Am. J. Pathol. 62:69a, 1971.

37. Rosenthal, R. L. Acute promyelocytic leukemia associated with hypofibrinogenemia. Blood 21:495–508, 1963.

38. Baker, W. G., Bang, N. U., Nachman, R. L., Raafat, F., Horowitz, H. I. Hypofibrinogenemic hemorrage in acute myelogenous leukemia treated with heparin. With autopsy findings of widespread intravascular clotting. Ann. Intern. Med. 61:116–123, 1964.

39. Shainoff, J. R., Page, I. H. Significance of cryoprofibrin in fibrinogen-fibrin conversion. J. Exp. Med. 116:687–707, 1962.

40. Lipinski, B., Wegrzynowicz, Z., Budzynski, A. Z., Kopec, M., Latallo, Z. S., Kowalski, E. Soluble unclottable complexes formed in the presence of fibrinogen degradation products (FDP) during the fibrinogen-fibrin conversion and their potential significance in pathology. Thromb. Diath. Haemorrh. 17:65–77, 1967.

41. Hawiger, J., Collins, R. D., Horn, R. G. Precipitation of soluble fibrin monomer complexes by lysosomal protein fraction of polymorphonuclear leukocytes. Proc. Soc. Exp. Biol. Med. 131:349–353, 1969.

42. Niewiarowski, S., Poplawski, A., Lipinski, B., Farbiszewski, R. The release of platelet clotting factors during aggregation and platelet viscous metamorphosis. Exp. Biol. Med. 3:121–128, 1968.

43. Horn, R. G., Spicer, S. S. Sulfated mucopolysaccharide and basic protein in certain granules of rabbit leukocytes. Lab. Invest. 13:1–15, 1964.

44. Fedorko, M. E., Morse, S. I. Isolation, characterization, and distribution of acid mucopolysaccharides in rabbit leucocytes. J. Exp. Med. 121:39–48, 1965.

45. Zeya, H. I., Spitznagel, J. K. Antibacterial and enzymic basic proteins from leukocyte lysosomes: separation and identification. Science 142:1085–1087, 1963.

46. Zeya, H. I., Spitznagel, J. K. Cationic protein-bearing granules of polymorphonuclear leukocytes: separation from enzyme-rich granules. Science 163: 1069–1071, 1969.

47. Horn, R. G., Collins, R. D. Fragmentation of granulocytes in pulmonary capillaries during development of the generalized Shwartzman reaction. Lab. Invest. 19:451–459, 1968.

48. Horn, R. G., Spicer, S. S. Sulfated mucopolysaccharide in fibrinoid glomerular occlusions of the generalized Shwartzman reaction. Am. J. Pathol. 46: 197–213, 1965.

49. Horn, R. G., Hawiger, J., Collins, R. D. Electron microscopy of fibrin-like precipitate formed during the paracoagulation reaction between soluble fibrin monomer complexes and protamine sulphate. Br. J. Haematol. 17:463–466, 1969.

Biochemical Mechanisms of Generation of Bradykinin by Endotoxin

Russell L. Miller, Michael J. Reichgott, and
Kenneth L. Melmon

*From the Division of Clinical Pharmacology,
Departments of Medicine and Pharmacology, University
of California San Francisco Medical Center, San
Francisco, California*

The vasodilatory properties of bradykinin are compatible with a pathogenetic role in man and subhuman primates during endotoxemia. The peptide is generated during endotoxemia when peripheral vascular resistance is decreased. A number of mechanisms of kinin generation can be recruited by endotoxin. In plasma containing complement and 19S antibody to endotoxin, the polysaccharide moiety of endotoxin activates plasma kallikreins. Similar quantities of bradykinin are generated in all species tested. Therefore, species-related variation in production of the peptide in vivo depends on the interaction of endotoxin with other tissues. Granulocytes of primates (as opposed to those of rabbits) contain a cytoplasmic kallikrein or kallikrein activator. Granulocytes release their cytoplasmic enzymes (and presumably kallikreins) when phagocytizing endotoxin. In-vitro phagocytosis requires complement and lipid-rich endotoxin. Polysaccharide fractions of endotoxin do not produce effects on granulocytes, their kinin generation, or the cardiovascular system. We suggest that differences in endotoxin effect among species are largely related to effects of the lipid moiety of endotoxin and the chemical machinery of the cells this moiety penetrates.

Endotoxemia may result from bacteremia, from entry of endotoxin into the circulation from a localized infection, or from injection of endotoxin into experimental animals or volunteer human subjects. Apparently, there are no significant differences between cardiovascular and metabolic effects of viable gram-negative organisms and those of gram-negative bacterial endotoxin [1–4]. Therefore, endotoxemia in animals has become a commonly used model for the study of septic shock. Since the cardiovascular effects of endotoxin vary among species [5], the subhuman primate is the model of endotoxemia most applicable to man. In the monkey as well as in humans, the severe hypotension of the early phase of septic shock is often associated with reduced peripheral vascular resistance [6–11].[1] In man the early phase is characterized by warm skin, bounding "hyperdynamic" pulses [11], and lowered peripheral resistance acompanied by normal or elevated cardiac output [9–11]; these findings suggest that peripheral vascular beds are highly if not effectively perfused. A second, species-specific response of the primate (both man and monkey) to endotoxin is the abnormally high generation of bradykinin [12–15][2] that may account for some of these cardiovascular effects.

After injection of endotoxin, kinin appears in the circulation during the period of vasodilatation [6, 12];[3] peak concentrations of peptide in plasma are reached when peripheral resistance is lowest. The polypeptide alone, however, has not even been proven to be necessary for such phenomena [16] and is not sufficient to account for all of the cardiovascular phenomena of endotoxemia [17].

This project was supported by research grant no. HL-09964 and training grant no. GM-01791 from the National Institutes of Health.

Please address requests for reprints to Dr. Kenneth L. Melmon, Division of Clinical Pharmacology, Moffitt Hospital, Room 1089, University of California San Francisco Medical Center, San Francisco, California 94122.

[1] M. J. Reichgott, R. P. Forsyth, D. K. Greineder, and K. L. Melmon, "Cardiovascular and Metabolic Effects of Whole or Fractionated Gram-Negative Bacterial Endotoxin in the Unanesthetized Rhesus Monkey," manuscript in preparation.
[2] See footnote 1.
[3] See footnote 1.

The peptide itself is probably not responsible for the lethality of endotoxin [18].

This presentation will summarize the known mechanisms of kinin generation and examine the influences of endotoxin on them. We will present considerations on the pathophysiologic importance of the endotoxin-bradykinin relationship to the cardiovascular abnormalities of endotoxemia.

General Mechanisms of Production of Kinin in Blood

Kinins are a group of polypeptides with similar pharmacologic activities. Three kinins—bradykinin (a nonapeptide), kallidin (a decapeptide), and methionyl-lysyl-bradykinin (an undecapeptide)—occur naturally in man and have been extensively characterized. Bradykinin may be considered the prototype of these three. It is normally present in concentrations below 3 ng/ml of blood [19], but in slightly greater concentrations it causes noticeable vasodilatation and increases vascular permeability.

Several plasma and tissue enzymes (kallikreins) function as specific proteases of plasma α_2 globulins (kininogens) from which they release kinins. There is an abundance of kininogen in plasma (and perhaps extracellular fluid); potentially 4–11 mg of bradykinin can be derived from 1 liter of human plasma [20]. Kallikreins derived from plasma and tissues differ chemically and have different affinities for the various molecular species of plasma kininogen [21, 22]. Kallikreins are most abundant in glandular tissues, plasma, granulocytes, and urine. Other plasma enzymes also contribute nonspecifically to production of kinin. Most often they activate kallikreins, which then release kinin. Occasionally, nonspecific proteases (e.g., trypsin) produce kinin directly from kininogen.

Hageman factor and plasmin. Kallikrein circulates in plasma as an inactive precursor, prekallikrein. Prekallikrein must be enzymatically activated to function. Hageman factor (clotting factor XII), which must itself be activated, usually by contact with charged surfaces, can enzymatically convert prekallikrein to kallikrein [22–26]. Factor XII can be activated by damaged vascular endothelium, blood cells, collagen, or tissues, and by colloid suspensions, crystals, salts of fatty acids, and a large number of inorganic and organic materials, including endotoxin [27–31]. The exact mechanism by which these substances activate factor XII is unknown, but the activators generally are negatively charged at *p*H 7.4 and are at least 50 A in diameter [25–26]. Relatively small quantities of kinin are formed when prekallikrein is converted to kallikrein by active Hageman factor. Apparently much more kinin is generated if plasmin is first activated [32, 33].

Plasmin specifically converts fibrinogen to fibrin but nonspecifically affects several other plasma proteins, including components of complement, several procoagulants, γ-globulin, and even the hormones glucagon and adrenocorticotropic hormone [34–37]. Plasmin ordinarily is in an inactive form, plasminogen, which is converted to plasmin by activators found in most tissues and body fluids; these activators are also usually in an inactive form, plasminogen proactivators [38]. Plasminogen proactivators are activated by a variety of substances and maneuvers, including exposure to plasmin itself [39].

As with Hageman factor, large amounts of plasmin are necessary for even a slow release of small amounts of kinin from kininogen [40–42]. However, when Hageman factor and plasmin interact, impressive generation of kinin results [32, 33, 43, 44]. Activated factor XII activates a plasminogen proactivator [45, 46]. Plasmin then releases several peptide fragments from factor XII. Some fragments of Hageman factor, which retain a specific active site, have six times the prekallikrein-activating potency of the parent factor XII molecule. Other fragments may lack this site but retain clot-promoting activity [32, 33].

Complement. Activation of complement and formation of kinin are also linked. Kallikrein is a relatively weak activator of C1, the first component of complement [47], while plasmin is a potent activator of C1 [48]. Hageman factor, by activating plasmin, also contributes to activation of complement. Activation of C1 by enzymes or by interaction with immune complexes or sensitized cells leads to the formation of C1 esterase (C$\overline{1}$). An α-globulin in plasma specifically inhibits C$\overline{1}$ [49, 50] and nonspecifically inhibits activated kallikrein and plasmin [51, 52]. A state of deficiency in C$\overline{1}$-esterase inhibitor has been described in hereditary angioedema [52, 53], and the activity of the disease may be associated with

138

elevated blood levels of bradykinin and depletion of kininogen [19, 54]. Epsilon-aminocaproic acid has been used to diminish the number and severity of attacks of the disease in some patients [55–57], perhaps by inhibition of plasmin and, therefore, formation of kinin.

Activation of the complement cascade also leads to generation and release of substances that can promote chemotaxis, immune adherence, phagocytosis, and cytolysis. These factors stimulate migration and activity of polymorphonuclear cells, which may contribute to kinin generation by several pathways (see below).

Antigen-antibody reactions and immune complexes. Immune complexes can activate kallikrein in plasma of various animal species [58, 59]. Other studies have suggested that, in many species, kinin is produced during anaphylaxis [60–62]. It is possible that immune complexes can fix complement and activate factor XII, perhaps as a result of the surface configuration of the immune complex or of the complement-related catalytic enzymes [63]. In addition, phagocytic leukocytes attracted by immune complexes can release cytoplasmic and lysosomal substances relevant to formation of kinin.

Leukocytes. Leukocytes influence both generation and destruction of kinins. They contain kininogenase, or kininogenase-activator activity, and kininase enzymes [64–66]. These substances are found primarily in neutrophils and eosinophils. Kininogenase activity or kininogenase-activating factors are in both the cell sap and the granular fractions of the polymorphonuclear cells. The cytoplasmic substances act rapidly, generate kinin from both whole plasma and partially purified kininogen, and are most active at pH 7.4–7.6. The granular fraction contains optimal kinin-forming enzyme activity at pH 6 or less. This latter activity probably is a nonspecific effect of lysosomal cathepsins previously described by Greenbaum et al. in rabbit leukocytes [67].

Hageman factor is required for generation of kinin by the leukocytic cytoplasmic factors. Both Hageman-deficient plasma and kininogen prepared from Hageman-deficient plasma are unable to support generation of kinin by whole granulocytes or by the cytoplasmic fraction of sonicated or homogenized cells.

Kininase activity has the same intracellular distribution as kininogenase activity. Both cyto-

plasmic and granule-fraction activity are found at pH 7.0. As the pH is lowered, cytoplasmic activity rapidly diminishes, and below pH 5.5, granular kininases are most active. Similar kininase activity was described by Zachariae et al. [68].

Greenbaum and associates have studied extensively kininogenase and kininase activities in rabbit peritoneal leukocytes and macrophages [67, 69–71]. Cells harvested after instillation of glycogen or mineral oil into the peritoneal cavities of rabbits [72] contained kallikrein-like enzymes, primarily in the lysosomes, active at acid pH. These enzymes released kinin-like peptides from human and bovine preparations of kininogen but acted much more slowly than the enzymes in human leukocytes. One of the released peptides, polymorphonuclear leukocyte (PMN)-kinin [67, 70, 71], has chemical and pharmacologic characteristics clearly distinguishing it from bradykinin.

Other workers have demonstrated the presence of nonlysosomal proteases in human granulocytes. Janoff and Zeligs [73] demonstrated that one such enzyme, released by phagocytosis and active at neutral pH, was capable of causing injury to epithelial cells, disruption of capillary basement membrane, and hemorrhagic edema. Movat et al. [74] described an enzyme, released during phagocytosis of antigen-antibody complexes by leukocytes, that injured capillaries and caused edema. Activity of this enzyme was also poorly released during incubation of leukocytes in the absence of particles for phagocytosis. Proteases, with neutral pH optima, have been found sequestered in a distinct class of granules in human PMNs [75]. These granules were separated by sucrose-density-gradient centrifugation and also contained β-glucuronidase and β-glycerol phosphatase, enzymes commonly associated with lysosomes. No proteases with pH optima below 6.0 were demonstrated in this study, and the authors suggest that acid-active enzymes may be absent from human PMNs. In addition, human and canine leukocytes have been shown to release plasminogen-activating enzymes when placed on plasminogen-rich fibrin plates [76]. This is noteworthy since the interaction of plasmin with Hageman factor (see above) represents the most important mechanism leading to generation of kinin in plasma. Other studies [77] indicate that enzymes found in granulocytes and capable of activating fibrinolytic mechanisms are inactivated at temperatures above 56 C and are inhibited by soybean-trypsin inhibitor, the same inhibitor that blocks plasma kallikrein. The cytoplasmic protease discovered by Janoff and Zeligs [73] was similarly heat-sensitive and inhibited by soybean-trypsin inhibitor. These studies may actually have been dealing with either a single enzyme or a small group of neutral proteases, each capable of generating kinin. At the least, the results indicate that the leukocytes of humans and other species may directly or indirectly contribute to formation of kinin.

In the granular fraction of leukocytes, acid-active enzymes are a clearly defined second group of kininogenases, identifiable with the acid proteases [71], and are presumably nonspecific. The acid proteases of rabbit cells act so slowly as kininogenases that their contribution to generation of kinin is unlikely to be of pathophysiologic importance [78].

Other kinin-generating enzymes. Tissue kallikreins derived from many other sources also participate in generation of kinin. Kallikreins are found in the pancreas, salivary glands, sweat glands, gut, kidney, carcinoid-tumor tissue, and spleen [79–81]. Such kallikreins are chemically distinct from kallikreins in plasma and from one another; they vary in physical characteristics, proteolytic specificities, immunologic properties, and patterns of inhibition by natural substances [79, 82, 83]. They may even be organ-specific [84, 85]. Pancreatic and salivary-gland kallikreins are stored in the granular fractions of cells [86] and have been found in association with other enzymes, such as renin [87]. Various stimuli are known to release stored kallikreins from granules in the pancreas [88] and from other organs [89, 90].

Summary of mechanisms of kinin production and limitations on kinin activity. Production of kinin follows activation of several enzymes in plasma and tissue (figure 1). Each activated enzyme seems to be intimately related to one or more other enzymes, resulting in a variety of cycles or cascades that can intensify further formation of kinin. For example, factor XII can activate small amounts of prekallikrein as well as plasmin. Plasmin can directly release kinin from kininogen but can also degrade factor XII into fragments that activate large amounts of prekallikrein. Plasmin also activates complement, which may facilitate production of kinin by a number of mechanisms, including direct production of kinin and formation of chemotactic substances leading

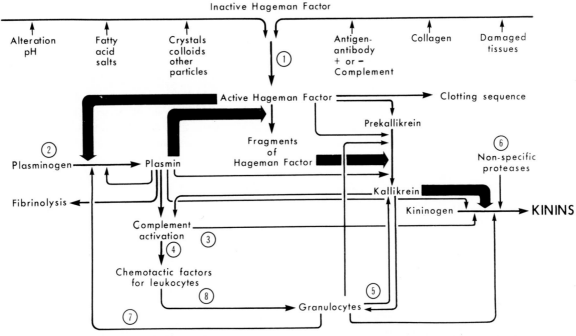

Figure 1. Possible mechanisms of kinin generation. Bold arrows represent the most potent pathways of generation. Circled numbers represent some of the steps at which endotoxin might initiate generation. (See text for explanation.)

to migration of leukocytes into the area. In addition, kallikrein itself has selective chemotactic activity for PMNs [91]. It is likely that when concentrations of kinins in blood are high, a cascade of events has contributed to the generation of kinin.

The ease of activating kinin-forming enzymes, the amount of potential kinin available, and the potency of the peptide make it necessary to have controls to prevent its accumulation in the circulation. Over-production of kinin in the body is partially prevented by inhibitors (e.g., $C\bar{1}$ inhibitor) that inactivate one or more of the components necessary for the activity of kallikrein.

Almost any modification of the kinin molecule by kininases causes it to lose its physiologic activity. Rapid metabolism limits accumulation of kinins. The half-life of kinins in the circulation is measured in seconds, due to kininases present in plasma, formed elements of the blood, and other tissues [36, 68, 78].

Possible Mechanisms by which Endotoxin Promotes Production of Kinin

Components of plasma. When endotoxin is added to normal plasma under appropriate conditions, bradykinin is released. There are several pathways by which endotoxin could initiate production of kinin: (*1*) direct formation by the action of endotoxin; (*2*) activation of factor XII; (*3*) direct or indirect activation of plasmin; and (*4*) interaction of endotoxin with specific immunoglobulins, with or without complement.

Direct formation of kinin. A few species of bacteria contain or elaborate kallikrein-like enzymes that apparently can release kinins from kininogens without the presence of kallikreins from plasma or tissue [79]. Endotoxin does not have kallikrein-like activity and requires the presence of endogenous plasma components for formation of kinin.

Activation of factor XII. Intravascular coagulation occurs during endotoxemia. Studies of endotoxin-induced coagulation have revealed that it activates factor XII in vitro (see figure 1, step 1) [92]. Another study found decreases in factor XII in experimental animals shortly after iv administration of endotoxin [31]. The Shwartzman reaction, **characterized** by disseminated intravascular coagulation and irreversible shock, is pro-

duced by injection of endotoxin. It can be reproduced experimentally in the absence of endotoxin by activation of factor XII with ellagic acid, sodium stearate, and other substances [31, 93]. However, some in-vitro experiments have shown that kinin is formed in endotoxin-treated plasma that is deficient in factor XII [94]. Therefore, activation of factor XII may be only one of many mechanisms by which endotoxin promotes production of kinin in plasma.

Plasmin. Bacterial activators, such as streptokinase or staphylokinase, activate plasminogen by their effects on a proactivator [95–97]. Other bacterial products, such as the purified lipopolysaccharide from *Salmonella abortus equi* (Pyrexal), release tissue activators into the circulation and promote disseminated intravascular coagulation (figure 1, step 2) [98]. It seems likely that other endotoxins could have similar indirect effects.

Complement and immune mechanisms. A macroglobulin antibody (directed toward the polysaccharide component of endotoxin) is demonstrable in sera of most animals shortly after birth [99]. Both complement and the 19S macroglobulin antibody [94] are required for generation of kinin in plasma incubated with endotoxin (figure 1, step 3). If complement is absent, kininogen is not depleted; complement-containing plasma exhibits a 42% depletion in kininogen and in production of kinin. Plasma of newborn humans, lacking in 19S antibody to endotoxin, does not support generation of kinin when incubated with endotoxin. Addition of 19S macroglobulin-containing antibody to deficient plasma of newborns allows depletion of kininogen when the plasma is incubated with endotoxin. Although Hageman factor is not required for production of kinin in this system, it may facilitate production by alternate mechanisms.

Complement may have a role in endotoxin-mediated production of kinin in addition to its interaction with immune complexes [100]. Endotoxin can react with complement and generate a substance chemotactic for PMNs (figure 1, step 4) [101]. In addition, a recent report suggests that bacterial endotoxins and other complement-activating substances initiate coagulation of blood through a complement-mediated pathway [102]. Although such pathways have not been fully characterized, they also may involve production of kinin. Finally, complement seems to be required for some endotoxin-leukocyte interactions [103, 104].

Endotoxin-leukocyte interactions. Endotoxin has profound effects on PMNs. In vivo, one of the earliest manifestations of endotoxemia is granulocytopenia [105–107] caused by adherence of the cells to vascular (capillary) epithelial surfaces [108]. Granulocytopenia persists for 1–3 hr and is followed both by return of the sequestered leukocytes to the circulating pool and by an outpouring of newly matured cells from the marrow to the periphery [107, 109].

Endotoxin is rapidly cleared from the circulation. It is initially bound to leukocytes and platelets [110] but is subsequently taken up and degraded by macrophages, primarily in the liver and spleen [111]. In-vitro experiments [112] have shown that endotoxin rapidly binds to rabbit polymorphoneutrophils. The binding is a passive phenomenon, occurs in physiologic salt solution, is dependent on divalent cations, but is independent of metabolic function or availability of glucose to the cell.

The interaction of endotoxin with leukocytes is more extensive than simple binding of the lipopolysaccharide to the cell surface, however. Addition of endotoxin to whole blood and incubation of the mixture in a Warburg flask resulted in increased consumption of O_2 in human or canine cells but not in cells of rabbits, cows, chickens, or rats. The increased respiratory activity was proportional to the dose of endotoxin, occurred about 10 min after addition of endotoxin [113], and could be inhibited by NaF or iodoacetate. Rough or excessive handling of white cells eliminated this respiratory response to endotoxin. Other investigators have demonstrated increased utilization of glucose and production of lactic acid and slightly decreased synthesis of glycogen in cells exposed to endotoxin [114]. Capability for phagocytosis increased in cells exposed to endotoxin within 20–30 min of incubation. The metabolic and phagocytic effects were obtained in physiologic saline solution and did not require serum. Nonetheless, the cells, although stimulated to phagocytize and kill microorganisms more effectively in the absence of serum, could do so only when serum opsonins were present.

In a study of the relationship between endotoxin and granulocytes, Cline et al. have also demonstrated augmented metabolic processes that are closely associated with phagocytosis [103]. Endotoxin stimulated incorporation of ^{3}H-uridine into RNA, incorporation of ^{14}C-amino acids into protein, and production of lactic acid and ${}^{14}CO_2$ from $1\text{-}{}^{14}$C-glucose. Lysosomal enzymes were released into the suspending medium, and cellular content of α-glucosidase (a lysosomal enzyme) and phosphorylase (a cytoplasmic enzyme) correspondingly decreased (figure 1, step 5). This system required complement for maximal activity, and no effects were seen when cells and endotoxin were incubated in a serumfree system. The same study examined the uptake of ^{14}C-labeled endotoxin by leukocytes. Only in complement-containing systems was there significant uptake of label. Autoradiography and electron microscopy confirmed that endotoxin was actually taken up into the granulocyte and not merely bound to the surface. Others have confirmed the central role of phagocytosis in endotoxin-leukocyte interaction when the system contains complement and opsonin [111, 115, 116]. In the absence of these serum factors, endotoxin binds only to the cell surface and is not phagocytized but still activates cellular metabolizing systems [112, 114, 116]. This has given rise to the term "sham phagocytosis" [112].

Phagocytosis of endotoxin is closely associated with the release of leukocytic lysosomal and cytoplasmic enzymes [103, 104, 117–119]. Endotoxin also releases lysosomal enzymes from other tissues (figure 1, step 6) [88–90]. The relationship of lysosomal acid proteases to generation of kinin has already been discussed. A recent paper [117] stated that phagocytosis of inert particles does not result in loss of cytoplasmic enzymes from cells. The direct or indirect effects (e.g., activation of complement) of endotoxin, however, might be expected to damage the cell extensively, allowing release of cytoplasmic proteases that could contribute to generation of kinin at neutral *p*H. A similar mechanism could be operating in other tissues.

The lipopolysaccharide portion of endotoxin stimulates fibrinolytic activity of leukocytes [76] by releasing plasminogen activators and lysosomal enzymes (figure 1, step 7). It is not known whether phagocytosis is required for these effects. In addition, endotoxin interacts with leukocytes to activate mechanisms of coagulation. The lipopolysaccharide can stimulate formation of leukocytic thromboplastic substances that directly activate factors VII and X, bypassing Hageman factor [120]. Endotoxin can also activate the intrinsic clotting system by releasing enzymes from leukocytes [121, 122].

In summary, interactions of endotoxin with leukocytes provide multiple possible mechanisms of kinin generation, including phagocytosis with release of lysosomal and cytoplasmic enzymes, initiation of coagulation via Hageman factor, generation of thromboplastin, and activation of plasminogen.

Species-related variability in endotoxin-induced generation of kinin. There seem to be multiple mechanisms in all species for endotoxin-induced generation of bradykinin. Kininogens and kininogenases are also universal. Nonetheless, no kinin has been detected in the plasma of cats receiving endotoxin (M. J. Reichgott, unpublished observations), and only low levels have been found in the dog [123, 124]. Even in the rabbit, in which generation of kinin by incubation of endotoxin with plasma is well documented [125–127], the concentration of peptide is not elevated in vivo, and in none of these animals are the cardiovascular abnormalities of endotoxemia consistent with elaboration of bradykinin or other vasodilators [5].

In an attempt to define the pathophysiologic significance of the kinin-endotoxin interrelationship in primates, we have examined the effects of chemically characterized fractions of endotoxin in vitro and in vivo in the monkey [104][4] (table 1). Our results allow us to examine the importance of each of these interactions to the development of the cardiovascular phenomena of endotoxemia. Initial studies compared the effects of crude endotoxin (LPS), the 165,000-*g* pellet (165P) obtained from a suspension of LPS, and the polysaccharide (PS) obtained by mild alkaline hydrolysis of 165P. In vitro, kininogen was depleted when monkey serum was incubated with LPS, 165P, and PS (table 2). Depletion was greatest with PS alone. Rabbit plasma also responded to incubation with PS by generation of kinin. Both primate [104] and rabbit models required complement and macroglobulin antibody.

Interaction of leukocytes from monkeys and humans with endotoxin fractions resembled phagocytosis and released lysosomal α-glucosidase into the suspending medium (table 3) [104]. It was demonstrable only with LPS and 165P, both of which have high content of fatty acids. PS alone did not affect the metabolism of the cell

[4] See footnote 1.

Table 1. Composition of lipopolysaccharide (endotoxin) preparations.

Preparation	Fraction	μmoles/mg of lyophilized substance*								
		Glc[†]	Gal	Rham	GlyManHep	KDO	HexNH$_2$	FA	Bound P$_I$[‡]	Rib[§]
A	165P	0.48	0.51	0.40	0.18	0.21	0.53		0.42	0.11
B	165P	0.52	0.45	0.41	0.20	0.19	0.62	0.50	0.39	0.11
C	165P	0.42	0.47	0.45	0.23	0.23	0.54	0.58	0.38	0.09
D	165P	0.47	0.55	0.48	0.22	0.23	0.54	0.54	0.41	0.10
A	165S	0.98	0.70	1.30	0.03	0.05	0.63	0.13	0.75	0.36
B	165S	1.02	0.76	1.24	0.03	0.07	0.76	0.12	0.72	0.38
C	165S	0.86	0.68	1.27	0.05	0.09	0.57	0.14	0.78	0.32
D	165S	0.82	0.65	1.33	0.04	0.06	0.54	0.11	0.70	0.42
B	PS	0.45			0.24	0.21	0.28	0.01		
C	PS	0.33			0.23	0.22	0.28	0.02		
D	PS	0.38			0.26	0.23	0.25	0.01		

NOTE. This table is reprinted from [104] with permission of Microforms International Marketing Corporation.

* Note similarity of chemical composition among different batches of lipopolysaccharide.

† Glc = glucose; Gal = galactose; Rham = rhamnose; GlyManHep = glyceromannoheptose; KDO = 2-keto-3-deoxyoctulosonic acid; HexNH$_2$ = hexosamine; FA = fatty acid; P$_I$ = inorganic phosphate; Rib = ribose; 165P = 165,000-*g* pellet obtained from lipopolysaccharide suspension; 165S = 165,000-*g* supernatant obtained from lipopolysaccharide suspension; PS = polysaccharide.

‡ Bound P$_I$ = total P$_I$ − free P$_I$.

§ Ribose values represent maximal values, because some increase in absorbance is obtained from nonribose carbohydrate.

Table 2. Effect of various fractions of endotoxin on kininogen in monkey plasma.

Fraction	No.	Kininogen concentration* 5 min	10 min
Saline	13	99 ± 2	96 ± 2
Crude LPS†	6	85 ± 4‡	75 ± 5‡
165P	9	92 ± 4	86 ± 3‡
PS§	13	78 ± 5‖	63 ± 6‖

NOTE. This table is reprinted from [104] with permission of Microforms International Marketing Corporation.

* Expressed as a percentage of the zero-time values (± SE).

† LPS = lipopolysaccharide.

‡ Significantly different from saline control (P < .05).

§ PS = polysaccharide.

‖ Significantly different from saline control (P < .01).

or become phagocytized; presumably, the leukocyte required at least some lipid or a relatively intact endotoxin molecule for the interaction.

Recently, we have compared the effects of endotoxin fractions in the unanesthetized rhesus monkey.[5] Whole LPS and two PS fractions (PS$_1$ with 6.3% of the original content of lipid and PS$_2$ with 0.5% of the total of lipid in LPS) [128] were infused into monkeys. Generation of kinin and the early vasodilatation characteristic of endotoxemia occurred after whole LPS and PS$_1$ (table 4). We detected neither kinin in blood nor any cardiovascular change after infusion of PS$_2$. It is necessary to recall, however, that PS$_2$ is able to support generation of kinin by plasma and endotoxin in vitro. All three fractions caused granulocytopenia, but that caused by whole LPS or PS$_1$ lasted for more than 2 hr and closely mimicked that seen in endotoxemia. The PS$_2$ lipidfree frac-

Table 3. Effect of various fractions of endotoxin on human granulocytes.

Fraction (no.)	Increment in lysozyme activity (μg/ml ±SE)*
Crude (LPS) (5)	30 ± 5†
Pellet (165P) (3)	53 ± 17
Polysaccharide (PS) (3)	3 ± 2

NOTE. This table is adapted from [104] with permission of Microforms International Marketing Corporation.

* Increment of lysosome in the suspending medium after incubation of granulocytes with endotoxin fractions for 60 min.

† Significantly different from controls (P < .01).

[5] See footnote 1.

Table 4. Effect of fractions of endotoxin on selected measurements in the monkey.

Fraction* and dose (mg/kg)	Hr after infusion 2 PVR (%)	Kinin (mg/ml)	WBC	6 PVR (%)
LPS-10 (5)	− 7	32	−61	+113
LPS-2.5 (1.25)	−22	52	−53	+ 58
PS$_1$† (0.075)	−22	42	−57	0
PS$_2$† (0.006)	+35	0	+79	+ 15

* Fraction indicates: LPS-10 = lipopolysaccharide, 10 mg/kg; LPS-2.5 = lipopolysaccharide, 2.5 mg/kg; PS$_1$ and PS$_2$ = polysaccharide with different lipid content. Approximate molecular weight of polysaccharide, <100,000. PVR = peripheral arterial vascular resistance. WBC = percentage change from control of total leukocyte count in peripheral blood. (See text for details.)

† PS$_1$ and PS$_2$ were administered at 2.5 mg/kg, which corresponds to PS obtained from about 10 mg of LPS.

tion was associated with transient granulocytopenia that lasted for less than 1 hr.

We have interpreted these data as follows. Both plasma and leukocytes can interact with endotoxin to generate kinin. Both primates and rabbits can generate kinin by interaction of plasma with endotoxin in vitro. In vivo, however, only the primate generates amounts of kinin sufficient to be detected in peripheral venous blood, and then only when exposed to lipid-containing preparations of endotoxin. Leukocytes generate kinin only when exposed to lipid-containing endotoxin, and leukocytes of primates and rabbits differ in their ability to generate kinin spontaneously. Therefore, it seems likely that the interaction between leukocytes and endotoxin, which results in generation of large quantities of kinin only in the primate, is more important in the production of detectable levels of kinin in plasma and pharmacologic effects than the plasma-endotoxin interaction common to both species. The understanding of species-related variability in generation of kinin and cardiovascular responses to endotoxin apparently requires evaluation of species-dependent cellular and plasma-enzyme capabilities. Species-dependent differences in enzyme content and kinin-generating capacity of granulocytes have already been described. These differences may determine which species generate amounts of kinin large enough to produce in-vivo effects. Thus, the primate, with a rapidly acting, relatively specific neutral leukocyte cytoplasmic kininoge-

144

nase, produces large amounts of kinin, whereas the rabbit, which lacks this enzyme, does not. On this basis, we predict that the leukocytes of neither the dog nor the cat contain this specific neutral protease.

Discussion

The clinical significance of endotoxin-induced generation of kinin has not been defined. Despite the fact that generation of kinin and the cardiovascular events of endotoxemia have yet to be dissociated, the significance of the association, even with respect to the cardiovascular effects of an episode of septic shock, is unclear. Our recent studies of fractions of endotoxin in primates indicated that, in the early phases of endotoxemia, profound generation of kinin in subhuman primates was dependent on the presence of some lipid in the LPS preparation. Late stages of toxicity were clearly related to the amount of lipid on the molecule of endotoxin (table 4) but seemed to be independent of the magnitude of kinin generation.

We feel that our studies with fractions of endotoxin provide an approach for understanding some differences among species in generation of kinin. Although we can determine species-related differences in endotoxin-induced kinin-generating capacity of leukocytes, we also must ask (1) whether there are species-related differences in tissues other than leukocytes to account for variations in the endotoxin-induced generation of kinin, and (2) whether species-related variation in the cellular contents of tissues is predominantly responsible for variability in noncardiovascular manifestations of endotoxemia. We should extend our understanding to endotoxin-induced phenomena in various tissues that do not generate kinin. There is at least one additional fundamental question, the answer to which is now elusive: Is there cause-and-effect relationship between early cardiovascular instability produced by endotoxin and ultimate mortality from endotoxemia? It could be that attention to the pathogenesis of the spectacular cardiovascular instability will reap its greatest reward not in cure of the disease or in alteration of mortality but in serving as a model to allow understanding of the critical interrelationship of endotoxin with cells and tissues in general.

References

1. Thomas, C. S., Jr., Melly, M. A., Koenig, M. G., Brockman, S. K. The hemodynamic effects of viable gram-negative organisms. Surg. Gynecol. Obstet. 128:753–760, 1969.
2. Buckberg, G., Cohn, J., Darling, C. *Escherichia coli* bacteremic shock in conscious baboons. Ann. Surg. 173:122–130, 1971.
3. Guenter, C. A., Fiorica, V., Hinshaw, L. B. Cardiorespiratory and metabolic responses to live *E. coli* and endotoxin in the monkey. J. Appl. Physiol. 26:780–786, 1969.
4. Hinshaw, L. B. A comparison of the responses of canine and primate species to bacteria and bacterial endotoxins (abstract). *In* R. C. Lillehei [ed.] Symposium on shock in high and low flow states. Excerpta Medica 247:245–249, 1972.
5. Kuida, H., Gilbert, R. P., Hinshaw, L. B., Brunson, J. G., Visscher, M. B. Species differences in effect of gram-negative endotoxin on circulation. Am. J. Physiol. 200:1197–1202, 1961.
6. Wyler, F., Forsyth, R. P., Nies, A. S., Neutze, J. M., Melmon, K. L. Endotoxin-induced regional circulatory changes in the unanesthetized monkey. Circ. Res. 24:777–786, 1969.
7. Hinshaw, L. B., Emerson, T. E., Jr., Reins, D. A. Cardiovascular responses of the primate in endotoxin shock. Am. J. Physiol. 210:335–340, 1966.
8. Gilbert, R. P. Endotoxin shock in the primate. Proc. Soc. Exp. Biol. Med. 111:328–331, 1962.
9. Bell, H., Thal, A. The peculiar hemodynamics of septic shock. Postgrad. Med. 48:106–114, 1970.
10. Hermreck, A. S., Thal, A. P. Mechanisms for the high circulatory requirements in sepsis and septic shock. Ann. Surg. 170:677–695, 1969.
11. Siegel, J. H., Greenspan, M., DelGuercio, L. R. M. Abnormal vascular tone, defective oxygen transport and myocardial failure in human septic shock. Ann. Surg. 165:504–517, 1967.
12. Nies, A. S., Forsyth, R. P., Williams, H. E., Melmon, K. L. Contribution of kinins to endotoxin shock in unanesthetized rhesus monkeys. Circ. Res. 22:155–164, 1968.
13. Kimball, H. R., Melmon, K. L., Wolff, S. M. Endotoxin-induced kinin production in man. Proc. Soc. Exp. Biol. Med. 139:1078–1082, 1972.
14. Attar, S. M. A., Tingey, H. B., McLaughlin, J. S., Cowley, R. A. Bradykinin in human shock. Surg. Forum 18:46–47, 1967.
15. Mason, J. W., Kleeberg, U., Dolan, P., Colman, R. W. Plasma kallikrein and Hageman factor in gram-negative bacteremia. Ann. Intern. Med. 73:545–551, 1970.
16. Reichgott, M. J., Melmon, K. L. Does bradykinin play a pathogenetic role in endotoxemia. *In* R. C. Lillehei [ed.] Symposium on shock in high and low flow states. Excerpta Medica 247:59–64, 1972.
17. Reichgott, M. J., Forsyth, R. P., Melmon, K. L. Effects of bradykinin and autonomic nervous sys-

tem inhibition on systemic and regional hemodynamics in the unanesthetized rhesus monkey. Circ. Res. 29:367–374, 1971.

18. Webster, M. E., Clark, W. R. Significance of the callicrein-callidinogen-callidin system in shock. Am. J. Physiol. 197:406–412, 1959.

19. Talamo, R. C., Haber, E., Austen, K. F. A radioimmunoassay for bradykinin in plasma and synovial fluid. J. Lab. Clin. Med. 74:816–827, 1969.

20. Erdös, E. G. Release and inactivation of kinins. Gastroenterology 51:893–900, 1966.

21. Pierce, J. V. Structural features of plasma kinins and kininogens. Fed. Proc. 27:52–57, 1968.

22. Webster, M. E. Human plasma kallikrein, its activation and pathological role. Fed. Proc. 27:84–89, 1968.

23. Ratnoff, O. D., Colopy, J. E. A familial hemorrhagic trait associated with a deficiency of a clot-promoting fraction of plasma. J. Clin. Invest. 34:602–613, 1955.

24. Margolis, J. Activation of plasma by contact with glass. Evidence for a common reaction which releases plasma kinin and initiates coagulation. J. Physiol. (Lond.) 144:1–22, 1958.

25. Margolis, J. The interrelationship of coagulation of plasma and release of peptides. Ann. N.Y. Acad. Sci. 104:133–145, 1963.

26. Margolis, J. The effect of colloidal silica on blood coagulation. Aust. J. Exp. Biol. Med. Sci. 39:249–258, 1961.

27. Ratnoff, O. D. The biology and pathology of the initial stages of blood coagulation. Progr. Hematol. 5:204–245, 1966.

28. Wilner, G. D., Nossel, H. L., LeRoy, E. C. Activation of Hageman factor by collagen. J. Clin. Invest. 47:2608–2615, 1968.

29. Kellermeyer, R. W., Breckenridge, R. T. The inflammatory process in acute gouty arthritis. I. Activation of Hageman factor by sodium urate crystals. J. Lab. Clin. Med. 65:307–315, 1965.

30. Ratnoff, O. D. Activation of Hageman factor by L-homocystine. Science 162:1007–1009, 1968.

31. Müller-Berghaus, G. Pathophysiology of disseminated intravascular coagulation. Thromb. Diath. Haemorrh. Suppl. 36:45–61, 1969.

32. Kaplan, A. P., Austen, K. F. A pre-albumin activator of prekallikrein. J. Immunol. 105:802–811, 1970.

33. Kaplan, A. P., Austen, K. F. A prealbumin activator of prekallikrein. II. Derivation of activators of prekallikrein from active Hageman factor by digestion with plasmin. J. Exp. Med. 133:696–712, 1971.

34. Lepow, I. H., Pillemer, L., Ratnoff, O. D. The influence of calcium ions on the inactivation of human complement and its components by plasmin. J. Exp. Med. 98:277–289, 1953.

35. Donaldson, V. H. Effect of plasmin in vitro on clotting factors in plasma. J. Lab. Clin. Med. 56:644–651, 1960.

36. Janeway, C. A., Merler, E., Rosen, F. S., Salmon, S., Crain, J. D. Intravenous gamma globulin. Metabolism of gamma globulin fragments in normal and agammaglobulinemic persons. N. Engl. J. Med. 278:919–923, 1968.

37. Mirsky, I. A., Perisutti, G., Davis, N. C. The destruction of glucagon, adrenocorticotropin, and somatotropin by human blood plasma. J. Clin. Invest. 38:14–20, 1959.

38. Sherry, S., Fletcher, A. P., Alkjaersig, N. Fibrinolysis and fibrinolytic activity in man. Physiol. Rev. 39:343–382, 1959.

39. Astrup, T. Tissue activators of plasminogen. Fed. Proc. 25:42–51, 1966.

40. Lewis, G. P. Active polypeptides derived from plasma proteins. Physiol. Rev. 40:647–676, 1960.

41. Eisen, V. Kinin formation and fibrinolysis in human plasma. J. Physiol. (Lond.) 166:514–529, 1963.

42. Webster, M. E., Pierce, J. V. Studies on plasma kallikrein and its relationship to plasmin. J. Pharmacol. Exp. Ther. 130:484–491, 1960.

43. Vogt, W. Kinin formation by plasmin, an indirect process mediated by activation of kallikrein. J. Physiol. (Lond.) 170:153–166, 1964.

44. Burrowes, C. E. Activation of human prekallikrein by plasmin (abstract). Fed. Proc. 30:451, 1971.

45. McDonagh, R. P., Jr., Ferguson, J. H. Studies on the participation of Hageman factor in fibrinolysis. Thromb. Diath. Haemorrh. 24:1–9, 1970.

46. Ogston, D., Ogston, C. M., Ratnoff, O. D., Forbes, C. D. Studies on a complex mechanism for the activation of plasminogen by kaolin and by chloroform: the participation of Hageman factor and additional cofactors. J. Clin. Invest. 48:1786–1801, 1969.

47. Gigli, I., Mason, J. W., Coleman, R. W., Austen, K. F. Interaction of kallikrein with C′1 esterase inhibitor (C′1 a INH) (abstract). J. Immunol. 101:814–815, 1968.

48. Ratnoff, O. D., Naff, G. B. The conversion of C′1S to C′1 esterase by plasmin and trypsin. J. Exp. Med. 125:337–358, 1967.

49. Pensky, J., Levy, L. R., Lepow, I. H. Partial purification of a serum inhibitor of C′1-esterase. J. Biol. Chem. 236:1674–1679, 1961.

50. Ratnoff, O. D., Lepow, I. H. Some properties of an esterase derived from preparations of the first component of complement. J. Exp. Med. 106:327–343, 1957.

51. Ratnoff, O. D., Pensky, J., Ogston, D., Naff, G. B. The inhibition of plasmin, plasma kallikrein, plasma permeability factor, and the C′1r subcomponent of the first component of complement by serum C′1 esterase inhibitor. J. Exp. Med. 129:315–331, 1969.

52. Landerman, N. S., Webster, M. E., Becker, E. L., Ratcliffe, H. E. Hereditary angioneurotic edema; II. Deficiency of inhibitor for serum globulin permeability factor and/or plasma kallikrein. J. Allerg. 33:330–341, 1962.

53. Donaldson, V. H., Evans, R. R. A biochemical ab-

146

normality in hereditary angioneurotic edema: absence of serum inhibitor of C'1-esterase. Am. J. Med. 35:37–44, 1963.

54. Burdon, K. L., Queng, J. T., Thomas, O. C., McGovern, J. P. Observations on biochemical abnormalities in hereditary angioneurotic edema. J. Allerg. 36:546–557, 1965.

55. Lundh, B., Laurell, A. B., Wetterqvist, H., White, T., Granerus, G. A case of hereditary angioneurotic edema successfully treated with epsilon-aminocaproic acid. Clin. Exp. Immunol. 3: 733–745, 1968.

56. Nilsson, I. M., Andersson, L., Björkman, S. E. Epsilon-aminocaproic acid (E-ACA) as a therapeutic agent. Based on five years' clinical experience. Acta Med. Scand. 48 (Suppl.):1–46, 1966.

57. Frank, M. M., Sergent, J. S., Kane, M. A., Alling, D. W. Epsilon aminocaproic acid therapy of hereditary angioneurotic edema. N. Engl. J. Med. 286:808–812, 1972.

58. Movat, H. Z., Treloar, M. P., DiLorenzo, N. L., Robertson, J. W., Sender, H. B. The demonstration of permeability factors and two kinin-forming enzymes in plasma. *In* H. Z. Movat [ed.] Proceedings of the third international symposium of the Canadian Society for Immunology, Toronto, October 3–5, 1968. Cellular and humoral mechanisms in anaphylaxis and allergy. Karger, Basel/New York, 1969, p. 215–223.

59. Epstein, W. V., Tan, M., Melmon, K. L. Rheumatoid factor and kinin generation. Ann. N.Y. Acad. Sci. 168:173–187, 1969.

60. Beraldo, W. T. Formation of bradykinin in anaphylactic and peptone shock. Am. J. Physiol. 163: 283–289, 1950.

61. Brocklehurst, W. E., Lahiri, S. C. The production of bradykinin in anaphylaxis. J. Physiol. (Lond.) 160:15P–16P, 1962.

62. Brocklehurst, W. E., Lahiri, S. C. Formation and destruction of bradykinin during anaphylaxis. J. Physiol. (Lond.) 165:39P–40P, 1963.

63. Eisen, V., Vogt, W. Plasma kininogenases and their activators. *In* E. G. Erdös [ed.] Handbuch der Experimentellen Pharmakologie. Vol. 25. Springer-Verlag, New York, 1970, p. 82–130.

64. Cline, M. J., Melmon, K. L. Plasma kinins and cortisol: a possible explanation of the anti-inflammatory action of cortisol. Science 153:1135–1138, 1966.

65. Melmon, K. L., Cline, M. J. Interaction of plasma kinins and granulocytes. Nature (Lond.) 213:90–92, 1967.

66. Melmon, K. L., Cline, M. J. The interaction of leukocytes and the kinin system. Biochem. Pharmacol. 17(Suppl.):271–281, 1968.

67. Greenbaum, L. M., Kim, K. S. The kinin-forming and kininase activities of rabbit polymorphonuclear leucocytes. Br. J. Pharmacol. 29:238–247, 1967.

68. Zachariae, H., Malmquist, J., Oates, J. A. Kininase in human polymorphonuclear leukocytes. Life Sci. 5:2347–2355, 1966.

69. Greenbaum, L. M., Yamafuji, K. The role of cathepsins in the inactivation of plasma kinins. *In* E. G. Erdös, N. Back, and F. Sicuteri [ed.] Hypotensive peptides. Springer-Verlag, New York, 1966, p. 252–262.

70. Greenbaum, L. M., Carrara, M. C., Freer, R. Inflammatory response and bradykinin. Fed. Proc. 27:90–91, 1968.

71. Greenbaum, L. M., Freer, R., Chang, J., Semente, G., Yamafuji, K. PMN-kinin and kinin metabolizing enzymes in normal and malignant leucocytes. Br. J. Pharmacol. 36:623–634, 1969.

72. Cohn, Z. A., Hirsch, J. G. The isolation and properties of the specific cytoplasmic granules of rabbit polymorphonuclear leucocytes. J. Exp. Med. 112:983–1004, 1960.

73. Janoff, A., Zeligs, J. D. Vascular injury and lysis of basement membrane in vitro by neutral protease of human leukocytes. Science 161:702–704, 1968.

74. Movat, H. Z., Uriuhara, T., Macmorine, D. L., Burke, J. S. A permeability factor released from leukocytes after phagocytosis of immune complexes and its possible role in the Arthus reaction. Life Sci. 3:1025–1032, 1964.

75. Folds, J. D., Welsh, I. R. H., Spitznagel, J. K. Neutral proteases confined to one class of lysosomes of human polymorphonuclear leukocytes. Proc. Soc. Exp. Biol. Med. 139:461–463, 1972.

76. Goldstein, I. M., Wünschmann, B., Astrup, T., Henderson, E. S. Effects of bacterial endotoxin on the fibrinolytic activity of normal human leukocytes. Blood 37:447–453, 1971.

77. Matsuoka, M., Sakuragawa, N., Shimaoka, M. Studies on fibrinolytic activities in normal human leucocytes. Acta Med. Biol. (Niigata) 16:91–104, 1969.

78. Erdös, E. G., Yang, H. Y. T. Kininases. *In* E. G. Erdös [ed.] Handbuch der Experimentellen Pharmakologie. Vol. 25. Springer-Verlag, New York, 1970, p. 289–323.

79. Erdös, E. G. Hypotensive peptides: bradykinin, kallidin, eledoisin. Adv. Pharmacol. 4:1–90, 1966.

80. Oates, J. A., Melmon, K. L. Biochemical and physiologic studies of the kinins in the carcinoid syndrome. *In* E. G. Erdös, N. Back, and F. Sicuteri [ed.] Hypotensive peptides. Springer-Verlag, New York, 1966, p. 565–578.

81. Greenbaum, L. M., Yamafuji, K. In vitro inactivation and formation of plasma kinins by spleen cathepsins. Br. J. Pharmacol. 27:230–238, 1966.

82. Vogel, R., Werle, E. Kallikrein inhibitors. *In* E. G Erdös [ed.] Handbuch der Experimentellen Pharmakologie. Vol. 25. Springer-Verlag, New York, 1970, p. 213–249.

83. Webster, M. E., Pierce, J. V. The nature of the kallidins released from human plasma by kallikreins and other enzymes. Ann. N.Y. Acad. Sci. 104:91–107, 1963.

84. Moriya, H., Pierce, J. V., Webster, M. E. Purification and some properties of three kallikreins. Ann. N.Y. Acad. Sci. 104:172–185, 1963.

85. Webster, M. E., Emmart, E. W., Turner, W. A., Moriya, H., Pierce, J. V. Immunological properties of the kallikreins. Biochem. Pharmacol. 12:511–519, 1963.

86. Bhoola, K. D. The subcellular distribution of pancreatic kallikrein. Biochem. Pharmacol. 18:2279–2282, 1969.

87. Chiang, T. S., Erdös, E. G., Miwa, I.. Tague, L. L., Coalson, J. J. Isolation from a salivary gland of granules containing renin and kallikrein. Circ. Res. 23:507–517, 1968.

88. Glenn, T. M., Lefer, A. M. Role of lysosomes in the pathogenesis of splanchnic ischemia shock in cats. Circ. Res. 27:783–797, 1970.

89. Janoff, A. Alterations in lysosomes (intracellular enzymes) during shock: effects of preconditioning (tolerance) and protective drugs. Int. Anesth. Clin. 2:251–269, 1964.

90. Janoff, A., Weissmann, G., Zweifach, B. W., Thomas, L. Pathogenesis of experimental shock. IV. Studies of lysosomes in normal and tolerant animals subjected to lethal trauma and endotoxemia. J. Exp. Med. 116:451–466, 1962.

91. Kaplan, A. P., Kay, A. B., Austen, K. F. A prealbumin activator of prekallikrein. III. Appearance of chemotactic activity for human neutrophils by the conversion of human prekallikrein to kallikrein. J. Exp. Med. 135:81–97, 1972.

92. Rodriguez-Erdmann, F. Studies on the pathogenesis of the generalized Shwartzman reaction. III. Trigger mechanism for the activation of the prothrombin molecule. Thromb. Diath. Haemorrh. 12:470–483, 1964.

93. McKay, D. G., Müller-Berghaus, G., Cruse, V. Activation of Hageman factor by ellagic acid and the generalized Shwartzman reaction. Am. J. Pathol. 54:393–400, 1969.

94. Nies, A. S., Melmon, K. L. Mechanism of endotoxin-induced kinin production in human plasma. Biochem. Pharmacol. 20:29–37, 1971.

95. Tillett, W. S., Garner, R. L. The fibrinolytic activity of hemolytic streptococci. J. Exp. Med. 58:485–502, 1933.

96. Astrup, T. The biological significance of fibrinolysis. Lancet 2:565–568, 1956.

97. Lack, C. H. Staphylokinase: an activator of plasma protease. Nature (Lond.) 161:559–560, 1948.

98. Beller, F. K. The role of endotoxin in disseminated intravascular coagulation. Thromb. Diath. Haemorrh. Suppl. 36:125–149, 1969.

99. Landy, M., Weidanz, W. P. Natural antibodies against gram-negative bacteria. In M. Landy and W. Braun [ed.] Bacterial endotoxins. Rutgers University Press, New Brunswick, N.J., 1964, p. 275–290.

100. Neter, E. Endotoxins and the immune response. Curr. Top. Microbiol. Immunol. 47:82–124, 1969.

101. Snyderman, R., Gewurz, H., Mergenhagen, S. E. Interactions of the complement system with endotoxic lipopolysaccharide: generation of a factor chemotactic for polymorphonuclear leukocytes. J. Exp. Med. 128:259–275, 1968.

102. Zimmerman, T. S., Müller-Eberhard, H. J. Blood coagulation initiation by a complement-mediated pathway. J. Exp. Med. 134:1601–1607, 1971.

103. Cline, M. J., Melmon, K. L., Davis, W. C., Williams, H. E. Mechanism of endotoxin interaction with human leucocytes. Br. J. Haematol. 15:539–547, 1968.

104. Nies, A. S., Greineder, D. K., Cline, M. J., Melmon, K. L. The divergent effects of endotoxin fractions on human plasma and leukocytes. Biochem. Pharmacol. 20:39–46, 1971.

105. Bennett, I. L., Jr., Beeson, P. B. The properties and biological effects of bacterial pyrogens. Medicine 29:365–400, 1950.

106. Gow, A. E. A note on certain phenomena associated with the protein shock reaction and intravenous vaccine therapy. Quart. J. Med. 13:82–104, 1919.

107. Cluff, L. E. Effects of lipopolysaccharides (endotoxins) on susceptibility to infections. In S. Kadis, G. Weinbaum, and S. J. Ajl [ed.] Microbial toxins. Vol. 5. Academic Press, New York, 1971, p. 399–413.

108. Mullholland, J. H., Cluff, L. E. The effect of endotoxin upon susceptibility to infection: the role of the granulocyte. In M. Landy and W. Braun [ed.] Bacterial endotoxins. Rutgers University Press, New Brunswick, N.J., 1964, p. 211–229.

109. Quesenberry, P., Morley, A., Stohlman, F., Jr., Rickard, K., Howard, D., Smith, M. Effect of endotoxin on granulopoiesis and colony-stimulating factor. N. Engl. J. Med. 286:227–232, 1972.

110. Cooper, K. E. Some physiological and clinical aspects of pyrogens. In G. E. W. Wolstenholme and J. Birch [ed.] Symposium on pyrogens and fever. Churchill-Livingstone, London, 1971, p. 5–21.

111. Filkins, J. P. Comparison of endotoxin detoxification by leukocytes and macrophages. Proc. Soc. Exp. Biol. Med. 137:1396–1400, 1971.

112. Gimber, P. E., Rafter, G. W. The interaction of Escherichia coli endotoxin with leukocytes. Arch. Biochem. Biophys. 135:14–20, 1969.

113. Strauss, B. S., Stetson, C. A., Jr. Studies on the effect of certain macromolecular substances on the respiratory activity of the leucocytes of the peripheral blood. J. Exp. Med. 112:653–669, 1960.

114. Cohn, Z. A., Morse, S. I. Functional and metabolic properties of polymorphonuclear leucocytes. II. The influence of a lipopolysaccharide endotoxin. J. Exp. Med. 111:689–704, 1960.

115. Wiener, E., Shilo, M., Beck, A. Effect of bacterial lipopolysaccharides on mouse peritoneal leukocytes. Lab. Invest. 14:475–487, 1965.

116. Graham, R. C., Jr., Karnovsky, M. J., Shafer, A. W., Glass, E. A., Karnovsky, M. L. Metabolic and morphological observations on the effect of

148

surface-active agents on leukocytes. J. Cell. Biol. 32:629–647, 1967.

117. Weissmann, G., Zurier, R. B., Spieler, P. J., Goldstein, I. M. Mechanisms of lysosomal enzyme release from leukocytes exposed to immune complexes and other particles. J. Exp. Med. 134 (Suppl.):149S–165S, 1971.

118. Selvaraj, R. J., Sbarra, A. J. Relationship of glycolytic and oxidative metabolism to particle entry and destruction in phagocytosing cells. Nature (Lond.) 211:1272–1276, 1966.

119. Weissmann, G., Thomas, L. Studies on lysosomes. I. The effects of endotoxin, endotoxin tolerance and cortisone on the release of acid hydrolases from a granular fraction of rabbit liver. J. Exp. Med. 116:433–450, 1962.

120. Niemetz, J., Fani, K. Role of leukocytes in blood coagulation and the generalized Shwartzman reaction. Nature [New Biol.] 232:247–248, 1971.

121. Margaretten, W., McKay, D. G. The role of the platelet and leukocyte in disseminated intravascular coagulation caused by bacterial endotoxin. Thromb. Diath. Haemorrh. Suppl. 36:151–157, 1969.

122. Lerner, R. G., Goldstein, R., Cummings, G. Stimulation of human leukocyte thromboplastic activity by endotoxin. Proc. Soc. Exp. Biol. Med. 138:145–148, 1971.

123. Shah, J. P., Shah, U. S., Appert, H. E., Howard, J. M. Studies on the release of bradykinin by the splanchnic circulation during endotoxin shock. J. Trauma 10:255–259, 1970.

124. Carretero, O. A., Nasjletti, A., Fasciolo, J. C. Kinins and kininogen in endotoxin shock. Experientia 26:63–65, 1970.

125. Erdös, E. G., Miwa, I. Effect of endotoxin shock on the plasma kallikrein-kinin system of the rabbit. Fed. Proc. 27:92–95, 1968.

126. Urbanitz, D., Sailer, R., Habermann, E. In vivo investigations on the role of the kinin system in tissue injury and shock syndromes. Adv. Exp. Med. Biol. 8:343–353, 1970.

127. Nies, A. S., Melmon, K. L. Variation in endotoxin-induced kinin production and effect between the rabbit and rhesus monkey. Am. J. Physiol. 1973 (in press).

128. Greineder, D. K. Chemical characterization of E. coli lipopolysaccharide fractions (thesis, Case Western Reserve University, 1970). Diss. Abstracts 32[B]:463, 1971.

Treatment and Prevention of Intravascular Coagulation with Antiserum to Endotoxin

Abraham I. Braude, Herndon Douglas, and
Charles E. Davis

*From the Departments of Medicine and Pathology,
University of California (San Diego) School of Medicine,
San Diego, California*

Because of the limited success of antibiotics, we have turned to antiserum against endotoxin to prevent one of the most devastating effects of gram-negative bacteremia: disseminated intravascular coagulation (generalized Shwartzman reaction). This reaction was produced in rabbits by iv injection of endotoxin at 24-hr intervals. Rabbit antiserum to endotoxin injected iv 2 hr before the second dose of endotoxin (treatment of disseminated intravascular coagulation) or 96 hr before the first dose (prevention) decreased the reaction from 90% to 18%. Intravascular coagulation was also prevented by antiserum to both heterologous bacteria and side-chain-deficient mutants, by the 19S immunoglobulins of antisera, and by hyperimmune globulin treated with mercaptoethanol and iodoacetate. Passive protection lasted for at least 30 days. Antiserum prevented the precipitous drop in fibrinogen and platelets that accompanies deposition of fibrin in the glomerulus and elsewhere. These findings indicate that protection by antiserum is independent of antibody to O antigen, does not require complement, and prevents the coagulation disturbances induced by endotoxin.

Intravascular coagulation and shock are among the most serious complications of infections with gram-negative bacteria. These infections may not respond to antibiotics, because the bacteria are resistant and their endotoxins produce shock and intravascular coagulation even when the bacteria are killed. To control these devastating effects of endotoxin, we have turned from antibiotics to antitoxin (antiserum to endotoxin). We recently reported our success in preventing the tissue necrosis of the local Shwartzman reaction by passive immunization with antitoxin [1]. This ability of antitoxin to prevent local intravascular coagulation in the skin suggested that such serum might also prevent the generalized Shwartzman reaction. Accordingly, we did experiments to prevent generalized intravascular coagulation and renal cortical necrosis with antitoxin. The present report describes the remarkable success achieved in the use of antitoxin both for preventing and for arresting

intravascular coagulation after it has been initiated by the first dose of endotoxin.

Materials and Methods

Endotoxins. Endotoxins were extracted from smooth strains of *Escherichia coli* O:111 and *Salmonella typhimurium* by the phenol-water method of Westphal [2]. The RNA was removed by ultracentrifugation at 90,000 g in a preparative sucrose gradient for 18 hr; a SW 25.1 Spinco swinging bucket rotor was used [3]. Endotoxins were checked for lipid A by the method of Westphal and Lüderitz [4], for component sugars by gas chromatography [5], and for toxicity by lethality to mice [6].

Production of intravascular coagulation and the generalized Shwartzman reaction. Two successive doses of endotoxin were injected into the marginal ear vein of white New Zealand rabbits weighing 1 kg. The interval between the two injections was 24 hr. The doses of endotoxin are given below for each experiment and were selected on the basis of preliminary titrations that determined the amounts of endotoxin needed for 95%–100% positive Shwartzman reactions. The animals were killed 24

This work was supported by contract no. MED/DADA 17-69-C-9161 from the U.S. Army and by grant no. AI 10108-02 from the U.S. Public Health Service.

Please address requests for reprints to Dr. A. I. Braude, University Hospital, 225 W. Dickinson Street, San Diego, California 92103.

hr after the second dose of endotoxin, and reactions were recorded as negative in the absence of both renal cortical necrosis and deposits of fibrin in the glomerular capillaries. Cortical necrosis was found by gross examination of fresh kidneys and by microscopic examination of formalin-fixed kidneys. Serial sections of the kidneys were stained with hematoxylin and eosin and by the periodic-acid Schiff (PAS) method. In kidneys stained with PAS, the fibrin in the glomeruli produced striking bright red deposits.

Preparation of antiserum. Albino rabbits of both sexes weighing 3 kg were injected iv with boiled cells of *E. coli* O:111, with a mutant (J5) deficient in uridine diphosphate galactose-4 epimerase, or with *S. typhimurium*. The bacteria in the stationary phase of growth in an 18-hr trypticase-soy broth culture were removed by centrifugation, washed three times in sterile 0.9% saline, boiled for 2.5 hr, resuspended to a density allowing 70% transmission of light at 610 nm, and injected into the marginal vein of the ear in doses of 1 ml. Blood was removed by cardiac puncture from each immunized rabbit seven days after the last of six iv injections of boiled cells given three times weekly for two weeks. The blood was allowed to clot and was centrifuged for 15 min at 1,250 g at room temperature (24 C). The serum was decanted and centrifuged again. Sera were then pooled, cultured, and frozen at −20 C. Frozen sera were thawed at 4 C and brought to room temperature before injection into rabbits. Strict precautions were taken for prevention of contamination of sera by bacteria and pyrogens, as previously described [6].

Collection of immunoglobulin fractions. Forty-milliliter samples of pooled serum were separated into three fractions on 5.0 × 100-cm glass columns packed with sterile G-200 Sephadex to a height of 90 cm. The eluant was phosphate-buffered saline (PBS, *p*H 7.0); this was pumped from a reservoir by a peristaltic pump and flowed upward at 50–65 ml/hr at room temperature. Optical density was recorded at 280 nm with the LKB Uvicord II. Ten-milliliter fractions were collected in an automatic fraction collector equipped with a 15-watt germicidal ultraviolet lamp (General Electric) and were combined under sterile conditions. The combined fractions were concentrated in sterile dialysis tubing against polyethylene glycol and restored to the original volume with a solution

of physiologic saline. The 19*S* fraction was then run through the column a second time, and the single, sharply defined peak fractions were pooled. The original volume was restored again by concentration in dialysis tubes. Freedom from contamination of the 19*S* fraction with IgG immunoglobulins was assured by determination of antibody, after treatment with 2-mercaptoethanol (2-ME) by immunodiffusion and by immunoelectrophoresis. Special precautions to assure sterility and freedom from pyrogens were taken in accordance with the measures previously described [6, 7]. All sera and serum fractions used for prevention of the Shwartzman reaction were shown to be sterile and free of pyrogens.

Preparation of γ-globulin by ammonium sulfate fractionation [8]. Ammonium sulfate was dissolved by heating 770 g/liter of distilled water; the hot solution was filtered through Whatman no. 1 paper and cooled, and the *p*H was adjusted to 6.5–7.2 with concentrated NH_4OH. Cold saturated $(NH_4)_2SO_4$ was added drop by drop to an equal volume of *E. coli* O:111 antiserum or nonimmune serum at 4 C to give 50% $(NH_4)_2SO_4$, and the mixture was stirred constantly during precipitation. The precipitate was removed by centrifugation at 10,000 g for 20 min, washed with 50% $(NH_4)_2$-SO_4, dissolved in PBS (*p*H 7.0), and brought to one-half of the original volume of serum. The precipitation and washing were repeated twice, the $SO_4^=$ ions were removed by dialysis against PBS, and the solution was concentrated to one-tenth of the original concentration of serum in an Amicon diafiltration cell with a UM-10 filter under nitrogen pressure. The complete procedure was carried out at 4 C with sterile, nonpyrogenic equipment.

Treatment of ammonium sulfate fraction with 2-ME. Thirty milliliters of the $(NH_4)_2SO_4$ fraction was dialyzed first against 3,000 ml of 0.1 M 2-ME in PBS (*p*H 7.0) for 3 hr at room temperature, next against 5.55 g of 2-iodoacetamide dissolved in 1,500 ml of PBS for 4 hr at room temperature, and then overnight at 4 C against PBS.

Antibody determinations. Antibody titers in serum were determined by hemagglutination and by bacterial agglutination. Hemagglutinins were measured by human group O red cells sensitized with alkaline-treated endotoxin [9], and bacterial agglutinins by bacteria that had been boiled for 2.5 hr. The bacterial cell suspension (diluted to give

70% light transmission at 610 nm on a Coleman Junior Spectrophotometer) was added in 0.5-ml volumes to all tubes. The samples were incubated overnight at 37 C and then examined for agglutination.

Coagulation. Platelet counts were done by phase microscopy [10]. Concentrations of fibrinogen were determined by a modification of the method of Ratnoff and Menzie [11]. Citrated blood was centrifuged at 2,000 *g* to obtain platelet-poor plasma. One milliliter of this plasma was added to 0.1 ml (100 units) of bovine topical thrombin (Parke-Davis) in 1.0 ml of saline. The tube was immediately inverted and a wooden applicator stick inserted. When the clot was firm, it was gently wound around the stick, fluid was expressed against the side of the tube, and the clot was blotted and washed three times. The clot was resuspended in 1 ml of 1 N sodium hydroxide and incubated at 100 C for 2–5 min until the clot was completely dissolved. Six milliliters of biuret reagent (Biotrowski's) was added to the digested tubes. Transmittance of light was read against a biuret reagent blank in a Klett spectrophotometer with a 540-nm filter. A standard fibrinogen curve, constructed from standardized bovine albumin for reference, was used for calculation of the concentration of fibrinogen.

Fibrin-split products were determined by the method of Leavelle et al. [12]. The Newman D_2C variant of *Staphylococcus aureus* (obtained from Dr. Charles Owen, Mayo Clinic, Rochester, Minn.), which produces bound coagulase only, was used in these tests. A suspension of killed, lyophilized bacteria was suspended in imidazole-saline buffer at a concentration of 3 mg/ml. Two milliliters of freshly drawn blood was placed in a tube containing 1 mg of soybean-trypsin inhibitor (Sigma Chemical Co.) and 100 units of bovine thrombin. The clot was incubated at 37 C until clot retraction began. It was then centrifuged to obtain serum for tests. Serial dilutions of samples were made in imidazole-saline buffer (*p*H 7.4) in type U microtiter plates with a calibrated pipette dropper and 0.025 ml of bacterial suspension added to each well. After 10 min of gentle agitation, the end points were read. The end point was judged as the last well with easily visualized, definite clumping. These results were expressed in μg/ml by comparison to standards of rabbit plasma with known fibrinogen content.

Prevention of intravascular coagulation and the generalized Shwartzman reaction with antiserum or 19S fraction. Antiserum or 19S immunoglobulin fractions in volumes of 10–20 ml were injected iv 2 hr before the provocative (iv) dose of endotoxin. Control rabbits received equal volumes of nonimmune rabbit serum or nonimmune 19S fractions. The exact volumes were specified for each experiment.

Statistical methods. The Chi-square test was used to determine whether the incidence of Shwartzman reactions in antisera-treated groups was significantly less than that in controls. Yates correction for continuity was used as indicated for small-cell frequencies. Student's *t*-test was used to determine whether levels of fibrinogen and platelets in rabbits given antisera were significantly different from those in controls.

Results

Arrest of intravascular coagulation. Renal cortical necrosis was greatly reduced with antisera given iv 2 hr before the provocative dose of endotoxin.

(1) Antiserum against homologous endotoxin. Antiserum to endotoxin from *E. coli* O:111 reduced the incidence of bilateral renal cortical necrosis from 96% to 19% in animals challenged with *E. coli* O:111 endotoxin (table 1). Kidneys from protected animals contained no sign of glomerular fibrin deposits in sections stained by the PAS method. The hemagglutinin (HA) titer of the serum for *E. coli* O:111 was 1:32,000, and the bacterial agglutinin (BA) titer for *E. coli* O:111 was 1:4,000. This antiserum also prevented the dermal Shwartzman reaction [1] and prevented death from endotoxin in mice [3, 6, 7].

Similar results were obtained with antiserum to *S. typhimurium*, which lowered the rate of renal cortical necrosis and glomerular thrombi from 81.2% to 17.6% in rabbits given *S. typhimurium* endotoxin (table 1). The HA titer of the *S. typhimurium* antiserum was 1:16,000, and the BA titer 1:4,000 against *S. typhimurium*.

(2) Antiserum against heterologous endotoxin. In this experiment antiserum was used for prevention of renal cortical necrosis produced by endotoxin of a different bacterial genus from that used for preparation of the antiserum. In rabbits given antiserum to *S. typhimurium*, the frequency of the

Table 1. Arrest of intravascular coagulation and prevention of renal cortical necrosis with antiserum given 22 hr after the first dose of homologous endotoxin.

Endotoxin	Preparative dose (mg)	Provocative dose (mg)	Amount of homologous antiserum (ml)	Incidence of renal cortical necrosis	
				Antiserum	Nonimmune serum
Escherichia coli	0.125	0.03	15	1/8*	9/9
E. coli	0.125	0.06	10	4/18	14/15
Total				5/26 (19%)	23/24 (96%)
				($P < .001$)	
Salmonella typhimurium	0.2	0.03	10	3/17 (17.6%)	13/16 (81.2%)
				($P < .001$)	

NOTE. Animals without renal cortical necrosis were also free of discernible fibrin deposits.
* Number of rabbits with positive reactions/total number of rabbits.

generalized Shwartzman reaction and renal cortical necrosis due to *E. coli* O:111 was reduced from 85.7% to 46.9% ($P < .01$, table 2). No antibody to *E. coli* O:111 endotoxin could be detected by tests for HA or BA in the protective antiserum.

(3) Antiserum free of O antibody. In animals prepared with 0.125 mg and provoked with 0.06 mg of *E. coli* O:111 endotoxin, 15 ml of antisera to the epimerase-deficient mutant (J5) of

Table 2. Arrest of intravascular coagulation with heterologous antisera 22 hr after first dose of endotoxin from *Escherichia coli* O:111—prevention of renal cortical necrosis by antiserum to *Salmonella typhimurium*.

Amount of serum (ml)	Incidence of renal cortical necrosis	
	Salmonella typhimurium antiserum	Nonimmune serum
10	3/9*	5/8
20	12/23	19/20
Total	15/32 (46.9%)	24/28 (85.7%)
	($P < .01$)	

NOTE. Each rabbit received a preparative dose of 0.125 mg and a provocative dose of 0.06 mg of *Escherichia coli* O:111 endotoxin.
* Number of rabbits with positive reactions/total number of rabbits.

E. coli O:111 reduced the incidence of generalized Shwartzman reactions from nine of 11 in controls to two of 11 ($P < .01$) in those given antiserum. The absence of O-antigenic determinants in the epimerase-deficient mutant (J5) indicates that O antibody is unnecessary for protection.

Prevention of renal cortical necrosis by the immunoglobulin fractions. The 19S fraction reduced the incidence of generalized Shwartzman reactions from 94.7% to 42.1% ($P < .01$) and the 7S from 77.7% to 40% ($P < .01$), as shown in table 3. Treatment of the 19S fraction with 2-ME lowered the BA titer from 1:512 to zero. In other words, all the detectable antibody to *E. coli* O:111 in the 19S fraction was inactivated by 2-ME as would be expected if the antibody were all IgM.

Passive immunization with the $(NH_4)_2SO_4$ *fraction of E. coli O:111.* An iv dose of 1.0 ml of the $(NH_4)_2SO_4$ fraction of *E. coli* O:111 antiserum 72 hr before the first dose of endotoxin reduced the incidence of renal cortical necrosis from 90% to 4% in rabbits given *E. coli* O:111 endotoxin for preparation and provocation of the generalized Shwartzman reaction (table 4). Treatment of the $(NH_4)_2SO_4$ fraction with 2-ME and iodoacetate did not impair its ability to prevent the generalized Shwartzman reaction (table 4).

Table 3. Arrest of intravascular coagulation with immunoglobulin fractions given 22 hr after the first dose of *Escherichia coli* O:111 endotoxin: prevention of renal cortical necrosis with the 7S and 19S fractions from *E. coli* O:111 antisera.

Fraction	Dose of endotoxin (mg)		Incidence of renal cortical necrosis		
	Preparative	Provocative	Antiserum	Nonimmune serum	P
19S	0.125	0.06	8/19 (42.1)*	18/19 (94.7)	<.01
7S	0.125	0.06	4/10 (40)	14/18 (77.7)	<.01

NOTE. Each rabbit received 10 ml of each fraction.
* Number of rabbits with positive reactions/total number of rabbits (percentage of positive reactions).

Table 4. Prevention of generalized Shwartzman reaction by iv injection of 1.0 ml of $(NH_4)_2SO_4$ fraction of *Escherichia coli* O:111 antiserum 72 hr before first dose of *E. coli* O:111 endotoxin.

Gamma globulin	2-mercapto-ethanol	Incidence of renal cortical necrosis*
Nonimmune	0	18/20 (90) †
Immune	0	1/23 (4)
Nonimmune	+	11/15 (76)
Immune	+	1/15 (6.8)

NOTE. For all data, $P < .001$.

* Each rabbit received a preparative dose of 0.2 mg and a provocative dose of 0.06 mg of *E. coli* O:111 endotoxin.

† Number of rabbits with positive reactions/total number of rabbits (percentage of positive reactions).

Duration of protection afforded by antiserum. The results in figure 1 show the incidence of generalized Shwartzman reactions when the intervals between 10 ml of *E. coli* O:111 antiserum and the preparative dose of 0.125 mg of *E. coli* O:111 endotoxin were three, seven, and 28 days, respectively. Ten milliliters of antiserum prevented 0.06 mg of *E. coli* O:111 endotoxin from provoking the generalized Shwartzman reaction for 28 days.

Effect of antiserum on fibrinogen, platelets, and fibrin-split products. *E. coli* O:111 antiserum, given iv 22 hr after the preparative injection of *E.*

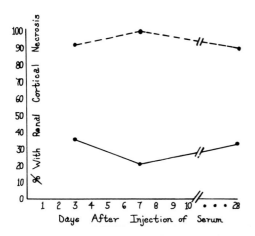

Figure 1. Duration of protection against the generalized Shwartzman reaction. Rabbits received either 10 ml of nonimmune serum (● – – – ●) or 10 ml of antiserum to *Escherichia coli* O:111 endotoxin (●——●). The preparative dose of *E. coli* O:111 endotoxin was 0.125 mg, and the provocative dose was 0.060 mg. The percentages at each point represent the number of animals with renal cortical necrosis.

coli O:111 endotoxin, reversed the drop in fibrinogen and retarded the fall in platelets induced by the provocative dose in animals developing bilateral renal cortical necrosis. As indicated in table 5, these differences were highly significant. The antiserum also reduced slightly the amount of fibrin-split products.

These results were obtained in two separate experiments. In the first, 5 ml of blood was drawn from the central ear artery before the first dose of endotoxin (0 hr), 24 hr later (just before the second dose), and again at 28 hr when the Shwartzman reactions were evident as bilateral renal cortical necrosis. Because removal of 5 ml of blood before the preparative dose accentuated the drop in platelets, the platelet counts before administration of endotoxin (at 0 hr) were done in the second experiment on a drop of blood from the central artery, and 0-hr determinations of fibrinogen and fibrin-split products were omitted. Otherwise the two experiments were the same.

The results confirm the observations of McKay and Shapiro [13] that each injection of endotoxin lowers the platelet count, while the level of fibrinogen rises after the first dose of endotoxin but drops sharply after the second, when glomerular thrombosis occurs. Our results show that antiserum to endotoxin allows the level of fibrinogen to continue its climb, despite a second dose of endotoxin, and markedly retards the second fall in platelets.

When antiserum was given 72 hr before the preparative first dose of endotoxin, the fall in fibrinogen and platelets was significantly less than that in control rabbits developing the generalized Shwartzman reaction. Antiserum also markedly prevented the rise in fibrin-split products (table 5).

Discussion

These studies show that antiserum to endotoxin can prevent not only renal cortical necrosis but also consumptive coagulopathy during the evolution of the generalized Shwartzman reaction. Since consumptive coagulopathy is associated with a high mortality rate in septic shock, the antiserum developed by this research may prove important in prevention and treatment of shock associated with gram-negative bacterial infection. Previous reports on the value of antiserum to endotoxin for prevention of death and skin necrosis (i.e., dermal Shwartzman reaction) due to endotoxin have al-

Table 5. Protection against consumptive coagulopathy by homologous iv antiserum to *Escherichia coli* O:111 endotoxin.

Time of antiserum	Clotting factors	Hr after preparative dose of endotoxin*		
		0	24	28
22 hr after preparative dose†	Fibrinogen (mg/100 ml)			
	Nonimmune serum	183 ± 33	350 ± 49 (+91%)	261 ± 55 (−25%)‡
	Antiserum	203 ± 19	385 ± 30 (+90%)	417 ± 41 (+8%)‡
Exp. no. 1	FSP§ (μg/ml)			
	Nonimmune serum	9 ± 2	37 ± 7	51 ± 12
	Antiserum	5 ± 1	31 ± 12	26 ± 8
	Fibrinogen (mg/100 ml)			
	Nonimmune serum		301 ± 72	214 ± 68 (−26%)‡
	Antiserum		393 ± 77	407 ± 75 (+14%)‡
Exp. no. 2	Platelets (per mm³ × 10³)			
	Nonimmune serum	596 ± 147	224 ± 76 (−51%)	84 ± 40 (−59%)‡
	Antiserum	837 ± 107	240 ± 54 (−69%)	171 ± 30 (−21%)‡
	FSP (μg/ml)			
	Nonimmune serum		13 ± 7	46 ± 26
	Antiserum		20 ± 21	22 ± 11
72 hr before preparative dose‖	Fibrinogen (mg/100 ml)			
	Nonimmune serum		347 ± 34	205 ± 27 (−41% ± 6)‡
	Antiserum		377 ± 31	336 ± 29 (−4% ± 6)‡
	Platelets (per mm³ × 10³)			
	Nonimmune serum	583 ± 110	332 ± 70	89 ± 30 (−73% ± 6)‡
	Antiserum	801 ± 138	418 ± 76	188 ± 43 (−55% ± 7)‡
	FSP (μg/ml)			
	Nonimmune serum		224	346 ± 94‡
	Antiserum		35	56 ± 40‡

* Mean ± 2 SE (%Δ); %Δ = percentage change from 0 to 24 hr and from 24 to 48 hr. Each rabbit received a preparative dose of 0.125 mg and a provocative dose of 0.060 mg of *E. coli* O:111 endotoxin.

† Fourteen rabbits were given 10 ml of nonimmune serum iv, and nine received 10 ml of *E. coli* O:111 antiserum iv in experiment no. 1. Seven were given 10 ml of nonimmune serum, and eight received 10 ml of immune serum in experiment no. 2.

‡ $P < .001$ for differences between protected and unprotected at 28 hr.

§ FSP = fibrin-split products.

‖ Thirteen rabbits were given 10 ml of nonimmune serum iv, and 21 received 10 ml of *E. coli* O:111 antiserum iv.

ready indicated the importance of passive immunity for protection against these effects of gram-negative bacterial infection [1, 3, 6, 7, 14, 15]. The present studies further clarify the nature of the protective action of antiserum and extend its range of potential value.

To explain how antiserum prevents intravascular coagulation, we must first analyze the coagulation disturbances in the generalized Shwartzman reaction. From the studies of Lee [16] and McKay [13], it appears that the first dose of endotoxin triggers intravascular conversion of fibrinogen to fibrin; but the reticuloendothelial system (RES) clears the fibrin before it can be deposited in the glomerular capillaries and other vessels. The first dose of endotoxin seems to condition the RES so

that it becomes blocked by the second dose. As a result the fibrin aggregates produced by the second dose are no longer removed and are filtered out by the glomeruli and occlude its capillaries. Antiserum could thus prevent the generalized Shwartzman reaction in two ways: (*1*) by stopping or slowing the clotting action of endotoxin, or (*2*) by protecting the RES from blockade by endotoxin. This question needs more study, but it is possible that antiserum might act in both ways. The sharp reduction in fibrin-split products when antiserum was given before the first dose of endotoxin supports the idea that antiserum slows the clotting action of endotoxin. This effect of antiserum is less marked when antiserum is given between the two doses, just before the provocative challenge with endotoxin. At this time there is only a slight difference between split-product levels, perhaps because normal serum given at this point may also slow clotting. Accordingly, it is likely that antiserum given just before the provocative dose acts through other mechanisms by protecting the RES from blockade so that it can continue to clear fibrin. Another factor to consider is that animals protected with antiserum did not show the signs of severe vascular collapse seen in those that went on to develop renal cortical necrosis; thus antiserum may have protected the kidneys from the sluggish glomerular flow that would favor deposition of fibrin.

Although more study is needed to establish the nature of the protective factor in antiserum against endotoxin, considerable evidence supports the idea that this factor is an antibody. Thus the protective factor increases after immunization, is concentrated in the γ-globulin fraction prepared by ammonium sulfate precipitation, is present in $19S$ and $7S$ fractions, and remains after treatment of the $(NH_4)_2SO_4$ fraction with 2-ME and iodo-acetate. The resistance of the protective property to 2-ME has other significant implications. It suggests that complement is not necessary for protection, since 2-ME destroys the complement-binding properties of gamma globulin [17]. For this reason it may be safe to administer the gamma globulin fraction of the antiserum iv after it is so treated, because the danger of iv gamma globulin resides in its ability to fix complement in vivo.

These findings on the effectiveness of antiserum against the generalized Shwartzman reaction resemble in several important respects the prevention of the local Shwartzman reaction with antiserum. In both cases, we were able, with antiserum against one species of bacteria, to prevent endotoxin of an unrelated species from producing the reaction. Such heterologous protection strengthens the concept that the toxic moiety of bacterial lipopolysaccharides is a molecule, composed of core sugars and lipid A common to many species of gram-negative bacteria [18–20]. It also provides new data that antibody to the polysaccharide units comprising the O antigens is not necessary for protection against endotoxin, even though it may contribute to such protection. Further evidence along this line comes from our finding that the *E. coli* rough mutant (J5) antiserum, lacking O-specific determinants, gave strong passive protection against local and generalized Shwartzman reactions due to either *E. coli* or *S. typhimurium* endotoxins [1].

It would appear from these results that antiserum against endotoxin offers a realistic approach to prophylaxis of intravascular coagulation in patients. In addition it may be of value in treatment during the initial stages of endotoxemia, before the disastrous effects of consumptive coagulopathy produce hemorrhage and before intravascular fibrin deposits obstruct vessels in the kidney and other vital organs. Such antisera have a broad range of activity against multiple endotoxins from unrelated bacterial genera and also protect against all serious effects of endotoxin: lethal shock [3, 6, 7, 14], local skin necrosis (the dermal Shwartzman reaction) [1], and generalized intravascular coagulation (the generalized Shwartzman reaction).

References

1. Braude, A. I., Douglas, H. Passive immunization against the local Shwartzman reaction. J. Immunol. 108:505–512, 1972.
2. Westphal, O., Lüderitz, O., Bister, F. Uber die Extraction von Bakterien mit Phenol/Wasser. Z. Naturforsch. [B] 7:148–155, 1952.
3. Tate, W. J., III, Douglas, H., Braude, A. I., Wells, W. W. Protection against lethality of *E. coli* endotoxin with "O" antiserum. Ann. N.Y. Acad. Sci. 133:746–762, 1966.
4. Kabat, E. A., Mayer, M. M. Experimental immunochemistry. 2nd ed. Charles C Thomas, Springfield, Ill., 1961, p. 835.
5. Davis, C. E., Freedman, S. D., Douglas, H., Braude, A. I. Analysis of sugars in bacterial endotoxins by gas-liquid chromatography. Anal. Biochem. 28:243–256, 1969.
6. Brown, K. R., Douglas, H., Braude, A. I. Prevention of death from endotoxin with antiserum: II. Elimination of the risk of anaphylaxis to endotoxin. J. Immunol. 106:324–333, 1971.
7. Davis, C. E., Brown, K. R., Douglas, H., Tate, W. J., Braude, A. I. Prevention of death from endotoxin with antisera: I. The risk of fatal anaphylaxis to endotoxin. J. Immunol. 102:563–572, 1969.
8. Kendall, F. E. Studies on serum proteins; I. Identification of single serum globulin by immunological means. J. Clin. Invest. 16:921–931, 1937.
9. Neter, E., Westphal, O., Lüderitz, O., Gorzynski, E. A., Eichenberger, E. Studies of enterobacterial lipopolysaccharides. Effects of heat and chemicals on erythrocyte-modifying, antigenic, toxic and pyrogenic properties. J. Immunol. 76:377–385, 1956.

156

10. Brecher, G., Cronkite, E. P. Estimation of the number of platelets by phase microscopy. *In* L. M. Tocantins [ed.]. The coagulation of blood: method of study. Grune and Stratton, New York, 1955, p. 41–44.

11. Ratnoff, O. D., Menzie, C. A. New method for the determination of fibrinogen in small samples of plasma. J. Lab. Clin. Med. 37:316–320, 1951.

12. Leavelle, D. E., Mertens, B. F., Bowie, E. J. W., Owen, C. A., Jr. Staphylococcal clumping on microtiter plates: a rapid simple method for measuring fibrinogen split products. Am. J. Clin. Pathol. 55:452–457, 1971.

13. McKay, D. G. Disseminated intravascular coagulation. Harper and Row, New York, 1965. 493 p.

14. Freedman, H. H. Passive transfer of protection against lethality of homologous and heterologous endotoxins. Proc. Soc. Exp. Biol. Med. 102:504–506, 1959.

15. Kim, Y. B., Watson, D. W. Modification of host responses to bacterial endotoxins. II. Passive transfer of immunity to bacterial endotoxin with fractions containing 19S antibodies. J. Exp. Med. 121:751–759, 1965.

16. Lee, L. Mechanisms involved in the production of the generalized Shwartzman. *In* M. Landy and W. Braun [ed.]. Bacterial endotoxins. Rutgers University Press, New Brunswick, N.J., 1964, p. 648–657.

17. Wiedermann, G., Miescher, P. A., Franklin, E. C. Effect of mercaptoethanol on complement binding ability of human 7S gammaglobulin. Proc. Soc. Exp. Biol. Med. 113:609–613, 1963.

18. Heath, E. C., Mayer, R. M., Edstron, R. D., Beaudreau, C. A. Structure and biosynthesis of the cell wall lipopolysaccharide of *Escherichia coli*. Ann. N.Y. Acad. Sci. 133:315–333, 1966.

19. Osborn, M. J. Biosynthesis and structure of the core region of the lipopolysaccharide in *Salmonella typhimurium*. Ann. N.Y. Acad. Sci. 133:375–383, 1966.

20. Nikaido, H., Naide, Y., Mäkelä, P. H. Biosynthesis of O-antigenic polysaccharides in *Salmonella*. Ann. N.Y. Acad. Sci. 133:299–314, 1966.

Effect of Endotoxin on Resistance of the Freshwater Crayfish (*Parachaeraps bicarinatus*) to Infection

D. McKay,* **C. R. Jenkin, and C. J. Tyson**

From the Department of Microbiology, University of Adelaide, Adelaide, South Australia

The resistance of the freshwater crayfish (*Parachaeraps bicarinatus*) to infection can be altered by prior injection of small amounts of lipopolysaccharide. Development and duration of the response is temperature-dependent and may be correlated with an increase in activity of phagocytic cells rather than an increase in titer of recognition factors free in the hemolymph. However, the increased activity of the cells may be measured only if the particle has been preopsonized with hemolymph.

During the past few years, we have been interested in the immune response of invertebrates and the mechanism by means of which phagocytic cells from these animals are able to discriminate between self and foreign material. We chose as our model the freshwater crayfish (*Parachaeraps bicarinatus*), which is widespread throughout the waterways of South Australia and can be obtained in large numbers throughout the year. Further advantages in using the animal are that it is easily maintained in the laboratory for many months or even years and that 20% of the weight of the animal is hemolymph containing about $2–5 \times 10^6$ cells/ml. This makes it an ideal animal for cellular and biochemical studies. Early in these studies we were fortunate to isolate a natural pathogen of the crayfish, a species of *Pseudomonas* that we have designated *Pseudomonas* CP. This organism produced an acute bacteremic infection; most animals died between two and four days after infection. The organism was transmissible only by injection; normal animals in the same tank as infected and dying animals showed no symptoms of disease [1, 2].

Effect of Killed Vaccines on the Resistance of the Crayfish to Infection

We attempted to immunize crayfish against this

infection with use of heat-killed or alcohol-killed vaccines prepared from the specific pathogen. Groups of 25 crayfish were immunized via the ventral hemal sinus; each animal received the equivalent of 2×10^7 bacteria/ml. The groups were challenged four days after the primary injection with 4 LD_{50} of *Pseudomonas* CP. The results indicated that a significant degree of resistance to infection was displayed by the treated animals. Further experiments showed that the degree of resistance or immunity was related to the dose of antigen given. Thus, four doses of vaccine (2×10^7 bacteria per dose) given at four-day intervals conferred better immunity against a challenge of 10 LD_{50} than did two doses. In view of these data, the specificity of this response was of interest. Alcohol-killed vaccines were prepared from a variety of bacteria, both gram-negative and gram-positive. Groups of 25 animals were injected via the ventral hemal sinus; each animal received 2×10^7 bacteria. All groups were callenged with 4 LD_{50} of *Pseudomonas* CP. The results (table 1) show that significant protection was achieved in only those animals receiving vaccines prepared from gram-negative bacteria. Further experiments involving larger and multiple doses of vaccines prepared from gram-positive bacteria failed to protect the animals against infection by this strain of *Pseudomonas*. Considering these results, the assumption was not unreasonable that the observed alteration in susceptibility to infection was due to the lipopolysaccharide endotoxin of the gram-negative bacteria. Lipopolysaccharide was prepared from *Salmonella typhimurium* by the method of Westphal et al. [3]. It was found that, for a given

This work was supported by a grant from the Australian Research Grants Committee.

Please address requests for reprints to Dr. C. R. Jenkin, Department of Microbiology, University of Adelaide, Adelaide, South Australia.

* Present address: National Biological Standard Laboratory, Canberra, A. C. T.

158

Table 1. Effect of immunization with vaccines prepared from various bacteria on the resistance of the crayfish (*Parachaeraps bicarinatus*) to infection with *Pseudomonas* CP.

Vaccine*	Percentage mortality	P
Staphylococcus aureus	85	NS†
Staphylococcus albus	89	NS
Micrococcus lysodiekticus	96	NS
Bacillus subtilis	95	NS
Pseudomonas CP	45	$0.001 < P < 0.005$
Salmonella typhimurium	35	<0.001
Escherichia coli	42	$0.001 < P < 0.005$
Pseudomonas fluorescens	37	<0.001
None	92	—

* All vaccines given at a dose equivalent to 2×10^7 bacteria.

† NS = not significant.

lipopolysaccharide preparation, crayfish were about 20–50 times more susceptible to the lethal effects of endotoxin than were mice. Preliminary studies [1] had indicated that a single dose of 2 µg of this preparation would give significant protection against 4 LD_{50} of the pathogen when animals were challenged four days after injection. The data expressed in figure 1 show that, as with

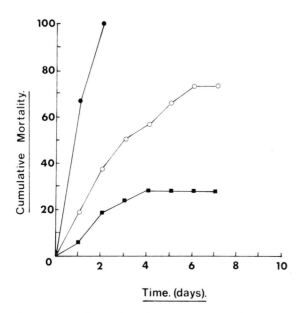

Figure 1. Effect of multiple doses of endotoxin on immunity of the crayfish (*Parachaeraps bicarinatus*) to infection with *Pseudomonas* CP. All animals were challenged with 10 LD_{50}. (●——●) = controls; (○——○) = one dose of 2 µg; (■——■) = four doses of 2 µg.

the alcohol-killed vaccine, the degree of immunity was dependent on the amount of lipopolysaccharide injected. Thus, animals receiving four separate 2-µg doses of lipopolysaccharide at four-day intervals showed a greater degree of resistance to 10 LD_{50} of this pathogen than did animals receiving only two doses. All animals were challenged four days after the last injection of lipopolysaccharide.

Effect of Temperature on Development and Duration of Immunity

In the area of South Australia in which these animals were caught, water temperatures may fluctuate from 10 C in winter to 26 C in summer. We were interested, therefore, in determining whether or not the expression and duration of immunity were related to the temperature at which the animals were maintained. Large numbers of crayfish were divided into groups of 25 and were maintained at three different temperatures (12 C, 19 C, and 26 C) for seven days. At certain intervals during the 14 days after this initial acclimatization, a group of crayfish at each of the above temperatures was injected with 2 µg of lipopolysaccharide. On day 15 the animals were returned to a temperature of 19 C and were challenged 6 hr later with 4 LD_{50} of *Pseudomonas* CP via the ventral sinus. Thus, all animals in this experiment had received a primary dose of lipopolysaccharide, but the time of immunization before challenge ranged from one to 14 days. The experiment was arranged so that all animals were challenged on the same day. The data illustrated in figure 2 show that the development and duration of immunity was temperature-dependent. Further experiments indicated that, even at 12 C, resistance to the pathogen developed (albeit more slowly), reached a peak eight weeks after the primary injection of endotoxin, and was still evident 16 weeks later.

In terms of resistance to infection after injection of lipopolysaccharide, these animals behave in a fashion similar to that in which the common vertebrate animals in the laboratory behave. The most noticeable difference is the duration of the response in these invertebrate animals, in contrast to the relatively short response of a few days in the vertebrates.

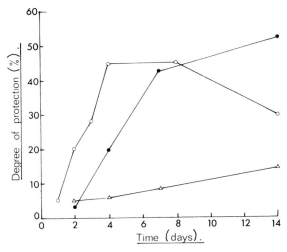

Figure 2. Effect of temperature on expression of immunity to infection with *Pseudomonas* CP in the crayfish (*Parachaeraps bicarinatus*) after vaccination with endotoxin (2 μg). Animals were maintained at 26 C (○———○), 19 C (●———●), and 12 C (△———△). Immunity is expressed as the degree of protection, which is the difference between the percentage of survivors in immunized groups and the percentage in control groups. All animals were challenged with 4 LD$_{50}$.

Mechanism of Resistance to Infection

In the vertebrates the increased resistance to infection with bacteria is dependent on two important physiologic changes: (*1*) an increase in the metabolism and activity of the phagocytic cells; and (*2*) a general increase in levels of specific antibody that the animal was making before the injection of lipopolysaccharide. Before we discuss the mechanism of this resistance in the crayfish, a pertinent question would be, "Is there any evidence to suggest that the recognition of bacteria by the phagocytic cells of the crayfish is mediated by factors (opsonins) free in the hemolymph of the crayfish?" The answer is yes (table 2). If one injects intravenously a known concentration of bacteria into vertebrates, the bacteria are cleared from the circulation at an exponential rate by the phagocytic cells of the liver and spleen. The rate of clearance, denoted by the phagocytic index K, varies with the strain of bacteria used and can be shown to be related to the titer of circulating antibody [4, 5]. If a second, similar dose of bacteria is injected into animals that have just eliminated a primary dose

Table 2. Clearance of bacteria from the circulation of normal and blockaded crayfish.

Treatment of animals	Challenge organism	Phagocytic index
Nothing	*Salmonella abortus equi*	0.15
S. abortus equi	*S. abortus equi*	0.035
S. abortus equi	*S. abortus equi* treated with normal hemolymph	0.15
S. abortus equi	*S. abortus equi* treated with hemolymph adsorbed with *S. abortus equi*	0.035
S. abortus equi treated with normal serum	*S. abortus equi*	0.11

of bacteria from the circulation, these bacteria are eliminated much more slowly due to the blockade of the animals by the first dose. However, if the second dose is treated (opsonized) with specific antibody, the rate of clearance may be returned to normal. This type of experiment is a sensitive assay for specific antibody. It may be seen from the data in table 2 that blockade in the crayfish may be reversed by treatment of the second dose of bacteria with hemolymph before injection. Furthermore, if the first dose of bacteria is opsonized before injection with hemolymph, then the animal is not blockaded, and the second dose of bacteria is cleared at the normal rate. This indicates that the elimination of bacteria from the circulation is dependent on the presence of factors (opsonins) in the hemolymph. Confirmation of these in-vivo results has been obtained in in-vitro studies concerned with the phagocytosis of erythrocytes and bacteria ([6] and C. J. Tyson and C. R. Jenkin, unpublished observations). By this technique, titrations of hemolymph obtained from lipopolysaccharide-stimulated animals, compared with titers of hemolymph from normal animals, showed that the dilution of hemolymph required to reverse blockade was similar in both groups of animals; this finding indicated that lipopolysaccharide had not increased the titer of opsonic factors. In view of these results, experiments were designed for measurement of the phagocytic activity of the cells derived from normal and lipopolysaccharide-stimulated animals. By a technique published elsewhere [6], ovine erythrocytes were phagocytosed in vitro by cells from crayfish that had received either an alcohol-killed vaccine or lipopolysaccharide and that were, therefore, immune to infection. The phagocytic activity of these cells was compared with that of cells obtained from normal animals. The data presented in figure 3 show that cells from immunized animals were much more active than those from normal ones. Also of importance is that the expression of phagocytic activity was dependent on the prior treatment of the erythrocyte with hemolymph. Since cells from immunized animals displayed increased activity, it was important that we determine whether or not there was any correlation between the development and duration of immunity and this cellular activity. Groups of 10 crayfish were maintained at 26 C for seven days. At intervals after this period for 24 days, different groups of crayfish were immunized with a single dose of 2.5 μg of endotoxin. Control animals received saline only. On day 25 the phagocytic activity of the hemocytes from animals in

each of the different groups was tested. For each test the hemocytes were pooled from four crayfish, and phagocytosis of opsonized erythrocytes was scored after incubation for 1.5 hr at 22 C.

Other groups (25 crayfish per group) were immunized by similar schedules and were maintained at the same temperature; these animals were challenged on day 25 with 4 LD$_{50}$ of *Pseudomonas* CP. Immunity was measured as the difference in percentages of survival between the control and immunized groups. The results (figure 4) show an apparent correlation between the appearance and duration of immunity and the increased activity of phagocytic cells. Similar experiments were done at 19 C and 13 C with similar results.

Correlation of Immunity with the Phagocytic Activity of Hemocytes

The preceding experimental data indicated the close relationship between the activity of the hemocytes, as measured in vitro by the uptake of opsonized ovine erythrocytes, and immunity to infection. These results were considered further for determination of the statistical relationship. The data presented in figure 4, those obtained at 19 C and 13 C, and those obtained in other

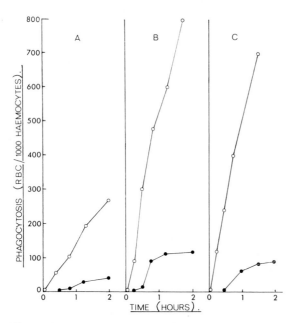

Figure 3. Phagocytosis of erythrocytes by hemocytes of the crayfish (*Parachaeraps bicarinatus*) in vitro. (**A**) = hemocytes from normal crayfish; (**B**) = hemocytes from crayfish immunized with four doses of endotoxin (2.5 μg per dose); (**C**) = hemocytes from crayfish immunized with four doses of killed *Pseudomonas* CP vaccine; (○——○) = erythrocytes opsonized with crayfish serum; (●——●) = nonopsonized erythrocytes.

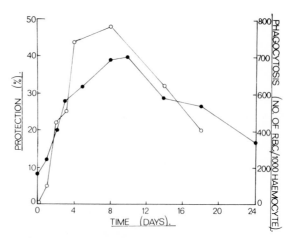

Figure 4. Relationship of immunity and phagocytic activity of hemocytes from the crayfish (*Parachaeraps bicarinatus*) after a single dose of endotoxin (2.5 μg) at 26 C. (○——○) = Immunity (percentage protection); (●——●) = phagocytic activity of hemocytes. Immunity is expressed as the degree of protection, which is the difference between the percentage of survivors in immunized groups and the percentage in control groups.

experiments were pooled. Before inclusion, data had to satisfy two requirements: the cells had to have been pooled from at least three animals, and the group tested for immunity had to have consisted of 20 or more animals. The combined data were plotted on a scatter diagram, and the best fitting line was calculated by the method of least squares (figure 5). The correlation coefficient ($r = 0.88$) and the coefficient of determination ($r^2 = 0.77$) were calculated. The latter figure indicates that 77% of the variation in immunity between groups of crayfish could be explained by variation in phagocytic activity of the hemocytes.

In summary, the data in this paper show that

endotoxin can alter the resistance of the freshwater crayfish (*P. bicarinatus*) to infection. Development and duration of immunity are dependent on the temperature at which the animal is maintained and, unlike the situation with the vertebrates, appear to be related solely to an increase in the activity of the phagocytic cells. However, the expression of this activity is dependent on contact of the particle with factors (opsonins) in the hemolymph.

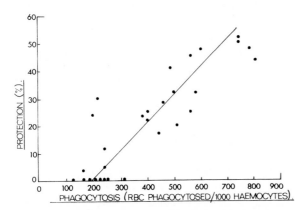

Figure 5. Relationship between phagocytic activity of hemocytes of the crayfish (*Parachaeraps bicarinatus*), as measured by the uptake of opsonized ovine erythrocytes, and immunity to infection with *Pseudomonas* CP. Immunity is expressed as the degree of protection, which is the difference between the percentage of survivors in immunized groups and the percentage in control groups.

References

1. McKay, D., Jenkin, C. R. Immunity in the invertebrates. II. Adaptive immunity in the crayfish (*Parachaeraps bicarinatus*). Immunology 17:127–137, 1969.
2. McKay, D., Jenkin, C. R. Immunity in the invertebrates. The fate and distribution of bacteria in normal and immunised crayfish (*Parachaeraps bicarinatus*). Aust. J. Exp. Biol. Med. Sci. 48: 599–607, 1970.
3. Westphal, O., Lüderitz, O., Bister, F. Über die Extraktion von Bakterien mit Phenol/Wasser. Z. Naturforsch. [B] 76:148–155, 1952.
4. Benacerraf, B., Sebestyen, M. M., Schlossman, S. A quantitative study of the kinetics of blood clearance of P32-labelled *Escherichia coli* and staphylococci by the reticuloendothelial system. J. Exp. Med. 110:27–48, 1959.
5. Jenkin, C. R. The immunological basis for the carrier state in mouse typhoid. *In* M. Landy and W. Braun [ed.]. Bacterial endotoxins, Rutgers University, New Brunswick, N. J., 1964, p. 263–274.
6. McKay, D., Jenkin, C. R. Immunity in the invertebrates. The role of serum factors in phagocytosis of erythrocytes by haemocytes of the freshwater crayfish (*Parachaeraps bicarinatus*). Aust. J. Exp. Biol. Med. Sci. 48:139–150, 1970.

Antibacterial Action of Antibody and Complement

D. Rowley

*From the Department of Microbiology, University
of Adelaide, Adelaide, South Australia*

Susceptibility to the lethal action of antibody and complement is a universal property of living cells, which are thereby attacked at their cell membranes. Certain cells (including some bacteria) may resist this action due to the impermeability of their cell walls to the activated components of complement. In susceptible bacteria there may be many cell-wall antigens that are capable of acting as receptors for antibody and of activating a lethal complement sequence. With *Salmonella typhimurium,* antibody against each of the Kauffman-White determinants is effective, as is antibody directed against rough-core polysaccharide and protein components. Little is known about the relative efficiency of antibodies against different determinants; such knowledge might lead to the recognition of the main protective antigens.

The fact that some gram-negative bacteria can be killed by fresh serum containing complement, in the presence of minute amounts of specific antibody, has been known for nearly 100 years. Work in the intervening years has persuaded us that this process is complex and obscure; the precise understanding of it still eludes us. I would like us to accept that, since most animal cells may be killed by specific antibody and complement, the final lethal blow is likely to fall on a common and basic cellular structure, such as the protoplasmic membrane. The relative or absolute resistance found in some bacteria is due either to lack of formation or to blocked access of the final lethal component to this basic, sensitive structure. This assertion is in keeping both with the general finding that rough, gram-negative organisms are sensitive, even though their parent, smooth strains are resistant [1], and with the particular finding of Muschel that protoplasts from resistant *Bacillus subtilis* are also sensitive [2]. One proposed sequence would involve fixation of antibodies to various cell-wall components of bacteria and activation of complement at those points. If the antibody allows the total sequence of complement activation and if the bacterial cell wall permits penetration of the final component as far as the protoplasmic membrane, then death of the cell can occur. This sequence implies that the reaction can be stated:

Please address requests for reprints to Dr. D. Rowley, Department of Microbiology, University of Adelaide, Adelaide, South Australia.

bacterial antigens + antibody + complement equals bacterial death; however, we find that this is an oversimplification and that each of these parameters is subject to considerable heterogeneity.

Heterogeneity of Antigens

Some years ago it was shown by Adler [3] that bacteria could be passively sensitized by coating with antigens (in a manner similar to red-cell sensitization) and that specific antibody against these coated antigens in the presence of complement killed the bacteria. This concept was expanded by Rowley and Turner [4], who found that the sensitivity of the coated organisms was roughly inversely proportional to the size of the antigen used for coating. This discovery led us to propose that an important factor determining the sensitivity of bacteria was the thickness of the cell wall, which affected the distance from the protoplasmic membrane at which complement was activated. There were similar differences in antigenic efficiency among native bacteria, and it was supposed that antibody against these bacteria might differ in efficiency; differences would be due more to position and other physical factors than to specificity. This idea led to attempts to determine which of the Kauffman-White antigens of *Salmonella typhimurium* (antigens 1, 4, 5, or 12) was most active as initial substrate in the bactericidal reaction. Murine antisera were prepared against various salmonellae and absorbed to render them monospecific with regard to the Kauffman-White antigens, then titrated in the bactericidal reaction against strains M206 (1, 4, 5, and 12) and ST11 (1, 4, and 12). The results (table 1) show no obvious pattern that might allow predictions about the relative importance of antigens 4, 5, or 12. On the other hand, antibody against antigen 9 was active against both strains of *S. typhimurium*, and antibody 5 was also active against strain ST11. In neither case did the strains possess the respective antigens. Moreover, antiserum against the living rough strain 11RX was also active against apparently unrelated organisms. When quantitative precipitin reactions were done with each antiserum against the respective lipopolysaccharides, the centrifuged supernatants at the equivalence points possessed undiminished bactericidal activity.

Similar experiments of Daguillard and Edsall [5] with antisera against *Salmonella typhi* 0901 (9 and 12) absorbed to and eluted from either *Salmonella haarlem* (9 and 46) or *Salmonella paratyphi* B (1, 4, 5, and 12) led them to the tentative conclusion that antibodies to antigen 12 were more effective than antibodies to antigen 9 in the bactericidal test. The authors made the reservation that unknown antigens shared between *S. typhi* and *S. paratyphi* B would vitiate their conclusions.

Our results, of the kind shown in table 1, led us to look for antigenic substrates other than the Kauffman-White oligosaccharide determinants. We soon concluded that, although antibodies against the smooth O-antigenic determinants can be bactericidal, they are clearly not the only active antibodies. In a representative experiment, a rabbit anti-*Citrobacter* (4 and 5) serum was absorbed with highly purified lipopolysaccharide from *S. typhimurium* and assayed against lipopolysaccharide-coated red cells and Boivin antigen-coated cells and in the bactericidal reaction against the sensitive *S. typhimurium* strain M206. The results (table 2) show that the antibody titer against lipopolysaccharide fell by 1,000–3,000, while the titer against Boivin antigen fell only fourfold, and the bactericidal titer fell about 50-fold. All of this remaining bactericidal antibody was removed by one more absorption with Boivin antigen. We must conclude that there was antigenic material in the Boivin antigen that was not present in the isolated lipopolysaccharide and that acted as a substrate in the bactericidal reaction. Similar experiments with *Vibrio cholerae* have led to the conclusion that this organism also contains antigens for the bactericidal reaction in its Boivin antigen. We have proposed without much good evidence that these antigens are protein in nature; but, even

Table 1. Bactericidal activity of antisera against several Kauffman-White (K. W.) antigens.

| | | Bactericidal titer vs. | |
| | | *Salmonella typhimurium* M206 | *Salmonella typhimurium* ST11 |
Antiserum vs.	K. W. antigens	(1, 4, 5,12)	(1, 4, 12)
Citrobacter	4, 5	10^6	10^3
Salmonella abortus equi	4, 12	10^4	10^4
Salmonella berta	9, 12	10^3	10^2
Salmonella berta absorbed with *Salmonella abortus equi*	9	10^3	10^2
Citrobacter absorbed with *Salmonella abortus equi*	5	10^4	10^2
Rough *Salmonella* 11RX	None	10^4	—

Table 2. Lack of correlation between antibodies to lipopolysaccharide (LPS) and bactericidal titer.

Antiserum	HA titer vs. *Salmonella typhimurium* LPS-coated cells	HA titer vs. *S. typhimurium* Boivin antigen-coated cells	Bactericidal titer vs. *S. typhimurium* M206 (1, 4, 5, 12)
Rabbit anti-*Citrobacter* (4, 5)	1:3,200	1:640	$1:10^6$
Absorbed 3 times with 200 μg of *S. typhimurium* LPS (1, 4, 5, 12)/ml	<1:2	1:160	$1:2 \times 10^4$
Absorbed with 1 mg of Boivin antigen (*S. typhimurium*)/ml	<1:2	<1:4	<1:10

Table 3. Amounts of various antigens inhibiting four vibriocidal doses of antiserum devoid of anti-lipopolysaccharide activity.

Antigen	Inhibiting amount (μg)	Protein content of inhibiting dose (μg)
Vibrio cholerae 569B cell walls (Inaba)	12.5	5.9
V. cholerae 569B Boivin antigen	1.6	0.1
V. cholerae 569B crude LPS*	0.8	0.012
V. cholerae 569B purified LPS	>1,400	>2.5
017 purified LPS (Ogawa)	700	13
Protein ex 569B Boivin antigen	38	30

* LPS = lipopolysaccharide.

without knowing its nature, one can assay the relative antigenic contents of *V. cholerae* accurately in a bactericidal-inhibition reaction (table 3).

There is a poor correlation between protein contents of these antigens and their activity in the inhibition test, so the belief that the unknown antigen is protein is still intuitive rather than logical. Further work is needed on these potentially important antigens; in fact, antibody against the unknown antigen in *V. cholerae* can be shown to be at least as protective against cholera in the baby-mouse model as is the antibody to lipopolysaccharide [6].

There is evidence from other sources that the bactericidal reaction may involve antigens other than smooth oligosaccharide determinants, in particular the common antigen of Neter [7], the rough-core antigens emphasized by Chedid [8], and the basic lipid A itself, which has recently been found to be antigenic (C. E. T. Galanos and O. Lüderitz, personal communication). In all of these cases, access of antibody to these antigens may be physically hindered in smooth organisms by the enormous excess of superficial O-antigen repeating units; in this sense, by prevention of effective antigen-antibody reactions, the smooth lipopolysaccharide coating may be regarded as virulence antigens.

Reynolds has clarified the way in which smooth lipopolysaccharide enables some bacteria to resist the effects of antibody and complement. Such resistant organisms become sensitive when incubated in Tris buffer [9, 10]. This resistance is due not to the stripping off of lipopolysaccharide from the cell wall but to the changed physical configuration of the lipopolysaccharide in the presence of Tris, which allows the activated complement molecules to penetrate to the inner protoplasmic membrane. Antibody attached to the bacterial cell wall in Mg^{++} and saline was ineffective, but if the bacteria plus the bound antibody were then transferred to Tris, rapid killing ensued [11].

Heterogeneity of Antibodies

An early observation in this field, the Neisser-Wechsberg prozone phenomenon, indicates heterogeneity of some kind in the antibody side of the reaction [12]. When concentrated, some sera are unable to kill bacteria, but further dilution in the presence of a standard complement system allows killing (table 4). Both sera are active up to a dilution of $1:10^6$, but serum A shows a marked concentration prozone. At one time I thought this could be explained by binding of only one combining site in each IgG molecule, which would prevent the cross-linking by complement of two f(c) ends of two adjacent IgG molecules (figure 1). This explanation now seems inadequate since, contrary to expectation, one can get prozone reactions in predominantly IgM sera. Kearney and Halliday [13] recently offered a well-supported explanation for similar prozones in hemolytic re-

Table 4. The prozone effect.

Antiserum to *Vibrio cholerae*	Dilutions of antisera								
	10^{-1}	10^{-2}	10^{-3}	10^{-4}	10^{-5}	10^{-6}	10^{-7}	10^{-8}	C*
A	cf†	cf	cf	10‡	5	8	100	cf	cf
B	0	3	2	6	10	3	80	cf	cf

* C = complement control.

† cf = confluent growth.

‡ Figures indicate the numbers of bacterial colonies obtained by plating 0.1 ml after incubation for 2 hr at 37 C.

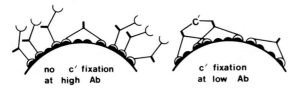

Figure 1. A possible explanation of the prozone effect. C′ = complement; Ab = antibody.

actions. This scheme depends on the greater avidity of noncomplement-fixing IgA, which at high concentrations in serum occupies most of the available antigenic sites to the exclusion of the active IgM and IgG molecules. This explanation is further supported by the work of Eddie et al., who purified different classes of antibody against *S. typhimurium* [14]. These authors found that IgA not only was inactive in the bactericidal reaction but also inhibited an otherwise effective IgM system.

Another simple experiment will further illustrate antibody heterogeneity in this system (table 5). At stage 2, we had a serum that behaved as antiserum to antigen 5 (anti-5). When this was absorbed onto C5 lipopolysaccharide containing antigen 5 and then eluted from the washed complex with glycine-HCl buffer, the eluate reacted with lipopolysaccharide-coated cells like anti-5 but was nonbactericidal. This eluted antibody was noncomplement-fixing, either because of damage during elution or more likely because it had always been so; therefore, the antibody in stage two was a mixture of noncomplement-fixing antibody with antibody having a high ratio of bactericidal-to-

hemagglutinating activity (although both had the same antigenic specificity). Obviously, the two antibodies had different avidity and elution characteristics.

The main heterogeneities of antibodies, apart from specificity, are those of class and complement-fixing ability. It might be helpful to know which antibody class is functionally the most efficient in terms of killing gram-negative bacteria, particularly since IgA predominates in the intestine, where most of these organisms begin to cause their pathogenic effects. We have been trying to purify IgG, IgM, and IgA containing anti-*V. cholerae* specificity from the secretions or sera of immunized dogs, mice, or rabbits. We have experienced extraordinary difficulty in inducing formation of specific antibody in the IgA class, even by prolonged administration of live organisms either by mouth or locally into the mammary gland, etc. In general, the more our IgA fractions have been purified, the lower the specific antibody activity, much of which originally belonged to contaminating IgM. It has not been as difficult to achieve immunoglobulin purity by gel-diffusion tests, but the IgA preparations would usually reveal their impurities by possessing vibriocidal activity; this is an extraordinarily sensitive test. It seems probable that when these immunoglobulins have been further purified by immunoadsorbents, IgA will be devoid of all complement-mediated bactericidal properties, just as Eddie et al. found with their material [14]. Pure IgG and IgM are not so difficult to obtain, and in table 6 the ratio of vibrio-

Table 5. Heterogeneity of antibodies in a rabbit antiserum to *Citrobacter*.

Antiserum	HA titer vs. RBC* sensitized with		Bactericidal titer vs. *Salmonella typhimurium* M206
	Native C5 LPS† (4, 5)	C5 LPS alk. sensitized (4)	
(1) *Citrobacter* (4, 5)	1:1,280	1:320	1:10^5
(2) Absorbed 2 times with 200 μg of *Salmonella typhimurium* ST11 LPS (1, 4, 12)/ml	1:640	<1:10	1:2 × 10^4
(3) Absorbed 2 times with 200 μg of C5 LPS (1, 4, 5, 12)/ml	<1:2	<1:2	1:2 × 10^3
(4) Eluate from C5 LPS of (3) adjusted to same volume	1:256	<1:2	<1:20

* RBC = red blood cell.
† LPS = lipopolysaccharide.

Table 6. Variable efficiency of immunoglobulin classes.

Animal species	Vibriocidal:hemagglutinating ratio in		
	IgG	IgM	IgA
Dog	...	−10,000	4–10 (Heddle, R. J.)
Rabbit	200	− 1,000	2–4 (Steele, T.)
Mouse	...	−10^6	<1–50 (Horsfall, D.)

cidal to hemagglutinating activities in our purified fractions is assembled. Incidentally, in our experience, the presence of lysozyme makes very little difference to the final bactericidal titer of any of the immunoglobulin classes (J. Knop, personal communication); this contradicts the results of Glynn and Adinolfi with human IgA [15].

It may be interesting to point out that in the protection test with baby mice, IgA is the most effective immunoglobulin class when evaluation is based on the hemagglutination unit.

In table 6 the IgM of mice is seen to be remarkably more active than that from the other two animal species. We have no explanation for this interspecies heterogeneity.

Heterogeneity of Complement Source

Some of the differences among species in antibody efficiency may be due to incompatibilities with the particular complement system used. Most people working on the bactericidal reaction have chosen to use guinea pig or fetal-calf serum as sources of complement, mainly because they work better than others. It is obvious that interactions between serum proteins of one animal and antibodies of others do occur, and these interactions might lead to fixation and deviation of complement from the bactericidal system. Great variations have been noted with different species of complement, particularly in the hemolytic reaction. In general, if one's interests are ultimately in reactions of the host against bacteria, it would seem best to use autologous serum as a source of complement.

I have stressed possible variations in antigens, antibodies, and complement, but I must mention an additional one. Bacteria are not inanimate bundles of antigens in fixed proportions; on the contrary, their composition can vary enormously depending on growth conditions. These phenotypic variations can exert an influence on the bacterial sensitivity to antibody and complement. In figure

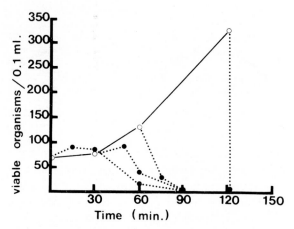

Figure 2. Changes in sensitivity of *Escherichia coli* to bactericidal action of serum at different phases in the growth cycle. (○—○) = increase in viable count after inoculation of the organisms into heat-inactivated serum and incubation at 37 C. (●···●) = death-time curves of samples taken from this culture and immediately transferred to unheated serum. (Reprinted with permission from the *Journal of General Microbiology* 18:529–533, 1958.)

2, sensitivities to a standard antibody-complement system, determined at various times in the growth curve, are shown [16]. It is clear that the bacteria used were most sensitive when rapidly growing. I have proposed elsewhere [17] that this fact could be due to rapidly growing bacteria that possess bare areas of R polysaccharides, which are left behind and exposed because the sugar transferases could not keep pace and link onto the oligosaccharide repeating units of the smooth O antigen.

Of all the variables of the bactericidal reaction, the one that most requires consideration in future work is the specificity of the antiserum used. Absorption to remove known antibody contaminants is clearly unacceptable, and the belief that one is working with monovalent antisera can probably be guaranteed only by absorption and elution of the antibody from an antigen-coupled immunoadsorbent column. The known structure of many Kauffman-White determinants should make it possible at least to compare precisely the efficiencies of antisera to antigens 4, 5, 12, etc.

References

1. Rowley, D. Some factors affecting the resistance of animals to bacterial infection. Ann. N. Y. Acad. Sci. 66:304–311, 1956.

2. Muschel, L. H. Immune bactericidal and bacteriolytic reactions. *In* G. E. W. Wolstenholme and J. Knight [ed.]. Ciba Foundation Symposium on Complement. Little, Brown, Boston, 1965, p. 155–174.

3. Adler, F. L. Studies on the bactericidal reaction. II. Inhibition by antibody, and antibody requirements of the reaction. J. Immunol. 70:79–88, 1953.

4. Rowley, D., Turner, K. J. Passive sensitization of *Salmonella adelaide* to the bactericidal action of antibody and complement. Nature (Lond.) 217:657–658, 1968.

5. Daguillard, F., Edsall, G. The agglutinating and bactericidal activity of IgM and IgG antibodies to the 9 and 12 factors of *Salmonella typhi* 0901. J. Immunol. 100:1112–1120, 1968.

6. Neoh, S. H., Rowley, D. The protection of infant mice against cholera by antibodies to three *Vibrio cholerae* antigens. J. Infect. Dis. 126:41–47, 1972.

7. Domingue, G. J., Neter, E. Opsonizing and bactericidal activity of antibodies against the common antigen of Enterobacteriaceae. J. Bacteriol. 91:129–133, 1966.

8. Chedid, L., Parant, M., Parant, F., Boyer, F. A proposed mechanism for natural immunity to enterobacterial pathogens. J. Immunol. 100:292–301, 1968.

9. Reynolds, B. L., Rowley, D. Sensitization of complement resistant bacterial strains. Nature (Lond.) 221:1259–1261, 1969.

10. Reynolds, B. L., Pruul, H. Sensitization of complement-resistant smooth gram-negative bacterial strains. Infect. Immun. 3:365–372, 1971.

11. Reynolds, B. L., Pruul, H. Protective role of smooth lipopolysaccharide in the serum bactericidal reaction. Infect. Immun. 4:764–771, 1971.

12. Neisser, M., Wechsberg, F. Münch. Med. Wschr. 48:697, 1901.

13. Kearney, R., Halliday, W. J. Immunity and paralysis in mice. Serological and biological properties of two distinct antibodies to type III pneumococcal polysaccharide. Immunology 19:551–560, 1970.

14. Eddie, D. S., Schulkind, M. L., Robbins, J. B. The isolation and biologic activities of purified secretory IgA and IgG anti-*Salmonella typhimurium* 'O' antibodies from rabbit intestinal fluid and colostrum. J. Immunol. 106:181–190, 1971.

15. Adinolfi, M., Glynn, A. A., Lindsay, M., Milne, C. M. Serological properties of gamma-A antibodies to *Escherichia coli* present in human colostrum. Immunology 10:517–526, 1966.

16. Rowley, D., Wardlaw, A. C. Lysis of gram-negative bacteria by serum. J. Gen. Microbiol. 18:529–533, 1958.

17. Rowley, D. Endotoxins and bacterial virulence. J. Infect. Dis. 123:317–327, 1971.

Contributions of the Classical and Alternate Complement Pathways to the Biological Effects of Endotoxin

Michael M. Frank, Joseph E. May, and Michael A. Kane

From the Laboratory of Clinical Investigation, National Institute of Allergy and Infectious Diseases, National Institutes of Health, Bethesda, Maryland

The interactions of endotoxin with sera from normal guinea pigs and sera from C4-deficient (C4D) guinea pigs that had a complete block in function of the classical complement sequence were examined. Endotoxin fixed the late components of complement (C3-C9) in both normal and C4D sera, demonstrating the existence of an alternate pathway that bypasses the early components of complement. Activation of the alternate pathway generates normal quantities of chemotactic factors, and the kinetics of chemotactic-factor generation is normal. The alternate pathway can opsonize bacteria and mediate the bactericidal reaction, although a difference in kinetics was observed. However, injection of endotoxin into normal and C4D animals revealed that activation of the classical pathway was an absolute requirement for thrombocytopenia and for induction of the hypercoagulable state.

At the First International Symposium on Endotoxin, Dr. Ivan Bennett summarized the difficulties encountered in defining the mechanisms involved in endotoxin action in vivo [1]. Almost every system one might care to study is altered by endotoxin, and the assignment of responsibility for one or another effect of endotoxin is therefore difficult. A productive approach to this problem is afforded by the study of animals in which a single physiologic or biochemical pathway is blocked and in which the consequences of endotoxin administration can be studied directly. The simplest of these is the study of animals with a biochemical block due to a genetically controlled defect in the synthesis of a specific protein. Such an approach was used in the present studies.

Some of the physiologic and biochemical consequences of the interaction of endotoxins with the complement system have been explored in a strain of guinea pigs with a genetic defect in the synthesis of C4. C4 deficiency is inherited as a simple mendelian recessive trait [2]. Heterozygous animals have 5%–50% of the normal levels of C4. Homozygous, deficient animals have none

(figure 1). Absence of C4 in serum has been shown to be due to a defect in synthesis of the C4 protein [3]. In-vivo and in-vitro studies of the sera of these animals have provided new insight into the mechanisms of endotoxin action. Some of these studies are reviewed here.

Materials and Methods

Normal, multipurpose NIH guinea pigs of the strain from which the C4-deficient (C4D) animals were derived, were used as controls for studies with heterozygous and homozygous deficient animals. Methods for the titration of C1, C2, and the C3–C9 complement-component complex are described elsewhere [4, 5]. Two methods were used for titration of C4. A classical method involved the sequential interaction of the components in the complement sequence. A newly developed method used C4-deficient serum as a reagent containing an excess of all complement components except C4. Details of this method will be reported elsewhere.[1] For various studies preparations of the endotoxin from *Escherichia coli* 0127:B8 and 011:B4 (Difco Laboratories, Detroit, Mich.) dissolved in sterile saline were used. All groups

Please address requests for reprints to Dr. Michael M. Frank, Bldg. 10, Rm. 11B-13, National Institutes of Health, Bethesda, Md. 20014.

[1] M. M. Frank and T. Gaither, "A New Method for the Titration of C4," manuscript in preparation.

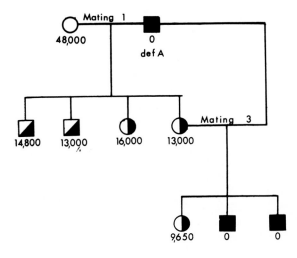

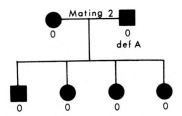

Figure 1. Mating studies with C4-deficient (C4D) animals: ■ = C4D male; ● = C4D female; ◖ = heterozygous female; ◪ = heterozygous male; and ○ = normal female. The serum C4 titer of each animal is indicated below the appropriate symbol.

of animals were simultaneously injected with the same lot of dissolved material. Details of the generation and assay of chemotactic preparations are to be reported elsewhere,[2] as are details of the bactericidal and phagocytic assays [6].

In studies with animals treated with cobravenom factor (CVF), CVF (Cordis Corp., Miami, Fla.) was injected ip into guinea pigs; the dose was 20 units/100 g body weight. Animals were used for experiments 20 hr later. Before experimental use, a sample of serum was obtained from each animal for titration of C3–C9. Only animals with less than 1% of the normal levels of C3–C9 were used. Studies in this laboratory have confirmed previous observations that administration of CVF leads to depletion of the

[2] R. A. Clark, M. M. Frank, and H. R. Kimball, "Generation of Chemotactic Factors in Normal and C4-deficient Guinea Pig Serum by Activation with Endotoxin and Complexes," manuscript in preparation.

late complement components but spares C1, C4, and C2 [7].

Results

Incubation of *E. coli* endotoxin with fresh sera of normal animals, heterozygous C4D animals, and homozygous C4D animals led to fixation of the late components of complement (C3–C9), with relative sparing of the components C1, C4, and C2 (figure 2). Thus, a mechanism existed in C4D serum for activation of the late components of complement in the absence of function of the classical C1, C4, and C2 pathway.

Addition of endotoxin to C4D serum led to generation of chemotactic activity. As shown in figure 3, kinetics of generation of chemotactic factor in sera from C4D animals were identical with kinetics in sera from normal guinea pigs, and equal amounts of activity were generated. Although not shown in figure 3, analysis of chemotactic activity has shown that in both C4D and normal sera, the activity resides in two small fragments with molecular weights of about 10,000 and 17,000, respectively; these fragments derive

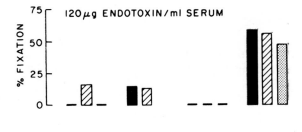

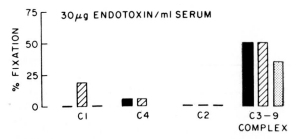

Figure 2. Fixation of complement components by endotoxin from *Escherichia coli*. Comparison of pattern of component fixation in sera from normal (■), heterozygous C4-deficient (▨), and homozygous C4-deficient (▧) guinea pigs.

170

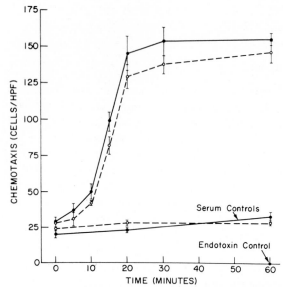

Figure 3. Generation of chemotactic factor by endotoxin in sera from normal (●——●) and C4-deficient (○– – –○) guinea pigs. The endotoxin control contained no serum; serum controls contained no endotoxin; activated samples contained 300 µg of *Escherichia coli* endotoxin/ml of serum. Incubation time at 37 C was varied from 0 to 60 min. All samples were then heated at 56 C for 30 min. None of the differences between normal and C4-deficient serum was significant.

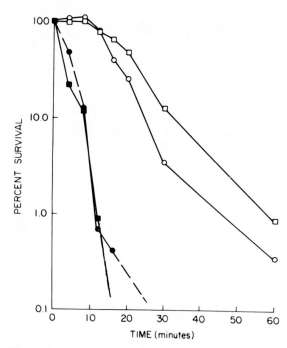

Figure 4. Bactericidal activity of sera of 100% normal (■——■), 50% normal (●——●), 100% C4-deficient (□——□), and 50% C4-deficient (○——○) guinea pigs. Effect of increasing the concentration of normal or C4-deficient serum from 50% to 100% and of prolonging the incubation time to 60 min on the survival of *Escherichia coli* C.

from cleavage of the complement component C5.[3]

As shown in figure 4, C4D serum is capable of destroying gram-negative bacteria. Unlike normal guinea-pig serum, which can kill 99% of the complement-sensitive strain of *E. coli* used in this study within 15 min, C4D serum demonstrated a marked lag before bactericidal action was evident, and a 60-min incubation period was required before 99% of the bacteria were nonviable. A less marked defect has been noted in studies of the opsonizing properties of C4D serum [6].

The experiments listed above demonstrate that endotoxins activate the alternate complement pathway in C4D serum in vitro. The next series of experiments examined some of the biologic consequences of administration of endotoxin into C4D guinea pigs.[4] When endotoxin was adminis-

[3] See footnote 2.
[4] M. A. Kaue, J. E. May, and M. M. Frank, "Interactions of the Classical and Alternate Complement Pathway with Endotoxin Lipopolysaccharide: Effect on Platelets and Blood Coagulation," manuscript in preparation.

tered intravenously to normal guinea pigs in doses of 1 mg/kg, there was a prompt fall in the platelet count. Fifteen minutes after injection, platelet levels were 5% ± 1% SE of the initial values. Neither C4D nor CVF-treated guinea pigs sustained a fall in platelet count (figure 5).

The effect of endotoxin administration on production of the hypercoagulable state was also explored. Both C4D and CVF-treated guinea pigs had normal clotting times in raw glass and silicon-coated glass tubes (table 1). When normal guinea pigs were injected with endotoxin, there was a prompt shortening of the clotting time, evidence for the induction of the hypercoagulable state. This was not observed with either CVF-treated normal guinea pigs or C4D guinea pigs.

Despite the failure of C4D animals to develop thrombocytopenia or the hypercoagulable state, injection of endotoxin did lead to activation of the alternate complement pathway, as evidenced by a fall in serum titer of C3–C9, although the

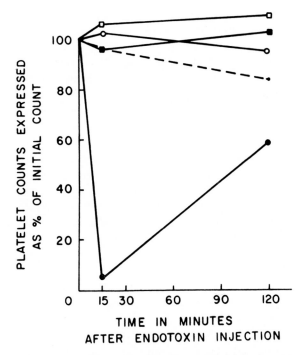

Figure 5. Effect of endotoxin on platelets in normal (●——●), C4-deficient (■——■), and cobra-venom factor (CVF)-treated (●– – –●) guinea pigs; (○——○) = normal controls; (□——□) = C4-deficient controls (all controls received saline). Shown in the figure are geometric means obtained at each point. Standard errors are not shown to increase clarity. The values at 15 min and the number of animals studied are as follows: endotoxin-treated normal animals, 5% ± 1% [8]; endotoxin-treated C4-deficient animals, 96% ± 14% [8]; saline-treated normal animals, 103% ± 8% [8]; saline-treated C4-deficient animals, 106% ± 11.6% [8]; normal animals treated with cobra-venom factor and given endotoxin, 95% ± 5% [5].

Table 2. Consumption of C4 and C3-C9 after injection of endotoxin.

Animals (no.)	% of C4 consumed	% of C3-C9 consumed
Normal (10)	12 ± 2*	29 ± 4
Normal control (10)	0 ± 2	0 ± 0.7
C4D (10)	—	16 ± 3
C4D control (10)	—	0 ± 0.6

NOTE. Blood for complement studies was taken just before and 15 min after injection of endotoxin. Control animals were bled at zero time and at 15 min without receiving endotoxin.

* Mean consumption ± SE of mean.

fall was somewhat less than that noted in normals (table 2).

Discussion

Addition of bacterial endotoxins to fresh serum leads to fixation of the late components of complement with relative sparing of C1, C4, and C2 [8]. When this observation was made with use of modern molecular assays, it raised the possibility of the existence of an alternate pathway for complement activation that could bypass the early components (C1, C4, and C2) and enter the complement sequence at the level of C3. However, another explanation of the data was that endotoxins, perhaps in concert with other serum factors (for example, after interaction with a particular subclass of antibody), used C1, C4, and C2 in very small amounts and thus used the clas-

Table 1. Alteration by endotoxin of clotting time in normal, C4-deficient (C4D), and cobra-venom factor (CVF)-treated animals.

Animals	Clotting time (min) before endotoxin		Clotting time 15 min after injection of endotoxin	
	Glass	Silicon	Glass	Silicon
Normal	12.53 ± 0.46 (15)*	29.2 ± 1.71 (15)	11.18 ± 0.46 (12)	19.25 ± 1.49 (12)
C4D	12.60 ± 0.49 (15)	30.2 ± 1.81 (15)	13.0 ± 0.52 (12)	33.91 ± 2.25 (12)
CVF-treated normal	13.37 ± 0.56 (8)	27.5 ± 1.83 (8)	11.66 ± 0.21 (6)	32.0 ± 3.34 (6)
C4D with restored C4†	12.16 ± 0.54 (6)	34.1 ± 2.21 (6)	10.83 ± 0.49 (6)	22.66 ± 0.71 (6)

* Mean ± SE (number of animals).

† See text.

172

sical pathway, but with great efficiency. It has been demonstrated that different classes of immunoglobulins use complement components with various degrees of efficiency [9], and data was presented to support the possibility that activation of C3–C9 by endotoxin did proceed through activation of small amounts of C1, C4, and C2 [10]. The availability of C4D guinea pigs allowed this point to be tested directly, because C4D animals have a complete block in the classical complement pathway. As shown in figure 2, endotoxin fixed the late components of complement in C4D serum, clearly demonstrating that an alternate pathway exists for activation and cleavage of the late components in the complement sequence. This mechanism can lead to production of normal amounts of chemotactic factors, which are thought to be essential to the inflammatory response (figure 3). The alternate pathway can opsonize bacteria and participate in the bactericidal reaction (figure 4), although the kinetics of the latter reaction are unlike those seen with normal serum: bacterial cell killing proceeds slowly and after a latent period. As a result of these experiments and those of other workers in the field, endotoxins have come to be considered potent activators of the alternate complement pathway.

The administration of endotoxin to animals of many species is followed by a precipitous fall in platelet count and by the development of the hypercoagulable state. There has been considerable controversy in the literature as to whether or not complement is responsible for these effects; in-vitro conditions for the reproduction of some of these events have not been those that maximize complement fixation or complement-dependent cytotoxic reactions [11]. Nevertheless, factors present in fresh serum are required for this reaction, as are divalent cations. It was of considerable interest to determine the effects of the injection of endotoxin into C4D animals that had an intact alternate pathway but a nonoperative classical complement pathway.

Normal guinea pigs given endotoxin sustained a marked fall in platelet count and induction of the hypercoagulable state (figure 5 and table 1). Neither C4D nor CVF-treated guinea pigs experienced a fall in blood platelets; thus, this effect is clearly complement-related and, moreover, proceeds through activation of the early components of complement (C1, C4, and C2) to activa-

tion and cleavage of the late components (C3–C9). If activation of the early components of complement alone were sufficient to induce this effect, CVF-treated animals with normal levels of C1, C4, and C2 should have developed thrombocytopenia. Neither CVF-treated normal nor C4D animals had shortening of their clotting times after injection of endotoxin. These effects, which parallel those noted above, indicate that the hypercoagulable state produced in guinea pigs on the injection of endotoxin requires the participation of the classical pathway and activates the late components of complement to produce its effect. That the alternate pathway was activated by endotoxin in C4D animals was evidenced by a reduction in titer of C3–C9 in these animals (table 2).

It is likely that the fall in platelet count, with attendant damage to the platelet membrane, is causally associated with the production of the hypercoagulable state. Data from our laboratory clearly indicate that antibodies directed at antigens localized on intact mammalian membranes (that is, antibodies directed at cell-surface antigens) are unable to activate directly the alternate complement pathway [12]. When endotoxin is injected iv, a portion of the material attaches to platelets [13]. We suggest that (1) this association forms the equivalent of membrane-associated antigen, and (2) damage to these membranes can be mediated only through the participation of natural antibody and the sequential action of C1, C4, and C2, and thus not by the alternate pathway.

References

1. Bennett, I. L. Introduction. *In* M. Landy and W. Braun [ed.] Bacterial endotoxins. Institute of Microbiology, Rutgers, New Brunswick, N.J., 1964, p. xiii–xvi.
2. Ellman, L., Green, I., Frank, M. Genetically controlled total deficiency of the fourth component of complement in the guinea pig. Science 170: 74–75, 1970.
3. Colten, H. R., Frank, M. M. Biosynthesis of the second (C2) and fourth (C4) components of complement in vitro by tissues isolated from guinea pigs with genetically determined C4 deficiency. Immunology 22:991, 1972.
4. Mayer, M. M. Complement and complement fixation. *In* E. A. Kabat and M. M. Mayer. Experimental immunochemistry. 2d ed. Charles C Thomas, Springfield, Ill., 1961, p. 133–240.

5. Rapp, H. J., Borsos, T. Molecular basis of complement action. Appleton-Century-Crofts, New York, 1970, p. 75–109.

6. Root, R. K., Ellman, L., Frank, M. M. Bactericidal and opsonic properties of C4 deficient guinea pig serum. J. Immunol. 109:477, 1972.

7. Shin, H. S., Gewurz, H., Snyderman, R. Reaction of a cobra venom factor with guinea pig complement and generation of an activity chemotactic for polymorphonuclear leukocytes. Proc. Soc. Exp. Biol. Med. 131:203–207, 1969.

8. Gewurz, H., Shin, H. S., Mergenhagen, S. E. Interactions of the complement system with endotoxic lipopolysaccharide: consumption of each of the six terminal complement components. J. Exp. Med. 128:1049–1057, 1968.

9. Frank, M. M., Gaither, T. Evidence that rabbit gamma-G haemolysin is capable of utilizing guinea-pig complement more efficiently than rabbit gamma M haemolysin. Immunology 19:975–981, 1970.

10. Snyderman, R., Gewurz, H., Mergenhagen, S. E., Jensen, J. Effect of C4 depletion on the utilization of the terminal components of guinea-pig complement by endotoxin. Nature (Lond.) 231:152–154, 1971.

11. Des Prez, R. M., Bryant, R. E. Effects of bacterial endotoxin on rabbit platelets IV. The divalent ion requirements of endotoxin induced and immunologically induced platelet injury. J. Exp. Med. 124:971–982, 1966.

12. May, J. E., Green, I., Frank, M. M. The alternate complement pathway in cell damage-antibody mediated cytolysis of erythrocytes and nucleated cells. J. Immunol. 109:595, 1972.

13. Herring, W. B., Herion, J. C., Walker, R. I., Palmer, J. G. Distribution and clearance of circulating endotoxin. J. Clin. Invest. 42:79–87, 1963.

Factors in Normal Human Serum that Promote Bacterial Phagocytosis

Ann B. Bjornson and J. Gabriel Michael

From the Department of Microbiology, University of Cincinnati College of Medicine, Cincinnati, Ohio

Heat-stable and heat-labile factors of normal human serum involved in opsonophagocytosis and intracellular killing of *Pseudomonas aeruginosa* by human polymorphonuclear leukocytes were investigated. Specific natural IgG antibodies, properdin, and a heat-labile factor (presumably C3 proactivator) were all required for phagocytosis and bactericidal activity. This finding suggested that an alternate pathway of the complement system was activated. Addition of immune IgG antibodies to the system eliminated the requirement for properdin and the heat-labile factor. Thus, two different mechanisms of complement activation can participate in the opsonophagocytosis of *P. aeruginosa*.

The capacity of normal animals to resist microbial infection depends on numerous humoral and cellular factors, which, when acting in concert, provide adequate antimicrobial protection in the absence of acquired specific immunity. Both heat-labile and heat-stable factors in normal serum aid in the ingestion and killing of extracellular microorganisms by phagocytic cells [1]. The heat-labile factors in normal serum that enhance phagocytosis are components of the complement system [2, 3]. The heat-stable natural antibodies of both classes IgG and IgM are present in sera of all normal adult animals and are directed against a wide array of microbial species [4]. It has been well documented that the bacteriolytic activity of normal serum is dependent on the synergistic action of these antibodies and complement [5, 6]. However, the contribution of these factors to the opsonic activity of normal serum remains controversial. In the present study, the factors of normal human serum that promote opsonophagocytosis and intracellular killing of bacteria by human polymorphonuclear leukocytes (PMNs) were investigated. A strain of *Pseudomonas aeruginosa* was selected for our experiments. This bacterium

causes infection in debilitated patients but is usually not pathogenic for normal individuals.

Materials and Methods

Preparations of immune IgG were kindly provided by Dr. M. Fisher, Parke-Davis, Detroit, Mich. These preparations were purified by Cohn fractionation of plasma obtained from volunteers immunized with a polyvalent pseudomonas vaccine consisting of seven serotypes of *P. aeruginosa*. Normal human IgG was prepared by Cohn fractionation of plasma from healthy adult donors and was provided by Ortho Research Foundation, Raritan, N.J. According to immunoelectrophoretic analysis, immune and normal gamma globulin preparations contained only IgG [7]. Drs. Fisher and Devlin also kindly provided us with *P. aeruginosa* strains and antigens. The methods for bacterial classification and preparation of antigens have been previously described [8]. The strain of *P. aeruginosa* used was classified as serotype 1. Highly purified human properdin [9] was kindly provided by Dr. I. Lepow, University of Connecticut Medical Center, Farmington, Conn. Goat antiserum to human C3 A and D [10] was provided by Dr. R. Spitzer, Children's Hospital Research Foundation, Cincinnati, Ohio.

A modification of the in-vitro system of Maaloe was used for the measurement of opsonophagocytosis [11]. Human peripheral leukocytes from dextran-sedimented blood were washed once and resuspended in Hanks' balanced salt solution (HBSS) containing 10 units of heparin/ml. Re-

This study was supported in part by Parke-Davis & Co., Detroit, Michigan.

We thank Drs. M. Fisher, I. Lepow, and R. Spitzer for preparations kindly donated to us and used in this investigation.

Please address requests for reprints to Dr. J. Gabriel Michael, Department of Microbiology, University of Cincinnati College of Medicine, Cincinnati, Ohio 45219.

action mixtures consisted of opsonized bacteria and leukocytes. Bacteria were opsonized by incubation of equal volumes of opsonins and bacterial suspension at 37 C for 1 hr and at 4 C overnight. The bacteria were then washed three times with sterile saline. The reaction mixtures were tumbled end over end on a rotator for 3 hr at 37 C; samples were then removed, diluted in distilled water for disruption of the leukocytes, and plated on brain-heart-infusion agar. The plates were incubated at 37 C overnight, and the colonies were enumerated for determination of the total killing that had occurred in each reaction mixture during the period of incubation.

The number of bacteria in each reaction mixture at zero time varied from 2.0×10^5 to 1.0×10^6 cells/ml. The number of leukocytes in each reaction mixture was 1.0×10^6 cells/ml. The killing effect at these ratios of leukocytes to bacteria was constant. At the end of the incubation period, smears of the reaction mixtures were prepared, stained with Giemsa, and examined microscopically. In all instances in which there was bactericidal activity, the bacteria were inside the PMNs.

Results

As is shown in table 1, fresh normal human serum promoted phagocytosis and bactericidal killing by human PMNs. Heating of the serum at 56 C for 30 min abolished its phagocytosis-promoting ability, indicating that the opsonins were heat-labile. Absorption of the normal serum with antiserum to human C3 for 15 min at 37 C before opsonization reduced its opsonic activity, providing evidence that components of complement were involved in the opsonization process (table 2).

Table 1. Requirement for natural IgG antibodies and complement for the opsonization of *Pseudomonas aeruginosa* by normal human serum (NHS).

Bacteria opsonized with	Leukocytes	No. of bacteria recovered after 3-hr incubation (as $\log_{10}$)	Reduction in bacterial counts (as $\log_{10}$)
NHS	Present	3.74	1.71
	Absent	5.45	
Heated NHS (56 C, 30 min)	Present	6.11	−0.10
	Absent	6.01	
NHS absorbed with antiserum to C3	Present	5.81	0.29
	Absent	6.10	
NHS absorbed with *Pseudomonas* (serotype 4)	Present	3.78	1.97
	Absent	5.75	
NHS absorbed with *Pseudomonas* (serotype 1)	Present	5.81	0.13
	Absent	5.94	
NHS absorbed with *Pseudomonas* (serotype 1) plus normal IgG (3.30 mg/ml)	Present	3.74	2.16
	Absent	5.90	
NHS absorbed with *Pseudomonas* (serotype 1) plus normal IgG absorbed with serotype 1 (3.30 mg/ml)	Present	5.78	−0.08
	Absent	5.70	
NHS absorbed with *Pseudomonas* (serotype 1) plus heated normal IgG (56 C, 30 min; 3.30 mg/ml)	Present	3.77	2.12
	Absent	5.89	
NHS absorbed with *Pseudomonas* (serotype 1) plus properdin (22 μg/ml)	Present	5.56	0.19
	Absent	5.75	

Table 2. Requirement for properdin and heat-labile factors for the opsonization of *Pseudomonas aeruginosa* by normal human serum (NHS).

Bacteria opsonized with	Leukocytes	No. of bacteria recovered after 3-hr incubation (as $\log_{10}$)	Reduction in bacterial counts (as $\log_{10}$)
NHS	Present	3.74	1.71
	Absent	5.45	
RP*	Present	5.64	0.04
	Absent	5.68	
RP plus properdin (22 µg/ml)	Present	3.91	1.69
	Absent	5.60	
RP plus normal IgG (3.30 mg/ml)	Present	5.78	−0.01
	Absent	5.77	
RP plus immune IgG (3.30 mg/ml)	Present	3.56	2.74
	Absent	6.30	
Heated NHS (50 C, 30 min)	Present	5.48	−0.06
	Absent	5.42	
Heated NHS plus properdin (22 µg/ml)	Present	5.77	−0.05
	Absent	5.72	
Heated NHS plus normal IgG (3.30 mg/ml)	Present	5.11	0.23
	Absent	5.34	
Heated NHS plus immune IgG (3.30 mg/ml)	Present	4.18	2.06
	Absent	6.24	

* Normal serum absorbed twice with 2 mg of zymosan/ml at 17 C.

For assessment of the role of natural antibodies in the opsonization of pseudomonas, normal serum was absorbed at 17 C with homologous (serotype 1) and heterologous (serotype 4) bacteria. Only absorption of the serum with homologous bacteria reduced its opsonic activity. This activity could be restored by addition of normal IgG. If, however, normal IgG was first absorbed with homologous bacteria, it did not restore opsonic activity. Thus, it became evident that, in addition to complement, specific antibodies to the bacteria were required for opsonophagocytosis of *P. aeruginosa*.

Another factor that has long been implicated in natural resistance to infection but remains a source of controversy is properdin. Recently, a renewed interest in properdin has been generated, since this protein has been purified and shown to be distinct from known immunoglobulins and complement components [9].

We attempted to determine whether properdin plays any role in opsonophagocytosis. Normal serum was absorbed twice at 17 C with 2 mg of zymosan/ml [12]. As is shown in table 2, opsonic activity of the absorbed serum (RP) was

eliminated, but it was restored by addition of purified properdin or by immune IgG. Antibody present in normal IgG did not restore the opsonic activity of properdin-depleted serum. These results suggest that, in addition to natural IgG antibodies, properdin was required for the activation of complement on the bacterial surface.

Since it has been suggested that the properdin system [13] and the recently described C3-activator system [14] represent the same pathway of complement activation [15], we considered the possibility that activation of complement on the bacterial surface might involve this alternate pathway. As a test of this hypothesis, we heated serum at 50 C for 30 min to inactivate C3 proactivator (C3PA), which is known to be required for C3 activation. As is shown in table 2, this heated serum did not promote opsonophagocytosis; its opsonic activity was not restored by properdin or natural IgG antibodies. Opsonic activity of this serum was restored by immune IgG antibodies. These results provided indirect evidence that the alternate pathway of complement activation might be operative in the opsonization of *P. aeruginosa* by normal human serum.

Studies are now in progress to determine if the heat-labile factor is indeed C3PA.

Discussion

Two different mechanisms of complement activation in opsonophagocytosis of *P. aeruginosa* have been demonstrated. The first requires properdin, natural IgG antibodies, and heat-labile factors and most probably occurs via the alternate pathway; the second is initiated by immune IgG antibodies and does not require these factors. The latter mechanism most probably represents the classical pathway, since studies by other investigators have shown that C1, C4, C2, and C3 are required for opsonophagocytosis in the presence of immune IgG antibodies [16, 17]. Activation of the alternate complement pathway on the bacterial surface would not be unexpected, since *P. aeruginosa* contains endotoxin, an activator of the C3PA system. The apparent requirements for antibody and properdin in the alternate pathway of complement activation demand further investigation.

The complement-fixing activities of natural and immune IgG antibodies to *Pseudomonas* have also been measured by conventional hemolytic tests [18]. In contrast to immune antibodies, which fixed complement very efficiently, natural IgG antibodies did not fix complement. Therefore, we suggest that natural and immune IgG antibodies to *Pseudomonas* initiate complement activation by different mechanisms. Sandberg et al. [19] have recently shown that guinea-pig gamma 1 antibodies activate only terminal components of complement, whereas guinea-pig gamma 2 antibodies activate all nine components. Experiments are now in progress to determine whether natural and immune IgG antibodies to *Pseudomonas* belong to different subclasses. The source of antigenic stimulation for the production of natural antibodies is uncertain, and it is possible that these antibodies are produced as a response to pseudomonas antigens. The term natural merely indicates that these antibodies are present in plasma of individuals who were not knowingly exposed to the homologous antigen. In contrast, immune IgG antibodies were prepared from plasma of volunteers hyperimmunized with a pseudomonas vaccine.

The restoration of opsonophagocytic activity

to RP with purified properdin endorses the role of this protein in natural resistance. In the past, it was suggested that properdin functioned in bacteriolysis [20], viral neutralization [21], and protozoal killing [22]. Landy showed that injection of bacterial endotoxins into experimental animals evoked a rise in antimicrobial resistance accompanied by an elevation of properdin levels in serum [23]. Later, Michael et al. [24] demonstrated that, under similar experimental conditions, levels of specific natural bactericidal antibodies were also substantially increased. Experiments by Pensky et al. [9] showed that both natural antibodies and properdin were required for lysis of bacteria. Although bacteriolysis may play a part in antimicrobial protection, the host's major defense mechanism against microbial infection depends on phagocytosis and intracellular killing.

The significance of reported low levels of properdin in debilitated individuals requires re-evaluation. In early studies Pillemer associated decreased properdin levels in serum with increased susceptibility of irradiated animals to bacterial infection [25]. Low levels of properdin were also found in humans in the terminal stages of cancer [26] and in burned patients [27]. With the availability of highly purified properdin and a newly developed radioimmunoassay for its measurement [28], it is now essential to examine again the role of properdin in these pathologic conditions.

References

1. Ward, H. K., Enders, J. F. Analysis of the opsonic and tropic action of normal and immune sera based on experiments with pneumococcus. J. Exp. Med. 57:527–547, 1933.
2. Smith, M. R., Wood, W. B., Jr. Heat labile opsonins to pneumococcus. I. Participation of complement. J. Exp. Med. 130:1209–1227, 1969.
3. Shin, H. S., Smith, M. R., Wood, W. B., Jr. Heat labile opsonins to pneumococcus. II. Involvement of C3 and C5. J. Exp. Med. 130:1229–1241, 1969.
4. Michael, J. G. Natural antibodies. Curr. Top. Microbiol. Immunol. 48:43–62, 1969.
5. Michael, J. G., Rosen, F. S. Association of "natural" antibodies to gram-negative bacteria with gamma-1-macroglobulins. J. Exp. Med. 118:619–626, 1963.
6. Muschel, L. H., Chamberlin, R. H., Osawa, E. Bactericidal activity of normal serum against

178

bacterial cultures. I. Activity against *Salmonella typhi* strains. Proc. Soc. Exp. Biol. Med. 97: 376–382, 1958.

7. Scheidegger, J. J. Une micro-méthode de l'immuno-électrophorèse. Int. Arch. Allerg. Appl. Immunol. 7:103–110, 1955.

8. Fisher, M. W., Devlin, H. B., Gnabasik, F. J. New immunotype schema for *Pseudomonas aeruginosa* based on protective antigens. J. Bacteriol. 98:835–836, 1969.

9. Pensky, J., Hinz, C. F., Jr., Todd, E. W., Wedgwood, R. J., Boyer, J. T., Lepow, I. H. Properties of highly purified human properdin. J. Immunol. 100:142–158, 1968.

10. Spitzer, R. E., Stitzel, A. E., Pauling, V. L., Davis, N. C., West, C. D. The antigenic and molecular alterations of C3 in the fluid phase during an immune reaction in normal human serum: demonstration of a new conversion product, C3x. J. Exp. Med. 134:656–680, 1971.

11. Maaloe, O. On the relation between alexin and opsonin. Einar Munksgaard, Copenhagen, 1946. 186 p.

12. Todd, E. W., Pillemer, L., Lepow, I. H. The properdin system and immunity. IX. Studies on the purification of human properdin. J. Immunol. 83:418–427, 1959.

13. Pillemer, L., Blum, L., Lepow, I. H., Ross, O. A., Todd, E. W., Wardlaw, A. C. The properdin system and immunity. I. Demonstration and isolation of a new serum protein, properdin, and its role in immune phenomena. Science 120: 279–285, 1954.

14. Götze, O., Müller-Eberhard, H. J. The C3-activator system: an alternate pathway of complement activation. J. Exp. Med. 134(Suppl.):90S–108S, 1971.

15. Lepow, I. H., Rosen, F. S. Pathways to the complement system. N. Engl. J. Med. 286:942–943, 1972.

16. Gigli, I., Nelson, R. A., Jr. Complement dependent immune phagocytosis. I. Requirements for C′1, C′4, C′2, C′3. Exp. Cell Res. 51:45–67, 1968.

17. Johnston, R. B., Jr., Klemperer, M. R., Alper, C. A., Rosen, F. S. The enhancement of bacterial phagocytosis by serum: the role of com-

plement components and two cofactors. J. Exp. Med. 129:1275–1290, 1969.

18. Bjornson, A. B., Michael, J. G. Contribution of humoral and cellular factors to the resistance to experimental infection by *Pseudomonas aeruginosa* in mice. II. Opsonic, agglutinative, and protective capacities of immunoglobulin G anti-*Pseudomonas* antibodies. Infec. Immun. 5:775–782, 1972.

19. Sandberg, A. L., Osler, A. G., Shin, H. S., Oliveira, B. The biologic activities of guinea pig antibodies. II. Modes of complement interaction with gamma 1 and gamma 2-immunoglobulins. J. Immunol. 104:329–334, 1970.

20. Wardlaw, A. C., Pillemer, L. The properdin system and immunity. V. The bactericidal activity of the properdin system. J. Exp. Med. 103:553–575, 1956.

21. Wedgwood, R. J., Ginsberg, H. S., Pillemer, L. The properdin system and immunity. VI. The interaction of Newcastle disease virus by the properdin system. J. Exp. Med. 104:707–725, 1956.

22. Feldman, H. A. The relationship of *Toxoplasma* antibody activator to the serum-properdin system. Ann. N. Y. Acad. Sci. 66:263–267, 1956.

23. Landy, M. Increase in resistance following administration of bacterial lipopolysaccharides. Ann. N. Y. Acad. Sci. 66:292–303, 1956.

24. Michael, J. G., Whitby, J. L., Landy, M. Increase in specific bactericidal antibodies after administration of endotoxin. Nature (Lond.) 191:296–297, 1961.

25. Ross, O. A. The properdin system in relation to fatal bacteremia following total-body irradiation of laboratory animals. Ann. N. Y. Acad. Sci. 66:274–279, 1956.

26. Southam, C. M., Pillemer, L. Serum properdin levels and cancer cell homografts in man. Proc. Soc. Exp. Biol. Med. 96:596–601, 1957.

27. Lowbury, E. J. L., Ricketts, C. R. Properdin and the defense of burns against infection. J. Hyg. (Camb.) 55:266–275, 1957.

28. Minta, J. O., Lepow, I. H. Solid phase radioimmunoassay of properdin [abstract]. Fed. Proc. 31:787, 1972.

Summary of Discussion

Sheldon M. Wolff

The discussion began when Dr. Edward J. Goetzl asked Dr. Robert Horn whether the increase in cationic peptides in the Shwartzman reaction was due to increased amounts of the peptides in the granulocytes or to increased release of preformed cationic peptides. Dr. Horn answered that there is an increased number of immature granulocytes in endotoxin-induced granulocytosis, and that these granulocytes probably have an increased concentration of such peptides. He was questioned by Dr. Jack S. C. Fong regarding the role of granulocytes in the coagulopathy of the Shwartzman reaction. Dr. Horn replied that the effect of granulocytes was probably to potentiate but not to initiate clotting. Dr. Raphael Shulman asked whether thrombocytopenic animals could develop a Shwartzman reaction. Dr. Horn responded that he did not believe that a reasonable model was presently available to answer this question, although he noted that evidence for the presence of thromboplastic activity in platelets was better than that for granulocytes.

Dr. Kenneth L. Melmon was asked by Dr. Otto Westphal to describe the preparation of the lipid-free polysaccharide he used. He answered that lipopolysaccharide (LPS) was obtained by extraction with phenol and was then deesterified by controlled alkaline hydrolysis. The final mixture contained less than 0.05% of the original lipid. Dr. Peter Ward asked whether kallikrein could be activated without Hageman factor; the answer was no, not at present.

The discussion then shifted to Dr. C. R. Jenkin's paper; he was asked by Dr. Edward H. Kass to give the LD$_{50}$ used in his studies. He answered that all animals survived an initial dose of endotoxin of 2 μg. Dr. Werner Braun asked Dr. Jenkin whether the endotoxin was found within phagocytic cells. Dr. Jenkin responded that, with use of whole unlabeled bacteria, 80%–90% appeared to be associated with circulating and marginating cells.

Dr. Kass asked Dr. Michael M. Frank whether the alternate complement pathway could be activated by gram-negative bacterial injections. Dr. Frank replied that he did not know; however, Dr. William McCabe answered, that in gram-negative bacteremia, levels of C3 were normal, but if shock occurred, the C3 levels were significantly depressed. Dr. Frank was asked by Dr. Peter Ward to explain the activation by LPS of the alternate complement pathway without promotion of coagulation. Dr. Frank answered that the studies showing activation of both the alternate complement pathway and coagulation were done in vitro, whereas clotting in his in-vivo system was not promoted, probably due to rapid clearance from the circulation of coagulation-promoting factors.

Dr. John B. Robbins inquired of Dr. Abraham Braude whether the protective effect of his antiserum to endotoxin could be absorbed with the immunizing agent. Dr. Braude answered that absorption with endotoxin always results in contamination with endotoxin.

Dr. Robert C. Skarnes was asked by Dr. Monto Ho whether the endotoxin-detoxifying factor could be related to the absence of interferon production during endotoxin tolerance. He replied that he did not know of any information on this point.

Dr. Louis Chedid was asked by Dr. Richard T. Smith to explain how endotoxin that affects bone-marrow (B) cells could protect against the graft-vs.-host reaction (GVHR) when it is known that the cells responsible for wasting disease in the GVHR are thought to be thymus (T) lymphocytes. Dr. Chedid mentioned two possibilites: firstly, the splenocytes might not be able to proliferate in vivo after incubation with endotoxin; secondly, T cells might be in some as yet unrecognized way influenced by endotoxin in vivo.

The final question asked by Dr. Loretta Leive of Dr. L. Joe Berry, was whether endotoxin might affect the rate of degradation of enzymes rather than just the rate of synthesis. He answered that all of the present evidence suggests no change in degradation rates.

Immunologic and Antineoplastic Effects of Endotoxin: Role of Membranes and Mediation by Cyclic Adenosine-3′,5′-Monophosphate

Werner Braun*

From the Institute of Microbiology, Rutgers University, New Brunswick, New Jersey

The role of endotoxin as it affects immunologic and antineoplastic activities mediated by cyclic adenosine-3′,5′-monophosphate (cAMP) is reviewed. Although evidence suggests that some effects of endotoxin may involve changing levels of cAMP, the mechanism by which endotoxin acts on the adenyl cyclase system is unknown.

The fascination and difficulty of dealing with the biological effects of bacterial endotoxins (LPS) are largely due to the multitude of events that can be triggered in animals exposed to LPS; such a variety of possible effects of LPS reflects the unusually large number of target cells and cellular sites that can be affected directly or indirectly by these practically ubiquitous molecules. In our attempts to dissect and to understand these complex events, we can be aided significantly by developments in related areas that furnish clues about some aspects of LPS activity without elucidating all of the known parameters.

In the last couple of years, there have been some significant developments in our understanding of factors and cells involved in the regulation of immune responses; these developments have permitted better identification of the nature of membrane-associated events and of ensuing intracellular changes in relation to modulations of the magnitude of specific immune responses. In particular, they have permitted us (1) to relate many adjuvant effects (i.e., enhanced antibody formation and enhanced cell-mediated immune responses to alterations in levels of cyclic adenosine-3′,5′-monophosphate (cAMP) within cells involved in the immune responses, and (2) to associate adjuvant-induced changes in endogenous levels of cAMP with altered rates of proliferation of tumor cells. This new information has been obtained in studies with agents other than LPS, but since LPS also are potent adjuvants and possess antitumor activity, I shall relate much of this recent information to certain parameters of endotoxin activity. I shall also present some data, obtained in tests with LPS, that indicate the justification of relating certain aspects of LPS activity to membrane effects that result in altered levels of endogenous cAMP.

Cyclic AMP, which has been recognized for quite some time as the second messenger in hormone-dependent activations of secretory cells, more recently has been found to play an important role in regulating many cell functions, including synthesis of nucleic acids and proliferation of cells [1–7], functions of muscles and nerves, behavioral activities, vision, etc. [1, 2, 8]. In view of this widespread role of endogenous cAMP as a regulator of cellular interactions, it is not surprising that the functions of cells involved in the immune response are also modified by agents that modify levels of endogenous cAMP.

Before we review various studies, let us take a brief, general look at the events that lead to

These studies were supported by grant no. AI 09343 from the National Institutes of Health, by grant no. GB-20162 from the National Science Foundation, and by a grant from the New York Cancer Research Institute.

Please address requests for reprints to Mr. Edward R. Isaacs, Institute of Microbiology, Rutgers University, New Brunswick, New Jersey 08903.

* Deceased, November 19, 1972.

alterations of intracellular levels of cAMP. Interactions between specific receptor sites on cell membranes and various extracellular membrane-active agents can lead to the activation of a membrane-associated enzyme, adenyl cyclase, which converts intracellular adenosine-5′-triphosphate (ATP) into cAMP (figure 1). Cyclic AMP, in turn, will combine with a specific, intracellular, cAMP-binding protein. It thereby activates a variety of kinases (mostly phosphorylating enzymes) that can influence macromolecular biosyntheses, including transcription and translation; it also activates preexisting enzymes [1, 2, 8]. Cyclic AMP, once formed, is quite rapidly converted by phosphodiesterases into inactive AMP.

Among the stimulators of adenyl cyclase activity are the catecholamines, including epinephrine, norepinephrine, and isoproterenol; among the inhibitors of phosphodiesterase activity (and thus the stabilizers of endogenous cAMP levels) are theophylline, caffeine, and papaverine. Our studies have identified double-stranded, synthetic polynucleotides as stimulators of adenyl cyclase activity [9–15].

Recent studies by several groups have shown that cAMP itself, as well as both its more potent dibutyryl derivative and known modifiers of cAMP formation and degradation, can alter the functions of the three major types of cells associated with the immune response: macrophages, thymus-derived (T) cells, and bone-marrow-derived (B) cells.

Macrophages

Unpublished studies by Bolis, Luly, and Braun have indicated the influence of repeated administration of cAMP to mice on the rate of carbon clearance from the circulation (figure 2). The iv injection of 200 µg of cAMP at 6-hr intervals resulted in an initial rise in rates of clearance, expressed as the phagocytic index (α) [16], followed by a slight decline in rates. However, a renewed increase to even higher levels results when injections of cAMP are given at 24-hr intervals. Enhancement of cell functions after modest stimulation of endogenous cAMP levels and reduced or inhibited responses at high levels of stimulation are recurrent themes in all cAMP-mediated events affecting cells involved in the immune response; these responses may account for the diminished effects after repeated injections of cAMP at 6-hr intervals and the renewed stimulation after injections at 24-hr intervals.

As a rule, stimulators of endogenous cAMP levels (including the polynucleotides, the catecholamines, and the methylxanthines) are better modifiers of cell functions than exogenous cAMP itself, presumably due to the instability of extracellular cAMP and the difficulty it has in getting into the cell. Accordingly, it is not surprising that

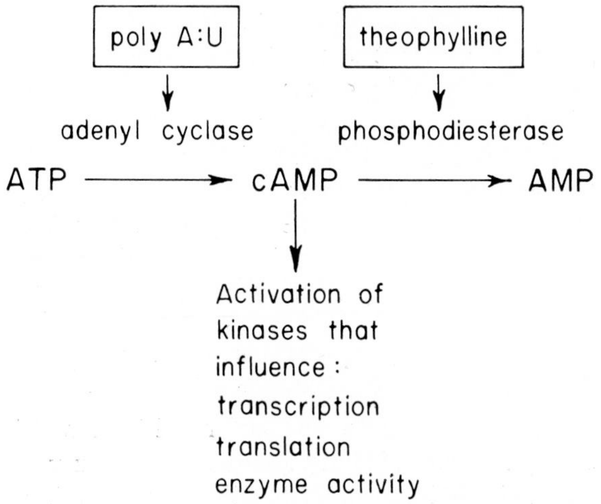

poly A:U → adenyl cyclase

theophylline → phosphodiesterase

ATP ⟶ cAMP ⟶ AMP

cAMP ↓

Activation of kinases that influence:

transcription

translation

enzyme activity

Figure 1. Steps in formation and degradation of cyclic adenosine 5′-monophosphate (cAMP). (Poly A:U = polyriboadenylic:polyribouridylic acid.)

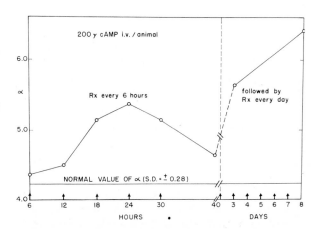

Figure 2. Effect of injections of cyclic adenosine-3′,5′-monophosphate (indicated by arrows), 200 µg/mouse iv per injection, on the clearance of carbon from the circulation. All values show the corrected phagocytic index α (data collected by L. Bolis and P. Luly).

polyriboadenylic-polyribouridylic acid (poly A:U) and polyriboinosinic-polyribocytidylic acid (poly I:C), both potent stimulators of adenyl cyclase, can stimulate or depress macrophage activities dependent on the time of administration [17, 18]. These studies on the influence of purposely altered cAMP levels on phagocytic activity are paralleled by the finding of Park et al. [19] that endogenous cAMP levels of macrophages are elevated during phagocytosis.

T Lymphocytes

The activation and functions of the thymus-dependent lymphocyte (T-cell) population, which includes cells that participate in cell-mediated immunity and act as helper cells in formation of antibody to many antigens, can also be stimulated or inhibited by agents altering endogenous cAMP levels. The direction of the alteration is a function of the extent of stimulation and the state of differentiation of the affected cells. Thus, phytohemagglutinin (PHA), which triggers transformation of T lymphocytes, has been recognized as a stimulator of endogenous cAMP levels [9, 20–23], and poly A:U enhances such effects [9]. Exogenous cAMP has been reported to inhibit and, occasionally, to stimulate PHA-elicited lymphocyte transformation [20, 24]; it is likely that the inhibitory effects were due to the use of an excessively high amount of cAMP. Poly I:C enhances graft rejection [25], and poly A:U enhances graft-vs.-host reactions ([26] and T. Matsumoto and W. Braun, unpublished observations) and delayed-type hypersensitivity responses, i.e., the activation of systems that are attributable to T-cell activities [10, 26]. There is also an increasing amount of information concerning the ability of cAMP and of stimulators of endogenous cAMP to inhibit antigen-triggered functions of already activated lymphocytes [27–29]. For example, the cytolytic activity of sensitized lymphocytes in the presence of target cells to which they were sensitized can be inhibited by isoproterenol, histamine, or prostaglandin E_1, and these inhibitory effects are directly correlated with an increase in endogenous cAMP levels [29]. Accordingly, the present indications are that a modest elevation of endogenous cAMP levels stimulates T-cell activation, whereas excessive stimulation inhibits activation. Once activated, however, the function of T cells is inhibited

only by agents that elevate endogenous cAMP levels.

B Lymphocytes

The activation of B cells, which are responsible for formation of antibody, is strikingly altered by administration of modifiers of endogenous cAMP levels at the time of immunization or shortly thereafter. The stimulatory effects of poly A:U and poly I:C on antibody formation in vivo and in vitro [10–15] can now be attributed to the capacity of the polynucleotides to enhance activity of endogenous adenyl cyclase [9–15]. This conclusion is supported by the fact that phosphodiesterase inhibitors (e.g., theophylline and caffeine), when employed at appropriate concentrations, can magnify the stimulatory effects of polynucleotides both in vivo and in vitro [14, 15]. Furthermore, we have recently reported [30, 31] that isoproterenol, norepinephrine, and epinephrine, all well-known stimulators of adenyl cyclase, can enhance activation and performance of antibody-forming cells of mice in vivo and in vitro; however, under in-vivo conditions isoproterenol stimulated only when administered in conjunction with low levels of poly A:U. We also showed that the effects of isoproterenol on formation of antibody are blocked by propranolol, a β-adrenergic blocking agent that competes with isoproterenol for the receptor site on the cell membrane; however, the effects of poly A:U are not antagonized by propranolol. These findings suggest that the immunoenhancing effects of different stimulators of adenyl cyclase can be mediated by different receptors on the cell membrane.

The role of cAMP in regulating the magnitude of antibody formation is also indicated by the observation that components of *Vibrio cholerae* as well as its exotoxin, which are potent stimulators of adenyl cyclase, possess strong adjuvant effects [32, 33].

Recent studies [31, 34, 35] have shown that antibody to T-cell–dependent antigens, such as ovine red blood cells, can be formed in vivo [34, 35] and in vitro [31] in the absence of the usually required number of T cells whenever normal, allogeneic spleen cells are injected into the T-cell–deficient animal or added to T-cell–deficient spleen-cell cultures. This helper effect of allogeneic spleen cells again can be attributed to alterations in endogenous cAMP levels, since studies in our laboratory (C. Shiozawa, unpublished observations) have shown that interactions between allogeneic, normal murine spleen cells lead to a rapid enhancement of adenyl cyclase activity.

On the basis of such information, it is possible to propose [31] that the productive activation of B-cell performance requires two signals to relevant B cells (figure 3): (*1*) a specific antigenic signal triggered by interactions between an antigenic determinant and a matching receptor site on appropriate B cells; and (*2*) a second, nonspecific membrane signal that activates a cAMP-mediated amplification system. The so-called T-cell–independent antigens (including endotoxin [36]) may have a molecular structure that permits the simultaneous triggering on B cells of the specific receptor site for antigen and the triggering of the second receptor site on the same cell, leading to the activation of the amplification system.

In the case of T-cell–dependent antigens (figure 4), triggering of detectable B-cell performance will depend on both interaction of the cell with antigen and an antigen-mediated contact with activated T cells, which triggers the amplification system. The latter natural process for the activation of the obligatory amplification system may be substituted for by contact with normal allogeneic cells or by agents capable of elevating endogenous cAMP levels (e.g., poly A:U [9], LPS, or materials released from other cells as the result of exposure to LPS). Thus, part of the adjuvant effects of LPS on T-cell–dependent antigens may reside in its ability to permit the bypass of the usually needed helper-cell functions of members of the T-cell population.

The data that have led to the conclusion expressed in figure 4 were obtained in studies in which the usually required number of T cells was reduced, either by thymectomy [26] or by anti-θ

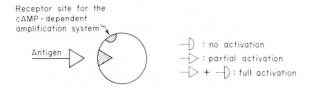

Figure 3. Presumed steps in the triggering of detectable B-cell functions (cAMP = cyclic adenosine 5′-monophosphate).

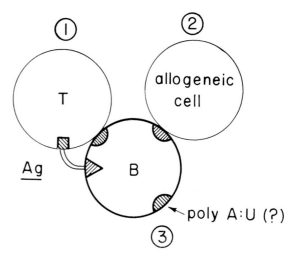

Figure 4. Three alternative mechanisms for the activation of the cAMP-dependent amplification system of antigen-exposed B cells (cAMP = cyclic adenosine 5′-monophosphate).

treatment in the presence of complement (D. Webb and W. Braun, unpublished observations). While such procedures reduce the number of T cells, they are unlikely to eliminate all T cells; therefore, it is impossible to rule out indirect effects of cAMP stimulation on B cells via effects on residual T cells that may, in turn, influence B-cell performance. Workers in several laboratories are currently attempting to resolve this matter.

We referred earlier to a biphasic dose-response relationship in regard to the influence of cAMP levels on macrophage and T-cell activation. A similar situation exists with B cells, in which biphasic responses have been analyzed most thoroughly so far [13–15]. In the process of lymphocyte activation, modest stimulation of endogenous cAMP levels will result in enhancement of antibody formation, whereas excessive stimulation will result in a lesser effect or even in an inhibition of antibody responses. An example of such biphasic relationships is the case in which increasing doses of theophylline are administered together with a low concentration of poly A:U at time of immunization of mice with ovine red blood cells (figure 5). Comparable dose-response curves have been obtained when increasing amounts of poly A:U alone are administered in murine spleen-cell cultures to which ovine red blood cells were added as immunogen [14] and when poly A:U is administered in the presence

184

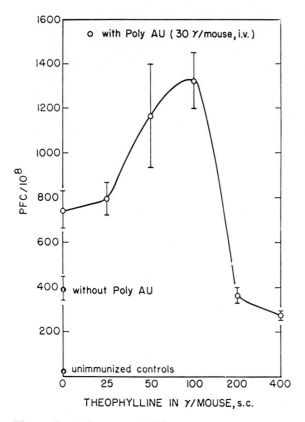

Figure 5. Influence of different concentrations of [theophylline]$_2$·ethylenediamine, given sc on antibody formation to ovine red blood cells in the absence or presence of polyriboadenylic-polyribouridylic acid. All assays were made 48 hr after immunization (from [15] reproduced with permission of Williams and Wilkins Co.).

of increasing levels of catecholamines both in vivo and in vitro [30]. Similar results have also been obtained in recent tests with interferon preparations, which have acted as modifiers of antibody responses and endogenous cAMP levels [37].

There are indications that the spectrum of the biphasic-response curve may differ for T and B cells [15]; the activation of B cells is apparently more readily inhibited by an excessive elevation of endogenous cAMP levels than is that of T cells.

Thus, studies with known modifiers of endogenous cAMP levels have shown that these substances can alter the activation and functions of all types of cells involved in immune responses and (depending on the time and intensity of the effects) can either stimulate or inhibit responses.

Effects of Endotoxin

Since LPS can produce modifications of macrophage and of lymphocyte activation and functions [38–41], we must ask to what extent the new information on the regulatory effects of modifiers of the cAMP system is not only parallel to many known LPS effects but actually part of its mode of action.

LPS is known to affect release of catecholamines and histamine and to alter cellular reactions to these agents [38, 42, 43]; some of these effects have been associated with a blocking of β-adrenergic receptors [44]. In view of the influence of catecholamines and histamine on cAMP formation [1, 29], one could suspect that adjuvant effects of LPS might operate partially through such LPS–elicited alterations. Supporting this conclusion are recent studies of Covert and Zarkower [43] on the adjuvant effects of early increases in histamine-forming capacity after the injection of LPS into mice immunized with ovine erythrocytes. Also, their finding that late increase in histamine depresses the antibody response may reflect the inhibited response of cells that have already been activated when subjected to agents that elevate levels of cAMP.

A second possible mode of action of LPS on antibody formation may be via its cytotoxicity, which may release from LPS-reacted cells some as yet unidentified modifiers of endogenous cAMP levels; these modifiers will then influence activities of cells involved in the immune response.

Last, but not least, components of the LPS molecule may have direct effects on the receptor sites that trigger the cAMP-dependent amplification system. In this regard, it is conceivable that the lipid moiety of LPS may resemble critical portions of prostaglandins and thus react with prostaglandin receptors; certain prostaglandins have been recently recognized as incredibly potent stimulators of the cAMP system [45]. Endotoxin receptors have been isolated in the studies of Springer et al. [46], and it would be interesting to learn whether or not such receptors would interact with prostaglandin.

We have collected some direct evidence indicating that the adjuvant effects of LPS operate via the cAMP system. As indicated earlier, one criterion for the involvement of cAMP-mediated events is the alteration of the response by theophylline or other inhibitors of the enzyme phosphodiesterase, which degrades cAMP. We previously showed that the effect of poly A:U on immune responses of mice to ovine erythrocytes could be potentiated by the simultaneous administration of theophylline. We have now found that the adjuvant effects of LPS also can be potentiated by theophylline, provided that low levels of LPS are used (table 1). Such potentiation was demonstrable not only in vivo but also in vitro (table 2).

The foregoing data show that cAMP-mediated events play a role in the adjuvant effects of LPS, but they do not reveal which of the several types of cells participating in formations of antibody to ovine erythrocytes is the principal target of the combined effects of LPS and theophylline. From in-vitro tests with poly A:U or LPS as stimulators of antibody formation to ovine red cells [10, 26, 47], one suspects that altered macrophage activity may be an important contributing factor.

It is noteworthy that theophylline did not alter the ability of LPS to cause, in the absence of any other administered antigen, a nonspecific activation of formation of antibody to various erythrocytic and bacterial antigens. In this regard, the immunoenhancing effects of LPS differ strikingly from those of polynucleotides and catecholamines, insofar as the latter can modify immune responses only when administered in the presence of antigen [10]. The exceptions are altered macrophage

Table 1. Influence of theophylline on the effects of endotoxin (LPS) from *Serratia marcescens* on formation of antibody to ovine red blood cells (oRBC) in CFW mice.

Treatment of spleen donors	Average no. of AFC† per 10⁸ spleen cells (±SE) 48 hr after immunization
Unimmunized controls	44 ± 1
oRBC (10⁸ iv)	674 ± 68
oRBC (10⁸ iv) + LPS (0.01 μg iv)	596 ± 145
oRBC (10⁸ iv) + LPS (0.01 μg iv) + theophylline*	1,024 ± 39
oRBC (10⁸ iv) + theophylline*	516 ± 157

NOTE. There were four animals per group.
* 100 μg of aminophylene (=[theophylline]₂·ethylenediamine) sc.
† AFC = antibody-forming cells.

Table 2. Influence of theophylline on the effects of endotoxin (LPS) from *Serratia marcescens* on formation of antibody to ovine red blood cells (oRBC) in vitro.

Experiment	RBC	LPS (μg)	Theophylline (μg)	Average no. of AFC per 10⁶ cells four (exp. A) or five (exp. B) days after start of cultures*
A	−	0	0	7.7
	+	0	0	460.4*
	+	0.01	0	508.0
	+	0.01	0.1	535.8
	+	0.01	0.001	706.1
	+	0.0001	0	444.3
	+	0.0001	0.1	672.2
	+	0.0001	0.001	495.2
	+	0	0.1	389.0
	+	0	0.001	493.8
B	−	0	0	12.1
	+	0	0	480.2*
	+	0.1	0	837.1
	+	0.1	1.0	977.8
	+	0.1	0.1	1,234.2
	+	0.01	0	501.8
	+	0.01	0.001	658.7

NOTE. Spleen cells from C57 B1 mice were used. All figures are averages from triplicate cultures.
* Antibody-forming cells (AFC) developed more slowly in experiment B, presumably because of differences in CO_2 content of the air during incubation.

activities that can result after administration of polynucleotides in the absence of antigen [10, 26] and the ability of polynucleotides to convert B-memory cells into antibody-secreting cells if polynucleotides alone are administered some time after specific priming with antigen [31]. In contrast, endotoxin alone can initiate specific antibody responses to many antigens, including those of gram-negative bacteria and red blood cells; however, the present consensus appears to be that this capacity of LPS may reflect the presence of cross-reacting antigens, as well as the productive activation of memory cells, and may involve the nonspecific, mitogenic effect of LPS on B cells discussed by Möller (see Session 1).

We have made an unsuccessful preliminary attempt to detect alterations in adenyl cyclase activity after exposure of murine spleen cells to LPS in vitro. In contrast, supernatants derived from LPS-exposed peritoneal cell populations

186

produced some stimulation, which, in a very preliminary fashion, suggests that the influence of LPS on cAMP-mediated events in cells involved in the immune response might be indirect, operating through materials released by other cells.

The problem of tolerance to LPS will be dealt with more thoroughly by others at this conference but, in view of my present discussion, one wonders to what extent some types of tolerance to LPS may not reflect the phase of inhibited responses that is typical after excessive stimulation of endogenous levels of cAMP. In this connection, one must also take note of Möller's report [36] that tolerance to LPS can express itself by a reduction in antibody-forming cells without concurrent reduction in rosette-forming cells; this phenomenon could be an expression of partial B-cell activation without the cAMP-mediated amplification required for secretion of antibody. One also wonders to what extent the negative phase in LPS effects may be a reflection of an initial, excessive stimulation of endogenous cAMP levels. The dose-dependent ability of prior injection of LPS to induce hyporeactivity to the stimulatory effects of polynucleotides (poly I:C) on antibody production and reticuloendothelial activity has been invoked by others as suggesting common mechanisms [18]; it is now tempting to attribute some of these to cAMP-mediated events.

Antitumor Effects

Finally, I wish to discuss briefly some recent results [48] that show that modifiers of the immune response can have antitumor effects; these effects are independent of the agents' effects on immunocompetent cells, a characteristic that is likely to apply to LPS also.

Both LPS and the synthetic polynucleotides have long been known to possess antitumor activity, which (specifically in the case of polynucleotides) was not always readily attributable to the immunoenhancing effects of these agents or to their capacity to induce interferon activity against virus-induced tumors [38, 49–54]. Even a very poor interferon inducer such as poly A:U possesses antitumor activity that is almost as potent as that of the good interferon inducer poly I:C. The tumor-retarding effects of poly A:U (450 µg/injection), theophylline (200 µg/injection), and poly A:U plus theophylline in Balb/C mice into which syngeneic Rauscher leukemia virus-

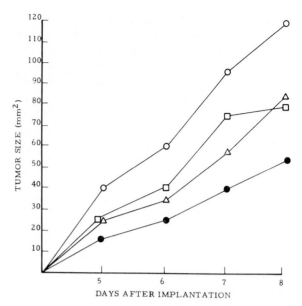

Figure 6. Effects of polyriboadenylic-polyribouridylic acid (poly A:U) and/or theophylline on the growth of Rauscher leukemia virus-induced tumor cells (MCDV-12) implanted intradermally (10^6 cells) into syngeneic Balb/C mice. Poly A:U (450 µg per injection) and/or theophylline (200 µg per injection) were given ip on days 0 and +1. (O———O) = controls; (□———□) = theophylline; (△———△) = poly A:U; (●———●) = poly A:U plus theophylline. There were five animals per group.

induced tumor ascites cells had been implanted intradermally are shown in figure 6. Treatment was given on the day of implantation and on the next day. As in the case of enhancement of antibody formation, poly A:U in combination with theophylline was more effective than poly A:U alone, a parallelism that suggested that altered host responses might indeed be responsible for the retardation of tumor growth. However, it turned out (table 3) that the effects of the polynucleotides and theophylline were just as good in irradiated animals as in nonirradiated hosts. Subsequent tests confirmed that the exposure of the tumor cells to polynucleotides in vitro, before the implantation of the tumor cells, sufficed to produce antitumor effects.

These observations indicate that agents that stimulate the activation of cells involved in the immune responses can also have a direct effect on proliferation of tumor cells or can at least evoke effects that are independent of alterations in lymphocyte-mediated immune responses. There is currently much interest in the role of histiocytes in

Table 3. Effects of polyriboinosinic:polyribocyti-dylic acid (poly I:C) and polyriboadenylic:polyribo-uridylic acid (poly A:U) (450 µg ip on days 0 and +1) on the growth of Rauscher leukemia virus-induced tumor-cell populations in irradiated and non-irradiated, syngeneic, Balb/C mice.

Group	Average tumor size (mm²) ± SE, day 7
Poly I:C, nonirradiated	32 ± 7.8
Poly I:C, irradiated	35 ± 2.0
Poly A:U, nonirradiated	58 ± 13.5
Poly A:U, irradiated	54 ± 11.1
Controls, nonirradiated	95 ± 6.2
Controls, irradiated	82 ± 8.0

NOTE. There were five animals per group.

the control of tumor growth [55], and we have collected data showing that the mere mixing in vitro of virus-induced tumor cells and syngeneic spleen cells elevates cAMP levels in both tumor and spleen cells (W. Braun and C. Shiozawa, unpublished observations). Thus, interactions between transformed cells and neighboring nontransformed cells may play an important role in regulating the proliferation of neoplastic cells.

We referred earlier to evidence supporting the conclusion that the functions of already activated, immunocompetent cells can be turned off, but cannot be stimulated, by agents that elevate endogenous cAMP levels. Proliferating tumor cells may be comparable to activated immunocompetent cells, insofar as their response to modifiers of endogenous cAMP levels is concerned; therefore, they may be inhibited in their proliferation after exposure to cAMP-elevating agents. Unfortunately, tumor-cell lines differ in their susceptibility to polynucleotides, catecholamines, and cAMP; this variability, which may reflect inherent differences in responsiveness to modifiers of cAMP levels as well as differences in base levels of endogenous cAMP, will probably complicate early clinical use of these findings.

In any event, it is now possible to suggest that the antitumor effects of LPS may also have two aspects: an effect on immune responses and more direct effects on the tumor cells themselves.

The foregoing review of progress in related areas so far has aided us only in recognizing that cAMP-mediated effects play a role in the adjuvant effects of LPS. It is likely that this is not the sole mechanism involved, since neither any known

stimulator of the cAMP system nor a combination of such stimulators has yielded effects as striking as those produced by LPS. It may be hoped, however, that the recognition of the role of cAMP in immune responses will enable us to determine to what extent similar mechanisms can account for the effects of LPS on immune responses and tumor growth.

References

1. Robison, G. A., Butcher, R. W., Sutherland, E. W. Cyclic AMP. Academic Press, New York, 1971. 531 p.
2. Robison, G. A., Nahas, G. G., Triner, L. [ed.]. Cyclic AMP and cell function. Ann. N. Y. Acad. Sci. 185:1–556, 1971.
3. Rasmussen, H. Cell communication, calcium ion, and cyclic adenosine monophosphate. Science 170:404–412, 1970.
4. Varmus, H. E., Perlman, R. L., Pastan, I. Regulation of lac messenger ribonucleic acid synthesis by cyclic adenosine 3',5'-monophosphate and glucose. J. Biol. Chem. 245:2259–2267, 1970.
5. Zubay, G., Schwartz, D., Beckwith, J. Mechanism of activation of catabolite-sensitive genes: a positive control system. Proc. Natl. Acad. Sci. U. S. A. 66:104–110, 1970.
6. Pastan, I., Perlman, R. Cyclic adenosine monophosphate in bacteria. Science 169:339–344, 1970.
7. Sheppard, J. R. Difference in the cyclic adenosine 3',5'-monophosphate levels in normal and transformed cells. Nature [New Biol.] 236:14–16, 1972.
8. Greengard, P., Paoletti, R., Robison, G. A. [ed.] Advances in cyclic nucleotide research. Vol. 1. Raven Press, New York, 1972.
9. Winchurch, R., Ishizuka, M., Webb, D., Braun, W. Adenyl cyclase activity of spleen cells exposed to immunoenhancing synthetic oligo- and polynucleotides. J. Immunol. 106:1399–1400, 1971.
10. Braun, W., Ishizuka, M., Yajima, Y., Webb, D., Winchurch, R. Spectrum and mode of action of poly A:U in the stimulation of immune responses. In R. F. Beers and W. Braun [ed.] Symposium on molecular biology, New York, 1970. Biological effects of polynucleotides. Springer-Verlag, New York, 1971.
11. Braun, W., Ishizuka, M., Winchurch, R., Webb, D. Cells and signals in immunological nonresponsiveness. Ann. N. Y. Acad. Sci. 181:289–298, 1971.
12. Braun, W., Ishizuka, M., Winchurch, R., Webb, D. On the role of cyclic AMP in immune responses. Ann. N. Y. Acad. Sci. 185:417–422, 1971.
13. Braun, W., Ishizuka, M. Antibody formation: reduced responses after administration of excessive amounts of nonspecific stimulators. Proc. Natl. Acad. Sci. U. S. A. 68:1114–1116, 1971.
14. Ishizuka, M., Braun, W., Matsumoto, T. Cyclic

188

AMP and immune responses. I. Influence of poly A:U and cAMP on antibody formation in vitro. J. Immunol. 107:1027–1035, 1971.

15. Braun, W., Ishizuka, M. Cyclic AMP and immune responses. II. Phosphodiesterase inhibitors as potentiators of polynucleotide effects on antibody formation. J. Immunol. 107:1036–1042, 1971.

16. Biozzi, G., Benacerraf, B., Stiffel, C., Halpern, B. N. Etude quantitative de l'activité granulopexique du système réticulo-endothélial chez la souris. C. R. Soc. Biol. (Paris) 148:431–435, 1954.

17. Berry, L. J., Smythe, D. S., Colwell, L. S., Schoengold, R. J., Actor, P. Comparison of the effects of a synthetic polyribonucleotide with the effects of endotoxin on selected host responses. Infec. Immun. 3:444–448, 1971.

18. Chester, T. J., DeClercq, E., Merigan, T. C. Effect of separate and combined injections of poly rI : poly rC and endotoxin on reticuloendothelial activity, interferon, and antibody production in the mouse. Infec. Immun. 3:516–520, 1971.

19. Park, B. H., Good, R. A., Beck, N. P., Davis, B. B. Concentration of cyclic adenosine 3′,5′-monophosphate in human leucocytes during phagocytosis. Nature [New Biol.] 229:27–29, 1971.

20. Hirschhorn, R., Grossman, J., Weissmann, G. Effect of cyclic 3′,5′-adenosine monophosphate and theophylline on lymphocyte transformation. Proc. Soc. Exp. Biol. Med. 133:1361–1365, 1970.

21. Smith, J. W., Steiner, A. L., Parker, C. W. Early effects of phytohemagglutinin (PHA) on lymphocyte cyclic AMP levels (abstract). Fed. Proc. 29: 369, 1970.

22. Smith, J. W., Steiner, A. L., Newberry, W. M., Jr., Parker, C. W. Cyclic adenosine 3′,5′-monophosphate in human lymphocytes. Alterations after phytohemagglutinin stimulation. J. Clin. Invest. 50:432–441, 1971.

23. Smith, J. W., Steiner, A. L., Parker, C. W. Human lymphocytic metabolism. Effects of cyclic and noncyclic nucleotides on stimulation by phytohemagglutinin. J. Clin. Invest. 50:442–448, 1971.

24. Gericke, D., Chandra, P., Haenzel, I., Wacker, A., Hoppe-Seyler, Z. Studies on the effect of nucleoside cyclic 3′,5′-monophosphates on antibody synthesis by spleen cells. Physiol. Chem. 351:305–308, 1970.

25. Turner, W., Chan, S. P., Chirigos, M. A. Stimulation of humoral and cellular antibody formation in mice by poly IR:CR. Proc. Soc. Exp. Biol. Med. 133:334–338, 1970.

26. Johnson, A. G., Cone, R. E., Friedman, H. M., Han, I. H., Johnson, I. G., Schmidtke, J. R., Stout, R. D. Stimulation of the immune system by homopolyribonucleotides. In R. F. Beers and W. Braun [ed.] Symposium on molecular biology, New York, 1970. Biological effects of polynucleotides. Springer-Verlag, New York, 1971, p. 157–177.

27. Bourne, H. R., Melmon, K. L. Adenyl cyclase in human leukocytes: evidence for activation by

separate beta adrenergic and prostaglandin receptors. J. Pharmacol. Exp. Ther. 178:1–7, 1971.

28. Bourne, H. R., Melmon, K. L., Lichtenstein, L. M. Histamine augments leukocyte adenosine 3′,5′-monophosphate and blocks antigenic histamine release. Science 173:743–745, 1971.

29. Henney, C. S., Bourne, H. R., Lichtenstein, L. M. The role of cyclic 3′,5′ adenosine monophosphate in the specific cytolytic activity of lymphocytes. J. Immunol. 1972 (in press).

30. Braun, W., Rega, M. J. Adenyl cyclase-stimulating catecholamines as modifiers of antibody formation (abstract). Fed. Proc. 31:803, 1972.

31. Braun, W. RNAs as amplifiers of specific signals in immunity. Ann. N. Y. Acad. Sci. 1972 (in press).

32. Winchurch, R. The effects of an immunoenhancing bacterial product on the adenyl cyclase activity of mouse spleen cells. J. Immunol. 1972 (in press).

33. Northrup, R. S., Fauci, A. S. Adjuvant effect of cholera enterotoxin on the immune response to sheep red blood cells. J. Infect. Dis. 125:672–673, 1972.

34. Katz, D. H., Paul, W. E., Benacerraf, B. Carrier function in anti-hapten antibody responses. V. Analysis of cellular events in the enhancement of antibody responses by the "allogenic effect" in DNP-OVA-primed guinea pigs challenged with a heterologous DNP-conjugate. J. Immunol. 107: 1319–1328, 1971.

35. Kreth, H. W., Williamson, A. R. Cell surveillance model for lymphocyte cooperation. Nature (Lond.) 234:454–456, 1971.

36. Möller, G. Triggering mechanisms for cellular recognition. In R. T. Smith and M. Landy [ed.] Immune surveillance. Academic Press, New York, 1970, p. 87–116.

37. Braun, W., Levy, H. Interferon preparations as modifiers of immune responses. J. Immunol. 1972 (in press).

38. Landy, M., Braun, W. [ed.]. Bacterial endotoxins. Rutgers University, New Brunswick, N. J., 1964. 691 p.

39. Franzl, R. E., McMaster, P. D. The primary immune response in mice. I. The enhancement and suppression of hemolysin production by a bacterial endotoxin. J. Exp. Med. 127:1087–1107, 1968.

40. Freedman, H. H., Nakano, M., Braun, W. Antibody formation in endotoxin-tolerant mice. Proc. Soc. Exp. Biol. Med. 121:1228–1230, 1966.

41. Peavy, D. L., Adler, W. H., Smith, R. T. Mitogenic effects of salmonella LPS and staphylococcal enterotoxin "B" on human peripheral lymphocytes and mouse spleen cells (abstract). Fed. Proc. 29: 370, 1970.

42. Schayer, R. W. Relationship of induced histidine decarboxylase activity and histamine synthesis to shock from stress and from endotoxin. Am. J. Physiol. 198:1187–1192, 1960.

43. Covert, J. B., Zarkower, A. Influence of Escherichia

coli lipopolysaccharide on histamine-forming capacity and antibody formation in lymphoid tissues. Infec. Immun. 4:452–455, 1971.

44. Keller, K. F., Fishel, C. W. In vivo and in vitro manifestations of adrenergic blockade in *Bordetella pertussis*-vaccinated mice. J. Bacteriol. 94:804–811, 1967.

45. Ramwell, P. W., Shaw, J. E. [ed.]. Prostaglandins. Ann. N. Y. Acad. Sci. 180:1–568, 1971.

46. Springer, G. F., Huprikar, S. V., Neter, E. Specific inhibition of endotoxin coating of red cells by a human erythrocyte membrane component. Infec. Immun. 1:98–108, 1970.

47. Ortiz-Ortiz, L., Jaroslow, B. N. Enhancement by the adjuvant endotoxin of an immune response induced in vitro. Immunology 19:387–399, 1970.

48. Webb, D., Braun, W., Plescia, O. J. Antitumor effects of polynucleotides and theophylline. Cancer Res. 32:1814–1819, 1972.

49. Bart, R. S., Kopf, A. W. Inhibition of the growth of murine malignant melanoma with synthetic double-stranded ribonucleic acid. Nature (Lond.) 224:372–373, 1969.

50. Gelboin, H. V., Levy, H. B. Polyinosinic-polycyti-

dylic acid inhibits chemically induced tumorigenesis in mouse skin. Science 167:205–207, 1970.

51. Larson, V. M., Clark, W. R., Hilleman, M. R. Influence of synthetic (poly I:C) and viral double-stranded ribonucleic acids on adenovirus 12 oncogenesis in hamsters. Proc. Soc. Exp. Biol. Med. 131:1002–1011, 1969.

52. Sarma, P. S., Shiu, G., Neubauer, R. H., Baron, S., Huebner, R. J. Virus-induced sarcoma of mice: inhibition by a synthetic polyribonucleotide complex. Proc. Natl. Acad. Sci. U. S. A. 62:1046–1051, 1969.

53. Braun, W., Plescia, O. J., Raskova, J., Webb, D. Basic proteins and synthetic polynucleotides as modifiers of immunogenicity of syngeneic tumor cells. Isr. J. Med. Sci. 7:72–82, 1971.

54. Gresser, I., Coppey, J., Fontaine-Brouty-Boye, D., Falcoff, R., Falcoff, E., Zajdela, A. Interferon and murine leukaemia. 3. Efficacy of interferon preparations administered after inoculation with Friend virus. Nature (Lond.) 215:174–175, 1967.

55. Zbar, B., Rapp, H. J., Ribi, E. E. Tumor suppression by cell walls of *Mycobacterium bovis* attached to oil droplets. J. Natl. Cancer Inst. 48:831–835, 1972.

Affinity of Endotoxin for Membranes

Joseph W. Shands, Jr.

From the Department of Immunology and Medical Microbiology, University of Florida College of Medicine, Gainesville, Florida

Morphologic studies have revealed that particles of endotoxin (LPS) associate with membranes by edge attachment. Also, under certain conditions endotoxin disks or lamellae may form vesicles on interaction with phospholipid. Edge attachment of LPS to membranes probably allows the lipids of LPS to interact with membrane receptors or lipids, and the vesiculation reaction provides a possible model for the incorporation of LPS into a membrane.

It is now well established that endotoxic lipopolysaccharide (LPS) is present in and can be extracted from the outer or L membrane of gram-negative bacteria [1–5]. When extracted and partially purified, LPS also resembles morphologically membrane fragments of various shapes, including ribbons, disks, lamellae, and vesicles [2, 6–10]. A model for the structure of endotoxin

Please address requests for reprints to Dr. Joseph W. Shands, Department of Immunology and Medical Microbiology, University of Florida College of Medicine, Gainesville, Florida 32601.

has been proposed (figure 1). Particles of LPS are envisioned as bilayers, each half of which is composed of polysaccharide covalently linked with lipid A. In aqueous solution two halves bond together, probably by hydrophobic interaction, to produce a bilayer with lipid buried in the center and the polysaccharide exposed to the surrounding water.

These membranous structures have a remarkable affinity for other biological membranes [11, 12], a phenomenon that has raised the intriguing possibility that the attachment of LPS to a mammalian membrane may cause some membrane

190

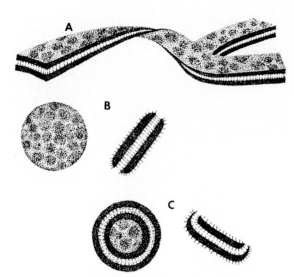

Figure 1. Proposed model for the shape of lipopolysaccharide. A, B, and C represent ribbon, disk, and donut, respectively, all depicted as bilayer structures with the fatty acids in the interior of the particle and the polysaccharide on the exterior surface. Illustration from [9].

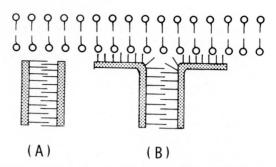

Figure 2. Model for the interactions of lipopolysaccharide (LPS) with a lipid bilayer membrane. **A**, edge attachment by which some fatty acids of the LPS might be "solubilized" into the bilayer. **B**, opening of the bilayer to allow interaction of the fatty acids.

alteration, which, in turn, may trigger some of its biological effects. In light of this possibility, we have examined the attachment of LPS to ovine erythrocytes to determine whether or not LPS caused gross alteration in membranes as measured by loss of hemoglobin. Experiments reported by Čižnár and Shands [12] showed that isolated LPS in high concentration had no disorganizing effects on the stability of erythrocyte membranes. However, mild hydrolysis with alkali increased the binding of LPS to erythrocytes 20-fold, and in high concentrations of this alkali-treated product, erythrocytes were lysed. These data indicate that alkali-treated LPS can cause conformational changes in membranes, but whether unaltered LPS is capable of doing so is presently unknown. Tests far more sensitive to subtle membrane changes will obviously have to be employed.

The mechanics of the attachment of LPS to membranes is somewhat difficult to envision. The reason for this difficulty is that the lipid moiety of LPS is essential for attachment to membranes [13], and if the model of the structure of LPS is correct and the lipid is buried in the interior of the particle, how does it become exposed to interact with membranes? Either the particles must attach by their edges, whereby some fatty acids may be "solubilized" into the membrane, or phase

transitions must take place in the particle. Figure 2 shows how edge attachment with membranes might occur.

We have taken two approaches toward the mechanics of LPS-membrane interaction. We have looked morphologically for the attachment of LPS to erythrocyte stromata, and we have studied by electron microscopy the interaction of LPS with crude and purified phosphatidyl choline, a membrane component known to inhibit the attachment of LPS to erythrocytes [14].

Materials and Methods

Lipopolysaccharide. The LPS used was extracted from a smooth strain of *Salmonella typhimurium* by the phenol-water technique [15]. The LPS was separated from contaminating RNA by centrifugation and washing.

Phosphatidyl choline. Crude phosphatidyl choline (90% pure, bovine origin) was obtained from Nutritional Biochemicals, Cleveland, Ohio. Chromatographically pure L-α-lecithin was obtained from General Biochemicals, Chagrin Falls, Ohio. The phospholipids were dispersed in distilled water or 0.1 M Tris buffer by sonication. Mixtures containing 0.1% phospholipid and 0.02% LPS in distilled water or 0.1 M Tris buffer were heated at 37 C or 56 C for 10 min. The mixture was then placed on a grid, negatively stained with vanatado-molybdate [16], and examined in an electron microscope.

Sensitization of erythrocytes. Fresh ovine erythrocytes (RBCs) were washed and suspended in isotonic saline containing LPS at a concentration of 1 mg per ml. After incubation for 1 hr at 37 C, the cells were washed three times in isotonic saline and lysed in 0.01 M Tris buffer (*p*H 6.8), and the stromata were pelleted. The stromata were fixed in 1% OsO₄ in 0.1 M Tris and embedded by the method of Shands [17]. Thin sections were stained with uranyl acetate and lead citrate and examined in an electron microscope.

Results

In figure 3 particles of LPS (arrows) attached to erythrocyte stromata are shown. All recognizable particles appear to attach by their edges in a manner consistent with the model shown in figure 2. We have been unable to find an opening of the bilayer structure (also shown in figure 2), although separation of the bilayer into monolayers has been demonstrated under some circumstances in vitro [18]. The interaction of endotoxin with a crude lecithin preparation is shown in figure 4. The usually long ribbons of LPS have been broken into smaller particles, and there is a notable stacking of LPS particles (arrows) with a remarkable periodicity and with particles separated by 205 ± 66 A of some amorphous material. The lamellar patterns of the lecithin (L) do not appear to be disturbed. In addition to the stacking, it is obvious that a vesiculation is also occurring; this is more clearly shown in figure 5. In figure 5 some lamellar or "disk-on-edge" structures can be seen (arrow); some of the particles appear in the shape of drumsticks, and some as vesicles. Possibly the "drumstick" particles represent the beginning of vesiculation in lamellar particles.

Discussion

The data show that particles of LPS can attach to membranes by their edges. With this orientation it is conceivable that hydrophobic groups in the interior of the LPS particle can react with hydrophobic groups or some receptor in the membrane. This is, therefore, one way in which LPS may attach to membranes. The interaction of LPS with lecithin suggests that there may be another mechanism for the interaction of LPS and membranes, and that this mechanism may lead to actual incorporation.

The stacking of LPS particles or L-membrane fragments has been reported previously by others [19, 20], and the transition of LPS particles from disks to vesicles has been elegantly demonstrated by De Pamphilis [19]. In our hands the stacking reaction occurred in both crude and purified preparations of phosphatidyl choline but did not occur with LPS alone. The phenomenon, therefore, indicates some sort of interaction between LPS and phospholipid, which under proper conditions produces a linear association. The role of metallic cations is undefined, but it is quite possible, as De Pamphilis has pointed out [19], that they are important. It seems that this interaction between LPS and phosphatidyl choline could explain the ability of the phospholipid to inhibit the attachment of LPS to cell membranes. This could be due either to the aggregating effect or to the covering of reactive groups.

The vesiculation of LPS with phospholipid was observed only with the crude phospholipid and not with the pure product. It seems, therefore, that some additional material is needed for initiation of the phase transition from disk to vesicle. A likely candidate for this material is, of course,

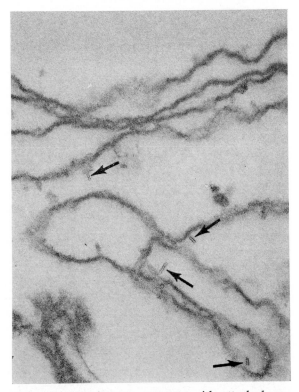

Figure 3. Erythrocyte stromata with attached particles of lipopolysaccharide. Those which can be identified (arrows) are attached by their edge (×100,000).

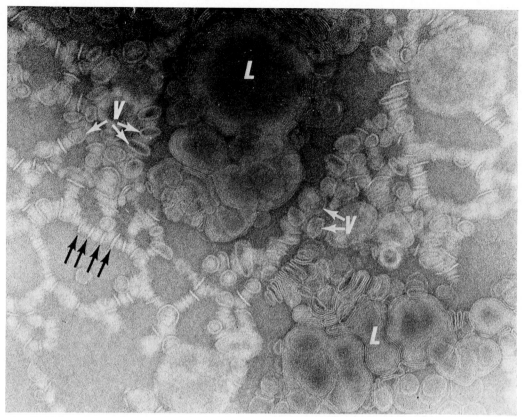

Figure 4. Stacking arrangement and vesiculation of crude lecithin and lipopolysaccharide (LPS) heated for 10 min at 37 C. LPS disks (arrows), vesicles (V), lecithin (L) (×100,000).

lysolecithin, which is known to induce membrane fusion, probably by the production of phase changes in membrane lipids [21]. In all probability the vesicle that results from the interaction of LPS and lecithin is composed of both substances, but how they are arranged relative to each other is unknown. In any event, this phenomenon may provide a model for the incorporation of LPS into mammalian membranes as well as into the L membrane of gram-negative bacteria. On attachment of a particle of LPS to a mammalian membrane, phospholipids interacting with the LPS particle may "solubilize" it and, by phase transitions similar to those of vesiculation, bring it into the membrane structure.

References

1. Bladen, H. A., Mergenhagen, S. E. Ultrastructure of *Veillonella* and morphological correlation of an outer membrane with particles associated with

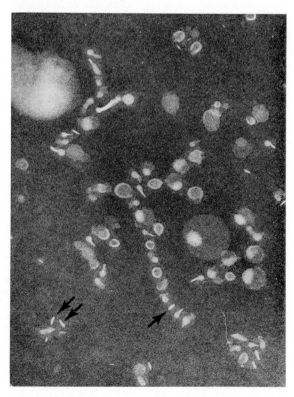

Figure 5. Same process as figure 4 but heated at 56 C for 10 min (×100,000).

endotoxic activity. J. Bacteriol. 88:1482–1492, 1964.

2. Burge, R. E., Draper, J. C. The structure of the cell wall of the gram-negative bacterium *Proteus vulgaris*. III. A lipopolysaccharide "unit membrane." J. Mol. Biol. 28:205–210, 1967.

3. De Petris, S. Ultrastructure of the cell wall of *Escherichia coli* and chemical nature of its constituent layers. J. Ultrastruct. Res. 19:45–83, 1967.

4. Frank, H., Dekegel, D. Electron microscopical studies on the localization of the different components of cell walls of gram-negative bacteria. Folia Microbiol. (Praha) 12:227–233, 1967.

5. Work, E. The chemistry and morphology of surface structures of lysine-limited *Escherichia coli*. Folia Microbiol. (Praha) 12:220–226, 1967.

6. Beer, H., Braude, A. I., Brinton, C. C., Jr. A study of particle sizes, shapes, and toxicities present in a Boivin-type endotoxic preparation. Ann. N.Y. Acad. Sci. 133:450–475, 1966.

7. Rothfield, L., Horne, R. W. Reassociation of purified lipopolysaccharide and phospholipid of the bacterial cell envelope: electron microscopic and monolayer studies. J. Bacteriol. 93:1705–1721, 1967.

8. Shands, J. W., Jr., Graham, J. A., Nath, K. The morphologic structure of isolated bacterial lipopolysaccharide. J. Mol. Biol. 25:15–21, 1967.

9. Shands, J. W., Jr., Graham, J. A. Morphologic structure and biological activity of bacterial lipopolysaccharides. *In* L. Chedid [ed.]. La structure et les effets biologiques des produits bactériens provenant de germes gram-negatifs. Colloques Internationaux du CNRS. 174:25–34, 1969.

10. Work, E., Knox, K. W., Vesk, M. The chemistry and electron microscopy of an extracellular lipopolysaccharide from *Escherichia coli*. Ann. N.Y. Acad. Sci. 133:438–449, 1966.

11. Lüderitz, O., Westphal, O., Sievers, K., Kröger, E., Neter, E., Braun, O. H. Über die Fixation von P³²-markiertem Lipopolysaccharide (Endotoxin) aus *Escherichia coli* an menschlichen Erythrocyten. Biochem. Z. 330:34–46, 1958.

12. Čižnár, I., Shands, J. W., Jr. Effect of alkali-treated lipopolysaccharide on erythrocyte membrane stability. Infect. Immun. 4:362–367, 1971.

13. Hämmerling, U., Westphal, O. Synthesis and use of O-stearoyl polysaccharides in passive hemagglutination and hemolysis. Eur. J. Biochem. 1:46–50, 1967.

14. Neter, E., Westphal, O., Lüderitz, O. Effects of lecithin, cholesterol, and serum on erythrocyte modification and antibody neutralization by enterobacterial lipopolysaccharides. Proc. Soc. Exp. Biol. Med. 88:339–341, 1955.

15. Westphal, O., Lüderitz, O., Bister, F. Über die Extraktion von Bakterien mit Phenol/wasser. Z. Naturforsch. 7b:148–155, 1952.

16. Callahan, W. P., Horner, J. A. The use of vanadium as a stain for electron microscopy. J. Cell. Biol. 20:350–356, 1964.

17. Shands, J. W., Jr. Embedding free-floating cells and microscopic particles; serum albumin coagulum-epoxy resin. Stain Technol. 43:15–17, 1968.

18. Shands, J. W., Jr. Evidence for a bilayer structure in gram-negative lipopolysaccharide: relationship to toxicity. Infect. Immun. 4:167–172, 1971.

19. De Pamphilis, M. L. Dissociation and reassembly of *Escherichia coli* outer membrane and of lipopolysaccharide, and their reassembly onto flagellar basal bodies. J. Bacteriol. 105:1184–1199, 1971.

20. Schnaitman, C. A. Effect of ethylenediaminetetraacetic acid, Triton X-100, and lysozyme on the morphology and chemical composition of isolated cell walls of *Escherichia coli*. J. Bacteriol. 108:553–563, 1971.

21. Lucy, J. A. The fusion of biological membranes. Nature (Lond.) 227:815–817, 1970.

Functional Aspects and Nature of the Lipopolysaccharide-Receptor of Human Erythrocytes

G. F. Springer, J. C. Adye, A. Bezkorovainy, and J. R. Murthy

From the Department of Immunochemistry Research, Evanston Hospital and the Department of Microbiology, Northwestern University, Evanston, Illinois, and the Department of Biochemistry, Rush-Presbyterian-St. Luke's Medical Center, Chicago, Illinois

We have isolated for the first time from human erythrocytes a physicochemically homogeneous lipo-glycoprotein that is rich in neuraminic acid, has a molecular weight of 228,000, and prevents attachment to erythrocytes of unheated and heated, smooth and rough endotoxin of all gram-negative bacteria tested. It did not interact with other bacterial antigens and, therefore, is named "lipopolysaccharide-receptor." The receptor activity is destroyed by proteases; lipid and neuraminic acid are not involved in the activity. We have established quantitatively with radioactive tracers that the receptor interacts with lipopolysaccharides and not with receptors on erythrocytes. The receptor blocks those lipopolysaccharide groupings that attach to erythrocytes. The action of the receptor is physical and reversible; it removes lipopolysaccharides fixed to red cells. While other compounds (namely, glycolipids, lipoproteins, and basic proteins) also inhibit attachment of lipopolysaccharide to cells, they are much less active, and those tested are unspecific.

Attachment of endotoxin (lipopolysaccharide or LPS, O antigen) to receptors in the host's tissue is a prerequisite for most, if not all, of its effects. LPS of gram-negative bacteria fixes to red cells in vitro [1] and, under extreme conditions, in vivo [2, 3]. A component of the human red-cell membrane to which all LPS preparations tested attach has been isolated and named lipopolysaccharide-receptor (LPS-receptor), since it fixes only LPS and the closely related Kunin antigen [4, 5].

Several other classes of compounds (e.g., glycolipids, aliphatic carbonyls, lipid-carrying proteins, basic proteins such as hemoglobin, protamine, and the basic antibiotic neomycin) inhibit the attachment of LPS to red cells as determined by passive hemagglutination [4, 6]. So far as it has been in-

vestigated, their action, in contrast to that of the LPS-receptor, does not seem to be specific [7]. Serologic experiments in this laboratory have shown that the inhibitors may prevent attachment of LPS to cells by blocking cell receptors, LPS, or both [4].

We now report on the isolation and nature of the LPS-receptor as well as its mode of action. We also compare it with other inhibitors of LPS attachment to red cells.

Materials and Methods

Isolation and characterization of LPS-receptor. Human erythrocyte stroma was prepared from donors of all blood groups and types as described previously [8]. Aqueous suspensions of stroma were homogenized in a Waring blender and extracted as follows:

Packed red blood cells
washed with 0.85% aqueous NaCl,
water hemolysis.
↓
(*1*) Erythrocyte stroma
| 1% aqueous suspension, homogenization Waring blender; butanol: H_2O (1:1) extraction, 16 hr, 4 C, *pH* 8.0. Activity in aqueous phase.

This study was supported by grants no. C72-28 and C72-10 from the Chicago Heart Association, by grant no. SD-340 from The John A. Hartford Foundation, Inc., and by the Susan Rebecca Stone Research Fund.

The authors acknowledge the excellent technical assistance of Miss Lynda Potzus, Mrs. Herta Tegtmeyer, and Mr. Thomas Burns.

Please address requests for reprints to Dr. Georg F. Springer, Department of Immunochemistry Research, Evanston Hospital, 2650 Ridge Avenue, Evanston, Illinois 60201.

(2) Crude LPS-receptor [4.4% of (1)]

Activity increase over (1) approximately 1,000%.
Centrifugation 33,000–151,000 g.
>90% activity in syrupy interphase.
Sepharose 4B column, 0.05 M Tris, pH 7.0; peak after void volume.

(3) Purified LPS-receptor [1.7% of (1)]

DEAE-Sephadex A-25 column. Elution 0.05 M Tris, pH 7.0, with NaCl. Stepwise increase, receptor recovered at 0.2 M NaCl. Sucrose density gradient 5%–20% in 0.01 M PO^{-3} 58,600 g Sephadex G-200 column, 0.05 M Tris, pH 7.0, receptor in void volume.

(4) Highly purified LPS-receptor [0.4% of (1)]

Activity increase over (2) approximately 400%.

Most of the receptor activity resided in the aqueous phase from which the receptor was isolated and purified as depicted above. The increase in activity of highly purified receptor over the stroma suspension was 40- to 50-fold.

Bacterial LPS. "Cold" as well as ^{32}P-labeled LPS was prepared from smooth, nonmotile *Escherichia coli* O_{86} B:7, which possesses high human blood-group B and low blood-group A activity [9]. *E. coli* was grown at 37 C on a fully defined glucose-salts medium devoid of blood-group–active substances [10]. When radioactive LPS was desired, the same medium containing 5 mCi of ^{32}P as $H_3{}^{32}PO_4$ per liter was used. After incubation for 48–72 hr at 37 C, the bacteria were collected by centrifugation, washed, and extracted with 45% aqueous phenol at 68 C [11]. The crude LPS was purified by ultracentrifugation, ethanol precipitation, and electrodialysis [8]. For comparative purposes, some *E. coli* O_{86} endotoxin was also prepared as the protein-LPS complex according to the procedure of Goebel et al. [12].

^{32}P-labeled or "cold" LPS was used for coating of red cells, either after heating in buffered saline (0.10 M sodium chloride and 0.05 M sodium phosphate, pH 7.4) at 100 C for 3 hr or after incubation at 37 C for 72 hr in the same solvent containing 0.1% NaN_3. All antigen solutions were used within 12 hr of dissolution. Immediately before use, insoluble material was removed by centrifugation at 2,000 g for 10 min.

Salmonella senftenberg group E LPS (Dr. D.A.L. Davies), *Serratia marcescens* protein-LPS (Dr. A. Nowotny), *S. marcescens* LPS (Difco, Detroit, Mich.), and Vi antigen from *Paracolobactrum ballerup* (Dr. E. Baker) were used. The two former preparations were purified before use as described above for *E. coli* O_{86} LPS. Purified culture fluids from *Salmonella typhimurium* H-K7 containing Kunin antigen and purified culture fluids from *Bacillus subtilis* and *Staphylococcus aureus* containing Rantz antigen were the gift of Dr. E. Neter. Stearoyl derivatives of streptococcus group antigens A and E were donated by Dr. H. D. Slade.

Analytical procedures. Pure reagent-grade chemicals were used throughout. All samples were dried to constant weight at 23–25 C (room temperature) at 10^{-1}–10^{-2} mm Hg over P_2O_5 before analysis in this laboratory. Drying to constant weight (at 80 C) was also done by Huffman Microanalytical Laboratories, Wheatridge, Colo., for determination of water loss and before performance of phosphorus (Ma and McKinley procedure [13]) and wet-ash analyses. Absorption of ultraviolet and visible light was measured either with a Beckman DU spectrophotometer to which a Gilford indicating photometer and cuvet positioner were attached or with a Perkin-Elmer model 202 spectrophotometer. Optical activity of 0.2% aqueous solutions of LPS-receptor was determined in a Perkin-Elmer model 141 digital readout polarimeter at 29 C with a 1-ml cell.

Sedimentation and diffusion constants of LPS and preparations of LPS-receptor were measured in a Beckman (Spinco) model E analytic ultracentrifuge and a model H electrophoresis apparatus as described recently [14]. Agar-gel electrophoresis was performed as by Wiemer [15]; 1.25% agar was poured onto lantern slides (8 × 10 cm), appropriate slots were made, and the electrophoresis was done at 20 V/cm and 0.2 mA/cm for 90 min at 23–25 C. Paper electrophoresis was performed for 60–75 min as previously described [16] on Whatman no. 1 paper at 8.3–11.8 V/cm and 0.11–0.15 mA/cm at 23–25 C. Cellulose acetate electrophoresis was done in a chamber for 25–60 min on Sepraphore III strips (Gelman Instrument Co., Ann Arbor, Mich.) at 13.3 V/cm and 0.17–0.2 mA/cm at 23–25 C. The following buffers were used at 0.01 M concentration unless stated otherwise: Tris-HCl, pH

9.0; 0.05 M veronal buffer, pH 8.6; monobasic-dibasic sodium phosphate, pH 7.1; sodium acetate-acetic acid, pH 5.0; sodium formate-formic acid, pH 3.1. Human serum and RBY dye (Gelman Instrument) served as controls. Paper and acetate strips were stained with 0.2% Ponceau S in 5% trichloroacetic acid, while the agar slides were stained with Amido Black 10B after fixation in acetic acid-water-methanol (2:9:9).

Sialic acid was estimated by the thiobarbituric acid [17] and resorcinol [18] procedures. For hexose determination, 0.05% receptor/ml was hydrolyzed in 1 N HCl for 10 hr under nitrogen, and the acid was removed by repeated freeze-drying and addition of water. Galactose (Gal) was measured with galactose dehydrogenase obtained from Worthington Biochemical Corp. (Freehold, N.J.), who also supplied the glucostat reagent for the glucose (Glc) assay. Fucose (Fuc) was estimated as by Dische [19], as was mannose (Man) after corrections were made for the Glc and Gal contents [20]. Total hexosamine was determined by the Gatt-Berman procedure [21].

Some of the quantitative amino acid analyses were carried out by Dr. P. Weber according to Spackman et al. [22]; 1 mg of receptor was hydrolyzed for 23 hr at 103 C under nitrogen with 4 ml of 6 N HCl or at 110 C for 24 hr with 5.8 N HCl. The hydrolysate was freeze-dried and dissolved in 1 ml of 0.2 N sodium citrate buffer, pH 2.20, and 0.25 ml of this solution was added to the column. Cysteic acid was measured after oxidation with performic acid in an ice bath before hydrolysis [23]. Determination of amino sugars in the analyzer differed from that of amino acids only in that the hydrolysis was in 4 N HCl for 4 hr at 100 C. Tryptophane (Trp) was determined as by Winkler [24], except that all volumes were reduced by four-fifths, and the absorbance was measured at 530 nm.

Lipid extraction. This procedure was based on that of Folch et al. [25]. Approximately 40 mg of LPS-receptor, dried to constant weight at 23–25 C, was suspended in predried chloroform-methanol 2:1 (v/v) in a glass-stoppered, spring-fastened pyrex tube and was stirred magnetically for 24 hr at 23–25 C. The phases were then separated by centrifugation. Reextraction was done twice, and the supernates were pooled and washed with 0.2 volume of aqueous 0.74% NaCl. The noncovalently bound lipids were recovered in the or-

ganic phase, freed of solvent in a stream of nitrogen, and dried to constant weight over paraffin, KOH, and P_2O_5 in vacuo. The residue was then hydrolyzed in a sealed tube at 100 C for 2 hr in 1% acetic acid in a nitrogen atmosphere [26] for separation of tightly bound lipid; after being dried to constant weight, the material was extracted with chloroform-methanol as described above. All fractions were soluble in buffered saline at concentrations up to at least 400 µg/ml after they had been mechanically shaken overnight at 4 C.

Enzymatic hydrolyses. The following enzymes were used (units/mg of substrate given in parentheses): insolubilized papain (1.8 units; Miles–Yeda, Rehovoth, Israel); crystallized mercuri-papain (0.8 units; Mann Research Laboratories, New York, N.Y.); twice-crystallized trypsin (500 µg/mg receptor; Mann); β-galactosidase from *E. coli* (27.4 units; Worthington); α-galactosidase from coffee beans (0.2 mg/mg of receptor [27]); neuraminidase from *Vibrio cholerae* (400 units; Behring Diagnostics, Woodbury, N.Y.); wheat germ lipase (0.02 units) and hog pancreas lipase (1 unit; Worthington); *Clostridium perfringens* phospholipase C (0.002 units; Calbiochem, La Jolla, Calif.); and phospholipase D from cabbage (0.002 units; Calbiochem). The enzymes were used in the buffers specified by the supplier, except that to β-galactosidase, Mg^{2+} (10^{-3} M, final concentration) was added, and the buffer in which the neuraminidase was purchased was adjusted from pH 5.6 to pH 6.8 by addition of 0.05 M NaOH. The substrate concentration was 2–3 mg/ml. Incubation was for 72 hr and at 37 C throughout with the exception of the cabbage phospholipase, which was incubated at 23–25 C; toluene served as preservative. The experiments with enzymes were concluded by thorough dialysis at 4 C in all instances except for that of the insolubilized papain, which was removed by centrifugation. All samples were freeze-dried and then dried at 23 C–25 C to constant weight before they were used in measurements of activity. All enzyme experiments included the following controls in the same final volume: (*1*) substrate alone, (*2*) enzyme incubated alone, and (*3*) boiled enzyme plus substrate.

Erythrocytes and antisera. Buffered saline served as serum diluent and red-cell–suspending solution. Human blood-group O erythrocytes from three adult donors (T. Z., J. C., and E. S.) were stored for less than two weeks at 1 C. Acid-citrate-dextrose (ACD, formulation A) solution was used as anticoagulant. Human antisera to blood group B (9220, -26, -34, -38, -42, and -71) were obtained from Ortho, Raritan, N.J. Rabbit antisera to *E. coli* O_{86} LPS were prepared in this laboratory as described previously [3]. All sera of animal origin were absorbed once with an equal volume of packed blood-group O red cells [9].

Red-cell coating and inhibition of coating. These procedures were similar to those described earlier [7]. Red cells were coated by the addition of nine volumes of ^{32}P-labeled LPS or ordinary LPS at various concentrations (indicated in Results) to one volume (0.2 ml) of washed, packed cells. The tubes were agitated in a rotary water bath at 37 C for 45 min and were also frequently inverted by hand. The red cells were then sedimented at 625 *g* and washed four times with 50 volumes of buffered saline before assay for ^{32}P and serologic activity. The smallest quantity of antigen affording maximal agglutination of red cells by homologous antiserum (reciprocal titer of 128, also called "optimal coating") under standard conditions was defined as one coating unit (U) [5].

The procedure in the inhibition assay differed from that in the coating test, in that the desired amount of inhibitor was added to one optimal coating dose of antigen and incubated in a total volume of 1.80 ml for 45 min at 37 C. Thereafter, 0.2 ml of packed erythrocytes was added, and the coating procedure was followed. In each test, a standard consisting of an ordinary coating assay using the antigen to be inhibited was included. The smallest amount of inhibitor that inhibited one coating unit of antigen by >95% was defined as one unit of inhibitor [5]. It must be stressed that definitions of coating and of inhibitory units are based on serologic results. The serologic data do not strictly correspond with the more sensitive and accurate radiologic measurements, but the serologic units proved to be useful in comparative studies [5]; therefore, they will be maintained.

Blocking of receptors on red cells by inhibitors. Packed, washed red cells were incubated with putative inhibitors for 45 min at 37 C; thereafter, they were washed four times. The inhibitor-treated red cells were then exposed to one unit of ^{32}P-labeled LPS.

Hemagglutination and hemagglutination-inhibition assays. The methods of titration and determination of hemagglutination titers (reciprocal) were the same as those used previously, as were the standards and the positive and negative controls included in all titrations [3, 9]. The tests were done at 23–25 C. The volume of all reagents in the tests was 0.05 ml; erythrocytes were used as 0.5% suspensions. A different pipette was used in each step of the titrations. Agglutinations were read microscopically and independently by at least two individuals about 1.5 hr after addition of red cells.

Hemagglutination-inhibition assays were carried out for measurement of the blood-group activity of the receptor. Methods and reagents have been described previously [8, 9].

Measurement of ^{32}P activity. Activity of red cells coated with ^{32}P-labeled LPS was assayed on planchets lined with lens tissue [28]. Supernatants and washes were placed on unlined planchets. Activity was measured with a gas-flow counter with automatic sample changer and printer (Nuclear Chicago, Des Plaines, Ill.). A uranium mock-^{32}P standard from International Chemical and Nuclear (Waltham, Mass.) was used for calibration of the counter. Decay was corrected for by inclusion of a ^{32}P-labeled LPS standard.

Competition between inhibitors and red cells for ^{32}P-labeled LPS. LPS and inhibitors were preincubated as in ordinary inhibition assays. Various quantities of inhibitor were incubated with 17 optimal coating units of LPS. The complexes [5] were exposed to erythrocytes at 37 C for periods of up to 5 hr under standard conditions, and duplicate samples of the mixtures were removed at intervals up to 5 hr. Thereafter, erythrocytes were washed, and the extent of their coating was determined. Red cells, LPS-coated and incubated in buffered saline only, served as standard. This procedure served to determine whether or not LPS, once complexed, was able to transfer to red cells.

Conversely, it was determined whether or not LPS already fixed to red cells could be removed under standard conditions (except for prolonged incubation; see Results).

Displacement of ^{32}P-labeled LPS from coated E. coli O_{86} LPS by other lipopolysaccharides. Red cells coated with ^{32}P-labeled LPS from *E. coli* O_{86} were incubated under standard conditions with other lipopolysaccharides and with Vi antigen. Thereafter, activity was measured on red cells after hemolysis as well as in the incubation medium. Radioactivity counts on red cells that had been coated with ^{32}P-labeled LPS and that had been incubated under the same conditions as the samples proper but with buffered saline only served as baseline, representing 100% coating.

Results

Inhibition of LPS coating of red cells. The inhibitory activity of the LPS-receptor (determined by serologic means) and the range of its specificity are shown in table 1. It can be seen in the last column of table 1 that the inhibitory activity of the LPS-receptor is quite specific, confined to O antigens and the related Kunin antigen. Closely similar receptor quantities (around 25 µg/ml) were needed for inhibition of coating by boiled bacterial suspensions and isolated LPS (table 1, column 3). That the ratio was lower for whole bacteria than for isolated LPS as shown in

Table 1. The lipopolysaccharide-receptor as inhibitor of antigen fixation to human erythrocytes.

Antigen fixed	(A) Smallest average amount of antigen affording optimal coating (µg/ml)	(B) Smallest average amount of receptor inhibiting coating by >95% (µg/ml)	B:A
Gram-negative bacteria			
Suspended bacteria (9)*	22	24	1.1
Isolated O antigens† (14)	3	27	9
Kunin antigens, dialyzed purified culture fluids (3)	85	125	1.5
Isolated Vi antigens (2)	0.6	700	1,170
Gram-positive bacteria			
Rantz antigens, dialyzed purified culture fluids (2)	65	4,500	69
Streptococcus pyogenes isolated group antigens stearoyl derivative (2)	4	1,750	440

* Figures in parentheses = numbers of different strains tested.
† Antigens in bold type are highly purified; those in regular type are crude preparations.

the last column is due to the contribution of inert material in coating experiments with boiled, whole bacteria.

The LPS-receptor exhibited activity not only towards all LPS preparations tested but also towards the protein-LPS antigens investigated. Although, in some instances, larger quantities of O antigen of a given bacterium were needed when protein-LPS was used instead of LPS, the amount of receptor needed to prevent coating of red cells was proportionally much smaller for the former. Thus the ratio of B to A depicted in table 1 was close to 1 or even <1 for protein-LPS complexes, as compared with an average of 9 for LPS. Coating of red cells by LPS activated by incubation at 37 C was also effectively inhibited by the receptor.

The inhibition given by the receptor was physical and not enzymatic, since incubation of LPS-receptor mixtures for 38 hr and subsequent receptor destruction by autoclaving (see below, *Physico chemical properties*) indicated no change in coating activity and serology of LPS as compared with LPS incubated and autoclaved without receptor. No decrease in the serologic activity of LPS coated onto red cells was observed after their exposure to 2 U of receptor under standard conditions for 45 min.

The activities of substances that we found to possess significant coating-inhibiting activity [4] are compared, on a strictly quantitative basis, in table 2. The inhibitors are listed in descending

Table 2. Inhibition of coating of red cells by ^{32}P-labeled lipopolysaccharide (LPS) from *Escherichia coli* O$_{86}$.

Inhibitor	Concentration (µg/ml)	Percentage inhibition of ^{32}P-labeled LPS uptake†
LPS-receptor	40	80
Serum albumin (human)	200	80
Phosphatidylethanolamine	500	80
Neomycin	1,000	75
Ganglioside	2,000	75
Asialoganglioside*	2,000	77
Human hemoglobin	2,000	82
Acrolein	10,000	20

* Prepared from bovine ganglioside (sialic acid content, 18.7%) in this laboratory; still contained 2.0%–2.5% sialic acid.
† 25 units of LPS.

order of efficiency, and the high activity of the LPS-receptor as compared with all other active substances is clearly indicated. Some poorly soluble substances such as cholesterol, which had previously been reported to inhibit erythrocyte coating [29], are not shown in table 2; they possessed <1% of the activity of the LPS-receptor. Acrolein showed only a very small effect; 10,000 µg/ml inhibited uptake of LPS by only 20% (table 2), even though its inhibitory effect as measured by serologic methods had appeared to be very much higher [4]. The reason for this seeming discrepancy is indicated in table 3: preincubation of red cells with the various inhibitors and their subsequent removal shows that only acrolein, ganglioside, and neomycin attach to red cells and thus significantly interfere with coating by LPS thereafter. All of these results were strictly reproducible except those with neomycin, where the effects sometimes were considerably fainter or even stronger than recorded in the table.

Because of the blocking effect of acrolein on the red-cell sites to which LPS attaches, its effect was also determined on the isolated receptor by incubation under standard conditions of 20 U of receptor with 40 U of acrolein. This was followed by exhaustive dialysis. The isolated receptor tested in inhibition assays was found to have lost more than 90% of its inhibitory activity.

None of more than 200 compounds, including all amino acids, proteins such as casein, γ-globulins, phospho-proteins, egg albumin, homo- and heteropolysaccharides, monosaccharides, DNA, and RNA, possessed any inhibitory activity.

Mode of LPS-receptor action. It was shown in the foregoing that the receptor appeared to interact only with the LPS; its action was confined to prevention of fixation of LPS to the cell surface. Therefore, we investigated whether or not the receptor permanently blocked the combining sites on the LPS molecule. The binding of LPS to any of the three inhibitors listed is not irreversible, and some LPS is transferred from the inhibitors to the red cells (figure 1). This transfer amounted to about 40% per 3.8 U of receptor during incubation for 5 hr. The transfer of LPS from both hemoglobin and ganglioside was much faster and exceeded 40% after 1 hr. After 5 hr, the transfer exceeded 80% for hemoglobin but had curiously decreased for the ganglioside. This was in contrast to both LPS-recep-

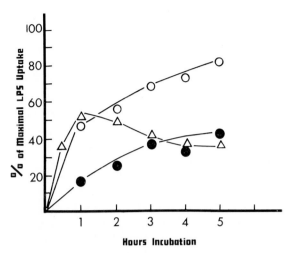

Figure 1. Transfer of ^{32}P-labeled lipopolysaccharide (LPS) of *Escherichia coli* O_{86} (17 U) from inhibitors to red cells. (●——●) = LPS-receptor, 3.8 units (35 µg); (○——○) = human hemoglobin, 2.5 units (500 µg); (△——△) = ganglioside, 3.8 units (1,000 µg).

tor and hemoglobin, where the transfer reached a plateau. This phenomenon is explained by the fixation of ganglioside itself to the red-cell surface and the concomitant release of LPS (table 3). Additional experiments, not listed in figure 2, showed that under standard coating conditions, an equilibrium was reached between LPS attached to LPS-receptor and LPS transferred to red cells. Furthermore, when incubated with receptor, LPS already fixed to red cells transferred to it, again until an equilibrium was reached. Thus, from 2 ml of a 2% red-cell suspension coated with 75 U of *E. coli* O_{86} LPS, 63 U of receptor removed 22.5% of all LPS after 1.5 hr (table 3) and approximately twice that amount after 6 hr. Some inhibitory lipids

Table 3. Removal by inhibitors of ^{32}P-labeled lipopolysaccharide (LPS) fixed to red cells.

Inhibitor	Concentration (µg/ml)	^{32}P-labeled LPS* removed by incubation for 1.5 hr (%)
LPS-receptor	625	22.5
Ganglioside	2,400	25.4
Phosphatidylethanolamine	2,500	21.5
Neomycin	3,000	2.3
Acrolein	10,000	2.3

* 25 units.

under the same conditions likewise removed coating LPS, although they were much less efficient on a weight basis; two other inhibitors, neomycin and acrolein, were virtually ineffective.

Displacement of red-cell–fixed LPS by bacterial antigens. LPS of gram-negative bacteria attaches to structures on the intact red-cell surface that are presumably identical with those on the isolated LPS receptor (see also [5]). We investigated, therefore, the possibility that one free bacterial antigen was capable of removing another already fixed to red cells. Our findings are shown in table 4, in which is demonstrated the ready displacement from the surface of the erythrocyte of one LPS by another but not by Vi antigen. The displacement on red-cell surfaces of one LPS by another has been described before [30].

Nature of the LPS-receptor. General properties. The electrodialyzed receptor was obtained as white, fluffy powder that was readily and immediately soluble at around pH 7 at concentrations up to at least 2%. Receptor solutions had a faintly yellow tinge. The receptor lost 8.48% water on drying at 80 C; it contained 0.49% ash and <0.02% phosphorus. Under standard conditions of incubation, the receptor retained full activity for at least 48 hr. It lost about 40% of its activity on incubation at 56 C for 6 hr. Boiling or autoclaving at 15 psi destroyed all receptor activity in less than 30 min; preincubation with LPS did not protect the receptor from inactivation. The receptor possessed no human blood-group A, B, M, and $Rh_o(D)$ activities at concentrations of at least 10 mg/ml. It did possess traces of H(O) activity at a concentration of 2.5–5.0 mg/ml, as measured with both human and eel antisera to H(O), and of N

activity with rabbit but not human antisera. The receptor was nonpyrogenic in rabbits at concentrations of 1 mg/kg of body weight [31].

Physicochemical properties. The procedures of extraction delineated in Materials and Methods resulted in a preparation that was electrophoretically homogeneous and migrated with a negative charge between *p*H 5.0 and *p*H 9.0. Its isoelectric point was close to *p*H 3.0. The properties of the receptor, which is either highly asymmetric or highly hydrated as indicated by the frictional ratio (f/f_o), are listed in table 5. A molecular weight of 228,000 was calculated from the data by the Svedberg equation, taking into account a 10% lipid content of the receptor (see below). Exhaustive succinylation, which acylated all lysyl residues, resulted in the release of 5%–10% of a smaller component. This changed the physical properties of the preparation as indicated in table 5. Succinylation followed by reduction-alkylation in urea [14] yielded three components: an apparent subunit, the starting material, and an apparent aggregate of the starting material. The light material had a molecular weight of 25,000. These findings suggest the involvement of noncovalent and/or -S-S-bonded subunits in the composition of the receptor; the latter possibility is remote (see below).

Chemical composition of LPS-receptor. The averaged results of two analyses of amino acids are listed in table 6. The peptide part of the receptor amounts to 61%. The receptor is rich in hydroxy- and mono- amino dicarboxylic acids. Glutamic acid is the predominant amino acid, followed closely by aspartic acid, serine and threo-

Table 4. Displacement from red cells of ^{32}P-labeled lipopolysaccharide (LPS) from *Escherichia coli* O_{86} by antigens from various bacteria.

Competing antigen (2×10^2–1.7×10^3 units)	Displacement (%) after	
	1.5 hr	6 hr
E. coli O_{86}, LPS	44.7	58.1
Salmonella senftenberg, LPS	46.2	ND*
Serratia marcescens, LPS	30.6	66.1
Paracolobactrum ballerup, Vi antigen	ND	0

NOTE. We used 1.6×10^2 units of ^{32}P-labeled LPS for coating.

* ND = not done.

Table 5. Physical data on highly purified lipopolysaccharide-receptor.

$S_{20^o, w}$*	6.5 (6.4) †
$D_{20^o, w}$	2.5 (2.8)
Mol wt	228,000 (202,000)
$\overline{v}$(ml/g)	0.718‡
dS/dC	1.6×10^{-4}
f/f_o	2.1
$[\alpha]_D^{29}$	−23.7 °
$A_{276nm}^{1\%}$	5.805

* Determined at 3–10 mg/ml.

† Numbers in parentheses indicate readings after exhaustive succinylation.

‡ Calculated from amino acid and carbohydrate composition. The molecule contained 9.61% lipid (see text) whose $\overline{v}$ (ml/g) was assumed to be 1.00.

Table 6. Amino acid composition of lipopolysaccharide (LPS)-receptor.

Amino acid	Percentage	Moles of amino acid per mole of LPS-receptor
Aspartic acid	5.78	99
Threonine	4.53	87
Serine	4.51	98
Glutamic acid	7.35	114
Proline	3.68	73
Glycine	2.20	67
Alanine	2.92	75
Cystine/2	trace	—
Valine	3.96	77
Methionine*	1.35	18
Isoleucine	3.03	53
Leucine	5.02	87
Tyrosine	2.12	28
Phenylalanine	3.41	47
Lysine	3.59	56
Histidine	2.82	41
Arginine	3.62	47
Tryptophan	1.18	13
Total	61.07	1,080

* As sulfoxide.

nine, while there is scarcity in Trp and sulfur-containing amino acids. Cystinyl-S- is demonstrable only in traces. The carbohydrate components of the receptor and its lipid content are listed in table 7. The N-acetylneuraminic acid (NANA) content of the LPS-receptor is more than three times that of the next common sugar, Gal. There is a low but significant content of Glc, Man, and Fuc.

The lipid content is quite high and amounts to slightly under 10% of the receptor molecule. The lipid is readily removed by organic solvents and, therefore, is not bound covalently. The lipid seems to play no role in the LPS-binding activity of the receptor (table 8). Extraction with organic solvents without prior hydrolysis did not significantly change the activity of the receptor. The activity of the

Table 8. Effect of lipid removal on lipopolysaccharide-receptor activity.

Preparation	Inhibitory activity as percentage of activity of intact receptor	
	Coating by serology*	Coating by ^{32}P uptake†
(1) Intact receptor	100	100
(2) CHCl$_3$:CH$_3$OH-extracted receptor	100	93.2
(3) CHCl$_3$:CH$_3$OH extract	<10	6.8
(4) Receptor (2) after hydrolysis and renewed CHCl$_3$:CH$_3$OH extraction	25–50	45.8
(5) CHCl$_3$:CH$_3$OH extract of (4)	<7.5	<1

* Standard conditions.
† Measured under standard conditions at 40 µg/ml *Escherichia coli* O$_{86}$ lipopolysaccharide concentration and 50 µg/ml receptor concentrations.

Table 7. Chemical composition of lipopolysaccharide-receptor.

Structural unit	Assay	Percentage	Moles of unit per mole of receptor
Carbohydrates			
Galactose	Galactose dehydrogenase	6.14	78
Glucose	Glucose oxidase	0.11	1
Mannose residue	Cysteine-H$_2$SO$_4$	1.46	18
Fucose residue	Cysteine-H$_2$SO$_4$	1.07	15
N-acetylneuraminic acid	Thiobarbituric acid	16.32*	120
Glucosamine	Amino acid analyzer	3.98	51
Galactosamine		4.42	56
Hexosamines	Gatt-Berman	10.33	131
Total carbohydrate		33.50	
Total peptide	Amino acid analyzer	61.07	
Lipids	Folch		
Noncovalent		9.61	
Covalent		< 0.30	
Total of all units		104.48	

* 16.67% by resorcinol assay.

extracted lipid was negligible. Even after hydrolysis at *p*H 3.67, which is known to damage the activity of the receptor severely [5], nearly 50% of its activity remained. The lipid extract after hydrolysis showed no activity.

Effect of enzymes on LPS-receptor activity. Proteases of animal and plant origin, trypsin and papain, inactivated the receptor by more than 95%. None of the other enzymes (namely, α- and β-galactosidase and neuraminidase) inactivated the receptor, although the latter enzyme removed over 95% of the receptor's NANA. The lipases likewise had no influence on the activity of the receptor.

Discussion

We describe here for the first time in detail the isolation and characterization of a cell component that specifically binds the endotoxins of all gram-negative bacteria investigated and also the related Kunin antigen, but none of the other antigens of either gram-negative or gram-positive bacteria tested. We have named this component of the red-cell membrane LPS-receptor because of the specificity of its reactivity. We have obtained it in physicochemically homogeneous form. It is a lipo-glycoprotein rich in neuraminic acid. Its protein part resembles those of the glycoproteins of the human blood groups ABH, Le, and MN [8, 32, 33]. The predominance of monoaminodicarboxylic acids and hydroxy-amino acids makes it reasonable to assume that the carbohydrate part of the molecule is attached to the protein part via glycosidic and amide linkages [34]. The NANA content of the LPS-receptor is quite similar to that of the antigens of the human blood group MN, but there is less Gal and somewhat more Fuc [8]. The small quantity of Glc may originate from the Sephadex and dialysis membranes used during the isolation of the receptor even though the appropriate blanks were subtracted.

The lipid part does not seem to have phosphorus-containing lipids. The lipids that could be removed were not involved in receptor activity. There is no proof as yet, however, that our procedures (even hydrolysis at about *p*H 3.7) rendered the receptor completely free of lipids.

Among the enzymes used, only the proteases proved to be capable of inactivating the receptor. Even though inactivation was virtually complete,

this does not prove that protein, peptides, or amino acids are involved specifically in receptor binding. It has been shown that proteases inactivate blood-group substances by destroying the peptide scaffolding that assures the appropriate arrangement in space of the immunodeterminant carbohydrate groupings [8]. The receptor had none or only traces of the blood-group specificities tested for. We have recently found it to be faintly immunogenic and to provoke antibodies specific for LPS-receptor in rabbits (G. F. Springer and J. C. Adye, unpublished observations).

Our results showed that LPS-receptor interacts only with LPS and not with the surface membrane of the red cell. That the structures on the isolated receptor and those on the red-cell surface that interact with the receptor appear to be the same is indicated by the inactivating effect of acrolein on the receptor groups in both locations and by the lack of any effect on either type of structure by neuraminidase. The attachment of LPS to red cells is reversible, as is the attachment to the receptor; the process of attachment seems to obey the laws of mass action [5]. The receptor also inhibits the cell-surface fixation of protein-LPS and LPS "activated" under physiologic conditions at 37 C.

We realize that the red-cell receptor that we have isolated may not be the same structure with which endotoxin interacts before it can begin its most noxious actions. Receptors on other cells may differ in nature, and there may be more than one kind of receptor. Nevertheless, we have isolated for the first time a cell-surface macromolecule that specifically prevents LPS from fixation to red cells. This macromolecule has a substantially higher affinity for endotoxin than any of the other substances investigated.

Our isolation of a cell-bound receptor substance that interacts specifically with the endotoxins of gram-negative bacteria has practical clinical implications in the treatment of endotoxic shock and septic abortion, especially since the receptor was also found to prevent the attachment to red cells of lipopolysaccharide that had been preincubated under physiologic conditions. Another outcome of such studies may be the isolation of active groupings of cell receptors and their use in prevention of attachment of endotoxins to cells, as well as in removal of LPS already bound to the cells.

From a basic scientific point of view, this receptor is likely to further the understanding of the mode of attachment of these and other toxic substances to cells and tissue components in addition to erythrocytes, since any toxic substance (or drug) has to attach itself to components of cells or tissues before it can begin to exert its effect.

References

1. Neter, E. Bacterial hemagglutination and hemolysis. Bacteriol. Rev. 20:166–188, 1956.
2. Buxton, A. The *in vivo* sensitization of avian erythrocytes with *Salmonella gallinarum* polysaccharide. Immunology 2:203–210, 1959.
3. Springer, G. F., Horton, R. E. Erythrocyte sensitization by blood group-specific bacterial antigens. J. Gen. Physiol. 47:1229–1250, 1964.
4. Springer, G. F., Wang, E. T., Nichols, J. H., Shear, J. M. Relations between bacterial lipopolysaccharide structures and those of human cells. Ann. N.Y. Acad. Sci. 133:566–579, 1966.
5. Springer, G. F., Huprikar, S. V., Neter, E. Specific inhibition of endotoxin coating of red cells by a human erythrocyte membrane component. Infec. Immun. 1:98–108, 1970.
6. Neter, E., Gorzynski, E. A., Westphal, O. Lüderitz, O., Klumpp, D. J. The effects of protamine and histone on enterobacterial lipopolysaccharides and hemolysis. Can. J. Microbiol. 4:371–383, 1958.
7. Whang, H. Y., Neter, E., Springer, G. F. Specificity of an erythrocyte membrane receptor for bacterial antigens. Z. Immunitaetsforsch. Allerg. Klin. Immunol. 140:298–303, 1970.
8. Springer, G. F., Nagai, Y., Tegtmeyer, H. Isolation and properties of human blood-group NN and meconium-Vg antigens. Biochemistry 5:3254–3272, 1966.
9. Springer, G. F., Horton, R. E. Blood group isoantibody stimulation in man by feeding blood group-active bacteria. J. Clin. Invest. 48:1280–1291, 1969.
10. Springer, G. F. Inhibition of blood-group agglutinins by substances occurring in plants. J. Immunol. 76:399–407, 1956.
11. Westphal, O., Lüderitz, O. Chemische Erforschung von Lipopolysacchariden gramnegativer Bakterien. Angew. Chem. 66:407–417, 1954.
12. Goebel, W. F., Binkley, F., Perlman, E. Studies on the Flexner group of dysentery bacilli. I. The specific antigens of *Shigella paradysenteriae* (Flexner). J. Exp. Med. 81:315–330, 1945.
13. Ma, T. S., McKinley, J. D. Determination of phosphorus in organic compounds: a new microprocedure. Mikrochimica Acta 1-2:4–13, 1953.
14. Bezkorovainy, A., Springer, G. F., Desai, P. R. Physicochemical properties of the eel anti-human blood-group H(O) antibody. Biochemistry 10:3761–3764, 1971.
15. Wiemer, R. J. Studies on agar gel electrophoresis, techniques, applications. Thesis. Ascia, Brussels, 1959.
16. Springer, G. F., Takahashi, T., Desai, P. R., Kolecki, B. J. Cross reactive human blood-group H(O) specific polysaccharide from *Sassafras albidum* and characterization of its hapten. Biochemistry 4:2099–2112, 1965.
17. Warren, L. The thiobarbituric acid assay of sialic acids. J. Biol. Chem. 234:1971–1975, 1959.
18. Svennerholm, L. Quantitative estimation of sialic acids. III. An anion exchange resin method. Acta Chem. Scand. 12:547–554, 1958.
19. Dische, Z., Shettles, L. B. A new spectrophotometric test for the detection of methylpentose. J. Biol. Chem. 192:579–582, 1951.
20. Dische, Z., Shettles, L. B., Osnos, M. New specific color reactions of hexoses and spectrophotometric micromethods for their determination. Arch. Biochem. Biophys. 22:169–184, 1949.
21. Gatt, R., Berman, E. R. A rapid procedure for the estimation of amino sugars on a microscale. Anal. Biochem. 15:167–171, 1966.
22. Spackman, D. H., Stein, W. H., Moore, S. Automatic recording apparatus for use in the chromatography of amino acids. Anal. Chem. 30:1190–1206, 1958.
23. Hirs, C. H. W. The oxidation of ribonuclease with performic acid. J. Biol. Chem. 219:611–621, 1956.
24. Winkler, S. Über die Tryptophanreaktion von Adamkiewicz-Hopkins. Z. Physiol. Chem. 228:50–60, 1934.
25. Folch, J., Lees, M., Sloane Stanley, G. H. A simple method for the isolation and purification of total lipides from animal tissues. J. Biol. Chem. 226:497–509, 1957.
26. Galanos, C., Lüderitz, O., Westphal, O. Preparation and properties of antisera against the lipid-A component of bacterial lipopolysaccharides. Eur. J. Biochem. 24:116–123, 1971.
27. Springer, G. F., Nichols, J. H., Callahan, H. J. Galactosidase action on human blood group B active *Escherichia coli* and on red cell substances. Science 146:946–947, 1964.
28. Comar, C. L. General procedures in radioassay. *In* Radioisotopes in biology and agriculture. McGraw Hill, New York, 1955, p. 158–200.
29. Neter, E., Zalewski, N. J., Zak, D. A. Inhibition by lecithin and cholesterol of bacterial (*Escherichia coli*) hemagglutination and hemolysis. J. Immunol. 71:145–151, 1953.
30. Lüderitz, O., Westphal, O., Sievers, K., Kröger, E., Neter, E., Braun, O. H. Über die Fixation von P^{32}-markiertem Lipopolysaccharid (Entodoxin) aus *Escherichia coli* an menschlichen Erythrocyten. Biochem. Zeit. 330:34–46, 1958.
31. U.S. Pharmacopoeia. Vol. XVIII, 1970, p. 886.
32. Carsten, M. E., Kabat, E. A. Immunochemical

studies on blood groups. XIX. The amino acids of blood group substances. J. Am. Chem. Soc. 78:3083–3087, 1956.

33. Pusztai, A., Morgan, W. T. J. Studies in immunochemistry. 22. The amino acid composition of

the human blood-group A, B, H and Le[a] specific substances. Biochem. J. 88:546–555, 1963.

34. Springer, G. F. Role of human cell surface structures in interactions between man and microbes. Naturwissenschaften 57:162–171, 1970.

Effects of Endotoxin on Macrophages and Other Lymphoreticular Cells

A. C. Allison, P. Davies, and R. C. Page

From the Clinical Research Centre, Harrow, Middlesex, England

In addition to exerting an adjuvant effect by direct stimulation of bone-marrow (B) lymphocytes, *Escherichia coli* endotoxin appears to have an adjuvant effect involving macrophages and thymus (T) lymphocytes. In cultured macrophages endotoxin stimulates the synthesis of the nonlysosomal enzymes lactic dehydrogenase and leucine-2-naphthylamidase (which are retained in the cytoplasm) and of the lysosomal enzymes acid phosphatase and β-N-acetylglucosaminidase (which are released into the medium). Properties of antibody-dependent cytotoxic lymphoid cells are discussed in relation to possible effects in endotoxin-mediated, immunopathologic reactions.

Although adjuvants have long been used as stimulants of immune responses, little is known about their mode of action. As a first approach to the analysis of this problem, it seemed worthwhile to define which of the cell types involved in immune responses are required for adjuvant effects. This involved experiments in which it was possible to treat separately with adjuvants each of the components of the complete system. Experiments in which antigen-containing peritoneal-exudate cells (PEC) were transferred to syngeneic recipients, and in which immune responses were reconstituted in irradiated recipients by transfers of syngeneic lymphoid cells, were appropriate. The macrophages or lymphocytes could be treated in vitro with adjuvants and washed before transfer into recipient mice; the use of radioactive antigens and adjuvants made it possible to ensure comparability of uptake in the system under investigation.

Adjuvant Effects

Our first experiments were carried out with PEC containing *Maia squinada* hemocyanin (MSH) in

CBA mice [1]. Antibody responses to MSH were markedly increased by injections of *Bordetella pertussis* when this adjuvant was administered at the same time as or a few hours before antigen. Antigen-containing PEC treated in vitro with *B. pertussis* elicited much higher antibody titers when injected into mice than they did in the absence of the adjuvant. Lymph-node cells treated in vitro with adjuvants and injected into irradiated mice gave the same antibody titers as they did in the absence of adjuvants; hence, adjuvants neither increased the response of immunocompetent cells nor prevented their capacity to reconstitute an immune response in irradiated recipients. In this system adjuvants exerted their stimulatory effects on antibody-producing cells only after uptake by PEC. Adjuvant-treated PEC injected at the same time as antigen-containing PEC also led to increases in the immune response; action of the adjuvant did not require the presence of adjuvant and antigen in the same cell. The overall catabolism and retention of ^{131}I-labeled hemocyanin in PEC was not significantly altered in the presence of adjuvants. In both cases most of the labeled antigen was broken down rapidly (4–5 hr), but 8%–14% remained in the cells during the second phase of slow catabolism.

In further experiments [2] it was found that

Please address requests for reprints to Dr. A. C. Allison, Clinical Research Centre, Harrow, Middlesex, England.

transfer of PEC that had taken up *Escherichia coli* lipopolysaccharide (LPS) and bovine serum albumin (BSA) consistently gave higher antibody responses than did transfer of cells containing approximately the same amount of BSA but no LPS. Again, treatment of lymphocytes with adjuvants before their use for reconstitution of irradiated recipients had no detectable effect on immune responses. In these and previous experiments, exposure to LPS of PEC that had taken up labeled BSA had no detectable effect on the rate of catabolism of the antigen. The BSA content of the cells fell rapidly to about 20% of the original level and then fell slowly or remained constant. The immunogenicity of the BSA during this period was unchanged.

Higher immune responses were found in mice inoculated with BSA associated with PEC than in mice inoculated with comparable amounts of free, soluble BSA. One interpretation of the higher antibody responses elicited by antigen in PEC is that PEC simply carry antigen to a site where it is likely that an immune response will be elicited. If this were so, it would be expected that the dose-response curves of antibody formation after exposure to antigen either in cells or free would be parallel, although the former would be higher. We found remarkably consistent linear dose-response curves when the logarithm of the primary dose of BSA was plotted against the logarithm of antigen-binding capacity after a secondary response, but the slopes of the two curves (antigen free or cell-associated) were significantly different. Moreover, when free BSA was administered together with cell-associated BSA, the immune responses were significantly less than those obtained with cell-associated BSA alone.

Thus it seems clear that PEC do more than act as passive carriers of antigen. The immune response to cell-associated BSA was equalled but not exceeded when the same amount of free BSA was administered together with the most potent known adjuvants. The results suggest that, with most adjuvants, there is a ceiling response to a given dose of antigen; this corresponds well to that observed with the same dose of antigen in cells. It seems reasonable to conclude that in both cases the balance betwen immunization and induction of tolerance is strongly altered in favor of the former, and that within certain limits the amount of antibody formed is dependent on the dose of antigen. The role of PEC in increasing immunogenicity can be explained if we suppose that they can present antigen to immunocompetent cells in a way that strongly favors immunity; moreover, if we limit the amount of BSA free to diffuse through the tissues, PEC reduce induction of tolerance.

Requirement of T (Thymus) Lymphocytes for Adjuvant Effects

If it is accepted that the initial reaction of at least some adjuvants is with nonimmunocompetent cells of the macrophage type, and that antibody is formed by B (bone-marrow) lymphocytes and cells derived from them, two hypotheses can be considered. According to the first, adjuvants can increase the efficiency of interaction of macrophages with B cells, by-passing the requirement for T-cell helper effects. According to the second, interaction of adjuvant with cells of the macrophage type increases the efficiency of their stimulation of T cells, so that the helper effect of T cells is augmented. We can test these hypotheses by ascertaining whether adjuvants can stimulate immune responses in animals deprived of T cells. Unanue [3] found that a dose of beryllium sulfate, which enhances 10-fold or more the immune response to keyhole limpet hemocyanin in normal mice, could not restore antibody formation in mice that were thymectomized, irradiated, and reconstituted with B cells. Allison and Davies [4] found that thymectomized mice either treated with anti-lymphocytic serum or irradiated and reconstituted with B cells gave very low immune responses to BSA. Administration of several adjuvants failed to increase immune responses in such animals, unless they had been reconstituted with grafts of T as well as B cells.

As has been reviewed by Möller at this meeting (see Session 2), endotoxins can have direct stimulating effects on B cells, so it is conceivable that some increase in antibody formation results from this mechanism. However, much of the adjuvant effect of endotoxin and of organisms such as *B. pertussis* (which may be due to endotoxin-like materials) appears to be mediated through macrophages (or analogous cells in peritoneal exudates) and T lymphocytes. Gery and Waksman [4a] have found that macrophages treated with endotoxin release a soluble factor that stimulates T lymphocytes.

Effects of Endotoxin and Other Bacterial Constituents on Enzymes in PEC

Lipopolysaccharide endotoxins possess a wide range of pharmacologic effects, some of which involve lysosomes, despite their lack of effect on the release of hydrolases from isolated granules. Endotoxin stimulates the bactericidal activity of both polymorphonuclear leukocytes [5] and macrophages [6]; acid phosphatase activity is also increased in the latter cell. These effects are explained on the basis of cellular injury, which, however, results in increased functional capacity.

Weissmann and Thomas [7] found that injection of *Enterobacter aerogenes* endotoxin into young rabbits increased the lability of lysosomes isolated immediately afterwards; such treatment also increased the serum levels of lysosomal enzymes. Both of these effects were reduced by prior induction of tolerance to endotoxin [8].

Endotoxin from *Salmonella abortus equi* stimulates the differentiation of murine peritoneal macrophages in vitro [9]. There is an increase in cell size and in phase-dense, acid phosphatase-positive granules, and electron microscopy reveals hypertrophy of the Golgi region with an increase in the number and size of lysosomes [10]. Increases in phagocytosis and pinocytosis of colloidal gold and in phagocytosis of bacteria by differentiating equine monocytes were observed in the presence of 1 μg/ml of bacterial LPS [11]. It is therefore probable that the activation of the reticuloendothelial system by endotoxin in vivo is in part mediated by an effect on the endocytic and digestive capacity of the lysosomal system.

The role of the macrophage lysosomal system in detoxification of endotoxin was investigated recently [12, 13]. It was found that sonicates of rat peritoneal, alveolar, and hepatic (Kupffer cells) macrophages effectively detoxified *Salmonella enteriditis* endotoxin (3 μg/ml) at protein concentrations of 0.5 mg/ml. On the other hand, sonicates of peripheral blood or peritoneal-exudate polymorphonuclear leukocytes were ineffective at concentrations of up to 5 mg/ml [14]. Sonicates of large-granule fractions from peritoneal-exudate macrophages rendered endotoxin inactive at protein concentrations as low as 12 μg/ml [14]. This suggests an important role for macrophage lysosomes in the detoxification of bacterial LPS.

When bacterial constituents are injected into an intact animal, they have multiple effects that are difficult to analyze. We have therefore examined their effects on cultures of murine PEC, which we took to be macrophages; however, it now appears that a second type of cell is also present, as described below. In this closed system, effects on macrophages and A (antibody-dependent) cells (see next section), both of which we believe to be important in certain subacute and chronic inflammatory reactions, are revealed. Most of our studies have been with a cell-wall component of group A streptococci, shown by Schwab and his colleagues [15] to produce, after a single injection, lesions persisting for many months in the skin, joints, and heart of recipient animals. The lesion contains many mononuclear cells, and in vitro studies have shown remarkable effects of the cell-wall preparation on murine PEC [16]. The cell-wall preparation, in concentrations of 1–50 μg/ml, causes an increase in size and spreading of macrophages on the surface of the culture vessel. The cells have a greater concentration of protein, which is reflected by increased concentrations of both lysosomal and nonlysosomal enzymes. In the presence of relatively low doses (up to 10 μg/ml) of cell-wall component, the acid hydrolases remain within the cells, but at higher concentrations selective exocytosis of a high proportion of certain acid hydrolases occurs (figure 1). The loss of lysosomal enzymes is not accompanied by death of the cell, and intracellular levels of nonlysosomal enzymes such as lactate dehydrogenase and leucine-2-naphthylamidase remain elevated. These in-vitro phenomena may be a reflection of the events in the intact animal after injection of the cell-wall component; selective exocytosis of acid hydrolases from viable cells would account for both the resultant damage to tissues and the chronicity of the lesion. This concept may serve as a model for events occurring in certain diseases such as rheumatic fever and periodontal disease, in which plaque containing bacterial constituents clearly plays a major role.

Endotoxin (LPS B, *E. coli* 055:B5, Difco) was added to cultures of murine PEC and incubated for 72 hr. Effects on enzyme levels are

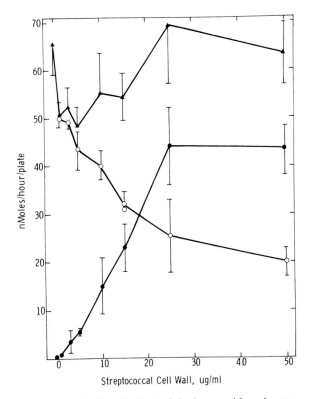

Figure 1. Redistribution of β-glucuronidase in murine peritoneal cells incubated with various concentrations of group A streptococcal cell wall for 24 hr. (▲——▲) = total activity, (●——●) = activity in cells, and (○——○) = activity found in the culture medium.

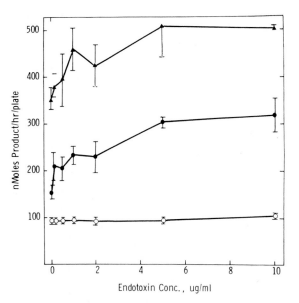

Figure 2. Effect of increasing concentrations of endotoxin on the levels of lysosomal enzymes in cultures of murine peritoneal cells. Values represent total enzyme activity after 72 hr in the presence of endotoxin. Activities are expressed as nmoles of product/plate per hr for β-glucuronidase (○——○) and N-acetyl-β-glucosaminidase (▲——▲), or as ΔOD_{560nm}/plate per hr for acid naphthylphosphatase (●——●).

shown in figure 2 and table 1. Endotoxin has no significant effect on production of the lysosomal hydrolases β-glucuronidase and β-galactosidase, but significantly stimulates the synthesis of two other lysosomal hydrolases, acid phosphatase and β-N-acetylglucosaminidase, and the selective release of these enzymes into the medium. Treatment with endotoxin leads to a marked increase

(about double) in activities of the nonlysosomal enzymes, lactic dehydrogenase and leucine-2-naphthylamidase, with no loss into the medium.

These and other differential results that we have obtained with several bacterial products imply that there is not one but a series of mechanisms that switch on enzyme synthesis in macrophages; one or more of these mechanisms may be activated by any particular stimulus. Endotoxin is one of the stimulating agents, and enzyme synthesis in and release from macrophages must be accepted as

Table 1. Effects of endotoxin on murine peritoneal cells in culture.

Enzyme measured	Control culture		Endotoxin (10 µg/ml)	
	Cells	Media	Cells	Media
Lactate dehydrogenase	189 (8)*	. . .	356 (14)	. . .
Leucine naphthylamidase	483 (47)	. . .	721 (51)	. . .
β-glucuronidase	81 (4)	13 (2)	84 (4)	17 (1)
β-N-acetylglucosaminidase	82 (6)	273 (24)	94 (19)	411 (23)
Acid naphthylphosphatase	0.744 (.045)	0.013 (.021)	1.182 (.196)	0.395 (.016)

* All values reported as nmol of product/hr per plate (±SD), except acid phosphatase, which is reported as OD_{560nm}/hr per plate. Cultures were continued for 72 hr.

208

one of the biological effects of endotoxin. Alexander and Evans [17] have reported that endotoxin activates PEC so that they nonspecifically kill target cells with which they come into contact; enzymatic changes in macrophages may possibly play a role in this phenomenon.

A New Cell Type and the Biological Effects of Endotoxin

A new mechanism of specific killing of target cells by effector cells from blood or lymphoid organs has been established in recent years. In this system the effector cells are not themselves specifically sensitized; they are nonspecific cells triggered into close attachment and cytotoxic activity by antibody bound to antigens on the target cell. The cytotoxic cells, however, kill only appropriately sensitized target cells and are not activated to indiscriminate killing. This system has been studied particularly in the laboratories of Perlmann in Stockholm and MacLennan in Oxford, and background information will be found in excellent reviews by Perlmann and Holm [18] and MacLennan [19].

The main features of antibody-dependent, cell-mediated cytotoxicity are summarized in figure 3. Most experiments have involved killing of foreign cells, especially chicken erythrocytes or tumor cells, but Perlmann and Holm [20] have shown that antiserum from guinea pigs that are sensitized

with BCG or heat-killed tubercle bacilli are able to render chicken red cells coated with purified protein derivative of tuberculin susceptible to cytotoxic lysis by normal spleen cells. Hence, it seems likely that any new antigen on the surface of target cells, e.g., endotoxin on host cells, could become coated with antibody and make the cells liable to cytotoxic attack by effector cells in the circulating blood. A brief account of the properties of the effector cells and sensitizing antibody will now be given.

Activity of effector cells can be assayed by addition to target cells sensitized with antibody of various numbers of cells from blood or lymphoid organs and measurement of ^{51}Cr release under standard conditions. Antibody-dependent effector-cell activity is greatest in PEC, second greatest in spleen cells, moderate in peripheral-blood lymphocytes, relatively low in lymph-node cells, and undetectable in thoracic-duct cells. The distribution immediately provides useful information about the effector cells. It is improbable that they are T lymphocytes: no activity is found in thoracic-duct cells, and rats or mice deprived of T lymphocytes have normal or increased numbers of effector cells [21, 22]. The effector cells are usually described as B lymphocytes [18, 19]. However, it seems equally unlikely that they are conventional, antibody-producing B lymphocytes. Effector-cell activity is not demonstrable in tho-

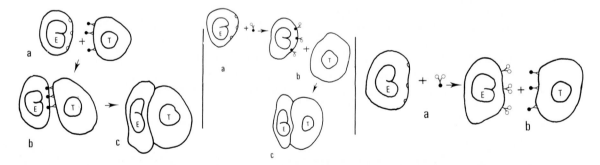

Figure 3. Diagrammatic representation of the main features of antibody-dependent, cell-mediated cytotoxicity. **Left**: Antibody (symbolized by a Y) becomes attached to a target cell and undergoes a configurational change in its Fc region (symbolized by a knob on the end of the Y). The effector cell (E) has receptors for modified Fc and becomes attached to the target cell (b). This is followed by close contact over a wide area (c) and lysis of the target cell. **Center**: Alternatively, antigen-antibody complexes in antibody excess can become attached to effector cells, making them able to kill specifically target cells with the same antigen but not antibody. **Right**: However, complexes in excess of specific or an unrelated antigen (or aggregated IgG) block the cytotoxic reaction, even when target cells have attached antibody.

racic-duct cells, which are able to restore fully primary and secondary antibody-forming capacity when transferred to lethally irradiated rats [23]. There is disproportionately low effector-cell activity in lymph nodes as compared with spleen, although lymph-node cells show at least equal capacity to synthesize antibody, including antibody against target cells. Pudifin et al. [24] found that in rats that had received 850 rads of whole-body irradiation, effector-cell activity returned before antibody-forming capacity. Indeed, all available evidence suggests that the effector cells are not themselves immunocompetent but are able to recognize antibody synthesized by other cells when the antibody has undergone a configurational change after contact with antigen. Because the cells are involved in antibody-dependent reactions, we propose the name A cells for them.

An important property of A cells is the presence on the plasma membrane of receptors for modified Fc. This can be shown in several ways, including rosette formation by red cells with attached IgG. Rosettes so formed sediment rapidly, and this reaction can be used to eliminate from populations of PEC or spleen cells those that can carry out antibody-dependent, cell-mediated cytotoxicity. If antigen-antibody complexes near equivalence or heated IgG are added to A cells, and if they are thoroughly washed and treated with a fluorescein-labeled antiglobulin serum, membrane fluorescence is obtained, showing that the modified immunoglobulin has been bound to the cells. After treatment with undenatured IgG, no such fluorescence is seen. The inhibition of antibody-dependent, cell-mediated cytotoxicity by aggregated IgG (described below) would be explained by prior attachment of the modified immunoglobulin to receptors on the effector cells, thereby preventing their reaction with antibody on target cells.

The properties of the bridging antibody in this system will now be considered. Holm and Perlmann [29] and MacLennan et al. [30] showed that attachment to target cells of amounts of antibody far too low to cause lysis in the presence of complement are sufficient to render the target cells susceptible to cell-mediated cytotoxicity. The antibody usually responsible for such lysis is IgG, and the reaction can be inhibited with aggregated IgG but not with native IgG [31]. Inhibition has likewise been obtained with soluble immune complexes [19]. MacLennan states that all subclasses of human IgG can react with the receptor on effector cells. In contrast, phagocytosis is induced in monocytes by a group of IgG1 or IgG3 or by a separate group on C3, but not by IgG2 or IgG4. These results suggest that the effector cells in antibody-dependent, cell-mediated cytotoxicity have receptors for modified Fc of all four subclasses of human IgG, including those that fix complement poorly (IgG2 and IgG4). Thus the receptors on the effector cells are different from those on cells of the monocyte-macrophage series.

Effector cells cannot be made selectively cytotoxic by treatment with antiserum against target cells alone, but this can be achieved by exposure of the effector cells to complexes of target-cell antigen and specific antibody [32]. After washing, the effector cells kill the same target cells not coated with antibody, but not other cells. From figure 3 it is clear that this type of sensitization is likely to depend critically upon the relative proportions of antigen and antibody; excess antigen will block the reaction. The relationship of this reaction to the blocking of cell-mediated cytotoxicity by complexes of antigen and antibody [33] deserves further consideration.

These cells might be involved in the effects of endotoxin in a number of ways. Endotoxin might stimulate the release of these cells from the bone marrow (as it does with granulocytes) or their local concentration in footpads and other sites. This could be a factor in the adjuvant action of endotoxin, increasing both formation of antibody and immunity against tumors. Some of the stimulation by endotoxin of enzyme synthesis (described in the previous section) may be attributable to A cells that were present in the populations of PEC used. Finally, some of the in-vivo effects of endotoxins might be due to attachment of endotoxin and antibody onto host cells and their sensitization for interaction with A cells. Endotoxin-antibody complexes themselves could become attached to A cells and release mediators of pharmacologic activity from them. The questions of what mediators are present in these cells and the role of these mediators in shock syndromes require thorough study.

210

References

1. Unanue, E. R., Askonas, B. A., Allison, A. C. A role of macrophages in the stimulation of immune responses by adjuvants. J. Immunol. 103: 71–78, 1969.

2. Spitznagel, J. K., Allison, A. C. Mode of action of adjuvants: effects on antibody responses to macrophage-associated bovine serum albumin. J. Immunol. 104:128–139, 1970.

3. Unanue, E. R. Thymus dependency of the immune response to hemocyanin: an evaluation of the role of macrophages in thymectomized mice. J. Immunol. 105:1339–1343, 1970.

4. Allison, A. C., Davies, A. J. S. Requirement of thymus-dependent lymphocytes for potentiation by adjuvants and antibody formation. Nature (Lond.) 233:330–332, 1971.

4a. Gery, I., Waksman, B. H. Potentiation of the T-lymphocyte response to mitogens. I. The responding cell. J. Exp. Med. 136:143, 1972.

5. Cohn, Z. A., Morse, S. I. Fuctional and metabolic properties of polymorphonuclear leucocytes. J. Exp. Med. 111:689–704, 1960.

6. Auzins, I., Rowley, D. On the question of the specificity of cellular immunity. Aust. J. Exp. Biol. Med. Sci. 40:283–291, 1962.

7. Weissmann, G., Thomas, L. Studies on lysosomes. I. The effects of endotoxin, endotoxin tolerance, and cortisone on the release of acid hydrolases from a granular fraction of rabbit liver. J. Exp. Med. 116:433–450, 1962.

8. Janoff, A., Weissmann, G., Zweifach, B. W., Thomas, L. Pathogenesis of experimental shock. IV. Studies on lysosomes in normal and tolerant animals subjected to lethal trauma and endotoxemia. J. Exp. Med. 116:451–466, 1962.

9. Cohn, Z. A., Benson, B. The differentiation of mononuclear phagocytes. Morphology, cytochemistry and biochemistry. J. Exp. Med. 121:153–170, 1965.

10. Cohn, Z. A., Hirsch, J. G., Fedorko, M. E. The *in vitro* differentiation of mononuclear phagocytes. IV. The ultrastructure of macrophage differentiation in the peritoneal cavity and in culture. J. Exp. Med. 123:747–756, 1966.

11. Bennett, W. E., Cohn, Z. A. The isolation and selected properties of blood monocytes. J. Exp. Med. 123:145–160, 1966.

12. Filkins, J. P. Bioassay of endotoxin inactivation in the lead-sensitized rat. Proc. Soc. Exp. Biol. Med. 134:610–612, 1970.

13. Filkins, J. P. Hepatic lysosomes and inactivation of endotoxin. J. Reticuloendothel. Soc. 9:480–490, 1971.

14. Filkins, J. P. Comparison of endotoxin detoxification by leukocytes and macrophages. Proc. Soc. Exp. Biol. Med. 137:1396–1400, 1971.

15. Ohanian, S. H., Schwab, J. H. Persistence of group A streptococcal cell walls related to chronic inflammation of rabbit dermal connective tissue. J. Exp. Med. 125:1137–1148, 1967.

16. Page, R. C., Davies, P., Allison, A. C. The role of macrophages in periodontal disease. *In* I. T. McPhee [ed.] Symposium of periodontal disease. Churchill Livingstone, London, 1972 (in press).

17. Evans, B., Alexander, P. Mechanism of immunologically specific killing of tumour cells by macrophages. Nature (Lond.) 236:168–170, 1972.

18. Perlmann, P., Holm, G. Cytotoxic effects of lymphoid cells *in vitro*. Adv. Immunol. 11:117–193, 1969.

19. MacLennan, I. C. M. Cytotoxic "B" cells. *In* A. J. S. Davies and R. Carter [ed.] Thymus dependency. Plenum, New York, 1972 (in press).

20. Perlmann, P., Holm, G. Studies on the mechanism of lymphocyte cytotoxicity. *In* P. A. Miescher and P. Grabar [ed.] Mechanism of inflammation induced by immune reactions. Schwabe, Basel, 1968, p. 325–341.

21. Harding, B., Pudifin, D. J., Gotch, F., MacLennan, I. C. M. Cytotoxic lymphocytes from rats depleted of thymus processed cells. Nature [New Biol.] 232:80–82, 1971.

22. Van Boxel, J. A., Stobo, J. D., Paul, W. E., Green, I. Antibody-dependent lymphoid cell-mediated cytotoxicity: no requirement for thymus-derived lymphocytes. Science 175:194–196, 1972.

23. Ellis, S. T., Gowans, J. L., Howard, J. C. The origin of antibody forming cells from lymphocytes. Antibiot. Chemother. (Basel) 15:40–55, 1969.

24. Pudifin, D. J., Harding, B., MacLennan, I. C. M. The differential effect of γ irradiation on the sensitizing and effector stages of antibody dependent lymphocyte mediated cytotoxicity. Immunology 21:853–860, 1971.

25. Miller, R. G., Phillips, R. A. Separation of cells by velocity sedimentation. J. Cell. Physiol. 73:191–201, 1969.

26. Allison, A. C., Young, M. R. Vital staining and fluorescence microscopy of lysosomes. *In* J. T. Dingle and H. B. Fell [ed.] Lysosomes in biology and pathology. Vol. 2. North Holland Publishing Co., Amsterdam, 1969, p. 600–628.

27. van Furth, R. Mononuclear phagocytes. Blackwell, Oxford, 1969. 654 p.

28. Thiéry, J.-P. Mise en évidence des polysaccharides sur coupes fines en microscopie électronique. J. Microscopie 6:987–1018, 1967.

29. Holm, G., Perlmann, P. Cytotoxicity of lymphocytes and its suppression. Antibiot. Chemother. (Basel) 15:295–309, 1969.

30. MacLennan, I. C. M., Loewi, G., Harding, B. The role of immunoglobulins in lymphocyte-mediated cell damage, *in vitro*. I. Comparison of the effects of target cell specific antibody and normal serum factors on cellular damage by immune and non-

immune lymphocytes. Immunology 18:397–404, 1970.

31. MacLennan, I. C. M., Howard, A. Evidence for correlation between charge and antigenic specificity of human IgG. Immunology 1972 (in press).

32. Perlmann, P., Perlmann, H., Biberfield, P. Specifically cytotoxic lymphocytes produced by pre-incubation with antibody complexed target cells. J. Immunol. 1972 (in press).

33. Sjögren, H. O., Hellström, I., Bansal, S. C., Hellström, K. E. Suggestive evidence that the "blocking antibodies" of tumor-bearing individuals may be antigen-antibody complexes. Proc. Natl. Acad. Sci. U.S.A. 68:1372–1375, 1971.

Mechanism of Interferon Induction by Endotoxin

Monto Ho, Yang H. Ke, and John A. Armstrong

From the Department of Epidemiology and Microbiology, Graduate School of Public Health, and the Division of Infectious Diseases, Department of Medicine, School of Medicine, University of Pittsburgh, Pittsburgh, Pennsylvania

When 1-kg rabbits received iv 500 μg of Boivin *Escherichia coli* endotoxin and were sacrificed 15 min later, production of interferon, as monitored by incubation of tissue slices in vitro, occurred primarily in the spleen, lung, liver, and thymus. When the temperature of incubation was 23 C, more interferon was produced than at 37 C. If the animals received 1 mg of actinomycin/kg 1 hr before administration of endotoxin, interferon production was accentuated in the spleen and lung. However, if the dose was increased to 5 mg/kg, no detectable interferon was produced. Production of interferon in slices was reversibly inhibited by cyclo-heximide or puromycin and was marginally accentuated after removal of these substances. These findings suggest that induction of interferon production by endotoxin, like induction by polyriboinosinic:polyribocytidylic acid (poly I:C) or virus, requires synthesis of messenger RNA and protein. Accentuated production by a marginal dose of actinomycin is explained by inhibited synthesis of messenger RNA for a control protein that inhibits interferon synthesis, as has been postulated for induction of interferon by poly I:C and by viruses.

Until now, it has been assumed that the mechanisms for induction of interferon production by endotoxin are basically different from mechanisms of production of interferon by viruses [1]. This concept is based primarily on the different effects of antimetabolites on these two inducers in experimental animals. Endotoxin-induced interferon, in contrast to virus-induced interferon, was not inhibited by inhibitors of synthesis of messenger RNA [2] and protein [3, 4]. These findings suggested that endotoxin-induced interferon, unlike virus-induced interferon, did not require synthesis of messenger RNA or protein and was released rather than synthesized. Other differences between these two inducing systems were observed, although the reasons for them were less apparent. For example, the production of endotoxin-induced interferon in the rabbit, in contrast to that of virus-induced interferon, was exquisitely inhibited

by cortisol and enhanced by adrenalectomy [5]. It was not inhibited by a reduction in body temperature of the animal, as was the production of virus-induced interferon [6].

On the other hand, work with endotoxin in cell cultures has consistently cast doubt on the interpretation that endotoxin-induced interferon is simply released. Unstimulated production of interferon has not been convincingly proved; the reported spontaneous production of interferon by peritoneal macrophages may actually have been stimulated by endotoxin [7]. But even such spontaneously produced interferon, or interferon produced by peritoneal macrophages after addition of endotoxin, was reportedly inhibited by actinomycin and puromycin, and hence presumably required synthesis of protein [7]. Finkelstein et al. [8] also found that production of interferon by endotoxin-stimulated macrophages was inhibited by actinomycin D and puromycin, although only in higher doses than those that inhibited virus-stimulated cells. Kobayashi et al. [9] maintained that there was an actinomycin-sensitive (early) and actinomycin-resistant (late) phase of interferon production under stimulation by endotoxin, while both phases were inhibited by puromycin.

A similar dichotomy of results obtained in vitro

This study was supported by grant no. 5R01-A1 02953 from the U.S. Public Health Service.

The authors thank Lucille Ople for excellent technical assistance.

Please address requests for reprints to Dr. Monto Ho, 427 Crabtree Hall, University of Pittsburgh, Pittsburgh, Pennsylvania 15213.

and in the animal pertains to polyriboinosinic: polyribocytidylic acid (poly I:C). Youngner and Hallum [10] first reported that the production of interferon induced by poly I:C in mice was not inhibited by cycloheximide, but work in cell cultures and tissue slices shows conclusively that production does indeed require synthesis of RNA as well as protein [8, 11–14]. However, under conditions of partial inhibition by actinomycin or cycloheximide, or of late addition of actinomycin, the production of interferon was actually accentuated [12, 13]. These apparently conflicting results were explained by the postulate that a labile control protein, capable of terminating interferon production in the course of the process, was formed [13]. The problem remains as to why antimetabolites such as actinomycin [15] and cycloheximide [10] did not inhibit the production of poly I:C-induced interferon in the animal. Perhaps effective doses of the antimetabolite were not delivered to interferon-producing tissues. Another partial explanation may be that accentuated interferon production in different tissues in response to antimetabolites varies. For example, accentuated interferon production by cycloheximide was more evident in slices of lung and kidney than in slices of liver [16]. Perhaps the most convincing evidence that induction of interferon by poly I:C and that by virus are identical with respect to requirement for RNA and protein was the finding that production of interferon induced by a virus in cell and slice cultures could also be markedly accentuated, under appropriate conditions, by cycloheximide or actinomycin [13]. Viruses have been the prototype of interferon inducers that require protein synthesis for their action. More recently, DeClercq and Merigan [17] found that cycloheximide administered late after the injection of an inducer into mice could accentuate interferon production irrespective of whether endotoxin, virus, or poly I:C was used.

In view of these findings with poly I:C, we decided to look again at the production of interferon in response to endotoxin. Progress in this area has been slower than that with poly I:C or with viruses, because (1) there is no consistent readily quantifiable cell-culture system for production of interferon in response to endotoxin; and (2) the amount of serum interferon produced in response to endotoxin in the animal or in tissue slices injected with endotoxin is usually low and often

variable. Nevertheless, the work reported below supports the hypothesis that, as far as metabolic requirements are concerned, there is no essential difference between endotoxin and the other two prototype inducers.

Materials and Methods

Endotoxin. Lipopolysaccharide from *Escherichia coli* 0127:B8 prepared by the Boivin method (Difco, Detroit, Mich.) was used. Another preparation of Boivin-type endotoxin from *E. coli* K-235, prepared by the Abbott Laboratory, Chicago, Ill. and obtained through courtesy of Dr. Warren R. Stinebring, was also used in a few of the experiments.

Cell cultures and interferon titration. The preparation of primary weanling-rabbit kidney cultures and their use in interferon assay has been previously described [18]. Samples containing interferon were incubated with these cells, after which they were challenged with vesicular-stomatitis virus; viral CPE was quantitated by uptake of methylrosaniline dye as measured in the spectrophotometer at 550 nm after elution with 2-methoxyethanol. All titers of interferon were standardized according to reference rabbit interferon (National Institutes of Health, Bethesda, Md.) and corrected to units/100 mg of tissue.

Tissue-slice method. The method of using tissue slices for study of the production of interferon has been described [14]. Since we wanted to follow the production process, we induced this process by injecting into animals 100–500 μg of endotoxin iv. Fifteen minutes after injection, each animal was sacrificed rapidly by exsanguination, after which the organs were removed and slices were made. These usually weighed 50–150 mg except for the thymus, which, owing to its fragility, was cut into chunks of about 300 mg. These slices or chunks were put into a small Erlenmyer flask (25 ml) containing 4 ml of oxygenated Ringer's balanced salt solution. Various additives (as indicated in the text) were included for the periods of time indicated. These cultures were then incubated in a Dubnoff shaking incubator at 37 C unless otherwise indicated. Substances added were in the following concentrations unless otherwise indicated: cycloheximide (Nutritional Biochemicals), 50 μg/ml; puromycin (Nutritional Biochemicals), 100 μg/ml; and actinomycin D

(Merck), 2 μg/ml. When incubation was to be continued without the antimetabolites after a period of exposure, the tissue slices were removed from the flask, washed three times in 50 ml of saline, transferred to a new flask containing fresh medium, and incubated. After collection, the culture fluids were centrifuged for removal of cell debris and were dialyzed before titration.

Occasionally animals received 1 or 5 mg of actinomycin/kg iv 30 min to 1 hr before injection of endotoxin.

Radioactivity-incorporation technique. For monitoring of RNA synthesis, tissue slices were incubated in Ringer's solution containing 1 μCi of ^{14}C-uridine/ml for 45 min. Cold trichloroacetic acid (TCA) solution containing Celite as a filter aid was then added to make the final concentration 10% with respect to TCA and 2% to Celite. The slices were homogenized in the cold and filtered onto filter paper with a Celite cushion. The precipitates were washed twice with 5% TCA, once with 0.5% TCA, and twice with methanol; they were then air-dried, transferred into test tubes, and dissolved in 0.1 N NaOH with heating. Aliquots were assayed for ^{14}C by the liquid-scintillation method and for RNA content by the orcinol method [19]. Specific activity expressed by count per min (cpm)/mg of RNA was used for calculation of the effect of treatment with antimetabolites.

For monitoring of protein synthesis, tissue slices were similarly incubated in Ringer's solution containing 1 μCi of ^{14}C-leucine or of ^{14}C-reconstituted protein hydrolysate/ml for 60 min. TCA precipitates were washed and redissolved, and radioactivity was counted as described above. Protein content was determined by Lowry's method [20], and specific activity was expressed as cpm/mg of protein.

Results

Tissues producing interferon. The distribution of production of interferon among various tissues was studied as follows. Fifteen minutes after injection of 100 mg of endotoxin, rabbits were sacrificed. Tissue slices were prepared and placed in flasks containing Ringer's solution. The production of interferon in a 12-hr period was assayed and is plotted in figure 1. Two sets of flasks were made; one was incubated at 37 C in a Dubnoff shaking waterbath and the other at room tempera-

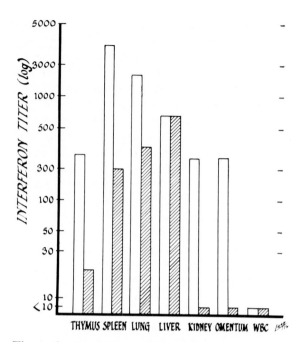

Figure 1. Production of interferon in tissue slices from rabbits injected with 100 mg of endotoxin. Temperatures of incubation: □ = 23 C; ▨ = 37 C.

ture (23 C) on an orbital shaker. In almost every tissue, the production of interferon at 23 C was greater than at 37 C. This is consistent with the experience of Laskovic et al. [21]. Secondly, tissues of the reticuloendothelial or phagocytic system were the ones that produced the greatest amount of interferon. This is consistent with the hypothesis of Kono and Ho [22] that the reticuloendothelial system is the main system forming interferon upon systemic induction. The peripheral white cells do not form interferon, despite the fact that these are the cells that take up large amounts of endotoxin when animals are injected iv [23]. It is interesting that the thymus produced significant amounts of interferon, particularly at 23 C, confirming the experience of Kojima [24].

Effect of antimetabolites in tissue slices from rabbits. The effect of antimetabolites on synthesis of RNA and protein in tissue slices was monitored by incorporation of ^{14}C-uridine or ^{14}C-leucine (see Materials and Methods). It can be seen from table 1 that 1 mg of actinomycin D/kg given iv 1 hr before sacrifice did not inhibit synthesis of RNA in organs by more than 77%. Inhibition was particularly slight in the lung, perhaps reflecting the distribution of actinomycin D after iv injection. On the other hand, if 5 mg/kg were

Table 1. Effect of antimetabolites on synthesis of RNA and protein in slices of rabbit tissues.

Agent	System	Dose per kg or ml	Percentage inhibition*		
			Spleen	Liver	Lung
Actinomycin D	In vivo	1 mg	70	77	47
Actinomycin D	In vivo	5 mg	86	82	75
Puromycin	In vitro	100 µg	89	79	88
Cycloheximide	In vitro	50 µg	99	95	99

* Percentage inhibition in slices incubated for 20–45 min with ^{14}C-uridine or ^{14}C-leucine, as compared with untreated slices (see Materials and Methods).

given, all tissues were inhibited at a higher level, although the lung again showed the least inhibition (75%). Puromycin (100 µg/ml) inhibited these three tissues by more than 79%, and 50 µg of cycloheximide/ml caused more than 95% inhibition.

Effect of actinomycin D on interferon production. The effect of actinomycin D on production of interferon by tissues of rabbits injected with three types of inducers is presented in table 2. Actinomycin D was injected in two doses, the metabolic effect of which has already been described (table 1). It is seen that when 1 mg of actinomycin/kg was administered, production of interferon induced by poly I:C, Newcastle disease virus, or endotoxin showed various patterns of inhibition or accentuation. For example, production of poly I:C- and virus-induced interferon was inhibited in the spleen and liver but accentuated in the lung. The production of endotoxin-induced interferon was accentuated in the spleen and lung but inhibited in the liver. We believe that this varying pattern of interferon production is explained by the fact that this dose of actinomycin is not equally distributed in various organs and is insufficient to inhibit completely synthesis of

messenger RNA. Partial inhibition permits accentuation of interferon production.

This interpretation was supported by results from animals that received 5 mg of actinomycin D/kg. Production of interferon induced by the three inducers was uniformly inhibited in all three tissues. This would suggest that synthesis of messenger RNA is indeed required for interferon production. If sufficient actinomycin D is given to inhibit interferon messenger RNA in the first place, production of interferon can no longer be accentuated during the later stages of the process.

Effect of inhibition of protein synthesis on endotoxin-induced interferon. This effect was studied with two groups of rabbits, both of which received 500 µg of endotoxin/kg iv 15 min before sacrifice. In one group no actinomycin was injected, while in the other group 1 mg of actinomycin/kg was injected iv 1 hr before endotoxin. A summary of our experience is presented in table 3. Results from slices of the spleen and the lung are presented because they consistently produce interferon in flasks.

When cycloheximide or puromycin was added for 4 hr of incubation, production of interferon was completely inhibited. On the other hand, during

Table 2. Effect of actinomycin D on production of interferon.

Actinomycin D (mg/kg)*	Inducer	Interferon produced/100 mg†		
		Spleen	Liver	Lung
0	Poly I:C‡	150,000	56,000	32,000
1	Poly I:C	900	700	72,000
5	Poly I:C	8	4	16
0	NDV§	950,000	31,000	3,100
1	NDV	9,000	300	11,000
5	NDV	4	4	16
0	Endotoxin	230	204	74
1	Endotoxin	1,660	42	501
5	Endotoxin	<4	4	4

* Actinomycin D was injected into 1-kg rabbits 1 hr before inducer. Slices were taken 15 min later.

† Each figure represents the mean from at least two animals; the figures in the endotoxin category represent at least four animals.

‡ Polyriboinosinic:polyribocytidylic acid.

§ Newcastle disease virus.

Table 3. Effect of cycloheximide and puromycin on endotoxin-induced interferon in tissue slices.

In vivo	In vitro	Spleen		Lung	
		0–4 hr	4–12 hr	0–4 hr	4–12 hr
None	None	128	64	32	16
None	Cyclo (0–4 hr)	<4	14	<4	180
None	Puro (0–4 hr)	<4	60	<4	70
None	Puro (0–12 hr)	<4	<4	<4	<4
Act D	None	127	120	160	430
Act D	Cyclo (0–4 hr)	<4	<4	<4	84
Act D	Puro (0–4 hr)	<4	<4	<4	64

NOTE. Actinomycin D (Act D, 1 mg/kg) was injected 1 hr before endotoxin (100 µg/kg). Fifteen minutes later, the animal was sacrificed, and tissue slices were incubated with cycloheximide (Cyclo, 50 µg/ml) or puromycin (Puro, 100 µg/ml) in flasks for the duration indicated in parentheses. Each figure represents the mean of results from two to six animals.

the 4–12-hr period when cycloheximide or puromycin was removed, interferon production resumed, and, in the case of lung slices, production was somewhat accentuated (table 3, lines 2 and 3). This is consistent with previous experience with poly I:C or Newcastle disease virus [16]. On the other hand, if puromycin was permitted to remain throughout the period of incubation (0–12 hr), production of interferon remained inhibited.

Slices from animals treated with 1 mg of actinomycin/kg produced similar or increased amounts of interferon, which could be similarly inhibited by cycloheximide or puromycin when these antimetabolites were added within 0–4 hr. In the case of the spleen, there was no reversal of the effect of puromycin or cycloheximide after these substances were removed. At this time we have no ready explanation for this observation. On the other hand, after removal of antimetabolites from lung slices, production of interferon resumed, as it did in tissues obtained from animals that did not receive actinomycin D.

It appears from these experiments that the production of endotoxin-induced interferon from either normal or actinomycin-treated animals requires synthesis of protein. We previously showed that the effect of puromycin and cycloheximide was reversible in rabbit kidney-cell cultures [13], thus explaining resumption of interferon production after these agents are removed. Accentuated production is consistent with the theory that interferon messenger RNA accumulates during the period when an antiprotein inhibitor acts, and that it is translated when the postulated control protein is absent or if it is inhibited.

Kinetics of interferon production under the effect of actinomycin D. Vilček showed that pro-

duction of interferon under partial inhibition by cycloheximide occurred late [12]. Here we studied the kinetics of interferon production in tissues from animals treated with actinomycin D and in controls. In figure 2, the kinetics, in terms of cumulative percentages of total amount of interferon produced, are shown. It is well known that endotoxin-induced interferon appears early [25, 26], as shown in the figure for lung slices. Interferon production is largely complete (more than 70%) by 4 hr, and by 6–8 hr has almost ceased. On the other hand, interferon production in the lung and spleen slices from animals injected with actinomycin D lags behind. This is consistent with the interpretation that low doses of actinomycin only partially inhibit interferon messenger RNA, thus decreasing production of interferon in the early phase. However, during the later phase, while interferon messenger RNA has accumulated with

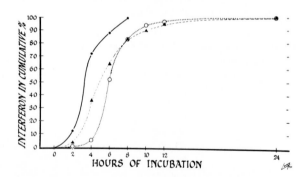

Figure 2. Kinetics of production of endotoxin-induced interferon and the effect of actinomycin D. Slices were obtained from animals injected with endotoxin alone or with 1 mg of actinomycin/kg 1 hr before endotoxin. (●——●) = lung slices, no actinomycin D; (▲– – –▲) = lung slices, actinomycin D; (○ ∙ ∙ ∙ ∙ ○) = spleen slices, actinomycin D.

time, the accumulation of messenger RNA for control protein or the control protein itself is preferentially inhibited, perhaps because of its lability, thus accounting for accentuated production of interferon.

Discussion

We believe that, in terms of synthesis of messenger RNA and protein, production of interferon can be explained by a single model, whether the inducer is poly I:C, virus, or endotoxin. The model that we postulate has already been described for virus and poly I:C [13] and is schematized in figure 3. We believe that the inducer, whether it be poly I:C, virus, or endotoxin, is able to activate an interferon gene in an as yet undefined manner. This gene transcribes interferon messenger RNA, which, in turn, is translated, perhaps through various intermediary gene products, such as "preinterferon" [15]. One of these translation products, either interferon or a precursor, activates the control-protein gene. The control protein depresses synthesis of interferon by inhibiting activity of interferon messenger RNA after transcription. We believe that the messenger RNA for control protein or its translational products is more labile than messenger RNA for interferon or than interferon itself; this accounts for the fact that, given a dose of actinomycin or antiprotein substance, the accumulation of control protein is preferentially inhibited. This also accounts for continued and even accentuated production of interferon under partial inhibition by these antimetabolites.

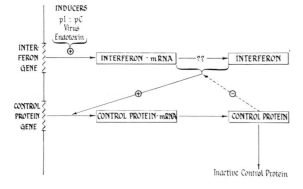

Figure 3. Scheme of induction and regulation of interferon. (⊕) designates facilitation of process; (⊖) designates inhibition.

There is one problem that this model does not entirely explain. Why is it that antimetabolites do not inhibit very well the in-vivo production of interferon induced by poly I:C or by endotoxin? We probably do not have all the answers yet, but the following considerations may be pertinent. (1) Antimetabolites probably cannot be delivered in adequate doses to various tissues making interferon. This is particularly true of the lung with respect to actinomycin D. Interferon can be produced when over 95% of protein synthesis is inhibited as measured by the usual methods [16]. (2) Different tissues and different cells are involved in responding to each inducer. (3) Different tissues have been shown to respond differently to antimetabolites in accentuating interferon production [16], which suggests that the level of production of control protein also varies in different cells or tissues. (4) When looked at carefully, early production of interferon in response to poly I:C and endotoxin is actually inhibited in animals [10, 17].

Other differences in the mode of interferon production in response to endotoxin and in response to virus [27] do not contradict the model. Why does the type of interferon induced by endotoxin or poly I:C appear more rapidly, or why do these interferons have different molecular weights and physicochemical stability [26, 28]? Perhaps different inducers activate different cell populations that produce the different types of interferon. Molecular weights and physicochemical ability may be explained by aggregation or polymerization of basic interferon molecules [29]. Another reason for the delay in the production of virus-induced interferon may be that the virus itself is not an inducer but that an effective inducer, perhaps double-stranded RNA, must first be made [30].

Finally, this model leaves unanswered the question of whether induction by other mechanisms, particularly by intracellular parasites and by immune stimulation of lymphocytes [31], requires synthesis of messenger RNA and protein. Undoubtedly, the answers will be forthcoming, and there is no reason to believe that induction by other mechanisms is different. Perhaps the crux of the induction problem is no longer whether synthesis of messenger RNA or protein is required, but by what process a genetic function is turned on. The effects of endotoxin and immune stimulation suggest an effect via membrane functions.

References

1. Ho, M., Fantes, K. H., Burke, D. C., Finter, N. B. Interferons or interferon-like inhibitors induced by non-viral substances. *In* N. B. Finter [ed.] Interferons. North-Holland, Amsterdam, 1966, p. 181–201.

2. Ho, M., Kono, Y. Effect of actinomycin D on virus and endotoxin-induced interferonlike inhibitors in rabbits. Proc. Natl. Acad. Sci. U.S.A. 53:220–224, 1965.

3. Youngner, J. S., Stinebring, W. R., Taube, S. E. Influence of inhibitors of protein synthesis on interferon formation in mice. Virology 27:541–550, 1965.

4. Ke, Y. H., Singer, S. H., Postic, B., Ho, M. Effect of puromycin on virus and endotoxin-induced interferonlike inhibitors in rabbits. Proc. Soc. Exp. Biol. Med. 121:181–183, 1966.

5. Postic, B., DeAngelis, C., Breinig, M. K., Ho, M. Effects of cortisol and adrenalectomy on induction of interferon by endotoxin. Proc. Soc. Exp. Biol. Med. 125:89–92, 1967.

6. Postic, B., DeAngelis, C., Breinig, M. K., Ho, M. Effect of temperature on the induction of interferons by endotoxin and virus. J. Bacteriol. 91: 1277–1281, 1966.

7. Smith, T. J., Wagner, R. R. Rabbit macrophage interferons. 1. Conditions for biosynthesis by virus-infected and uninfected cells. J. Exp. Med. 125: 559–577, 1967.

8. Finkelstein, M. S., Bausek, G. H., Merigan, T. C. Interferon inducers *in vitro:* difference in sensitivity to inhibitors of RNA and protein synthesis. Science 161:465–468, 1968.

9. Kobayashi, S., Yasui, O., Masuzumi, M. Studies on early-appearing interferon *in vitro.* Production of endotoxin-induced interferon by mouse spleen cells cultured *in vitro.* Proc. Soc. Exp. Biol. Med. 131:487–494, 1969.

10. Youngner, J. S., Hallum, J. V. Interferon production in mice by double-stranded synthetic polynucleotides: induction or release? Virology 35:177–179, 1968.

11. Field, A. K., Tytell, A. A., Lampson, G. P., Hilleman, M. R. Inducers of interferon and host resistance. V. *In vitro* studies. Proc. Natl. Acad. Sci. U.S.A. 61:340–346, 1968.

12. Vilček, J. Metabolic determinants of the induction of interferon by a synthetic double-stranded polynucleotide in rabbit kidney cells. Ann. N.Y. Acad. Sci. 173:390–403, 1970.

13. Tan, Y. H., Armstrong, J. A., Ke, Y. H., Ho, M. Regulation of cellular interferon production: enhancement by antimetabolites. Proc. Natl. Acad. Sci. U.S.A. 67:464–471, 1970.

14. Ho, M., Ke, Y. H. The mechanisms of stimulation of interferon production by a complexed polyribonucleotide. Virology 40:693–702, 1970.

15. Ho, M., Breinig, M. K., Postic, B., Armstrong, J. A. Effect of pre-injections on the stimulation of interferon by a complexed polynucleotide, endotoxin and virus. Ann. N.Y. Acad. Sci. 173:680–693, 1970.

16. Ke, Y. H., Ho, M. Patterns of cycloheximide and puromycin effect on interferon production stimulated by virus or polyribonucleotide in different tissues. Proc. Soc. Exp. Biol. Med. 136:365–368, 1971.

17. DeClercq, E., Merigan, T. C. Stimulation or inhibition of interferon production depending on time of cycloheximide administration. Virology 42: 799–802, 1970.

18. Armstrong, J. A. Semi-micro, dye-binding assay for rabbit interferon. Appl. Microbiol. 21:723–725, 1971.

19. Schneider, W. C. Determination of nucleic acids in tissues by pentose analysis. *In* S. P. Colowick and N. O. Kaplan [ed.] Methods in enzymology. Vol. 3. Academic Press, New York, 1957, p. 680–684.

20. Lowry, O. H., Rosebrough, N. J., Farr, A. L., Randall, R. J. Protein measurement with the Folin phenol reagent. J. Biol. Chem. 193:265–275, 1951.

21. Lackovič, V., Borecky, L., Sikl, D., Masler, L., Bauer, S. The temperature requirement for interferon production in cells stimulated by Newcastle disease virus or microbial agents *in vitro.* Acta Virol. (Praha) [Eng.] 11:500–505, 1967.

22. Kono, Y., Ho, M. The role of the reticuloendothelial system in interferon formation in the rabbit. Virology 25:163–166, 1965.

23. Braude, A. I. Absorption, distribution, and elimination of endotoxin and their derivatives. *In* M. Landy and W. Braun [ed.] Bacterial endotoxins. Rutgers University Press, New Brunswick, N.J., 1964, p. 98–109.

24. Kojima, Y. Sites of interferon production in rabbits induced by bacterial endotoxin. Kitasato Arch. Exp. Med. 43:35–44, 1970.

25. Stinebring, W. R., Youngner, J. S. Patterns of interferon appearance in mice injected with bacteria or bacterial endotoxin. Nature (Lond.) 204:712, 1964.

26. Ho, M. Interferon-like viral inhibitor in rabbits after intravenous administration of endotoxin. Science 146:1472–1474, 1964.

27. Ho, M. The induction of interferons and related problems. Jap. J. Exp. Med. 37:169–182, 1967.

28. Hallum, J. V., Youngner, J. S., Stinebring, W. R. Interferon activity associated with high molecular weight proteins in the circulation of mice injected with endotoxin or bacteria. Virology 27:429–431, 1965.

29. Carter, W. A. Interferon: evidence for subunit structure. Proc. Natl. Acad. Sci. U.S.A. 67:620–628, 1970.
30. Tytell, A. A., Lampson, G. P., Field, A. K., Hilleman, M. R. Inducers of interferon and host resistance. III. Double-stranded RNA from reovirus type 3 virions (REO 3-RNA). Proc. Natl. Acad. Sci. U.S.A. 58:1719–1722, 1967.
31. Green, J. A., Cooperband, S. R., Kibrick, S. Immune specific induction of interferon production in cultures of human blood lymphocytes. Science 164:1415–1417, 1969.

Involvement of a Chemical Moiety of Bacterial Lipopolysaccharide in Production of Interferon in Animals

Julius S. Youngner, David Sidney Feingold, and Joseph K. Chen*

From the Department of Microbiology, School of Medicine, University of Pittsburgh, Pittsburgh, Pennsylvania

Interferon is released in mice that receive an injection of the lipopolysaccharide (LPS) of *Salmonella typhimurium* (strain LT-2), glycolipids from organisms of Rb and Re chemotypes, or lipid A. No interferon is released by O-antigenic hapten lacking the glycolipid moiety. The interferon-releasing ability of glycolipid from the Re chemotype is reduced by partial hydrolysis with acid, which removes 2-keto-3-deoxy-octonic acid and reduces the solubility of the hydrolyzed product. The ability of the glycolipid to release interferon is increased by partial hydrolysis with alkali. This treatment removes some of the fatty acids but little or no 2-keto-3-deoxy-octonic acid and increases the solubility of the hydrolyzed product; interferon-releasing ability is progressively reduced with more extensive hydrolysis and concomitant loss of fatty acids. The interferon-releasing ability of unhydrolyzed glycolipid is markedly increased by complexing with bovine serum albumin. These results show that the interferon-releasing property of LPS resides in the lipid A portion of the molecule. Furthermore, the relative solubility and the presence of fatty acids in LPS and its derivatives play a decisive role in their ability to release interferon.

Since the discovery of interferon by Isaacs and Lindenmann in 1957, it has become apparent that the production of this protein inhibitor of viral replication is not a response of host cells solely to infection by viruses. It has been well documented that interferon is produced by animals or cell cultures exposed to a wide variety of nonviral substances. These nonviral substances include materials of microbial origin [1], phytohemagglutinins [2], synthetic polymers [3, 4], and low-molecular-weight organic compounds [5].

Bacterial lipopolysaccharides (LPS) have been the most intensively studied interferon stimuli of microbial origin; in this presentation our efforts to determine which chemical structure of LPS is responsible for the production of interferon in mice are described.

This investigation was conducted under the sponsorship of the Commission on Influenza, Armed Forces Epidemiological Board, and was supported by the U.S. Army Medical Research and Development Command, Department of the Army, under research contract no. DADA 17-67-C7046, and in part by Public Health Service research grant no. AI-06264 from the National Institute of Allergy and Infectious Diseases. Dr. Feingold is a Research Career Development Awardee of the National Institutes of Health, U.S. Public Health Service.

Please address requests for reprints to Dr. J. S. Youngner, Department of Microbiology, School of Medicine, University of Pittsburgh, Pittsburgh, Pa. 15213.

* Present address: Department of Biochemistry, School of Medicine, University of Pittsburgh, Pittsburgh, Pa. 15213.

Materials and Methods

Bacteria. The growth and properties of wild-

220

type *Salmonella typhimurium* (strain LT-2) and rough mutant strains TV-161 (Rb chemotype) and G-30/C21 (Re chemtoype) have been described elsewhere [6]. A culture of *Salmonella minnesota* (strain R595) was kindly provided by Dr. Otto Lüderitz.

LPS. LPS and core glycolipid were prepared from *S. typhimurium* strains LT-2 and TV-161, respectively, by phenol-water extraction [7]. Glycolipid (GLP) from *S. typhimurium* (strain G-30/C21) and *S. minnesota* (strain R595) were prepared by the phenol:chloroform:petroleum ether extraction method of Galanos et al. [8]. The purified cell-wall LPS and GLP contained 1% or less protein when tested by the Lowry method [9].

Interferon. Interferon was produced by iv injection of the material being tested into female Swiss Webster mice weighing 25–30 g (Taconic Farms, Germantown, N.Y.). At least 10 animals were used per sample. At appropriate times, blood was obtained by cardiac puncture, and pooled plasmas were tested for interferon by the plaque-reduction method [10]. Cultures of murine L cells (clone 929) in 60-mm petri dishes were exposed for 20 hr at 37 C to 3 ml of twofold serial dilutions of plasma in growth medium (Eagle's Minimal Essential Medium plus 4% calf serum) and then were challenged with 40–60 pfu of vesicular-stomatitis virus. Three dishes were used for each dilution. Titers of interferon were determined by plotting on probit paper the percentage of inhibition against the different dilutions and were expressed as the reciprocal of the dilution of the sample that reduced the plaque count to 50% of the control plaque count.

Lethality of LPS, GLP, and their derivatives was determined by injection into the tail vein of mice of 0.1 ml of twofold or 10-fold dilutions of the materials in pyrogenfree saline; at least 10 mice were used per dilution. Deaths were recorded over a 72-hr period, and the LD_{50} was calculated by the method of Reed and Muench [11].

Results

Our initial experiments were done with suspensions of heat-killed mutants of *S. typhimurium* with known blocks in the biosynthesis of LPS. It was found that the interferon response elicited in mice was not dependent on the presence of a complete LPS [6]. A rough mutant (G-30/C21), which contains in its cell wall a GLP consisting only of 2-keto-3-deoxyoctonate (KDO) and lipid, was indistinguishable in its interferon-stimulating ability from the wild-type (LT-2), which possesses a complete O antigen with polysaccharide side chains.

When these experiments were repeated with LPS and GLP extracted from the different mutants, the data showed that purified cell-wall GLP from the rough mutant G-30/C21 and LPS from the wild-type organism did not differ significantly in their interferon-releasing activity in mice [12].

Additional experiments demonstrated that the O-antigenic side chains of LPS do not in themselves elicit production of interferon in mice. Core GLP from strain TV-161 (Rb chemotype) was prepared by phenol extraction, and pure O-specific hapten from the same strain was isolated and purified by the method of Kent and Osborn [13]. The chemical analysis of the purified core GLP and the O-specific hapten are given in table 1. These data indicate that the maximal contamination of the hapten by core GLP (as measured by KDO content) is less than 2%. In addition, no fatty acids could be detected in the hapten preparation by chromotography.

From table 2 it can be seen that, while core GLP was lethal for mice treated with 80 mg of cycloheximide/kg [14], 10-fold greater doses of the pure hapten had no demonstrable lethality for mice. With regard to interferon, the response to core GLP was similar to that which would be seen after injection of complete LPS (which contains O side chains) or after injection of GLP of Re chemtoype. However, no evidence of interferon production by hapten was observed. Even in mice treated with 244 mg of cycloheximide/kg, a dose which markedly enhances release of interferon by core GLP [15], injection of hapten did not increase the level of interferon significantly above that released by cycloheximide alone.

Table 1. Characterization of *Salmonella typhimurium* (TV-161) glycolipid and hapten fractions.

Preparation	Total carbohydrate (%)	KDO* (%)	Abequose (%)	Heptose (%)	Nucleic acid (%)
Core glycolipid	26	5.3	<0.4	8.7	<1
Hapten	100	0.07	8.7	1.0	<1

* 2-keto-3-deoxyoctonate.

Table 2. Effect in mice of *Salmonella typhimurium* (TV-161) glycolipid and hapten fractions.

Preparation	LD$_{50}$ in mice treated with		Dose (µg)	Interferon titer of plasma of mice treated with	
	Saline (µg)	Cycloheximide (80 mg/kg) (µg)		Saline*	Cycloheximide† (244 mg/kg)
Core glycolipid	>500	39	100	180	7,200
			10	210	4,000
Hapten	>500	>500	100	<32	400
			10	<32	100
Saline	. . .	. . .	. . .	<32	340

* Mice bled at 2 hr.
† Mice bled at 8 hr.

The essential role of the lipid portion of salmonella LPS in the production of interferon in mice was demonstrated in the following manner. GLP from the heptoseless mutant R595 of *S. minnesota* was hydrolyzed by treatment with acid, and thus a series of lipids containing decreasing amounts of KDO were produced [14]. The lipids were isolated by ultracentrifugation, washed with water, and lyophilized. The conditions of hydrolysis and the minimal interferon-producing dose of a number of materials are presented in table 3. In this and succeeding tables, the interferon-producing potency of materials tested is expressed as the minimal amount of material that, when injected into mice, is capable of eliciting a titer of interferon of ≥ 64. In every case, this value was determined from a dose-response experiment.

It can be seen from table 3 that the minimal interferon-producing dose for native GLP is 1 µg per mouse. When 90%–100% of the KDO is removed by hydrolysis with acetic acid or H$_2$SO$_4$

Table 3. Effect of treatment with acid on production of interferon by glycolipid from *Salmonella minnesota* (R595).

Treatment	Duration (min)	KDO* (%)	Minimal interferon-producing dose (µg)
None		23.6	1
Acetic acid (0.05 N, *p*H 3.1, 100 C)	10	13.1	1
	30	1.1	10
H$_2$SO$_4$ (0.02) N, *p*H 1.7, 100 C)	20	0.0	10
HCl (0.1 N, *p*H 1.0, 100 C)	30	0.2	100

* 2-keto-3-deoxyoctonate.

under the conditions shown, the minimal interferon-producing dose is 10 times higher. With more drastic treatment with HCl, a 100-fold increase in minimal interferon-producing dose is observed. Analysis has shown that, in addition to cleavage of KDO, increasing hydrolysis by acid causes other changes in the structure of the lipid portion of the GLP.

In addition to acid hydrolysis, hydrolysis in the presence of sodium hydroxide was done. After alkaline treatment, the mixture was neutralized, extracted with hexane, and dialyzed against water, and the product was recovered by lyophilization (table 4). Graded alkaline hydrolysis with 0.1 N NaOH at 37 C for periods of time ranging from 20 min to 24 hr removed up to 67% of the fatty acids originally present (expressed as µ-equivalents of ester per mg of GLP). On the other hand, the loss of KDO after alkaline hydrolysis was minimal. The minimal interferon-producing dose of the GLP subjected to this mild alkaline hydrolysis was 10-fold less than that of the native GLP, an effect we ascribe to the increased solubility of the GLP caused by loss of hydrophobic fatty acids.

However, as can be seen at the bottom of the table, when alkaline hydrolysis was performed in 0.25 N NaOH at 100 C for 60 min, a fatty acid-free product was obtained with a 1,000-fold increase in the minimal interferon-producing dose. These results show that there is a dramatic loss of biological activity when all of the fatty-acid residues are removed from the GLP.

In view of the demonstration by Galanos et al. [16] that complexing with bovine serum albumin (BSA) increased the anticomplementary and toxic activity of lipid A, GLP-BSA complexes were prepared and tested as stimuli of interferon. Table 5

Table 4. Effect of graded hydrolysis with NaOH on production of interferon by glycolipid from *Salmonella minnesota* (R595).

Treatment	Duration	Ester (μ-equivalent per mg)	KDO* (%)	Minimal interferon-producing dose (μg)
None		1.20	23.6	1
NaOH	20 min	0.84	25.6	1
(0.1 N,	100 min	0.74	25.6	0.1
37 C)	7.5 hr	0.56	28.5	0.1
	24 hr	0.38	31.0	0.1
NaOH (0.25 N, 100 C)	60 min	0.00	NT†	1,000

* 2-keto-3-deoxyoctonate.
† Not tested.

shows that complexing of GLP with an equal weight of BSA causes a 100-fold reduction in the minimal interferon-producing dose, representing a significant enhancement of biological activity.

Discussion

Several conclusions can be reached on the basis of these data.

(*1*) The O-antigen haptenic polysaccharide alone does not elicit an interferon response in mice.

(*2*) The interferon-producing activity of LPS and GLP resides in the lipid-A portion of the molecule. This conclusion is consistent with the views of other workers that lipid A plays a key role in the biological activities of LPS [17].

(*3*) The interferon-producing capacity of GLP is affected by acid or alkaline hydrolysis. Partial mild hydrolysis with acid, which removes KDO from the molecule, results in some loss of activity because of a reduction in the solubility of the material. The opposite effect is seen when up to 67% of the fatty-acid residues is removed by partial hydrolysis with NaOH. This treatment increases the interferon-producing activity of GLP, probably as a result of the increased solubility of the altered molecule. However, more drastic alkaline hydrolysis, which causes the complete removal of fatty acids, results in a virtually total loss of interferon-producing activity.

(*4*) Interferon-producing activity of GLP is significantly enhanced by complexing with BSA, a procedure that increases the solubility (or dispersion) of the GLP.

(*5*) From our results thus far, it is possible to conclude that solubility and the presence of fatty acids are major factors that determine the interferon-producing capacity of LPS and its derivatives.

Table 5. Effect of complexes of glycolipid from *Salmonella minnesota* (R595) and bovine serum albumin (BSA) on production of interferon and lethality in mice.

Treatment	Minimal interferon-producing dose (μg)	LD$_{50}$ (μg)
None	1	>2,000
BSA complex	0.01	1,320

References

1. Youngner, J. S. Interferon production by nonviral stimuli of microbial origin. J. Gen. Physiol. 56 (Suppl.):25S–40S, 1970.

2. Wheelock, E. F. Interferon-like virus-inhibitor induced in human leukocytes by phytohemagglutinin. Science 149:310–311, 1965.

3. Merigan, T. C. Induction of circulating interferon by synthetic anionic polymers of known composition. Nature (Lond.) 214:416–417, 1967.

4. Field, A. K., Tytell, A. A., Lampson, G. P., Hilleman, M. R. Inducers of interferon and host resistance. II. Multistranded synthetic polynucleotide complexes. Proc. Natl. Acad. Sci. U.S.A. 58: 1004–1010, 1967.

5. Krueger, R. F., Mayer, G. D. Tilorone hydrochloride: an orally active antiviral agent. Science 169: 1213–1214, 1970.

6. Youngner, J. S., Feingold, D. S. Interferon production in mice by cell wall mutants of *Salmonella typhimurium*. J. Virol. 1:1164–1167, 1967.

7. Westphal, O., Jann, K. Bacterial lipopolysaccharides. *In* R. L. Whistler [ed.] Methods in carbohydrate chemistry. Vol. V. Academic Press, New York, 1965, p. 83–91.

8. Galanos, C., Lüderitz, O., Westphal, O. A new method for the extraction of R lipopolysaccharides. Eur. J. Biochem. 9:245–249, 1969.

9. Lowry, O. H., Rosebrough, N. J., Farr, A. L., Randall, R. J. Protein measurement with the Folin phenol reagent. J. Biol. Chem. 193:265–275, 1951.

10. Youngner, J. S., Stinebring, W. R. Comparison of interferon production in mice by bacterial endotoxin and statolon. Virology 29:310–316, 1966.

11. Reed, L. J., Muench, H. A simple method of estimating fifty per cent end points. Am. J. Hyg. 27: 493–497, 1938.

12. Feingold, D. S., Youngner, J. S., Chen, J. Inter-

feron production in mice by cell wall mutants of *Salmonella typhimurium*. II. Effect of purified glycolipids from S and Re chemotypes. Biochem. Biophys. Res. Commun. 32:554–557, 1968.

13. Kent, J. L., Osborn, M. J. Properties of the O-specific hapten formed *in vivo* by mutant strains of *Salmonella typhimurium*. Biochemistry 7:4396–4408, 1968.

14. Feingold, D. S., Youngner, J. S., Chen, J. Interferon production in mice by cell wall mutants of *Salmonella typhimurium*. III. Role of lipid moiety of bacterial lipopolysaccharide in interferon production in animals. Ann. N. Y. Acad. Sci. 173: 249–254, 1970.

15. Youngner, J. S., Stinebring, W. R., Taube, S. E. Influence of inhibitors of protein synthesis on interferon formation in mice. Virology 27:541–550, 1965.

16. Galanos, C., Rietschel, E. T., Lüderitz, O., Westphal, O. Interaction of lipopolysaccharides and lipid A with complement. Eur. J. Biochem. 19: 143–152, 1971.

17. Mergenhagen, S. E., Gewurz, H., Bladen, H. A., Nowotny, A., Kasai, N., Lüderitz, O. Interactions of the complement system with endotoxins from a *Salmonella minnesota* mutant deficient in O-polysaccharide and heptose. J. Immunol. 100: 227–229, 1968.

Comparative Biological Activity of Endotoxin and Synthetic Polyribonucleotides

Inh H. Han, Arthur G. Johnson, Joseph Cook, and Seong S. Han

From the Departments of Microbiology and Anatomy, University of Michigan, Ann Arbor, Michigan

A comparison of the biological activity of endotoxin with that of synthetic polyribonucleotides was undertaken in search of a nontoxic, practical adjuvant to the immune response. Although both products were effective adjuvants, the action of polynucleotides differed from that of endotoxin in that the former appeared to be mediated through thoracic-duct lymphocytes, causing a decrease in uptake of antigen by this cell, with little effect on formation of germinal centers. In addition, polyadenylic-polyuridylic acid complexes were found to be essentially nontoxic to mice, nonpyrogenic, and incapable of preparing for the local Shwartzman reaction.

Although the response of the immune system to vaccination generally is subliminal, lengthy search for an effective nontoxic adjuvant has proved frustrating [1]. Despite the overt toxicity of bacterial endotoxins, their potent adjuvant action [2] deserves mechanistic definition as a possible lead to less harmful materials. Thus, our laboratory has been seeking to understand the means by which endotoxins, when given with unrelated antigens, induce a higher peak titer and a shortening of the time necessary for free antibody to appear in the circulation. The action of immunocompetent cells is hastened to a remarkable degree by endotoxin; the induction period in mice is reduced by three to five days with a variety of antigens [3].

We postulated that the effect of endotoxin might be mediated through its cytotoxic action on small lymphocytes, i.e., the liberation of nucleic acids in the vicinity of potentially immunocompetent cells, which causes these cells to divide more quickly [4]. Support for this hypothesis was gained when nucleic acids per se were shown to function as adjuvants [5]. Recently, synthetic double-stranded polyribonucleotides have been shown to heighten the activity of antigens in sev-eral species [6–8]; their action is attributed to an effect on helper cells of the thymus gland [9] and possibly a phagocytic cell [10]. Systematic measure of any toxic properties of one of the active polynucleotide complexes, polyriboadenylic-polyribouridylic acid (poly A:U), has not been reported, although the toxic attributes of another complex, polyriboinosinic-polyribocytidylic acid (poly I:C), have been described [11]. Accordingly, we performed the following study to compare the relative toxicity and adjuvant action of poly A:U and poly I:C complexes with the well-known properties of endotoxins derived from gram-negative bacteria.

Materials and Methods

Animals. Sprague-Dawley rats (250 g) were used. Twelve-week-old New Zealand white rabbits were used in tests for pyrogenicity and the Shwartzman reaction. For tests of toxicity, five- to six-week-old inbred Balb/aj mice of both sexes were used. Deaths were tabulated daily for four days.

Antigen. Cadmium-free equine ferritin, six times crystallized, was obtained from Pentex, Kankakee, Ill. Ferritin was chosen for this study because of its strong antigenicity in rats, its uniform molecular size, its inability to induce phagocytosis in nonphagocytic cells, and its unmistakable identification in the electron microscope.

Adjuvants. Gram-negative bacterial endotoxin

This study was supported by research grants no. AM 14273 and DE 02734 from the U. S. Public Health Service.

Please address requests for reprints to Dr. Arthur G. Johnson, Department of Microbiology, University of Michigan, Ann Arbor, Michigan 48104.

(*Serratia marcescens,* SM-TCA-AA-1) was prepared by Dr. A. Nowotny of Temple University, Philadelphia, Pa. The polynucleotides used in this study were freeze-dried preparations free of nucleotides, nucleosides, and other low-molecular-weight materials and were obtained from Miles Laboratories, Elkhart, Ind. Their molecular weight was in excess of 100,000 with an average of 2.29 µmoles of phosphate per mg of poly A and 2.50 µmoles of phosphate per mg of poly U.

Iodination of equine ferritin. Ten milligrams of equine ferritin was iodinated with 2 mCi of ^{131}I according to the method of McConahey and Dixon [12].

Collection of thoracic-duct cells (TDC). Rats were anesthetized by an intraperitoneal injection of 35 mg of chloralhydrate/100 g of body weight. The upper left quadrant of the dorsal skin was shaved and disinfected by topical application of 2% Zephiran chloride in 70% alcohol. An incision one-half inch long was then made along the costal margin. The left kidney was loosened from the perirenal fat of the retroperitoneal region and pushed towards the median for exposure of the thoracic duct lying under the abdominal aorta. Blunt forceps were used to separate the thoracic duct from the aorta. The thoracic duct was tightly ligated, and a polyethylene tube filled with heparin in physiological buffered saline was inserted. Animals were then immobilized in a restraining cage and were provided ad lib with regular pellet food and a pyrogen-free mixture of 5% dextran and physiological saline. Thoracic-duct lymph was collected in Eagle's basal medium containing heparin at 24 C (room temperature). For experimentation as many as three rats were prepared on the same day, and TDC used in experiments on uptake of antigen were collected within a 60-min period between 4 hr and 12 hr of cannulation. The lymph was centrifuged at 100 *g* for 10 min, and the pellet was resuspended to make 2×10^7 TDC/ml. A model B Coulter counter was used for establishment of the desired concentration of cells.

In-vitro uptake of antigen. Various doses of iodinated ferritin were added to 2×10^7 TDC in 1 ml of Eagle's medium and incubated at 37 C for 30 min. This was followed by three washings of the reacted TDC with 2 ml of Eagle's medium each. Fluids from the washings were collected as well as the final TDC pellet, which was resus-

pended in 2 ml of the medium. Radioactivity of the supernatant and the resuspended cells was counted in a well-type gamma counter (Nuclear Chicago). The various concentrations of the adjuvants were added to the TDC mixture just before the addition of antigen.

Preparation of cells for electron microscopy. Pellets of TDC reacted in vitro with ferritin were fixed in an equal mixture of 2% paraformaldehyde and 2% osmium tetroxide in 0.1 M cacodylate buffer (*pH* 7.4) at 0 C, rinsed in the same buffer, dehydrated through a series of graded ethanol, and embedded in a mixture of epoxy resin. Sections were made on a LKB ultramicrotome, stained with uranyl acetate and lead citrate, and studied in a Phillips model 300 electron microscope.

Histologic study. Spleens from two-week-old Balb mice were fixed in Bouin's solution, embedded in paraffin, stained with methyl green–pyronin or Azure II–hematoxylin, and observed for size, number, and formation of germinal centers.

Results

Adjuvant action. The adjuvant action of the polyribonucleotide in rats, as in other species, approximates that of endotoxin (table 1).

The capabilities of the two adjuvants differed considerably, however, when they were tested in rats whose thoracic ducts had been cannulated and drained of approximately one billion lymphocytes during a 24-hr period. It may be seen in table 2, which presents the geometric mean of data from 10 experiments, that poly A:U was unable to stimulate such animals to respond with increased titers of antibody to ferritin, whereas the effect of endotoxin remained intact. Similarly, elevation of immunologic memory by poly A:U was negated by the drainage procedure, but this property of endotoxin was not affected.

It was shown previously that small amounts of equine ferritin were localized within lymphocytes that were in the lymph node and were draining the site of antigen injected into the footpads of rats [13]. Although the functional significance of this observation is not known, the effect of adjuvants on the uptake of antigenic ferritin by thoracic-duct lymphocytes (TDL) in vitro might serve to elucidate it. To this end, poly A:U and

Table 1. Adjuvant effect of polyribonucleotides or endotoxin on titers of antibody to equine ferritin in Sprague-Dawley rats.

Ferritin	Adjuvant	Antibody titer* on day		
		4	7	14
1 mg	None	10 (1, 1, 1)†	6,500 (10, 10, 11)	6,500 (10, 10, 11)
1 mg	Poly A:U‡	230 (5, 6)	20,500 (11, 13)	14,500 (11, 12)
1 mg	Endotoxin	160 (3, 6, 6)	16,300 (11, 11, 13)	12,900 (10, 11, 13)
0.1 mg	None	10 (0, 1, 2)	160 (4, 5, 6)	0 (0, 0, 0)
0.1 mg	Poly A:U	200 (4, 6, 6)	1,600 (8, 8, 9)	250 (4, 5, 8)
0.1 mg	Endotoxin	100 (3, 5, 5)	1,600 (7, 9, 9)	50 (0, 4, 5)

* Geometric mean titer of three rats as determined by passive hemagglutination.

† Parentheses enclose highest positive $\log_2$ dilution for each individual rat. Original dilution = 1:10.

‡ Polyribodadenylic-polyribouridylic acid.

Table 2. Effect of 24-hr drainage of thoracic-duct lymph on adjuvant action of polyribonucleotides or endotoxin.

Adjuvant for primary injection*	Reciprocal HA titer† (days after inoculation)							
	Primary response				Secondary response‡			
	7	15	21	28	3	7	12	21
None	190	113	43	10	2,150	4,300	7,240	8,610
250 µg of poly A:U§	166	113	23	37	260	2,037	2,032	4,560
20 µg of endotoxin	508	2,280	2,280	5,120	10,240	84,000	186,000	92,000

* All injections included 0.1 mg of ferritin.

† Geometric mean titer of six Sprague-Dawley rats.

‡ All rats were given a second injection of 0.1 mg of equine ferritin without adjuvant one month after primary injection.

§ Polyriboadenylic-polyribouridylic acid.

endotoxin were tested with regard to their effect on the uptake of ^{131}I-labeled ferritin by TDL. It is evident from the results (table 3) that a substantial, dose-dependent decrease was observed in the presence of poly A:U, whereas endotoxin did not appear to affect this binding significantly.

Table 3. Effect of polyribonucleotides or endotoxin on uptake of ^{131}I-labeled ferritin by thoracic-duct cells (TDC) of rats.

TDC incubated with	Uptake (%)*
Ferritin	100
Ferritin plus poly A:U† (18 µg)	10
Ferritin plus poly A:U (1.8 µg)	19
Ferritin plus poly A:U (0.18 µg)	48
Ferritin plus endotoxin (1 µg)	95
Ferritin plus endotoxin (0.1 µg)	95
Ferritin plus endotoxin (0.01 µg)	85

NOTE. 2×10^7 TDC plus 10 µg of ^{131}I-labeled ferritin plus adjuvant were incubated for 30 min at 37 C.

* Percentage of radioactivity as compared with that on cells incubated with ^{131}I-labeled ferritin alone.

† Polyriboadenylic-polyribouridylic acid.

These results, representative of seven experiments, further hint at a difference in mechanism between the two adjuvants.

The type of cell within the population of TDL affected by poly A:U might be a small to medium-sized lymphocyte, inasmuch as over 90% of the TDL possessed these structural characteristics. Less than 5% of the cells observed during the first 24 hr of cannulation were large blastoid cells, cells with plasmacytic tendencies, or other leukocytic elements. The predominant cell had a large nucleocytoplasmic ratio, and the cytoplasm contained numerous free ribosomes both in polymeric and monomeric configurations, a few scattered profiles of rough-surfaced endoplasmic reticulum, and several small mitochondria, particularly around the cytocentrum. The nuclear chromatin was dense and demonstrated various patterns of clumping along the nuclear membrane. A small nucleolus was usually embedded in dense clumps of nucleolus-associated chromatin. A number of perichromatin granules and occasional ring-shaped fibrillar structures were also seen.

The basic morphology of the small lymphocyte did not change in cells fixed after incubation for 30 min in vitro with the antigen. Now, however, there were occasional ferritin granules along the surface of such lymphocytes that appeared to be bound to the plasma membrane. In addition, several to a dozen or more of these usually could be seen in each of the pinocytic vesicles.

A second major difference between the adjuvant action of endotoxin and that of poly A:U appeared in a histologic study of the spleens of two-week-old Balb mice. Under ordinary conditions of immunization, bovine gamma globulin is a poor antigen in this young mouse. Nevertheless, under stimulus by either endotoxin or poly A:U, antibody was readily and regularly observed (table 4). In the spleens of endotoxin-treated mice, marked hyperplasia of the white pulp and lymphoid nodules was observed. Germinal centers were prominent, peaking in activity on day 8. In contrast, in poly A:U-treated mice, only a few small germinal centers were visible, with little or no hyperplasia. Yet, titers of antibody were as elevated as in mice stimulated with endotoxin.

Toxicity. With poly A:U exhibiting an adjuvant action quantitatively similar to that of endotoxin, it was important to determine its toxicity relative to that of endotoxin and that reported for a like adjuvant, poly I:C. The following parameters were tested.

(1) Lethality in mice treated with actinomycin D. Treatment of mice with actinomycin D has been shown to reduce by 3–4 $\log_{10}$ the LD_{50} of endotoxin [14], and this procedure has been useful in the detection of minute amounts of this toxin. Accordingly, mice were injected intraperitoneally with various amounts of poly A:U, poly I:C, or endotoxin together with 12.5 µg of acti-

Table 4. Adjuvant effect of polyribonucleotides or endotoxin in two-week-old Balb/aj mice.

Bovine gamma globulin (1 mg) plus	Reciprocal HA titer (days after inoculation)			
	4	6	8	12
None	0	20	160	160
Poly A:U* (250 µg)	40	2,560	2,560	640
Endotoxin (10 µg)	20	1,280	1,280	320

  * Polyriboadenylic-polyribouridylic acid.

Table 5. Comparative LD_{50} of polyribonucleotides and endotoxin in Balb mice treated with actinomycin D.

Substance tested	LD_{50} (µg)	
	Experiment 1	Experiment 2
Poly A:U*	50	40
Poly I:C†	1	0.4
Endotoxin	0.1	0.04

  * Polyriboadenylic-polyribouridylic acid.
  † Polyriboinosinic-polyribocytidylic acid.

nomycin D. The results of two experiments are shown in table 5. It may be seen that poly A:U was relatively nonlethal in this exquisitely sensitive assay, as compared with the acute toxicity exhibited by poly I:C and endotoxin.

(2) Pyrogenicity. Comparison in 10 rabbits each of the relative pyrogenicity of adjuvant-active doses of poly A:U and poly I:C revealed a mean maximal rise in degrees Fahrenheit of 3.7 for 300 µg of the complexed poly I:C, as compared with 1.6 for a like amount of the complexed poly A:U.

(3) Shwartzman reaction. The abilities of poly A:U, poly I:C, and endotoxin to prepare for or to provoke the inflammatory features of the local Shwartzman reaction were compared in New Zealand white rabbits. As is apparent from table 6, poly I:C and endotoxin were capable of both preparing for and provoking this reaction, but

Table 6. Capacity of polyribonucleotides or endotoxin in preparing for or provoking the local Shwartzman reaction.

Preparing agent	Provoking agent	Reactivity
Poly A:U* (400 µg)	Poly A:U (1,000 µg)	−
Poly I:C† (400 µg)	Poly A:U (1,000 µg)	+
Endotoxin (50 µg)	Poly A:U (1,000 µg)	+
Poly A:U (400 µg)	Poly I:C (1,000 µg)	−
Poly I:C (400 µg)	Poly I:C (1,000 µg)	+
Endotoxin (50 µg)	Poly I:C (1,000 µg)	+
Poly A:U (400 µg)	Endotoxin (100 µg)	−
Poly I:C (400 µg)	Endotoxin (100 µg)	+
Endotoxin (50 µg)	Endotoxin (100 µg)	+

NOTE. New Zealand white rabbits were given the preparing agent intradermally at two sites on the abdomen. The provoking agent was given intravenously 24 hr later. Reactivity was measured at 18 hr and designated negative if edema, erythema, and/or induration was absent.

  * Polyriboadenylic-polyribouridylic acid.
  † Polyriboinosinic-polyribocytidylic acid.

228

once again poly A:U failed to exhibit any significant signs of toxicity.

Discussion

These limited results document the lack of toxicity of poly A:U, a potent stimulator of both antibody synthesis and cell-mediated immunity [7]. In addition, they confirm and extend previous data [15–17] on the endotoxin-like activity of poly I:C. After a broader array of toxicity tests, additional study of poly A:U in man may be indicated. However, while freedom from toxicity is important for any future use with human vaccines, additional investigative parameters are warranted when consideration is given to the usefulness of poly A:U in the immunotherapy of cancer. Here, the balance between stimulation of cell-mediated immunity and production of tumor-enhancing antibody is crucial. A note of optimism along this line was obtained recently by Lacour et al. [18], who showed poly A:U to be a potent adjunct to surgical treatment in reducing by 50% the appearance of metastases of a transplantable melanoma in hamsters and in increasing the mean survival time of mice bearing spontaneous mammary adenocarcinomas.

The difference in capability of endotoxin and poly A:U as adjuvants in rats drained for 24 hr of TDL may be either quantitative or qualitative. For example, in additional experiments where drainage before challenge with antigen was extended to two days rather than one, neither endotoxin nor poly A:U exhibited adjuvant action. Thus (as an explanation of the data in table 2), endotoxin might have been more capable than poly A:U of stimulating the residual TDC remaining after 24 hr. On the other hand, a qualitatively different cell may have played the key role in endotoxin action, since it is known that, after drainage of the circulating cells, the sessile lymphocytes of the lymphoid tissues leave these areas and become available to the drainage system. Indirect support for the latter hypothesis was found in the data in table 3, in which only poly A:U was found to diminish the uptake of antigen by TDC. Whether or not this phenomenon is associated with the adjuvant action of poly A:U is under investigation in this laboratory. Our early data suggest that the direct addition and uptake of ferritin by TDC does not result in tolerance to this antigen, the diminution of which could have explained the adjuvant action of poly A:U.

References

1. Regamey, R. H. [ed.]. International symposium on adjuvants of immunity, Utrecht, 1966. S. Karger, Basel, 1967. 375 p.
2. Johnson, A. G., Gaines, S., Landy, M. Studies on the O antigen of *Salmonella typhosa*. V. Enhancement of antibody response to protein antigens by the purified lipopolysaccharide. J. Exp. Med. 103: 225–246, 1956.
3. Merritt, K., Johnson, A. G. Studies on the adjuvant action of bacterial endotoxins on antibody formation. V. The influence of endotoxin and 5-fluoro-2-deoxyuridine on the primary antibody response of the Balb mouse to a purified protein antigen. J. Immunol. 91:266–272, 1963.
4. Kind, P., Johnson, A. G. Studies on the adjuvant action of bacterial endotoxins on antibody formation. I. Time limitation of enhancing effect and restoration of antibody formation in X-irradiated rabbits. J. Immunol. 82:415–427, 1959.
5. Merritt, K., Johnson, A. G. Studies on the adjuvant action of bacterial endotoxins on antibody formation. VI. Enhancement of antibody formation by nucleic acids. J. Immunol. 94:416–422, 1965.
6. Braun, W., Ishizuka, M., Yajima, Y., Webb, D., Winchurch, R. Spectrum and mode of action of poly A:U in the stimulation of immune responses. *In* R. F. Beers and W. Braun [ed.]. Biological effects of polynucleotides. Springer-Verlag, New York, 1971, p. 139–156.
7. Johnson, A. G., Cone, R. E., Friedman, H. M., Han, I. H., Johnson, H. G., Schmidtke, J. R., Stout, R. D. Stimulation of the immune system by homopolyribonucleotides. *In* R. F. Beers and W. Braun [ed.]. Biological effects of polynucleotides. Springer-Verlag, New York, 1971, p. 157–178.
8. Schmidtke, J. R., Johnson, A. G. Regulation of the immune system by synthetic polynucleotides. I. Characteristics of adjuvant action on antibody synthesis. J. Immunol. 106:1191–1200, 1971.
9. Cone, R. E., Johnson, A. G. Regulation of the immune system by synthetic polynucleotides. III. Action on antigen-reactive cells of thymic origin. J. Exp. Med. 133:665–676, 1971.
10. Johnson, H. G., Johnson, A. G. Regulation of the immune system by synthetic polynucleotides. II. Action on peritoneal exudate cells. J. Exp. Med. 133:649–664, 1971.
11. Philips, F. S., Fleisher, M., Hamilton, L. D., Schwartz, M. K., Sternberg, S. S. Polyinosinic-polycytidylic acid toxicity. *In* R. F. Beers and W. Braun [ed.]. Biological effects of polynucleotides. Springer-Verlag, New York, 1971, p. 259–274.

12. McConahey, P. J., Dixon, F. J. A method of trace iodination of proteins for immunologic studies. Int. Arch. Allerg. Appl. Immunol. 29:185–189, 1966.

13. Han, S. S., Johnson, A. G. Radioautographic and electron-microscopic evidence of rapid uptake of antigen by lymphocytes. Science 153:176–178, 1966.

14. Pieroni, R. E., Broderick, E., Bundeally, A., Levine, L. A simple method for the quantitation of submicrogram amounts of bacterial endotoxin. Proc. Soc. Exp. Biol. Med. 133:790–794, 1970.

15. Hilleman, M. R. Prospects for the use of double-stranded ribonucleic acid (poly I:C) inducers in man. J. Infect. Dis. 121:196–211, 1970.

16. Stinebring, W. R., Absher, M. The double-bitted axe: a study of toxicity of interferon releasers. *In* R. F. Beers and W. Braun [ed.]. Biological effects of polynucleotides. Springer-Verlag, New York, 1971, p. 249–257.

17. Berry, L. J., Smythe, D. S., Colwell, L. S., Schoengold, R. J., Actor, P. Comparison of the effects of a synthetic polyribonucleotide with the effects of endotoxin on selected host responses. Infec. Immun. 3:444–448, 1971.

18. Lacour, F., Spira, A., Lacour, J., Prade, M. Polyadenylic-polyuridylic acid, an adjunct to surgery in the treatment of spontaneous mammary tumors in C3H/He mice and transplantable melanoma in the hamster. Cancer Res. 32:648–649, 1972.

Patterns of Tolerance to Endotoxin

Kelsey C. Milner

From the Rocky Mountain Laboratory, National Institute of Allergy and Infectious Diseases, Hamilton, Montana

Induced resistance (tolerance) to the pyrogenic effect of endotoxins has been studied in rabbits by means that permit a clear differentiation of at least two mechanisms that produce the same clinical signs. Within 24 hr after a single sublethal dose of endotoxin, animals showed marked tolerance with no inter-endotoxin specificity. This early tolerance did not seem to be associated with any humoral factor and could not be transferred passively. If the endotoxin was strongly immunogenic, a late form of tolerance, specific for the endotoxin employed, developed coincidentally with the production of O antibody; sera from animals with this type of resistance could passively transfer tolerance to normal recipients. These findings suggest that late tolerance to pyrogen may be a manifestation of anaphylaxis, but evidence for such a mechanism has not been obtained.

In 1942, Favorite and Morgan [1] inoculated human subjects with a toxic antigen (endotoxin) derived from the typhoid bacillus. They observed that repeated injections resulted in a state of "tolerance" to toxic effects, and they reported that this tolerance appeared to bear little relationship to the titers of antibodies that also

The collaboration of Dr. J. A. Rudbach in certain of these experiments is acknowledged. A fuller treatment of our joint work will be offered as a separate communication. I am also grateful to Dr. A. M. Pappenheimer, Jr., for bringing the work of Uhr and Brandriss [15] to my attention.

Please address requests for reprints to Dr. Kelsey C. Milner, Rocky Mountain Laboratory, Hamilton, Montana 59840.

developed. These findings have been abundantly verified, rediscovered, or buttressed by succeeding investigators, many of whom, like Beeson [2], adopted "tolerance" as the most suitable familiar designation for the phenomenon of induced resistance to the pathophysiologic effects of endotoxin.

More recently, the signs by which tolerance is recognized have been shown to result from more than one mechanism. The first clear demonstration of this fact was by Greisman et al. [3, 4], who differentiated the ordinary form of tolerance from a "pyrogenic refractory state," which developed rapidly during continuous infusion of endotoxin. Formerly, tolerance was always induced by a series of injections given over a period of days or weeks. Obviously, animals prepared in this manner had the opportunity to respond in several ways, e.g., with cellular effects, metabolic disturbances, release of natural antibodies or other humoral factors, and production of antibody to O-antigenic determinants and, perhaps, to unknown immunogens. Milner and Rudbach tried to disentangle some of these factors and presented a preliminary report [5] distinguishing between an early form of tolerance without inter-endotoxin specificity and a late form specific to the O antigens. They also demonstrated a role for O antibody in the passive transfer of tolerance. Meanwhile, Greisman and colleagues [6, 7] had, by somewhat different methods, arrived at virtually identical conclusions.

Because the matter remains controversial, this paper has been prepared as a validation of earlier work from this laboratory and in support of the conclusions of Greisman et al. [6, 7]. The work reported deals exclusively with tolerance to the pyretic effect of endotoxins in rabbits, and extrapolation of these results to other effects or other animals should be avoided.

Materials and Methods

Endotoxins. Several endotoxins of different specificities were prepared in this laboratory by a variety of standard methods. All were biologically potent to the extent of producing significant fever in rabbits or killing chick embryos in doses of 5 ng or less.

Animals. Albino New Zealand rabbits were obtained from commercial suppliers. All weighed between 2 and 3 kg (average, about 2.3 kg). Doses were adjusted in terms of μg per animal, because preliminary observations indicated that, among adult rabbits, there was no correlation between body weight and fever response to a given total dose of endotoxin.

Antisera. As specified below, serum was sometimes collected either 24 hr or about one week after one or several injections of endotoxin. In other cases, rabbits were given intensive courses of immunization with endotoxin or boiled whole cells for eight to 12 weeks and were bled seven days after the last inoculation. The rabbits were then rested four to six weeks and were bled again five to seven days after a recall dose. The two bleedings were shown to be equivalent in effect and quantity of antibody (more than 1 mg of

specifically precipitable antibody protein per milliliter, except after immunization with material from Re mutants). Sera prepared in this manner are hereafter referred to as hyperimmune sera. Sera from at least six (and usually from 12 to 24) rabbits were pooled in each lot. Rigorous precautions were taken against contamination, and all sera (as well as other fluids) were shown to be pyrogen free in doses larger than those used.

Measurement of fever. Tests for pyrogen were performed as described elsewhere [8, 9], with temperatures of preconditioned animals recorded automatically every 12 min. Temperature curves were plotted on a scale of one inch per hr or degree C, and the areas enclosed between 6-hr curves and baselines (fever index or FI) were measured in cm^2 by planimetry or computer. Challenge doses usually approximated 1 FI_{40} (the dose calculated to produce an FI of 40 cm^2 under the stated conditions) because, in the average case, this quantity has been shown to lie at the sensitive center of the straight-line portion of the dose-response curve [8]. This dose was also especially suitable because it did not induce a fatal Shwartzman reaction in any prepared rabbit, owing in part to its modest size and in part to the maturity of the rabbits. Fever curves in all figures are point-by-point averages from at least six, and usually from eight to 12, identically treated rabbits. All inoculations were given iv.

Results

Specificity of early and late tolerance. Study of tolerance that developed within 24 hr after one exposure to endotoxin was prompted by the reports of Greisman et al. [3, 4] on the rapid development of tolerance during continuous infusion with endotoxin, and by knowledge that levels of circulating antibody did not increase during that interval. In the first experiment (figure 1), marked tolerance was observed 24 hr after injection of 1.0 μg of endotoxin. (However, this was a large preparative dose, about 5 FI_{40}.) Slightly greater tolerance was produced by seven daily injections of the same size, as judged by depression of the first peak of fever, in addition to elimination of the second peak, and rapid fall of temperature to below the normal level. The early tolerance demonstrated by curve A in figure 1 was not associated with increased circulating antibody, but it is possible that

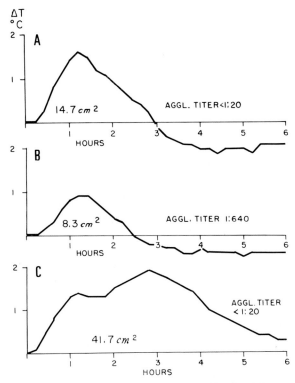

Figure 1. Tolerance to challenge with about 1 FI_40 (dose producing fever index of 40 cm²) of endotoxin induced by one or by seven prior exposures. (**A**) Rabbits challenged homologously 24 hr after one iv injection of 1.0 μg of endotoxin. (**B**) Rabbits challenged 24 hr after the last of seven daily iv injections of 1.0 μg of endotoxin. (**C**) Control rabbits with no prior treatment. The challenge dose in all cases was 0.2 μg of the same endotoxin used for preparation of groups A and B. Areas expressed in cm² are average fever indices.

the still greater tolerance shown by curve B might have been contributed to by the O antibody present after seven inoculations.

Figure 2 summarizes a titration of the amount of endotoxin required to induce measurable tolerance within 24 hr. Preparative doses graded from 1.0 μg by decimal decrements and the 0.1 μg challenge with the same material approximated 1 FI_40. Preparation with as little as 0.001 μg modified the response to challenge. It is also apparent from figure 2 and corroborated by much additional experience that the fever indices bear an inverse relationship to the size of the initial exposure and objectively reflect the degree of tolerance. For many purposes, therefore, one may indicate tolerance by numerical expression of the areas and estimate the significance of differences between

groups by statistical procedures. Where this has been done, I have used the median test in conjunction with the Fisher calculation of exact probability [10].

Another experiment (figure 3) confirms the previously reported persistence [11] and recall [12] of tolerance. Rabbits had been given an intensive schedule of inoculations of boiled cells of *Salmonella enteritidis* for production of antiserum. Five weeks after the last stimulation, some animals were challenged with 1 FI_40 of endotoxin from the same organism. As seen in part A of figure 3, they were still appreciably tolerant. Seven days after another stimulation, however, they gave the minimal response, seen in part B, to 1 FI_40 of endotoxin. Still other rested hyperimmune rabbits were stimulated again with homologous endotoxin and challenged seven days later with endotoxin from *Bordetella pertussis* (not shown). These animals responded to heterologous challenge exactly as did normal controls, thus revealing a remarkable degree of specificity under these circumstances.

As a further check on the specificity of tolerance with a simplified protocol, the experiment summarized in table 1 was performed. Rabbits given single preparing doses of endotoxins from three types of bacteria of different O specificities were challenged with endotoxin from one type; half of them were challenged after 24 hr and half after seven days. The early challenge demonstrated a marked degree of tolerance in all animals, with no species specificity. After late challenge, only those prepared homologously were tolerant.

One means for examining the role of O antigen in endotoxic phenomena is by use of extracts from a given wild-type culture and from its heptoseless Re (O-antigen deficient) mutant. Figure 4 represents an experiment in which rabbits were given single doses of one or the other of these kinds of endotoxin (prepared from *Salmonella typhimurium* and adjusted to equal toxicity) and were subsequently challenged with endotoxin from the wild-type culture, part of them after 24 hr and part after eight days. Either type produced strong early tolerance, but only the wild type produced late tolerance, presumably because the defective O antigen of the Re mutant is virtually nonimmunogenic. Similar results were obtained with an analogous pair of extracts from *Salmonella minnesota,*

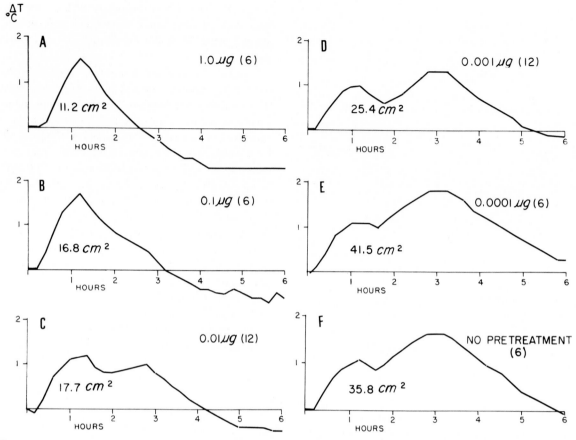

Figure 2. Titration of the inducing dose for early tolerance. Doses indicated at the upper right of each curve were given 24 hr before challenge with about 1 FI_{40} (dose producing fever index of 40 cm²) of the same endotoxin. Figures in parentheses specify number of rabbits in group. Areas expressed in cm² are average fever indices.

although late tolerance was not induced by a single injection of wild-type endotoxin, probably because this phenol-water extract was not sufficiently immunogenic.

Passive transfer of tolerance. The technique of passive transfer offers another means for study of the mechanism of tolerance. It is now well established (e.g., see [13]) that serum or plasma from tolerant rabbits can modify the febrile responses to endotoxin of normal rabbits. This phenomenon has been studied further, and evidence has been accumulated that antibody to specific O determinants is at least partially responsible for the passive transfer of tolerance. In numerous preliminary experiments, it was found that (*1*) serum and plasma could be used interchangeably; (*2*) serum was equally effective whether used fresh-frozen and thawed or heated at 56 C to inactivate com-

plement; and (*3*) serum could be given at any time from a few minutes to at least 24 hr before challenge with almost identical results, provided it was itself nonpyrogenic. The quantity of serum required to modify significantly the response of rabbits to about 1 FI_{40} of endotoxin was not very great in the cases of hyperimmune sera and others found to be effective. Table 2 shows a titration, still without end point, of one such pool of serum. It had been previously established that 10 ml and 15 ml gave no greater effect than 5 ml, so the latter was taken as the high dose. Amounts of specifically precipitable antibody nitrogen, as well as volumes of serum, are given as additional points of reference, although it is not asserted that precipitins are the most effective humoral mediators. As little as 0.1 ml of serum containing 17 µg of antibody nitrogen distinctly modified the fever

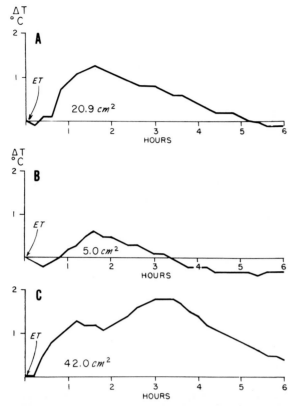

Figure 3. Febrile responses to homologous endotoxin of rested hyperimmune rabbits before and after recall. Areas expressed in cm² are average fever indices. (**A**) Hyperimmunized rabbits rested for five weeks. (**B**) Hyperimmunized rabbits rested for six weeks and challenged seven days after booster. (**C**) Normal controls.

curves, and the effect was graded within the range of 0.1–5.0 ml.

The belief that O antibody plays a major role

Table 1. Specificity of tolerance to endotoxin in prepared rabbits.

Source of endotoxin*	Fever indices (cm²)	
	24 hr	7 days
Serratia marcescens	10.8	47.2
Bordetella pertussis	10.8	47.5
Salmonella enteritidis	11.1	21.9
Saline diluent	36.5	47.2

* Normal rabbits, accustomed to restraint, were prepared with 1 FI_{40} (0.1–0.3 μg) of endotoxin from indicated bacterial source. Half were challenged with about 1 FI_{40} of endotoxin from *Salmonella enteritidis* 24 hr later, and the other half were given the same challenge seven days later. Fever indices are average responses of six rabbits each.

in late and persistent tolerance arose from the observed O specificity of this form of resistance and was supported by a large body of evidence similar to that appearing in figures 5 and 6 and tables 3 and 4. For each of groups A–E (table 3) eight donor rabbits were given single inoculations of the indicated doses of endotoxin. Blood was taken after 25 hr and again after seven days, and the recovered sera were pooled by groups. Additional rabbits (not shown in table) were prepared in the same way as were the donors and challenged after one or seven days. All gave tolerant responses, although animals challenged early were more markedly resistant. Donors for groups G and H were prepared with 1.0 μg of a different endotoxin and were bled after 24 hr. Donors for control groups F, I, and J were normal rabbits. Six to eight normal recipients per group were given 10 ml of the indicated serum or plasma and were challenged 24 min later with about 1 FI_{40} of homologous endotoxin.

Figure 5 (from A and B of table 3) illustrates the attempts at passive transfer and shows the constant result in this and similar experiments. Serum or plasma obtained from donors one day after exposure to endotoxin never conferred tolerance on normal recipients. Sera from later bleedings transferred tolerance to homologous challenge, provided that O antibody was present (see also table 3, C–J). Pertinent examples of passive transfer with hyperimmune sera, in which O antibody appears to play a role, are given in table 4.

A further indication that O antibody contributes to passive transfer of tolerance was obtained by absorption of an effective antiserum with nontoxic O-specific substance and demonstration that this reduced ability to transfer resistance (figure 6). A material isolated in this laboratory from various smooth cultures of *Escherichia coli* (originally called "native hapten") contained no heptose, ketodeoxyoctonates (KDO), phosphorus, or long-chain fatty acids, yet it would precipitate from an immune serum up to 80% of the antibody precipitable by whole endotoxin from the same organism [14]. Prior treatment with an immune serum to *E. coli* O113 had a modifying effect on the febrile response to homologous endotoxins (figure 6, part A). One absorption with native hapten, in slight antigen excess, removed a significant amount of the transfer factor ($P < .02$), as shown

234

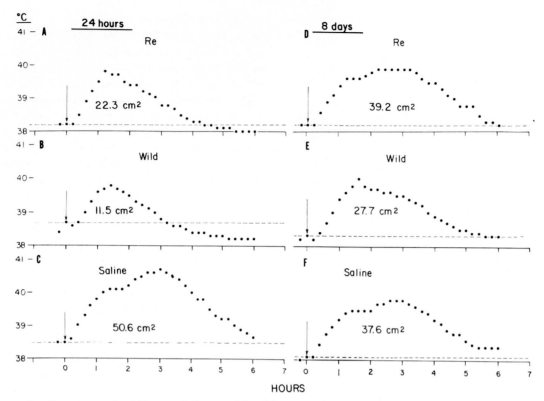

Figure 4. Responses of rabbits to challenge with wild-type endotoxin 24 hr and eight days after preparation with equipyrogenic doses of extracts from wild-type or Re-mutant cultures of *Salmonella typhimurium*. The type of preparation is indicated over each curve. Areas expressed in cm² are average fever indices.

Table 2. Quantity of antiserum or antibody required for passive transfer of tolerance to endotoxin.

Antiserum* (ml)	Antibody nitrogen (μg)	Endotoxin† (μg)	Average fever index (cm²)
5	830	None	0
5	830	0.4	11.9
2	332	0.4	21.0
1	166	0.4	25.5
0.4	66	0.4	28.2
0.1	17	0.4	30.6
None	0	0.4	46.8

* Donor rabbits were immunized with whole boiled cells of *Salmonella typhimurium*. Specifically precipitable antibody nitrogen (166 μg/ml) was determined with the homologous endotoxin used in this test.

† M NaCl extract of *S. typhimurium*.

at B; it appeared unlikely that the absorption would have removed anything but antibodies to O determinants.

Discussion

It is recognized that the data and conclusions presented here are at variance with many other reports. The findings are set forth because they have been consistent and are based on investigations in considerable depth. In searching for explanations of differences, it is most important to bear in mind that this paper is concerned solely with the febrile responses of rabbits. Other experiments have shown, for example, that tolerance to the lethal effect of endotoxin in mice does not follow a similar pattern. For clarity, therefore, it will be necessary to examine each effect in each animal.

Distinguishing between early and late tolerance to endotoxic pyrogen in the rabbit (the one without inter-endotoxic specificity and largely independent of humoral factors, the other specific to O antigen and mediated by antibody) makes it possible to reconcile these findings with many in which the two separate mechanisms were not recognized. Animals rendered tolerant in the traditional manner have both types of resistance, and consideration of the mixture without separation of the two types of which it is composed leads inevitably to confusion. Furthermore, the data are

Table 3. Failure of serum or plasma from donors in early active tolerance to transfer tolerance passively.

	Donors		Recipients		
Group	Preparation*	Bleeding (day)	Fluid transferred	Challenge†	Fever index (cm²)
A	1.0 µg Se ET	1	Serum	0.1 µg Se ET	40.7
B	1.0 µg Se ET	7	Serum	0.1 µg Se ET	18.5
C	0.1 µg Se ET	7	Serum	0.1 µg Se ET	27.2
D	10.0 µg Se ET	1	Serum	0.1 µg Se ET	. . .‡
E	10.0 µg Se ET	7	Serum	0.1 µg Se ET	15.6
F	None	. . .	Serum	0.1 µg Se ET	38.8
G	1.0 µg Ec ET	1	Serum	0.25 µg Ec ET	42.1;43.0
H	1.0 µg Ec ET	1	Plasma	0.25 µg Ec ET	41.9;41.7
I	None	. . .	Serum	0.25 µg Ec ET	44.2
J	None	. . .	Plasma	0.25 µg Ec ET	41.8

* Test donors were given a single dose of endotoxin. Se ET = aqueous-ether extract of *Salmonella enteritidis;* Ec ET = hot phenol-water extract of *Escherichia coli* O113.

† Normal recipients were given 10 ml of serum or plasma and challenged as indicated 24 min later.

‡ Pool of sera from donors was pyrogenic and extremely chylous.

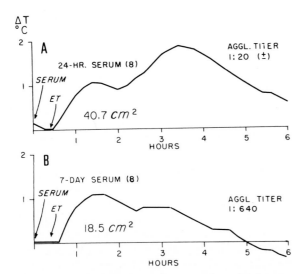

Figure 5. Attempted passive transfer of tolerance with serum obtained from donor rabbits one or seven days after a single injection of endotoxin. Preparation: 1.0 µg of aqueous-ether extract of *Salmonella enteritidis*. Areas expressed in cm² are average fever indices.

not so clear-cut as to establish a simple dichotomy: there may be still other mechanisms for producing the signs of tolerance.

An area of current interest is the possible production by animals of antibodies to the toxic groupings of endotoxin, as distinguished from antibodies to the saccharide O determinants. This paper enters that area at only one point, namely, the failure of rabbits, under these particular conditions, to produce an antibody directed specifically

Table 4. Passive transfer of tolerance with hyperimmune sera.

Donors			Average fever index (cm²)
Immune to*	Amount of serum transferred (ml)	Challenge†	
Stm wild	5	Stm wild ET	17.9
Stm Re	5	Stm wild ET	36.9
Nothing (NRS)	5	Stm wild ET	43.4
S. minn. wild	10	*S. minn.* wild ET	23.6
S. minn. Re	10	*S. minn.* wild ET	36.3
Nothing (NRS)	10	*S. minn.* wild ET	35.8
S. minn. wild	10	*S. minn.* Re ET	37.1
S. minn. Re	10	*S. minn.* Re ET	35.2
Nothing (NRS)	10	*S. minn.* Re ET	41.3

* Donor rabbits intensively immunized with boiled whole cells. Stm wild = *Salmonella typhimurium* 5609, a wild-type virulent strain; Stm Re = *S. typhimurium* G30/C21, a heptoseless (Re) mutant; *S. minn. wild* = *Salmonella minnesota* 1114; *S. minn.* Re = *S. minnesota* 1167; NRS = normal rabbit serum; ET = endotoxin.

† Challenges approximated 1 FI_{40} of endotoxins extracted from cultures used for immunization. Stm wild ET endotoxin extracted with 1 M NaCl; the other two were obtained by the hot phenol-water method. Animals were challenged 24 min after injection of serum.

against whatever component of endotoxin is responsible for pyrogenicity. The endotoxins from Re mutants used in these experiments were approximately as toxic and as soluble in water as those from wild-type cultures. They were authentically heptoseless and contained no saccharide determinants, except possibly KDO. Because the endotoxins

236

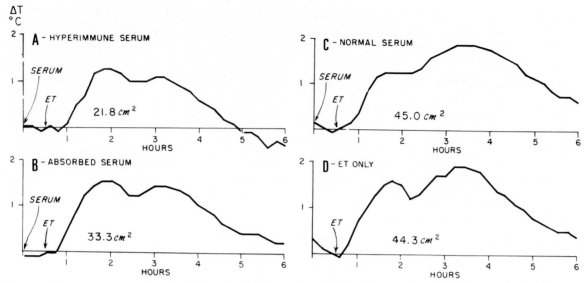

Figure 6. Effect on passive transfer of partial absorption of O antibodies. (**A**) Tolerant response of normal rabbits given 5 ml of hyperimmune serum before challenge with phenol-water extract of *Escherichia coli* O113. (**B**) Reduced passive transfer with hyperimmune serum absorbed once with nontoxic O-specific substance (native hapten). (**C, D**) Controls.

were extracted by the phenol-water method, which frequently (but not always) leads to a relatively nonimmunogenic product, attempts to hyperimmunize rabbits were usually conducted with whole boiled cells. However, when rabbits were immunized with the extracted endotoxins, the results were not different. Neither quantitative precipitation nor indirect hemagglutination revealed more than very low levels of antibody to Re determinants, and none of the antisera to Re determinants appreciably modified the fever response to challenge with any endotoxin. More intensive immunization, possibly with use of adjuvants, could lead to higher titers of antibody to Re determinants, but it does not appear likely that such antibodies are important in ordinary active tolerance to pyrogen.

Failure heretofore to discuss tolerance to endotoxin in connection with anaphylaxis may be attributed to the prevailing opinion that antibody is not much involved in tolerance. The present findings, in conjunction with recent papers by Greisman and colleagues, indicate a necessity for such consideration. Some of the schedules employed for production and challenge of tolerance might easily produce at least mild anaphylactic responses, and the administration of preformed antibody followed by specific antigen certainly meets basic conditions for passive anaphylaxis. It has

also been on record since the early years of this century that, in anaphylactic shock, rabbits suffer a drop in temperature of 0.6–1.7 C and guinea pigs a drop of 3-5 C [15]. Furthermore, Uhr and Brandriss [16] showed that passive administration of as little as 5 μg of antibody nitrogen would modify the fever normally provoked in a hypersensitive guinea pig by challenge with specific antigen, and larger amounts of antibody would abort the fever altogether. Their normal control animals, when given a small amount of antibody and challenged 24 hr later with specific nontoxic antigen, underwent a period of hypothermia. Thus, in light of these and other experiments, guinea pigs in a state of mild, nonfatal anaphylaxis, active or passive, undergo a period of hypothermia that can obliterate fevers of hypersensitivity and that very likely could also modify endotoxin fevers. The parallel is obvious. The question of whether or not late tolerance to endotoxic pyrogen in the rabbit, and especially the passive transfer of tolerance, could simply be manifestations of anaphylaxis must be answered experimentally.

The hypothesis is still considered tenable, although various attempts in this laboratory have failed to provide evidence in its favor. Numerical data are not given because all results to date have been negative. In four different systems in which nontoxic antigens and specific antibody of high titer have been available, schedules of doses approximating those used to demonstrate passive transfer of tolerance have failed to produce hypothermia in the rabbit. These were bovine γ-globulin, type 3 pneumococcal polysaccharide, normal murine serum, and the nontoxic O-specific polysaccharide, native hapten, from *E. coli* O113. Moreover, histamine, in a wide range of doses from lethal to very small, never produced hypothermia in a live rabbit; neither did serotonin or a combination of histamine and serotonin. Therefore, the mechanism of anaphylaxis, which would seriously complicate studies of tolerance to pyrogen in the guinea pig, may not be an important factor in the rabbit. Further work along these lines is indicated, however, at least to the extent of establishment of conditions that will produce anaphylaxis in the rabbit and determination of their effect on response to endotoxin.

References

1. Favorite, G. O., Morgan, H. R. Effects produced by the intravenous injection in man of a toxic, antigenic material derived from *Eberthella typhosa:* clinical, hematological, chemical and serological studies. J. Clin. Invest. 21:589–599, 1942.

2. Beeson, P. B. Development of tolerance to typhoid bacterial pyrogen and its abolition by reticulo-endothelial blockade. Proc. Soc. Exp. Biol. Med. 61: 248–250, 1946.

3. Greisman, S. E., Woodward, W. E. Mechanisms of endotoxin tolerance. III. The refractory state during continuous intravenous infusions of endotoxin. J. Exp. Med. 121:911–933, 1965.

4. Greisman, S. E., Young, E. J., Woodward, W. E. Mechanisms of endotoxin tolerance. IV. Specificity of the pyrogenic refractory state during continuous intravenous infusions of endotoxin. J. Exp. Med. 124:983–1000, 1966.

5. Milner, K. C., Rudbach, J. A. Relationship between presence of specific antibody and tolerance to endotoxin (abstract). Bacteriol. Proc. p. 96, 1968.

6. Greisman, S. E., Young, E. J., Carozza, F. A., Jr. Mechanisms of endotoxin tolerance. V. Specificity of the early and late phases of pyrogenic tolerance. J. Immunol. 103:1223–1236, 1969.

7. Greisman, S. E., Young, E. J. Mechanisms of endotoxin tolerance. VI. Transfer of the "anamnestic" tolerant response with primed spleen cells. J. Immunol. 103:1237–1241, 1969.

8. Keene, W. R., Silberman, H. R., Landy, M. Observations on the pyrogenic response and its application to the bioassay of endotoxin. J. Clin. Invest. 40:295–301, 1961.

9. Milner, K. C., Finkelstein, R. A. Bioassay of endotoxin: correlation between pyrogenicity for rabbits and lethality for chick embryos. J. Infect. Dis. 116:529–536, 1966.

10. Siegel, S. Nonparametric statistics for the behavioral sciences. McGraw-Hill, New York, 1956. 312 p.

11. Mulholland, J. H., Wolff, S. M., Jackson, A. L., Landy, M. Quantitative studies of febrile tolerance and levels of specific antibody evoked by bacterial endotoxin. J. Clin. Invest. 44:920–928, 1965.

12. Watson, D. W., Kim, Y. B. Modification of host responses to bacterial endotoxins. I. Specificity of pyrogenic tolerance and the role of hypersensitivity in pyrogenicity, lethality, and skin reactivity. J. Exp. Med. 118:425–446, 1963.

13. Freedman, H. H. Further studies on passive transfer of tolerance to pyrogenicity of bacterial endotoxin. The febrile and leucopenic responses. J. Exp. Med. 112:619–634, 1960.

14. Anacker, R. L., Finkelstein, R. A., Haskins, W. T., Landy, M., Milner, K. C., Ribi, E., Stashak, P. W. Origin and properties of naturally occurring hapten from *Escherichia coli*. J. Bacteriol. 88: 1705–1720, 1964.

15. Seegal, B. C. Anaphylaxis. *In* F. P. Gay [ed.] Agents of disease and host resistance. Charles C Thomas, Springfield, Ill., 1935, p. 36–78.

16. Uhr, J. W., Brandriss, M. W. Delayed hypersensitivity. IV. Systemic reactivity of guinea pigs sensitized to protein antigens. J. Exp. Med. 108: 905–924, 1958.

Antigenic Determinant Involved in Hypersensitivity of Infected Mice to Lipopolysaccharide of *Salmonella*

Masaya Kawakami

From the Department of Molecular Biology, School of Medicine, Kitasato University, Sagamihara, Kanagawaken, Japan

Sensitivity of mice to cell-wall lipopolysaccharide (LPS) extracted from certain strains of *Salmonella* was increased by infection with various strains of *Salmonella typhimurium, Salmonella minnesota,* and *Salmonella enteritidis*. Smooth or rough strains possessing a polysaccharide chain longer than that of the glucoseless mutant were capable of inducing this hypersensitivity. Infection with one strain sensitized animals to LPS from other strains. However, strains of *Escherichia coli* that had no antigen in common with rough-core polysaccharide could not induce hypersensitivity. Fractions containing either O side chains or LPS of the heptoseless mutants of *Salmonella* were ineffective in provoking hypersensitivity. Sensitization to smooth LPS by bacterial infection was prevented by prior administration to mice of rough LPS and not of the O side chain. These results indicate that the major antigenic determinant participating in this hypersensitivity is in the rough-core polysaccharide, presumably in the polysaccharide sequence from glucose-1 to glucose-2 of the rough core, but not in the O side chain or lipid A of LPS.

Various authors have speculated on the possible participation of allergy in certain biological activities of endotoxin [1]. They assumed that animals sensitized by natural infection with gram-negative bacteria that are nevertheless normal in appearance acquired a hypersensitivity to endotoxin, and that, thereby, these animals might have an increased response to endotoxin. Watson and Kim [2, 3] have observed a lower susceptibility of young rabbits to endotoxin. They presumed that the high sensitivity of adults to endotoxin was due to the hypersensitivity acquired as a result of subclinical infection or contact with intestinal flora, and that a common antigen in the endotoxin of various microorganisms participated in the hypersensitivity reaction because of its cross-reactivity.

A mild infection in mice with suitable strains of *Salmonella* causes development of a carrier state. The infecting bacteria persist for months in organs of mice (e.g., liver, spleen, and lymph nodes) without the appearance of severe symptoms in the host [4]. This sublethal infection confers on mice an immunity against infection with virulent *Salmonella*. On the other hand, this immunization with live bacteria also makes these mice more sensitive than uninfected mice to the lethal action of endotoxin extracted from the *Salmonella* species. Infection with one strain sensitizes animals to lipopolysaccharide (LPS) from other strains [5]. This increased sensitivity to endotoxin may also be explained by hypersensitivity based on the immunologically specific reaction. To prove this assumption, it is necessary to demonstrate an immunologically specific antigen in the endotoxin molecule, which participates in the hypersensitive reaction. If the assumption is correct, a common antigen must be involved in the cross-reaction among various species of *Salmonella*.

In a previous paper [5], we indicated the presence of such an antigenic determinant in the rough-core polysaccharide chain in LPS of salmonella cell walls. The present paper reports an extended study on the antigenic determinant that relates to hypersensitivity to LPS in mice infected with *Salmonella*.

Materials and Methods

Details of the methods have been given in a previous report [5]. A smooth strain of *Salmonella*

Please address requests for reprints to Dr. Masaya Kawakami, Department of Molecular Biology, School of Medicine, Kitasato University, Sagamihara, Kanagawaken, Japan.

typhimurium (LT2) and a series of rough mutants derived from this strain were donated by H. Nikaido, University of California, Berkeley. Strains of *Salmonella minnesota* and *Escherichia coli* were given to us by G. Schmidt, Max-Planck-Institut für Immunbiologie, Freiburg, Germany. Our stock strains of *Salmonella enteritidis* were also used. LPS was extracted from these bacteria with 90% phenol containing 0.001 M disodium EDTA. After precipitation with ethanol and treatment with nucleases, the LPS was purified by ultracentrifugation and ion-exchange cellulose chromatography. The O side chain and O side chain-lipid complex were prepared from strain TV160 [5].

Generally, 4×10^4 viable bacteria were injected iv into each specific-pathogenfree DDN mouse. In cases of infection with virulent strains LT2 and 116-54, iv inoculation was followed by treatment with kanamycin, which saved the host from death and maintained viable bacteria at a certain level in the host. Fourteen to 16 days after infection, the mice were given an iv injection with various amounts of antigenic material. According to the method of Litchfield and Wilcoxon [6], the LD_{50} and its confidence limits were calculated from the death rate recorded 48 hr after administration.

Results

In our previous paper [5], it was shown that hypersensitivity to LPS from smooth or rough strains of *Salmonella* was conferred by infection with *Salmonella* that was pathogenic for mice, whereas nonpathogenic bacteria such as *E. coli* and *Enterobacter aerogenes* did not sensitize the mice. The hypersensitive state developed during the second week and decreased within four weeks of infection with *Salmonella*. The reaction seemed to be delayed-type, because the mean survival time of mice infected and then given LPS was 8 hr. There was no significant increase in number of bacteria in the organs of infected mice after death due to LPS; thus death was thought not to be due to septicemia caused by the aggressive effect of LPS. We noted in our previous paper that heterologous *Salmonella* also caused hypersensitivity to LPS; i.e., mice infected with *S. enteritidis* were hypersensitive to LPS extracted from *S. typhimurium,* and vice versa. Rough strains pos-

sessing a complete rough-core polysaccharide were able to sensitize mice, and LPS from these strains were capable of evoking the reaction. From these results, it was thought that a common antigen must be participating in the hypersensitive reaction in the rough core of these salmonella strains.

To prove this assumption, we studied the cross-reaction, using a strain of *S. minnesota;* the rough-core polysaccharide structure in the LPS of this strain was known to be the same as that in LPS of *S. typhimurium* (figure 1). As shown in table 1, those mice sensitized by infection with rough (rouB) or smooth strains of *S. typhimurium* or *S. minnesota* were equally hypersensitive to LPS extracted from either *S. typhimurium* (strain his-rfb388, rouB) or *S. minnesota* (strain R60, rouB). In contrast, infection with the glucose-1-less mutant of *S. typhimurium* (strain SL1004), which lacked a part of the rough-core polysaccharide, was found to be incapable of sensitizing mice. Our strain of *E. coli* possessed an antigen in its rough LPS that was not cross-reactive serologically to the rough core of *Salmonella,* and this strain was also ineffective in sensitizing mice (table 1). Furthermore, mice sensitized by smooth or rough *S. typhimurium* or *S. enteritidis* to rough LPS were not sensitive to LPS of a heptoseless mutant lacking the rough-core polysaccharide. They rather became resistant to this LPS at a later stage of infection; the LD_{50} of heptoseless LPS in the uninfected control group was 273 μg/mouse, while it increased to 1,700 μg five weeks after the infection. Essentially the same results were obtained when LPS from a heptoseless strain of *S. minnesota* (R595) was used as antigen.

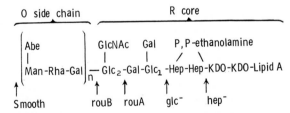

Figure 1. Structure of endotoxin (LPS) from *Salmonella typhimurium*. Gal = galactose; Rha = rhamnose; Man = mannose; Abe = abequose; Glc = glucose; GlcNAc = N-acetylglucosamine; Hep = glyceromannoheptose; KDO = 3-deoxyoctulosonate. Arrow indicates distal end of LPS chain of each class of mutants. The rough core of LPS from *Salmonella minnesota* has the same structure as does that of LPS from *S. typhimurium*.

Table 1. Sensitivity of infected mice to endotoxin (LPS) extracted from rough strains of *Salmonella*.

Mice infected with	LPS from	LD$_{50}$ (µg)	Confidence limits (P = .05)
Salmonella typhimurium LT2 (S)*	*S. typhimurium* his-rfb388	27	13–55
S. typhimurium his-rfb388 (rouB)	*S. typhimurium* his-rfb388	33	15–73
S. typhimurium SL1004 (glc$_1$$^-$)	*S. typhimurium* his-rfb388	>187	
Salmonella enteritidis SER (R)	*S. typhimurium* his-rfb388	31	14–65
Salmonella minnesota R60 (rouB)	*S. typhimurium* his-rfb388	34	22–54
Escherichia coli F862 (coliR$_1$)	*S. typhimurium* his-rfb388	232	120–392
Uninfected control	*S. typhimurium* his-rfb388	255	181–360
S. typhimurium his-rfb388 (rouB)	*S. minnesota* R60	27	14–52
S. minnesota R60 (rouB)	*S. minnesota* R60	29	16–52
Uninfected control	*S. minnesota* R60	198	125–314

* S = smooth; rouB = rouB class mutant (see [12]); glc$_1$$^-$ = glucose-1-less mutant; R = rough; coli R$_1$ = coli R$_1$ class mutant (see [13]). For the mutant of *S. minnesota*, the name corresponding to the class of *S. typhimurium* was used for the convenience of comparison.

The results of present and previous experiments are summarized in table 2. These results indicate that the antigenic determinant involved in the hypersensitive reaction is in the rough-core polysaccharide but not in the lipid A of LPS.

In our previous report [5], it was demonstrated that mice infected with smooth or rough strains were not sensitive to the O side chain of LPS extracted from *S. typhimurium* (strain TV160). The O side chain-lipid complex was also ineffective. Although the isolated O side chain or O side chain-lipid complexes are of sufficiently high molecular weight to be precipitated by O-specific antiserum, we thought that they might be too small to be involved in the hypersensitivity reaction. To eliminate this possibility, we tested their inhibitory effect on the development of hypersensitivity to smooth LPS. Mice were injected twice iv with 10 µg of rough LPS; the interval between injections was two weeks. Mice were infected with a smooth strain, LT2, four weeks after the second dose of LPS. Finally, the sensitivity of these mice to LPS was tested two weeks after the infection. As shown in table 3, prior treatment with rough LPS prevented the development of hypersensitivity caused by infection with smooth *Salmonella*. However, treatment with the same dose of O side chain was completely ineffective. From these re-

Table 2. Sensitivity of infected mice to endotoxin lipopolysaccharide (LPS) of various strains of *Salmonella*.

Mice infected with	Increase in sensitivity to LPS from				
	Salmonella enteritidis (S)*	*Salmonella typhimurium* (S)	*S. typhimurium* (rouB)	*Salmonella minnesota* (rouB)	*S. typhimurium* (hep$^-$)
S. enteritidis (S)	+	+	...	...	...
S. typhimurium (S)	+	+	+	...	...
S. enteritidis (R)	+	...	...	...	−†
S. typhimurium (rouB)	+	+	+	...	−†
S. typhimurium (rouA)	+	...	+	+	...
S. minnesota (rouB)	...	+	+	+	...
S. typhimurium (glc$_1$$^-$)	...	...	−	...	...
S. typhimurium (hep$^-$)	−	...	...	...	−†
Escherichia coli (S)	−	...	...	...	...
E. coli (coliR$_1$)	−	...	...	...	...
Enterobacter aerogenes (S)	...	−	...	...	...

NOTE. + = Significant increase in sensitivity; − = no increase in sensitivity (P = .05).

* S = smooth; R = rough; rouA and rouB = rouA and rouB classes of mutants, respectively (see [12]); glc$_1$$^-$ = glucose-1-less mutant; hep$^-$ = heptoseless mutant; coli R$_1$ = coli R$_1$ class mutant (see [13]). For the mutant of *S. minnesota*, the name corresponding to the class of *S. typhimurium* was used for the convenience of comparison.

† − = decrease in sensitivity.

sults, it appears to be most likely that the O side chain is not essential to the hypersensitivity reaction.

Discussion

It has been reported that hypersensitivity to LPS in rabbits, rats, and mice has been induced by multiple injections of LPS into these animals [7–9]. This hypersensitivity was characterized by intravascular coagulation in infected animals and has been distinguished carefully from classic hypersensitivity by Braude (see Session 3). Besides the common antigen, O-specific antigen was reported to participate in this reaction [8, 9]. It is thought that this hypersensitivity is similar to the immediate type. Hypersensitivity of mice to LPS, which is produced by infection with *Salmonella,* is thought to differ from that reported by Braude, because the former seems to be of the delayed type. Attempts to transfer sensitivity passively with serum failed (author's unpublished observation). Therefore, the antigenic groups involved in both types of hypersensitivity apparently differ from each other. Present studies indicate that the major antigenic component that participates in our hypersensitivity may be in the rough-core-polysaccharide sequence, because animals can be sensitized by infection with strains possessing LPS longer than that of the glucose-1-less mutant, and the fractions containing either O side chain or LPS of heptoseless mutants were ineffective in provoking the hypersensitivity reaction.

Watson and Kim have speculated that there are two different activities in the endotoxin molecule: the primary toxicity associated with the lipid portion of the molecule, and the secondary toxicity resulting from the acquisition of hypersensitivity of the host to some portion of the molecule [3]. According to their theory, hypersensitivity to endotoxin is caused by subclinical infections or contact with the intestinal flora that have a common antigen. The authors tried to explain the secondary toxicity of endotoxin from all species of gram-negative microorganisms. The hypersensitivity we observed is one caused only by infection with *Salmonella.* However, it is plausible that hypersensitivity to LPS from *Salmonella* and to LPS from common gram-negative bacteria is produced by the same immunologic mechanism. Hyperreactivity to endotoxin has also been observed in animals infected with heterologous bacteria such as *Brucella* [10] and *Mycobacterium tuberculosis* [11]. In these cases, the increased sensitivity was not thought to be a hypersensitivity based on the immunologically specific reaction [11]. However, these phenomena must be examined again from the standpoint of common antigens of various microorganisms.

References

1. Stetson, C. A. Role of hypersensitivity in reactions to endotoxin. *In* M. Landy and W. Braun [ed.] Bacterial endotoxins. Rutgers University Press, New Brunswick, N.J., 1964, p. 658–662.
2. Watson, D. W., Kim, Y. B. Modification of host responses to bacterial endotoxins. I. Specificity of pyrogenic tolerance and the role of hypersensitivity in pyrogenicity, lethality, and skin reactivity. J. Exp. Med. 118:425–446, 1963.
3. Watson, D. W., Kim, Y. B. Immunological aspects of pyrogenic tolerance. *In* M. Landy and W. Braun [ed.] Bacterial endotoxins. Rutgers Uni-

Table 3. Effect of prior treatment with endotoxin lipopolysaccharide (LPS).

| | | Sensitivity to LPS | | |
Mice treated with	Infected with	Source	LD_{50} (µg)	Confidence limits ($P = .05$)
LPS from *Salmonella typhimurium* his-rfb388	*S. typhimurium* LT2	*S. typhimurium* his-rfb388	66	39–112
None	*S. typhimurium* LT2	*S. typhimurium* his-rfb388	26	16–43
None	None	*S. typhimurium* his-rfb388	227	120–422
LPS from *S. typhimurium* his-rfb388	*S. typhimurium* LT2	*S. typhimurium* LT2	94	50–175
O side chain from *S. typhimurium* LT2	*S. typhimurium* LT2	*S. typhimurium* LT2	16	9–28
None	*S. typhimurium* LT2	*S. typhimurium* LT2	17	10–31
None	None	*S. typhimurium* LT2	152	71–323

versity Press, New Brunswick, N.J., 1964, p. 522–536.

4. Kawakami, M., Ishibashi, H., Mitsuhashi, S., Sakaino, K., Fukai, K. Experimental salmonellosis. Unstable L forms in liver of infected mice. Jap. J. Microbiol. 14:143–153, 1970.

5. Kawakami, M., Hara, Y., Osawa, N. Experimental salmonellosis: hypersensitivity to cell wall lipopolysaccharide and anti-infectious resistance of mice infected with *Salmonella*. Infec. Immun. 4:519–524, 1971.

6. Litchfield, J. T., Jr., Wilcoxon, F. A simplified method of evaluating dose-effect experiments. J. Pharmacol. Exp. Ther. 96:99–113, 1949.

7. Homma, Y., Sora, T., Sagehashi, T., Hosoya, S., Miyazaki, Y., Wakabayashi, Y. Anaphylaxis in mice against the endotoxin of *Pseudomonas aeruginosa* and distribution of antigens in anaphylactic shock. Allergy 3:177–183, 1954.

8. Braude, A. I., Siemienski, J. The influence of endotoxin on resistance to infection. Bull. N. Y. Acad. Med. 37:448–467, 1961.

9. Davis, C. E., Brown, K. R., Douglas, H., Tate, W. J., III, Braude, A. I. Prevention of death from endotoxin with antisera: I. The risk of fatal anaphylaxis to endotoxin. J. Immunol. 102:563–572, 1969.

10. Abernathy, R. S., Bradley, G. M., Spink, W. W. Increased susceptibility of mice with brucellosis to bacterial endotoxins. J. Immunol. 81:271–275, 1958.

11. Suter, E., Ullman, G. E., Hoffman, R. G. Sensitivity of mice to endotoxin after vaccination with BCG (Baccillus Calmette-Guérin). Proc. Soc. Exp. Biol. Med. 99:167–169, 1958.

12. Nikaido, H. Bacterial cell wall, deep layer. *In* Davis and Warren [ed.] The specificity of cell surfaces. Prentice-Hall, Englewood Cliffs, N.J., 1967, p. 3–30.

13. Schmidt, G., Fromme, I., Mayer, H. Immunochemical studies on core lipopolysaccharides of Enterobacteriaceae of different genera. Eur. J. Biochem. 14:357–366, 1970.

Escherichia coli Bacteremia in the Squirrel Monkey: Demonstration of a Complement-Dependent Neutrophil Response

David N. Gilbert, Jack A. Barnett, and Jay P. Sanford

From the Department of Internal Medicine, University of Texas Southwestern Medical School, Dallas, Texas

Squirrel monkeys were significantly depleted of complement by a nontoxic protein constituent of cobra venom, and the influence of complement depletion on the course of bacteremia due to *Escherichia coli* was studied. Striking neutropenia occurred rapidly in control animals, while the rate of occurrence of neutropenia was 20–30 times slower in animals treated with cobra-venom; this finding suggested that the neutropenia induced by *E. coli* was at least partially a complement-mediated response. In complement-depleted monkeys, sustained levels of *E. coli* bacteremia were consistently higher than in control animals. These observations are consistent with a hypothesis that complement-mediated neutrophilic-leukocyte function is an important defense mechanism of the host in gram-negative bacillary bacteremia.

Interaction between gram-negative bacteria and the complement (C) system has been demonstrated by electron-microscopic study of gram-negative bacteria after exposure to antibody and C [1, 2], by the in-vitro consumption of selected components of C after incubation with endotoxin [3–12], and by the in-vivo fall in serum titers of C in experimental animals that received iv endotoxin [13–16]. Studies were designed for determination of the influence of in-vivo depletion of C on the clearance of bacteremia due to *Escherichia coli* by inactivation of the C3 component of C in a subhuman primate with a nontoxic protein of cobra venom, termed cobra-venom factor (CoF) [17–21].

Materials and Methods

CoF was prepared from the venom of *Naja naja* by previously described methods [18]. Male Brazilian squirrel monkeys (*Saimiri sciureus*) were

During the tenure of these studies, Dr. Gilbert was supported by training grant no. 5 T01 AI 00030 from the National Institute of Allergy and Infectious Diseases, National Institutes of Health. His present address is: Director of Medical Education, Providence Hospital, Portland, Oregon.

Please address requests for reprints to Dr. Jay P. Sanford, Department of Internal Medicine, University of Texas Southwestern Medical School at Dallas, 5323 Harry Hines Boulevard, Dallas, Texas 75235.

given a total dose of 300 units of CoF per day (four equally divided ip injections during 24 hr). Control animals received no CoF. Bacteremia was induced in test animals 24–48 hr after the last injection. Baseline plasma samples for determinations of C were obtained immediately before the induction of bacteremia. Only local anesthetics were used.

Total hemolytic titers of C were determined in duplicate for each sample by the method of Mayer [22], modified by the reduction of all reagent volumes by 60%. The concentration of C3 was estimated by the radial-immunodiffusion technique with use of agar plates containing antibodies to human C3 [23].

All animals were given 4.0×10^9 *E. coli* per kg iv. Bacteremia was quantitated by means of serial 10-fold dilutions of 0.1-ml samples of blood on MacConkey agar.

Results

Depletion of C by CoF. The mean total hemolytic titer was 226 units in control animals compared with 31 units in CoF-treated monkeys (figure 1). The mean value of C3 as determined by radial immunodiffusion was 187 units in control animals and 36 units in C-depleted animals.

Effect of E. coli bacteremia on leukocyte counts. In control monkeys a dramatic decrease in leukocyte count was demonstrated after iv ad-

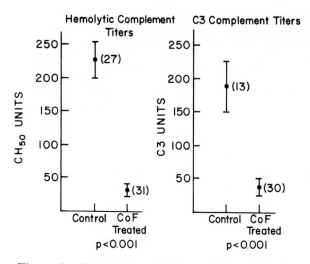

Figure 1. Hemolytic and C3 complement levels in monkeys treated with cobra-venom factor (CoF) and in controls. Hemolytic complement titers are expressed as 50% hemolytic units (CH_{50}), and C3 titers are expressed as units of human C3. Values represent means ± 2 SE. Numbers in parentheses indicate number of observations.

ministration of *E. coli* (figure 2). Maximal depression of the leukocyte count occurred within 1–2 min and occasionally was demonstrated as soon as 45 sec after injection. Both viable and heat-killed ^{32}P-labeled *E. coli* were used, and no dif-

ferences were noted between their effects on the leukocyte count. In C-depleted monkeys, leukocyte counts also decreased after iv administration of *E. coli*; however, the decrease occurred at a much slower rate (figure 2). In CoF-treated animals, the maximal decrease in leukocyte count usually occurred between 12 and 30 min after injection, but two animals required 2 hr to reach minimal values. Differential leukocyte counts demonstrated that the leukopenia was the result of neutropenia. Neutrophils reappeared in normal animals earlier than in CoF-treated animals. During the phase of rebound, many immature granulocytic forms were demonstrable in peripheral-blood smears of both groups.

Quantitation of bacteremia. Quantitation of *E. coli* in blood at 1, 2, 3, and 4 hr demonstrated consistently higher levels in C-depleted animals than in controls (figure 3). Differences between control and CoF-treated animals were statistically significant at all intervals studied beyond 30 min.

Discussion

These experiments provide additional data implicating the C system in the pathophysiologic consequences of *E. coli* bacteremia. Granulocytopenia after endotoxin has been well documented in man and other mammalian species [24–26]. Through in-vivo microscopic studies in animals having received bacteria iv, it has been shown

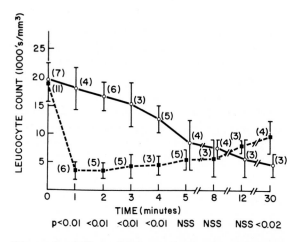

Figure 2. Effect of bacteremia due to *Escherichia coli* on leukocyte counts of monkeys treated with cobra-venom factor (O———O) and of controls (■– –■). Total leukocyte counts are plotted on the ordinate, and the interval in minutes after iv challenge with viable *E. coli* is plotted on the abscissa. Values given are mean ± 2 SE. The significance of differences was assessed by Student's *t* test. NSS = not statistically significant.

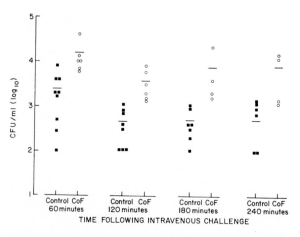

Figure 3. Level of bacteremia in monkeys treated with cobra-venom factor (CoF) (O) and in controls (■), expressed as number of colony-forming units (CFU) of *Escherichia coli* cultured from blood at intervals after iv challenge. (—) = mean.

that neutrophils rapidly adhere to capillary endothelium and engage in phagocytosis [27–29]. Similar rapid adherence of circulating neutrophils to endothelium (margination) was recently shown in volunteers given small iv doses of endotoxin [30, 31]. This phenomenon of margination may represent the C-mediated immune-adherence reaction [27, 28, 32]. Recently, granulocyte receptors for the immune-adherence reaction have been demonstrated [33].

Little is known of the neutrophil response to endotoxin in vivo when the C system is inhibited. In one study, treatment with CoF had no influence on the endotoxin-induced neutropenia of the Shwartzman reaction in rabbits, but events during the first 4 hr after injection were not reported [16]. Accumulation of neutrophils in inflammatory exudates after ip administration of endotoxin was markedly retarded in C5-deficient mice [34]. In the studies reported here, CoF-treated monkeys required 20–30 times longer to develop maximal neutropenia after bacterial challenge than did control animals. This blunted response was paralleled by the observation that the sequestration of neutrophils in the pulmonary circulation was decreased during the first few minutes after injection of E. coli into CoF-treated animals. Thus, the development of neutropenia during E. coli bacteremia may be at least partially mediated by C.

There are probably multiple factors involved that account for the observation of higher levels of bacteremia in CoF-treated monkeys. The function of the fixed phagocytic cells of the reticuloendothelial system (RES) may be important. The relation of the C system to bacterial clearance by the RES is controversial. Some investigators have found that the presence of C is necessary for clearance, while others have found that the presence of C has no influence on clearance of endotoxin or of gram-negative bacteria by the RES [35, 36]. Spielgelberg et al. demonstrated impaired clearance of E. coli in mice that were depleted of C by the injection of heat-aggregated human gamma globulin [37].

As suggested by the present studies, treatment with CoF may impair neutrophil margination and subsequent phagocytic efficiency. In-vitro studies by other investigators further support this hypothesis by demonstrating impaired phagocytosis of E. coli by neutrophils suspended in CoF-treated human serum as compared with control human serum [38]. Lastly, the C-dependent bactericidal reaction of plasma may contribute to the ultimate level of bacteremia. Studies reported elsewhere demonstrate that treatment of monkeys with CoF impaired the normal in-vitro bactericidal activity of plasma against E. coli [39]. Thus, the prolonged and higher levels of bacteremia in CoF-treated monkeys may have resulted from a combination of the impaired activities of the RES, circulating neutrophils, and bactericidal activity of plasma. While the roles of the RES and serum bactericidal activity have previously been emphasized in the clearance of bacteremia, neutrophilic leukocytes have not been considered to have an important role. The present observations suggest that C-mediated neutrophilic-leukocyte function does have a significant role in host defenses against gram-negative bacillary bacteremia.

References

1. Bladen, H. A., Evans, R. T., Mergenhagen, S. E. Lesions in *Escherichia coli* membranes after actions of antibody and complement. J. Bacteriol. 91:2377–2381, 1966.
2. Bladen, H. A., Gewurz, H., Mergenhagen, S. E. Interactions of the complement system with the surface and endotoxic lipopolysaccharide of *Veillonella alcalescens*. J. Exp. Med. 125:767–786, 1967.
3. Day, N. K., Good, R. A., Finstad, J., Johannsen, R., Pickering, R. J., Gewurz, H. Interactions between endotoxic lipopolysaccharides and the complement system in the sera of lower vertebrates. Proc. Soc. Exp. Biol. Med. 133:1397–1401, 1970.
4. Lichtenstein, L. M., Gewurz, H., Adkinson, N. F., Jr., Shin, H. S., Mergenhagen, S. E. Interactions of the complement system with endotoxic lipopolysaccharide: the generation of an anaphylatoxin. Immunology 16:327–336, 1969.
5. Gewurz, H., Shin, H. S., Mergenhagen, S. E. Interactions of the complement system with endotoxic lipopolysaccharide: consumption of each of the six terminal complement components. J. Exp. Med. 128:1049–1057, 1968.
6. Hook, W. A., Snyderman, R., Mergenhagen, S. E. Consumption of hamster complement by bacterial endotoxin. J. Immunol. 105:268–270, 1970.
7. Gewurz, H. Interaction between endotoxic lipopolysaccharides (LPS) and the complement (C) system: solubilization of C-consuming substances during brief absorption of 0°. Proc. Soc. Exp. Biol. Med. 136:561–564, 1971.
8. Marcus, R. L., Shin, H. S., Mayer, M. M. An alternate complement pathway: C3 cleaving activity not due to C4,2a, on endotoxic lipopolysaccharide after treatment with guinea pig serum; relation

to properdin. Proc. Natl. Acad. Sci. U.S.A. 68: 1351–1354, 1971.

9. Shin, H. S., Gewurz, H., Snyderman, R. Reaction of cobra venom factor with guinea pig complement and generation of an activity chemotactic for polymorphonuclear leukocytes. Proc. Soc. Exp. Biol. Med. 131:203–207, 1969.

10. Hook, W. A., Snyderman, R., Mergenhagen, S. E. Histamine-releasing factor generated by the interaction of endotoxin with hamster serum. Infec. Immun. 2:462–467, 1970.

11. Mergenhagen, S. E., Snyderman, R., Gewurz, H., Shin, H. S. Significance of complement to the mechanism of action of endotoxin. Curr. Top. Microbiol. 50:37–77, 1969.

12. Snyderman, R., Gewurz, H., Mergenhagen, S. E. Interactions of the complement system with endotoxic lipopolysaccharide: generation of a factor chemotactic for polymorphonuclear leukocytes. J. Exp. Med. 128:259–275, 1968.

13. Gilbert, V. E., Braude, A. I. Reduction of serum complement by endotoxin in vivo (abstract). Fed. Proc. 21:17, 1962.

14. Spink, W. W., Davis, R. B., Potter, R., Chartrand, S. The initial stage of canine endotoxin shock as an expression of anaphylactic shock: studies on complement titers and plasma histamine concentrations. J. Clin. Invest. 43:696–704, 1964.

15. Füst, G., Keresztes, M. Effect of endotoxin on the serum level of complement components. II. Effect of endotoxin on dog serum complement level in vivo and in vitro. Acta Microbiol. Acad. Sci. Hung. 16:135–147, 1969.

16. Fong, J. S. C., Good, R. A. Prevention of the localized and generalized Shwartzman reactions by an anti-complementary agent, cobra venom factor. J. Exp. Med. 134:642–655, 1971.

17. Nelson, R. A., Jr. A new concept of immunosuppression in hypersensitivity reactions and in transplantation immunity. Survey of Opthalmol. 11: 498–505, 1966.

18. Cochrane, C. G., Müller-Eberhard, H. J., Akin, B. S. Depletion of plasma complement in vivo by a protein of cobra venom: its effect on various immunologic reactions. J. Immunol. 105:55–69, 1970.

19. Müller-Eberhard, H. J., Fjellström, K.-E. Isolation of the anti-complementary protein from cobra venom and its mode of action on C3. J. Immunol. 107:1666–1672, 1971.

20. Götze, O., Müller-Eberhard, H. J. The C3-activator system: an alternative pathway of complement activation. J. Exp. Med. 134:905–1085, 1971.

21. Sandberg, A. L., Osler, A. G. Dual pathways of complement interaction with guinea pig immunoglobulin. J. Immunol. 107:1268–1273, 1971.

22. Mayer, M. M. Complement and complement fixation. In E. A. Kabat and M. M. Mayer [ed.] Experimental immunochemistry. 2nd ed. Charles C Thomas, Springfield, Ill., 1971, p. 133–240.

23. Shanbrom, E., Khoo, M., Lou, K. Simplified quantitation of C'3 by radial immunodiffusion (abstract). Clin. Res. 15:114, 1967.

24. Mechanic, R. C., Frei, E., III, Landy, M., Smith, W. W. Quantitative studies of human leukocytic and febrile response to single and repeated doses of purified bacterial endotoxin. J. Clin. Invest. 41: 162–172, 1962.

25. Sheagren, J. N., Wolff, S. M., Shulman, N. R. Febrile and hematologic responses of rhesus monkeys to bacterial endotoxin. Am. J. Physiol. 212:884–890, 1967.

26. Dinbar, A., Rapaport, S. I., Patch, M. J., Grant, W., Fonkalsrud, E. W. Hematologic effects of endotoxin on the macaque monkey. Surgery 70:596–603, 1971.

27. Smith, M. R., Perry, W. D., Berry, J. W., Wood, W. B., Jr. Surface phagocytosis in vivo. J. Immunol. 67:71–74, 1951.

28. Wood, W. B., Jr., Smith, M. R., Perry, W. D., Berry, J. W. Studies on the cellular immunology of acute bacteremia. J. Exp. Med. 94:521–533, 1951.

29. Mulholland, J. H., Cluff, L. E. The effect of endotoxin upon susceptibility to infection: the role of the granulocyte. In M. Landy and W. Braun [ed.] Bacterial endotoxins. Rutgers University, New Brunswick, N.J., 1964, p. 211–229.

30. Athens, J. W., Haab, O. P., Raab, S. O., Maver, A. M., Ashenbrucker, H., Cartwright, G. E., Wintrobe, M. M. Leukokinetic studies. IV. The total blood, circulating and marginal granulocyte pools and the granulocyte turnover rate in normal subjects. J. Clin. Invest. 40:989–995, 1961.

31. Athens, J. W., Raab, S. O., Haab, O. P., Maver, A. M., Ashenbrucker, H., Cartwright, G. E., Wintrobe, M. M. Leukokinetic studies. III. The distribution of granulocytes in the blood of normal subjects. J. Clin. Invest. 40:159–164, 1961.

32. Petz, L., Fudenberg, H. H., Fink, H. Immune adherence reactions of human erythrocytes sensitized with complement in vitro and in vivo. J. Immunol. 107:1714–1722, 1971.

33. VanLoghem, J. J., Von Dem Borne, A. E. G. K., Van Der Hart, M., Peetoom, F. Immune adherence and blood cell destruction. Vox Sang. 12: 361–373, 1967.

34. Snyderman, R., Phillips, J. K., Mergenhagen, S. E. Biologic activity of complement in vivo. Role of C5 in the accumulation of polymorphonuclear leukocytes in inflammatory exudates. J. Exp. Med. 134:1131–1143, 1971.

35. Howard, J. G., Wardlow, A. C. The opsonic effect of normal serum on the uptake of bacteria by the reticuloendothelial system. J. Immunol. 1:338–352, 1958.

36. Goodman, J. S., Rogers, D. E., Koenig, G. The hepatic uptake of bacterial endotoxin. I. The influence of humoral and cellular factors. Proc. Soc. Exp. Biol. Med. 132:372–375, 1969.

37. Spielgelberg, H. Z., Miescher, P. A., Benacerraf, B. Studies on the role of complement in the immune clearance of *Escherichia coli* and rat erythrocytes by the reticuloendothelial system in mice. J. Immunol. 90:751–759, 1963.

38. Johnson, F. R., Agnello, V., Williams, R. C., Jr. Opsonic activity in human serum deficient in C2. J. Immunol. 109:141–145, 1972.

39. Gilbert, D. N., Barnett, J. A., Sanford, J. P. *Escherichia coli* bacteremia in the squirrel monkey. I. Effect of cobra venom factor treatment. J. Clin. Invest. 52:406–413, 1973.

Summary of Discussion

Chandler A. Stetson, Jr.

Dr. L. Joe Berry began the question-and-answer period by commenting that he has been unable to induce tolerance against lethality with polyriboinosinic-polyribocytidylic acid (poly I:C); he asked Dr. Kelsey C. Milner about his experience with poly I:C tolerance. Dr. Milner replied that he has not examined tolerance to the lethal effect of poly I:C but has looked at tolerance to the pyrogenic effects. He reported that he could make rabbits tolerant to the pyrogenic effect of poly I:C with endotoxin, but that a larger dose was required than it takes to make them tolerant to endotoxin per se. Experiments attempting to make the animals tolerant to endotoxin with poly I:C could not be fully interpreted, because contamination of the poly I:C preparation with endotoxin could not be ruled out. Dr. Sheldon M. Wolff then referred to studies in which he was unable to induce tolerance to the pyrogenic effect of poly I:C with either poly I:C or endotoxin. Furthermore, his studies would suggest that poly I:C was not contaminated. Dr. Wolff concluded by stating that, as far as tolerance to endotoxin was concerned, there were probably a number of factors involved; perhaps the major factor was some immunologic response directed toward a specific antigen. This statement precipitated a number of additional comments tending to support the fact that tolerance to endotoxin is multifactorial. Dr. Berry noted studies in which mice made tolerant to endotoxin were also remarkably resistant to actinomycin D and ethionimide, which he claimed had to be antibody-independent and probably reflected some change in the cellular system of the mice. Dr. Louis Chedid then reminded the audience that hypercorticism, and even Drum shock, rendered animals resistant to endotoxin and, therefore, reaffirmed the point that there were other mechanisms for tolerance than those mediated by antibody. Finally, Dr. Richard A. Finkelstein revived the group's memories with regard to the natural acquisition of tolerance to endotoxin seen in the developing chick embryo between days 10 and 15. It was his feeling that this may represent a hormonal-mediated mechanism.

The discussion then shifted to the area of lipopolysaccharide (LPS)-receptors when Dr. Chedid asked Dr. J. W. Shands Jr. to what extent the presence of plasma prevents or inhibits the attachments of LPS to the erythrocyte membrane. Dr. Shands stated that, although a considerable amount is inhibited, there is still some LPS that does attach. Dr. Georg Springer elaborated on Dr. Shands' answer by pointing out that the plasma proteins responsible for this inhibition were the lipoproteins and albumin. Since these proteins may be decreased in certain disease states, it is quite possible that sera from these patients would not inhibit the attachment of LPS as well. Dr. Springer then referred to a patient with renal failure after trauma whose serum had very little effect in preventing the attachment of LPS to red cells. In reference to the specificity of the LPS-receptor, Dr. Erwin Neter reemphasized that the term specificity is applicable in only a limited sense, since these receptors could bind any gram-negative LPS. Furthermore, Dr. Springer stressed that these polysaccharides were freely exchangeable, in that, if the red-cell receptor was saturated with one LPS, it could be displaced by addition in excess of a totally different LPS. Dr. Constantin A. Bona concluded the questioning on LPS-receptors by asking for a definition of their biological significance. Dr. Springer said that this significance could not be defined at the present time, because it was not known what other types of cells may possess such receptors.

Dr. Peter Ward introduced the questioning on mechanisms of macrophage activation by asking Dr. Anthony Allison whether he had any evidence to indicate that the A cells to which he refers have any tissue-damaging activity after contact with endotoxin. Dr. Allison had no firm evidence for this at that time but commented that he intended to examine the question. Dr. Chandler A. Stetson, Jr. wanted to know what proportion of cells in the peritoneal cavity were of the A type. According to Dr. Allison, this proportion varied with the species and state of the animal. As an example, he claimed that relatively clean mice had about 4% of such cells in their peritoneal cavity, whereas chronically infected mice had about 10%–12%.

The final question dealt with induction of interferon, and Dr. Monto Ho was asked what tissues besides thymus could be induced to produce interferon with endotoxin. Dr. Ho replied that, other than thymus and peritoneal macrophages, there were no other tissues in which any consistent effect was observed.

SPECIAL ADDRESS

Role of the Federal Government in Support of Biomedical Research

Robert Q. Marston

*From the National Institutes of Health,
Bethesda, Maryland*

I have often described the biomedical research enterprise in this country as a consortium of agencies and institutions working toward a common goal. Today, I would like to examine the role of the federal partner in this enterprise, drawing primarily for illustration on the mission and work of the National Institutes of Health.

In the long run, the state of health in this country and the prospects of the individual citizen for a long and productive life depend on the initiative and imagination with which our scientists pursue the advancement of knowledge. Thus a basic strategy of the federal health effort is to emphasize those activities such as research that, in the long run, will fundamentally improve the health prospects of the nation.

The history of medical research in this country is, therefore, in large measure a history of the involvement of the federal government in seeking the solution to national health problems. Until late in the 19th century, American medicine was dependent on the advances made by such pioneers as Pasteur, Koch, Lister, and Ehrlich in the great European universities and medical schools.

Our government first became involved in medical research in the context of public health when a "laboratory of hygiene" was established at the U.S. Marine Hospital on Staten Island, New York, in 1900. The laboratory's primary concern was to deal with cholera and other infectious diseases arising from the increasing numbers of immigrants ariving in the port of New York. Out of this very specialized function, the laboratory and its successor, the National Institute of Health, gradually developed an expanding national program of research on infectious diseases during the late 19th and early 20th centuries.

It was not until the National Cancer Act of 1937 that a major effort was made to investigate the most complex of chronic diseases. This legislation created the National Cancer Institute, later incorporated into the National Institutes of Health (NIH), and helped to establish the precedent for research through grants-in-aid to nonfederal scientists.

The enactment of the Public Health Service Act of 1944 gave the Surgeon General broad authorization to foster, conduct, support, and cooperate

Dr. Marston is the former director of the National Institutes of Health.

Please address requests for reprints to Dr. Robert Q. Marston, National Institutes of Health, Bethesda, Maryland 20014.

in research relating to health and disease. Authority to award funds for nonfederal research and research training, until then limited to cancer, was extended to health problems in general.

The success of research in solving major problems, including medical and health problems, connected with the national effort during World War II did much to encourage postwar public investments in science and technology. Research activities in all sectors of the nation—academic, industrial, governmental—began to expand; research became more professionalized, and patterns of sponsorship began to shift.

The rapid growth in federal support for biomedical research is essentially a phenomenon of the past 25 years. In 1947 the total national investment in biomedical research and development was about $87 million, and today it is roughly $3.3 billion. There was a plateau in NIH funding from 1967 to 1970, but the downward drift was checked in 1971. In fact, the NIH budget for research institutes and divisions has increased by about 40% in two years to a total of almost $1.5 billion in 1972.

Today the NIH accounts for almost 60% of all federal support for biomedical research and development, and its role is larger than the figures indicate, because most of its support is to nonprofit institutions, particularly universities, which currently receive some $800 million each year.

The NIH presently supports (1) about 11,200 research grants ($676 million), (2) about 2,100 training grants ($131 million), and (3) about 4,000 fellowships and career development awards ($25 million).

The rapid and significant growth of the NIH and of biomedical research in general has been concurrent with a new concern for reordering of priorities, growing consumerism, changing economic relationships internationally and domestically, continuation of the civil rights struggle, and a sense of the urgency of problems of the ghetto, of population, of pollution, and of possible exhaustion of vital resources.

It is inevitable, then, that biomedical science, now a very sizable enterprise, would be a topic of concern for nonscientists and indeed for the political process.

Another compelling reason for its public visibility is the increasingly urgent problem of the delivery of health services to the American people.

Such societal pressures, expectations, and hopes predictably have had a significant impact on the ways in which the increased funds have been distributed. Thus, of the 40% increase I mentioned earlier (roughly $400 million), $200 million is invested in cancer research, $60 million in cardiovascular research, and $30 million in population research, earmarked to a greater degree than in the past for expenditure through centrally programmed, targeted efforts.

As significant as the increased targeting has been the great effort by professional and lay groups to gain additional visibility for the area of their particular interest. You all know about the pressures in the fields of cancer and heart disease. Campaigns have also been mounted and legislation introduced to confer NIH status on and to increase the amount of support in such areas as population, gerontology, gastroenterology, sickle-cell disease, and kidney disease.

This kind of concentration of purpose and resources runs head on against a history of evidence that major innovations in medicine have emerged largely as the result of the work of individuals and small groups pursuing ideas arising out of their own experience and reflecting their personal creativity. Yet we have already taken significant steps toward a new way of doing business: the targeted-program approach to research, reminiscent of some of the successful programs in technology and development. The advocates for the two positions of targeted research vs. individual initiative have become in some ways more highly polarized than was the case a few years ago. Fortunately, increased overall support has made it possible to accommodate the emphasis on targeted research without serious disruption of investigator-initiated research.

As director of the NIH, I supported both targeted programs and regular research projects on the grounds of what I believed to be sound philosophy. However, in recognition of the fact that undifferentiated research fares less well in nonscientific decision-making than the highly visible targeted programs, I swung my weight when possible toward the maintenance of investigator-initiated research projects.

The research programs for which additional funds are requested in the 1973 NIH budget—cancer, arteriosclerosis, lung diseases, sickle-cell anemia, and population research—are often referred to as "special-emphasis" programs. While these programs do, indeed, receive special emphasis, each of the institutes has, in addition, some half a dozen or more areas that it describes as "new initiatives," "high-priority programs," or "special-emphasis programs." These programs are highlighted because of the special health problems they represent or because of the special opportunities for progress they offer.

These concerns of the institutes are often overlooked by those who think of activities of the NIH mainly in terms of free-ranging research and who regard the regular research grant as the federal government's chosen instrument for supporting medical research in academic institutions and hospitals. Since 1948, the NIH has been highly successful in managing a rapidly expanding program in a way that has subjected the academic community to a minimum of bureaucratic interference with its conduct of research. Despite the very considerable competence and high repute of its own scientific staff and advisors, the NIH has never thought it appropriate (or, indeed, feasible) to attempt to direct the entire national medical research effort. On the contrary, the NIH has always felt that in medical research maximal freedom for the investigator is a condition for maximal creativity.

The effectiveness of the system is manifest, on the one hand, in the remarkable growth of American capability in the biomedical sciences and, on the other hand, in the tremendous advances in medical practice and in basic knowledge which that enhanced capability has brought about.

The free-enterprise philosophy of the NIH and the managerial system that has evolved from it have tended to obscure the fact that NIH is and always has been a mission-oriented agency with quite specific goals that it wishes to achieve. While its central task has been to build and sustain the national capability for medical research, its primary purpose, as an agency of the Department of Health, Education, and Welfare, is to improve the health of the American people.

In summary, the role of the federal government in support of research is to provide the tools and resources for the advancement of knowledge on behalf of human health and well-being and to broaden the base of that knowledge. It also helps to create and sustain a climate in which biomedical research can flourish.

SESSION V: CLINICAL ASPECTS

Biological Effects of Bacterial Endotoxins in Man

Sheldon M. Wolff

From the Laboratory of Clinical Investigation, National Institute of Allergy and Infectious Diseases; National Institutes of Health, Bethesda, Maryland

The studies reviewed in the present manuscript outline some of the biologic effects of endotoxins in man. The human being is the most sensitive animal to bacterial endotoxins. The sensitivity of the various parameters measured revealed that granulocytosis induced by endotoxin is the most sensitive (i.e., requires the smallest dose), fever the next most sensitive, an increase in plasma cortisol the third most sensitive, and the induction of a reproducible growth hormone response the least sensitive. Although many of our studies are descriptive, it is hoped that they will lead to better understanding of some of the pathophysiologic events that accompany gram-negative bacterial infections in man. Finally, when appropriately used, bacterial endotoxins can be useful tools in the diagnosis and management of certain hematologic and endocrinologic disorders of man.

Studies of the biological effects of bacterial endotoxins have had an unusual history. The earliest recognition of some of these effects (fever, for example) was noted after contaminated infusions [1] or fever therapy [2]. Despite these early observations, the overwhelming majority of careful studies characterizing the biological properties of bacterial endotoxins (lipopolysaccharides) have been done in animals. Some investigations in humans were done in the 1940s and 1950s, but the materials were often whole, killed bacterial vaccines. The preparation of some highly purified materials such as Lipexal (Pyrexal) in the laboratory of Dr. Otto Westphal [3, 4] and a *Salmonella typhosa* 0282 lipopolysaccharide in the laboratory of Dr. Maurice Landy [5] has resulted in purified, standardized materials for studies in patients and animals in many institutions. Unfortunately, Lipexal was removed from the American market during the mid1960s and the original *S. typhosa* endotoxin is, to our knowledge, available only in Dr. Sheldon Greisman's laboratory and in our own. The endotoxin Piromen, derived from pseudomonas organisms, is commercially available in this country, but this material is contaminated with protein, varies from lot to lot, and is not as chemically pure as the above-mentioned endotoxins.

Approximately 12 years ago, we began systematically to study some of the biological activities of bacterial endotoxins in normal human volunteers. Some of the effects of endotoxin in animals have been used as explanations of certain clinical situations, and one of our purposes was to compare the responses of man to those of experimental animals. We have tried, not always successfully, to define our studies in quantitative terms. Whenever possible, we have compared the effects of bacterial endotoxin with those we elicited with the pyrogenic hormone metabolite, etiocholanolone [6]. Although this paper will cover our own work, it should be emphasized that others have made many outstanding contributions in this area. Furthermore, I have selected four aspects of our work for review: fever, hematologic effects, humoral immunity, and endocrinologic effects. I

Throughout these studies, I have been fortunate to have a large number of able collaborators. There are far too many for specific mention, but, in particular, I should like to acknowledge the work of Drs. Harry R. Kimball and Ronald J. Elin. Mr. Stanley B. Ward has been an active participant in the studies, and his contributions are gratefully acknowledged. None of this work would have been possible without the extraordinary group of normal volunteers who so willingly participated.

Please address requests for reprints to Dr. Sheldon M. Wolff, National Institutes of Health, Building 10, Room 11N-232, Bethesda, Maryland 20014.

do this primarily because of the limitations of space, although our studies in such areas as the effect of endotoxins on drug metabolism [7, 8] and nonspecific resistance [9] may be of more interest to some readers. Finally, due to the increased sensitivity of man to endotoxin, many of the investigations remain descriptive and, in fact, raise more questions than provide answers.

Materials and Methods

Subjects and endotoxins. All subjects in the studies were healthy, normal volunteers between the ages of 18 and 40 years and of either sex. When endotoxin was given, they rested in bed for the day of the study and had rectal temperatures taken for at least 1 hr before and until they became afebrile after the injection of endotoxin. Volunteers were not studied if their baseline temperature was greater than 38 C. Blood pressure and pulse were obtained hourly for 1 hr before and for the duration of the study. Febrile responses were plotted as previously reported [10].

For many of the studies, the endotoxin derived from *Salmonella abortus equi* (Lipexal in the U.S., Pyrexal in Europe) was used, while in other studies the endotoxin obtained from *S. typhosa* 0282 was employed. Lipexal was made available by the Dorsey Company, and *S. typhosa* endotoxin was a generous gift of Dr. Maurice Landy.

The techniques and procedures used in the studies were standard laboratory methods and may be found in the references cited.

Results and Discussion

Fever. For many years, clinicians have recognized the exquisite sensitivity of man to the pyrogenic properties of bacterial endotoxins. An example of such sensitivity is the response of patients to contaminated intravenous infusions. In fact, man is the most sensitive of all animals studied to the effects of endotoxin. It takes far less endotoxin to produce fever in man than in other animals. For example, with Lipexal in humans, we routinely obtain a mean 7-hr fever index of 50 cm² (maximal rise in temperature of 1.9 C) with a dose of 0.002 µg/kg. In contrast, a dose of 0.005 µg/kg of the same material in rabbits results in a mean 7-hr fever index of 25 cm² and a mean maximal temperature of 1.3 C. Similarly, the same febrile response in rabbits and humans with the

S. typhosa endotoxin requires 10 times as much endotoxin (0.05 µg/kg) in the rabbit as in man (0.005 µg/kg). These studies are in general agreement with those of Greisman and Hornick, who reported that the ascending portion of the dose-response curve in man was much steeper than that in rabbits, although they found that the minimal pyrogenic dose on the basis of weight for both species was the same [11]. This extreme sensitivity of man obviously limits the studies that can be performed. It is fair to assume that the inability to reproduce in man many of the biological effects seen in experimental animals is related to the small dose we can administer to human beings.

The characteristics of the febrile responses of man to small (up to 5 ng/kg) doses of either Lipexal or *S. typhosa* endotoxins follow a dose-response relationship. The usual latent period is 90–120 min after injection (depending on dose), and the length of the fever (again depending on dose) ordinarily is not more than 3–4 hr. The fever curves in human beings are always monophasic; we have never seen a biphasic curve characteristic of rabbits, despite the fact that we have given thousands of injections of endotoxin to patients and volunteers. The pulse usually follows the course of the fever; i.e., tachycardia accompanies the fever. Only at higher doses does one see a change in blood pressure, ordinarily an increase in pulse pressure. At the doses we use, we have never observed a fall in systolic blood pressure.

A number of years ago, we studied the physiologic effects of small doses of Lipexal in normal volunteers [12]. In the subjects who developed fever, we observed two distinct patterns of response. In the largest group, it was noted that production of heat preceded a rise in core temperature. Surprisingly, we observed another group in whom a rise in core temperature preceded an increase in production of heat or metabolic activity. This effect in the second group was due to thermal redistribution [12]. We interpreted these studies to show that endotoxin could produce fever by alteration of either vasomotor control of heat redistribution or heat production.

The phenomenon of tolerance of endotoxin, the subject of considerable study in experimental animals, has also been investigated in man [13–15]. The human being develops pyrogenic tolerance to the same dose of endotoxin very rapidly, and differences can be noted within 24 hr. Unlike the rabbit, which maintains a monophasic fever curve when tolerant, the febrile response can disappear completely in man. Similar to that in the rabbit, the mechanism for development of tolerance in man is poorly understood. Reticuloendothelial-system function is not enhanced in man [14] as in the rabbit [16]. Development of classical humoral antibody is probably not the major means for the induction of tolerance. It appears to last a much shorter period of time [17] than is true of the rabbit [18], but this may be again a reflection of the doses we can use.

It has been reported that agammaglobulinemic humans develop pyrogenic tolerance [19]. This study has often been cited to support the thesis that circulating antibody is not responsible for the development of tolerance. In view of the importance of this observation, we repeated those experiments. Two adults, both with very low (less than 150 mg/100 ml) levels of IgG and undetectable IgM or IgA, were each given seven daily intravenous injections of S. typhosa 0282. A third patient was similarly studied who had low levels of all three immunoglobulins. All three patients developed pyrogenic tolerance without detectable levels of bentonite flocculating antibody. In addition, studies of levels of bactericidal antibody were no different after the series of injections than before. Thus, we were able to confirm that pyrogenic tolerance in man could occur in the absence of detectable humoral antibodies and, in fact, in the absence of detectable IgM or IgA immunoglobulins.

Hematologic effects. One of the earliest recognized biological effects of endotoxin in man was leukocytosis. During the last 15 years, it has been shown that leukocytosis may be the most sensitive indicator of sensitivity to endotoxin in humans, since it can occur at subpyrogenic doses [10, 20]. It has been shown that endotoxin-induced leukocytosis is secondary to release of granulocytes from the bone-marrow granulocyte reserves [21]. In fact, before this latter fact became known, Heilmeyer had suggested that endotoxins might be useful in the estimation of marrow granulocyte reserves, a fact soon verified and expanded upon [22, 23]. The estimation of bone-marrow granulocyte reserves with use of endotoxins has proved to be a useful adjunct to cytotoxic chemotherapy [24], to be of value in the study of various neutropenias [25], and to be helpful in the study of some

of the humoral aspects of granulocyte regulation [26].

Although granulocytosis is a usual and expected response to endotoxin in human beings, the neutropenia so common to other animals is not routinely seen in man [10, 20, 27]. It may or may not occur, and this probably reflects the dosage used, since smaller doses of endotoxin do not induce neutropenia in animals. It is also possible that neutropenia does occur, but that, due to mobilization of cells from the marginating to the circulating granulocyte pools, it might be obscured.

Thrombocytopenia is seen in certain experimental animals when high enough doses of endotoxin are given. We do not see it in our volunteers after a single dose of endotoxin, probably because the dose is small. However, recent studies in our laboratory[1] have detected some small but insignificant decreases in platelet counts 4 hr after a second injection of the same dose of Lipexal (5 ng/kg) given 24 hr after the first.

A diverse array of clotting-factor abnormalities (e.g., a decrease in fibrinogen followed by an elevation) have been reported in animals. Fibrinolysis has been reported to occur in humans after administration of endotoxin [4, 28], and hypofibrinogenemia has been reported after the infusion of typhoid vaccine [29]. We have recently conducted a variety of coagulation studies in normal volunteers in collaboration with Drs. Harvey Gralnick and Ronald Elin.[2] Factors V, VII, VIII, IX, and X, plasma fibrinogen, and euglobulin clot lysis were measured before and at 30, 120, and 240 min after each of two consecutive daily injections of 5 ng/kg of Lipexal. The only significant change was a mild but significant increase in factor VIII 4 hr after injection on both days. In addition, fibrinogen-survival studies were done in five volunteers before and after endotoxin, and no significant changes were observed.[3]

The effects of endotoxin on iron in sera of animals are well known and documented [30]. In fact, the fall in concentrations of serum iron is so reproducible that it has been suggested as a bioassay for endotoxin [31]. To our knowledge,

[1] H. Gralnick, R. J. Elin, and S. M. Wolff, manuscript in preparation.
[2] See footnote 1.
[3] See footnote 1.

studies of serum iron after administration of endotoxin to human beings have not been reported. We recently completed studies that clearly showed that the infusion of 5 ng/kg of Lipexal resulted in a highly significant decrease in concentration of serum iron [32].

Humoral immunity. There have been a large number of studies on the effects of bacterial endotoxins on such immunologically important organs as the thymus, lymph nodes, macrophages, etc., of animals. In fact, many immunologists have once again become interested in endotoxins because of their mitogenic effects on B lymphocytes (see papers of Möller and Peavy, session 2). However, such studies in human systems are few, and, in fact, in-vivo studies on the effect of endotoxin on cell-mediated immunity in man are almost non-existent. We have done a few such studies, but, since they have been generally negative and preliminary, the following discussion will be restricted to studies of humoral immunity.

A large literature exists concerning the effects of endotoxin on immunologic responses in animals. In contrast, the literature concerning human responses to purified materials is not very large. Most of the studies in man have employed typhoid vaccines not administered intravenously, and there is disagreement in the literature regarding the types of response elicited. Since we could not measure circulating antibodies to *S. abortus equi* endotoxin after its administration, all of our studies have employed the *S. typhosa* 0282 endotoxin given intravenously.

The antibody responses of a group of normal volunteers to single injections of *S. typhosa* endotoxin were initially studied. Circulating antibody, as measured by bentonite flocculating antibody [33], became detectable about 72 hr after injection and reached its peak at about six to seven days. The response was dependent on the dose of endotoxin administered. A similar pattern was seen in levels of bactericidal antibody measured in these same sera.

Many investigators have reported that the circulating O-antibody response to somatic antigens in man is predominantly of the 19S macroglobulin type. On the other hand, other investigators have demonstrated antibodies of the IgA and IgG immunoglobulin classes in addition to those of IgM specificity in response to killed typhoid bacilli [34]. In a series of experiments previously reported, we were able to demonstrate that, after 14 daily injections of *S. typhosa* endotoxin, normal volunteers had very high titers of circulating antibodies, which peaked about the eighth day of the study and remained up for the one month of the study [35]. These sera when fractionated revealed that the antibody response was heterogeneous [35]. Antibodies were found in three immunoglobulin classes (IgM, IgA, and IgG) in serum and in IgA and IgG classes in nasal-wash material. In fact, in separate experiments [36], we were able to detect another low-molecular-weight protein, with a sedimentation constant of approximately 2S and a molecular weight of approximately 20,000, which contained antibody activity. The significance of this low-molecular-weight antibody is unknown.

Endocrinologic effects. The reactivity of the hypothalamic-pituitary-adrenal axis to endotoxins is a well recognized and described phenomenon [37]. It has been suggested that many of the biological effects of endotoxin may be secondary to the release of adrenocortical hormones. For example, it has been stated that the granulocytosis after administration of endotoxin was due to release of these hormones. Some of our studies have focused on whether such a relationship exists.

The use of bacterial endotoxins as a diagnostic test in evaluation of pituitary-adrenal function has had wide application [37, 38]. In these tests, endotoxin is administered, and plasma cortisol or growth hormone is measured. It is quite clear that the cortisol response follows a dose-response relationship and is a rapid and convenient way of testing pituitary-adrenal function. Some authors have stated that there is a clear correlation between fever and adrenal responses [39], but our studies in volunteers repeatedly have shown that such a relationship does not exist. The following facts support the latter statement. (*1*) Volunteers given aspirin to prevent fever after administration of endotoxin have the expected rise in plasma cortisol [40]. (*2*) Volunteers given etiocholanolone develop far more fever than those given endotoxin but do not have any increase in plasma cortisol [41]. (*3*) Statistical analysis failed to reveal any correlation between fever and plasma cortisol [41]. Such data clearly separate the adrenal response from the fever and constitutional symptoms that follow infusion of endotoxin.

Recently, the effect of endotoxin on release of growth hormone from the pituitary gland has become a useful test in the diagnostic workup of patients with suspected hypothalamic or pituitary disorders [42]. It is clear that the patterns of increase in levels of growth hormone after administration of endotoxin are different from those seen for plasma cortisol. The peak concentration of plasma cortisol is achieved by 4 hr, while growth-hormone concentration peaks at 2 hr [41]. In addition, suppression of fever is accompanied by erratic changes in growth-hormone concentration. Therefore, for measurement of growth-hormone responses, the unpleasant constitutional symptoms such as fever must be allowed to occur. Also of interest is the fact that smaller doses of Lipexal are required for production of increased concentrations of plasma cortisol (in the order of 2 ng/kg), while to obtain reproducible results with growth-hormone concentrations, one must use at least 5 ng/kg [40].

References

1. Siebert, F. B. Fever-producing substances found in some distilled waters. Am. J. Physiol. 67:90–104, 1923.
2. Bennett, I. L., Jr., Cluff, L. E. Bacterial pyrogens. Pharmacol. Rev. 9:427–475, 1957.
3. Westphal, O., Lüderitz, O. Uber die chemische und biologische Analyse Hochgereinigter Bakterienpolysaccharide. Allergie (Deutsch. Med. Wschr.) 2:17, 1953.
4. Eichenberger, E., Schmidhauser-Kopp, M., Hurni, H., Fricsay, M., Westphal, O. Biologische Wirkungen eines Hochgereinigten Pyrogens (Lipopolysaccharids) aus *Salmonella abortus equi*. Schweiz. Med. Wschr. 85:1190–1196, 1955.
5. Webster, M. E., Sagin, J. F., Landy, M., Johnson, A. G. Studies on the O antigen of *Salmonella typhosa* I. Purification of the antigen. J. Immunol. 74:455–465, 1955.
6. Wolff, S. M., Kimball, H. R., Perry, S., Root, R., Kappas, A. The biological properties of etiocholanolone. Ann. Intern. Med. 67:1268–1295, 1967.
7. Song, C. S., Gelb, N. A., Wolff, S. M. The influence of pyrogen-induced fever on salicylamide metabolism in man. J. Clin. Invest. 51:2959–2966, 1972.
8. Blaschke, T. F., Elin, R. J., Berk, P. D., Song, C., Wolff, S. M. The effect of induced fever on BSP kinetics in man. Ann. Intern. Med. 78:221–226, 1973.
9. Elin, R. J., Wolff, S. M. Effect of endotoxin administration on the growth of *Candida albicans* in sera from various species. Canad. J. Microbiol. 1973 (in press).
10. Wolff, S. M., Rubenstein, M., Mulholland, J. H., Alling, D. W. Comparison of hematologic and febrile response to endotoxin in man. Blood 26:190–201, 1965.
11. Greisman, S. E., Hornick, R. B. Comparative pyrogenic reactivity of rabbit and man to bacterial endotoxin. Proc. Soc. Exp. Biol. Med. 131:1154–1158, 1969.
12. Buskirk, E. R., Thompson, R. H., Rubenstein, M., Wolff, S. M. Heat exchange in men and women following intravenous injection of endotoxin. J. Appl. Physiol. 19:907–913, 1964.
13. Neva, F. A., Morgan, H. R. Tolerance to the action of endotoxins of enteric bacilli in patients convalescent from typhoid and paratyphoid fevers. J. Lab. Clin. Med. 35:911–922, 1950.
14. Greisman, S. E., Wagner, H. N., Jr., Iio, M., Hornick, R. B. Mechanisms of endotoxin tolerance. II. Relationship between endotoxin tolerance and reticuloendothelial system phagocytic activity in man. J. Exp. Med. 119:241–264, 1964.
15. Greisman, S. E., Woodward, W. E. Mechanisms of endotoxin tolerance. III. The refractory state during continuous intravenous infusions of endotoxin. J. Exp. Med. 121:911–933, 1965.
16. Beeson, P. B. Tolerance to bacterial pyrogens. II. Role of the reticulo-endothelial system. J. Exp. Med. 86:39–44, 1947.
17. Rubenstein, M., Mulholland, J. H., Jeffery, G. M., Wolff, S. M. Malaria induced endotoxin tolerance. Proc. Soc. Exp. Biol. Med. 118:283–287, 1965.
18. Mulholland, J. H., Wolff, S. M., Jackson, A. L., Landy, M. Quantitative studies of febrile tolerance and levels of specific antibody evoked by bacterial endotoxin. J. Clin. Invest. 44:920–928, 1965.
19. Good, R. A., Varco, R. L. A clinical and experimental study of agammaglobulinemia. The Journal—Lancet 75:245–272, 1955.
20. Mechanic, R. C., Frei, E., III, Landy, M., Smith, W. W. Quantitative studies of human leukocytic and febrile response to single and repeated doses of purified bacterial endotoxin. J. Clin. Invest. 41:162–172, 1962.
21. Athens, J. W., Haab, O. P., Raab, S. O., Mauer, A. M., Ashenbrucker, H., Cartwright, G. E., Wintrobe, M. M. Leukokinetic studies. IV. The total blood, circulating, and marginal granulocyte pools and the granulocyte turnover rate in normal subjects. J. Clin. Invest. 40:989–995, 1961.
22. Heilmeyer, I. A test of leukopoietic function of the bone marrow. Deutsch. Med. Wschr. 2:683, 1957.
23. Craddock, C. G., Jr., Perry, S., Ventzke, L. E., Lawrence, J. S. Evaluation of marrow granulocyte reserves in normal and disease states. Blood 15:840–855, 1960.
24. Fink, M. E., Calabresi, P. The granulocyte response to an endotoxin (Pyrexal) as a measure of functional marrow reserve in cancer chemotherapy. Ann. Intern. Med. 57:732–742, 1962.
25. Dale, D. C., Wolff, S. M. Studies of the neutropenia of acute malaria. Blood 41:197–206, 1973.

26. Boggs, D. R., Marsh, J. C., Chervenick, P. A., Cartwright, G. E., Wintrobe, M. M. Neutrophil releasing activity in plasma of normal human subjects injected with endotoxin. Proc. Soc. Exp. Biol. Med. 127:689–693, 1968.

27. Kimball, H. R., Melmon, K. L., Wolff, S. M. Endotoxin-induced kinin production in man. Proc. Soc. Exp. Biol. Med. 139:1078–1082, 1972.

28. Von Kaulla, K. N. Intravenous protein-free pyrogen. A powerful fibrinolytic agent in man. Circulation 17:187–198, 1958.

29. Shulman, N. R. Studies on the inhibition of proteolytic enzymes by serum. III. Physiological aspects of variations in proteolytic inhibition. The concurrence of changes in fibrinogen concentration with changes in trypsin inhibition. J. Exp. Med. 95:605–618, 1952.

30. Kampschmidt, R. F., Schultz, G. A. Hypoferremia in rats following injection of bacterial endotoxin. Proc. Soc. Exp. Biol. Med. 106:870–871, 1961.

31. Baker, P. J., Wilson, J. B. Hypoferremia in mice and its application to the bioassay of endotoxin. J. Bacteriol. 90:903–910, 1965.

32. Elin, R. J., Wolff, S. M. Effect of fever on serum iron in man. Clin. Res. 21:598, 1973.

33. Wolff, S. M., Ward, S. B., Landy, M. The serologic properties of bentonite particles coated with microbial polysaccharides. Proc. Soc. Exp. Biol. Med. 114:530–536, 1963.

34. Turner, M. W., Rowe, D. S. Characterization of human antibodies to Salmonella typhi by gel filtration and antigenic analysis. Immunology 7:639–656, 1964.

35. Rossen, R. D., Wolff, S. M., Butler, W. T. The antibody response in nasal washings and serum to S. typhosa endotoxin administered intravenously. J. Immunol. 99:246–254, 1967.

36. Rossen, R. D., Wolff, S. M., Butler, W. T., Vannier, W. E. The identification of low molecular weight flocculating antibodies in the serum of the rabbit, monkey and man. J. Immunol. 98:764–777, 1967.

37. Carroll, B. J., Pearson, M. J., Martin, F. I. R. Evaluation of three acute tests of hypothalamic-pituitary-adrenal function. Metabolism 18:476–483, 1969.

38. Farmer, T. A., Jr., Hill, S. R., Jr., Pittman, J. A., Jr., Herod, J. W., Jr. The plasma 17-hydroxycorticosteroid response to corticotrophin, SU-4885 and lipopolysaccharide pyrogen. J. Clin. Endocrinol. Metab. 21:433–455, 1961.

39. Brinck-Johnsen, T., Solem, J. H., Brinck-Johnsen, K., Ingvaldsen, P. The 17-hydroxycorticosteroid response to corticotrophin, metopiron and bacterial pyrogen. Acta Med. Scand. 173:129–140, 1963.

40. Wright, L. J., Lipsett, M. B., Ross, G. T., Wolff, S. M. Effects of dexamethasone and aspirin on the responses to endotoxin in man. J. Clin. Endocrinol. Metab. 34:13–17, 1972.

41. Kimball, H. R., Lipsett, M. B., Odell, W. D., Wolff, S. M. Comparison of the effect of the pyrogens, etiocholanolone and bacterial endotoxin on plasma cortisol and growth hormone in man. J. Clin. Endocrinol. Metab. 28:337–342, 1968.

42. Kohler, P. O., O'Malley, B. W., Rayford, P. L., Lipsett, M. B., Odell, W. D. Effect of pyrogen on blood levels of pituitary trophic hormones. Observations of the usefulness of the growth hormone response in the detection of pituitary disease. J. Clin. Endocrinol. Metab. 27:219–226, 1967.

Mechanisms of Endotoxin Tolerance with Special Reference to Man

Sheldon E. Greisman and Richard B. Hornick

From the Departments of Medicine and Physiology, University of Maryland School of Medicine, Baltimore, Maryland

In this review of the mechanisms underlying pyrogenic tolerance to bacterial endotoxins, two earlier concepts require modification. (*1*) Granulocytes are the major source of endogenous pyrogen release by endotoxin. (*2*) Pyrogenic tolerance results simply from accelerated clearance of endotoxin by the reticuloendothelial system (RES), diverting the toxin from the granulocytes. Rather, it now appears that hepatic Kupffer cells are the major source of endogenous pyrogen release by endotoxin, and that pyrogenic tolerance results primarily from refractoriness of these cells to further release of endogenous pyrogen. Accelerated clearance by the RES appears to be an ancillary protective mechanism that brings toxin more rapidly to the refractory cells. Refractoriness to the release of endogenous pyrogen can be achieved both by direct cellular interaction with endotoxin and by antibodies to endotoxin. The direct cellular effect is transient and requires closely spaced injections of endotoxin for maximal maintenance; antibody-mediated protection is delayed but more enduring. The diverse, often conflicting observations on pyrogenic tolerance are explicable by the interplay of these protective mechanisms.

Assessment of the role of endotoxin in the pathogenesis of gram-negative bacterial infection of man requires an understanding of the mechanisms used by man to develop resistance to this toxin. In contrast to studies with animal models, however, few critical studies exist on human mechanisms of acquisition of tolerance to endotoxin. Moreover, such studies in man are by necessity restricted to parameters (e.g., pyrogenicity) that pose minimal risk to the subjects. The following report presents a review, drawn from investigations conducted in this and other laboratories, of the tolerant responses of man to bacterial endotoxin. Our experience is based upon studies in several hundred volunteers given repetitive intravenous (iv) or intradermal (id) inoculations of bacterial endotoxins from various sources. These studies were conducted for the prime purpose of developing improved prophylactic and therapeutic approaches to gram-negative bacterial infections. All volunteers were fully informed of the nature of these studies and were free to withdraw at any time. Experimental protocols were reviewed and approved by the University of Maryland Human Research Committee. Where pertinent, results of animal studies are included to supplement the human data.

General Considerations

Quantitation of reactivity. Man is one of the species most reactive to gram-negative bacterial endotoxin. When highly purified preparations, including those derived from rough mutant strains, are administered id to healthy adults, gross inflammatory responses are readily elicited with 10^{-3}–10^{-4} µg/0.1 ml. These responses become visibly greater than those at saline control sites after approximately 1 hr and are well developed by 3 hr. Histologically, the 3-hr lesions are characterized

These studies were supported by contract no. DA-49-193-MD-2867 from the United States Army Medical Research and Development Command and by research grant no. AI-07052 from the U.S. Public Health Service.

We express our deepest appreciation to the volunteers who participated in these studies and to the officials of the Maryland House of Correction, Jessup, Maryland, for their interest and support, without which these studies could not have been done.

Please address requests for reprints to Dr. Sheldon E. Greisman, Department of Medicine, University of Maryland School of Medicine, Baltimore, Maryland 21201.

by predominance of mononuclear cells. As the concentration of endotoxin is increased logarithmically, dermal inflammation increases in intensity, and the 3-hr infiltrate becomes progressively more polymorphonuclear leukocytic in nature. Polymorphonuclear predominance generally appears with concentrations of endotoxin approximately 100-fold above the minimal inflammation-inducing dose. When endotoxin is administered daily into the same site for one week, tolerance does not develop to its local inflammatory activity. Rather, the inflammatory lesions persist and become intensely infiltrated with lymphocytes and macrophages. The implications of the fact that tolerance to the local inflammatory activity of endotoxin fails to develop, at least in some tissue sites, will be considered later.

When endotoxin is administered to him iv, healthy man responds in a characteristic fashion. On a per-kg basis, rabbit and man are almost equally reactive to threshold pyrogenic quantities of endotoxin. In both species, unequivocal elevations of temperature can be evoked with less than 0.001 µg/kg of purified preparations. However, as the quantity of endotoxin is increased, the dose-response relationship becomes considerably steeper for man, and the intensity of the subjective human toxic responses increases in parallel with the pyrogenic response [1]. When endotoxin is administered to man in single, daily, iv injections, tolerance to both febrile and subjective toxic reactions of endotoxin becomes evident within 48 hr. It is of interest that, with certain endotoxins, the febrile and subjective toxic responses are actually increased 24 hr after the initial iv administration [2]. Such enhanced febrile responses at 24 hr are never seen in the rabbit. Thereafter, however, progressive tolerance ensues and is virtually maximal within the week. Tolerance in man, as in the rabbit, is relative, not absolute, and fever and subjective toxic responses can be elicited again by an increased dose of endotoxin [3].

Qualitative aspects of reactivity. When healthy man is given an iv injection of purified endotoxin, a latent period of 45–90 min generally elapses before any clinical reaction is seen. The shortest latent period encountered in this laboratory was 35 min and was observed in a patient during the hyperreactive phase of typhoid illness [3]. The febrile response characteristically peaks in about 3 hr. Unless the dose of endotoxin is very large, the temperature then begins to decline rapidly. During the phase of ascending temperature, accompanying subjective symptoms are chills, rigor, headache, myalgia, anorexia, nausea, and (at times) vomiting. For unknown reasons, despite equipyrogenic doses, nausea and emesis are consistently more pronounced with some endotoxins (e.g., *Escherichia coli* 0127B8) than with others (e.g., *Pseudomonas* sp., Piromen®).

The pyrogenic response of man differs from that of the rabbit in one important respect. The rabbit exhibits a rapidly appearing first-fever peak with a latent period of approximately 15 min and a summit at about 90 min. This early response is absent in man. With larger doses of endotoxin, a second phase of fever develops in the rabbit, whose peak and onset (as seen when the first phase is blocked by perfusion of the endotoxin directly through the liver [5]) closely parallel the time course of the human response. Morover, as tolerance is induced by constant, single, daily iv doses of endotoxin, the second febrile peak in the rabbit again parallels the human response in that it declines progressively, whereas the first peak is not reduced appreciably [4, 5].

Evidence for circulating endogenous pyrogen. If we assumed that man behaves as does the rabbit, tolerance to the pyrogenic activity of endotoxin would be based upon reduced generation of endogenous pyrogen in response to the toxin; the thermoregulatory centers (presumably the anterior hypothalamus) would remain fully responsive to this protein [6]. A major criterion for the activity of endogenous pyrogen in man is the onset of an unequivocal rise in temperature that commences within 40 min of administration of the pyrogen [7]. In addition, in contrast to endotoxin, tolerance to repetitive doses of endogenous pyrogen, or to its continuous iv infusion, does not develop [6]. Considerable effort has been devoted to proving the existence of a circulating intermediate responsible for the pyrogenic effects of endotoxin in man. Circulating endogenous pyrogen is extremely difficult to detect in humans, although this is not the case in experimental animals. Currently, there exist only two instances in which circulating endogenous pyrogen has been presumptively demonstrated in volunteers after administration of endotoxin [8, 9]. Large quantities of plasma must be transferred, and the resulting febrile response is small [9].

Cells responsible for generation of endogenous pyrogen. In 1948, Beeson extracted a pyrogenic substance from polymorphonuclear leukocytes [10]. For years thereafter, the polymorphonuclear leukocyte was regarded by most investigators as the major source of endogenous pyrogen release by endotoxins [6]. Indeed, studies by Herion et al. with nitrogen mustard in rabbits led to the conclusion that the granulocyte was the sole source of endogenous pyrogen after administration of endotoxin [11]. The general poisonous effects of mustard, however, make such studies difficult to interpret.

The important studies of Page and Good with respect to man, although carried out before many of the studies with rabbits, were largely overlooked and deserve great emphasis. These workers observed that some patients with cyclic neutropenia and others with agranulocytosis in the absence of circulating neutrophils developed a febrile response to endotoxin-containing preparations (typhoid vaccine) "at least as high as that produced when normal numbers of neutrophils were present in the blood" [12]. The authors stated: "It must be concluded from this study that injury to neutrophils with liberation of endogenous pyrogen plays no important role in the development of fever following the intravenous injection of pyrogens in man." Moreover, when tolerance was tested, it was concluded that ". . . refractoriness to endotoxin develops in the complete absence of neutrophils in the circulating blood or in the blood forming tissues and consequently demonstrates that these cells are not essential to this defense reaction" [12].

Later studies by Bodel and Atkins demonstrated that human monocytes could elaborate endogenous pyrogen when stimulated by pyrogens [13]. Moreover, a series of classic studies carried out by Dinarello et al. indicated that, in the rabbit, isolated hepatic Kupffer cells could be stimulated in vitro by endotoxin to generate endogenous pyrogen. Furthermore, the isolated Kupffer cells (but not blood leukocytes) from rabbits tolerant to endotoxin were found to be refractory to the release of endogenous pyrogen by this toxin [14]. Subsequent in-vivo studies in our laboratory, using direct hepatic perfusion with

endotoxin via chronic, indwelling, portal-vein cannulae, indicated that the second phase of the biphasic febrile response in the rabbit (the one paralleling the human response) could be related primarily to hepatic release of endogenous pyrogen, and tolerance to refractoriness of such release [5]. If these studies can be extrapolated to man, the findings of Page and Good in their agranulocytic patients become readily explicable.

The above discussion is not meant to imply that endotoxin stimulates the release of endogenous pyrogen only from hepatic Kupffer cells. Certainly, endogenous pyrogen is also released in vitro from granulocytes, lung and peritoneal macrophages, blood monocytes, and spleen cells [13–16]; endotoxin also appears to be capable of acting directly upon the thermoregulatory centers to evoke fever without the intermediate of endogenous pyrogen [17, 18]. However, since the hepatic Kupffer cells comprise the major segment of the reticuloendothelial system, and since this segment sequesters the major portion of iv-injected endotoxin [19], it is this segment that would be expected to play the major role in the generation of endogenous pyrogen. In this respect, it is of interest that the first phase of the biphasic febrile response in the rabbit, in contrast to the second phase, can be reduced or eliminated by direct perfusion of the liver with endotoxin; presumably, therefore, the first febrile phase in the rabbit (for which a human counterpart does not appear to exist) is not mediated by hepatic release of endogenous pyrogen [5] and may reflect release of neutrophil endogenous pyrogen or a direct cerebral effect, as postulated by Bennett et al. [17].

Mechanisms of Tolerance to Endotoxin

Based on the above considerations, and in complete agreement with the concept of Dinarello et al. [14], our current working hypothesis for the acquisition of pyrogenic tolerance to iv administration of endotoxin entails the development of refractoriness of hepatic Kupffer cells to the generation and release of endogenous pyrogen. The mechanisms whereby such refractoriness operates will now be considered.

Early refractory state to continuous iv infusion of endotoxin. In both rabbit and man, continuous iv infusions of endotoxin lead to pyrogenic refractoriness within hours. Volunteers express disbelief that the infusion of endotoxin is continuing during this refractory state, since the subjective toxic responses also subside. This early refractory state is relative and can be overcome by increasing the rate of infusion. It is not associated with increments in titers of antibody to endotoxin and is not transferable with plasma. Furthermore, it develops readily in splenectomized rabbits that exhibit depressed synthesis of antibody to endotoxin. This state is not associated with enhanced ability of the plasma to inactivate endotoxin. It is specific for endotoxins as a class (at least at low rates of infusion) but exhibits no interendotoxin specificity. It is associated with inability to generate endogenous pyrogen, even while endotoxin is demonstrable in the bloodstream.

Full responsiveness persists to preformed endogenous pyrogen. It appears, therefore, that a cellular state refractory to the generation of endogenous pyrogen in response to endotoxemia can develop rapidly in both rabbit and man [20, 21]. Since this refractory state is not reversible by infusion of large quantities of fresh whole blood and occurs despite an increased number of circulating granulocytes, the findings remain consistent with the concept of the hepatic Kupffer cell as the primary target cell responsible for endotoxin fever. Moreover, the data suggest that it is a direct interaction of these target cells with endotoxin, rather than depletion or production of humoral factors, that rapidly renders them refractory to the release of endogenous pyrogen. The cellular events leading to the refractory state, however, are unknown. There is no enhancement in the potency of liver homogenates to detoxify endotoxin [21]. The hypothesis proposed by Janoff et al. [22] that depletion of the more susceptible intracellular lysosomes may be responsible for resistance to endotoxin appears pertinent in this regard and may ultimately prove to be relatable to this form of early tolerance.

Specificity of tolerance. Initial classic studies in man by Morgan demonstrated that tolerance produced by repetitive, daily iv injections of one endotoxin evoked cross-tolerance to endotoxins from heterologous bacterial species, that tolerance waned within four to five weeks after discontinuance of the injections, and that O antibody titers did not correlate with tolerance [23]. It should be emphasized that, while tolerance to heterologous endotoxins was unequivocally demonstrated, the relative effectiveness of such "heterologous" tolerance as compared with "homologous" tolerance was not quantitated. Careful quantitative studies of this nature have yet to be carried out in man. When such studies have been performed in animals, tolerance has been found to be significantly greater to the homologous endotoxin preparation [24].

In any case, it became clear by 1948 that man behaved similarly to the experimental animal in that tolerance to one endotoxin conferred definite tolerance to all others tested. More recent studies demonstrated that this problem of specificity of tolerance is complex. By study of the induction of pyrogenic tolerance after a single iv injection of endotoxin at various intervals, it could be shown in the rabbit that two temporally distinct mechanisms were involved. An early phase of pyrogenic tolerance, i.e., that appearing within hours, was probably identical with that elicited by the continuous iv infusion of endotoxin. This tolerance was specific for endotoxins as a class (provided massive doses of toxin were not used) but exhibited no interendotoxin specificity. This early phase was not associated with increments in antibody to endotoxin and could be readily induced by endotoxins lacking O-specific polysaccharide side chains (i.e., endotoxins from rough mutants). The degree of early tolerance was directly proportional to the intensity of the initial response. This early tolerance after one injection of endotoxin waned rapidly and became minimal by 48 hr. However, tolerance reappeared by 72 hr and increased over the next several days. In contrast to early tolerance, this later phase was unrelated to the initial intensity of the febrile response but was related rather to the antigenicity of the immunizing endotoxin. Moreover, it was largely, though not completely, specific for the homologous endotoxin used for the initial injection when endotoxins from smooth bacteria were employed for immunization.

In the only studies in man currently available, this late phase of tolerance after a single iv injection of endotoxin was found, as in the rabbit, to be independent of the initial pyrogenic response and to be related to the O antigenicity of the endotoxin preparation [24]. These findings led to the conclusion that tolerance to endotoxin involved at least two distinct mechanisms: an early, transient, cellular, refractory state that extended to all endotoxins, probably mediated by a direct cellular effect of the toxin; and a later, antibody-mediated phase that was, at least in part, directed against O-related antigens. The concept was developed that the importance of antibody in the mediation of pyrogenic tolerance was dependent on the interplay of these mechanisms [21, 24]. Closely spaced injections or continuous infusions would provide optimal conditions for the direct cellular refractory mechanism, and pyrogenic unresponsiveness could thus be induced rapidly and maintained without the requirement for antibody. However, as the interval between challenges was lengthened, the direct cellular effect waned, and tolerance became increasingly dependent upon antibodies to endotoxin. The former tolerant state would exhibit no O specificity; the latter, a significant level. After repetitive daily injections, both states would coexist, and either O specificity or its absence could be emphasized, depending upon the prejudice of the investigator.

Passive transfer of tolerance to endotoxin. Although earlier investigators reported failure in transferral of tolerance to endotoxin with serum, there is no longer any question that this can be accomplished in the experimental animal (see review in [24]) and in man [2]. The importance of using serum and serum fractions collected with pyrogenfree precautions cannot be overemphasized, since contamination with endotoxin will itself evoke the characteristic early tolerant state, which will be proportional to the degree of contamination and which will possess no O specificity. Recent studies have been done in our laboratory on quantitation of the specificity of transfer of pyrogenic tolerance with serum and with pyrogenfree serum fractions (prepared by ammonium sulfate or by Sephadex G-200 fractionation) from rabbits immunized daily for one week with smooth gram-negative bacteria or endotoxins derived therefrom. These studies demonstrated that the humoral tolerance factors appear to be directed primarily, though not completely, towards O-specific antigens of the toxin. This O-specific protection was divided between the antibody fractions sensitive to 2-mercaptoethanol and those resistant to it; this finding suggests that both IgM and IgG antibodies are important.

Earlier studies on the specificity of tolerance transferable with spleen cells from rabbits immunized with smooth endotoxins support this concept by demonstrating that an "anamnestic" tolerant response could be transferred, but only with viable cells, and that this was largely (though not entirely) O-specific [25]. Protective antibodies in the rabbit have also been demonstrated in other laboratories with use of passive protection against lethality and against the Shwartzman reaction. Here, again, when pyrogen-free sera from animals immunized with smooth gram-negative bacteria or their endotoxins were used for transfer, protection was always significantly greater to the homologous toxin; indeed, at times protection extended only to the homologous toxin [26–29].

Of special importance is the fact that, when certain rough-mutant, gram-negative bacteria are used for immunization, the resulting antisera, carefully shown to be pyrogenfree, give high levels of cross-protection to heterologous endotoxins. The latter findings by Braude and Douglas were interpreted to indicate that, in the absence of the O-antigenic, polysaccharide side chains of endotoxin, common-core antigenic configurations are more readily exposed; these configurations allow the induction of high titers of protective antibody to these antigens [29]. These findings have recently been fully confirmed in our laboratory and are of potential therapeutic significance, since a method now appears available for development of an effective antiserum to endotoxin with broad specificity.

Minimal dose of endotoxin required to evoke tolerance. Studies in this laboratory have quantitated the relationship of tolerance of man toward endotoxin to the immunizing dose. Tolerance on day 7 was found to be similar in healthy volunteers, regardless of whether they were injected iv with 0.01 µg/kg of *E. coli* endotoxin on days 0 and 5 or with 0.001 µg/kg. Since the former immunizing doses evoked marked febrile and subjective toxic responses, whereas the latter elicited no clinical reactions, it is clear that the later phase of tolerance in man, as in the rabbit, is not dependent upon the initial reactivity to endotoxin. Rather, the factor common to both doses of endotoxin was their ability to evoke elevated titers of O antibody by the time of the test for tolerance. Only when the immunizing dose of endotoxin was reduced to 0.0001 μg/kg and antibody titers were minimal did tolerance become minimal.

It may be concluded that antigenicity, not toxicity, constitutes the major factor in evoking the late phase (i.e., day 7) of endotoxin tolerance in man, when the refractory state that follows each endotoxin dose is permitted to wane by appropriate spacing of injections; the pattern is similar in the rabbit [24].

Tolerance and reticuloendothelial system (RES) phagocytic activity. In 1947, Beeson reported that pyrogenic tolerance in rabbits was associated with accelerated clearance of endotoxin from the blood, and that the RES "blockade" retarded clearance and abolished tolerance. Since tolerance did not appear to be transferable with serum and extended to heterologous endotoxins, it was postulated that resistance was based on accelerated RES clearance of toxin from the blood and its consequent diversion from other target organs [30]. Subsequent studies in our laboratory [31], confirmed by Wolff et al. [32], demonstrated that tolerance to endotoxin was not actually abolished by the RES blockade. Rather, the reactivity of both normal and tolerant animals was increased markedly, while striking differences in reactivity between the normal and the tolerant states persisted.

The mechanism whereby the RES blockade enhances pyrogenic reactivity of both normal and tolerant animals, while leaving the tolerance mechanisms intact, is not entirely clear. However, data now available permit a reasonable working hypothesis. Since, as discussed previously, pyrogenic tolerance appears to be based primarily upon refractoriness of hepatic Kupffer cells to release of endogenous pyrogen by endotoxin, this refractoriness presumably persists after the RES blockade. Instead, the diversion of endotoxin to less refractory tissues after after blockade (as evidenced from studies on comparative rates of clearance of Cr51-labeled endotoxin in normal, tolerant, blockaded normal, and blockaded tolerant rabbits [33] and on reduced total endotoxin uptake by the liver of blockaded animals [34, 35]), could account for the observed responses.

This hypothesis is identical to that proposed by Dinarello et al. [14] and is extended to include the normal RES-blockaded animal (figure 1). According to this concept, the tolerant blockaded animal remains tolerant when compared with the normal blockaded control, because the former still (1) possesses refractory hepatic reticuloendothelial cells and (2) clears the endotoxin into these refractory cells at a faster rate (compare panels B and D, figure 1). That antibodies to endotoxin continue to contribute to these tolerant mechanisms after the RES blockade is suggested by the findings that sera from blockaded tolerant rabbits retain the ability to transfer protection to normal recipients [36] and that sera from tolerant rabbits transfer protection to blockaded normal rabbits [31]. The antiserum appears capable of contributing to both tolerant mechanisms (see below). The hypothesis outlined in figure 1 is also compatible with earlier observations that the RES blockade fails to enhance the pyrogenic reactivity of either the normal or the tolerant animal when endotoxin is administered as a slow iv infusion; this method of administration would not tax the phagocytic capacity of a blockaded RES as would sudden injection of the entire bolus of toxin [20].

The hypothesis presented in figure 1 states that the enhanced fever evoked in normal animals by the RES blockade results from diversion of endotoxin from hepatic to extrahepatic tissues (panel A versus panel B). The converse of this postulate is that enhanced uptake of toxin by normal Kupffer cells should result in tolerance. Such tolerance, if it occurred, could never be as impressive as that achieved in the tolerant animal whose Kupffer cells are simultaneously refractory to the generation of endogenous pyrogen. In the absence of such refractoriness, no reduction in pyrogenic response was observed after diversion of endotoxin to the Kupffer cells by means of injection of the toxin via portal-vein cannulae [5]. Studies in this and other laboratories indicate that RES clearance of endotoxin can be accelerated within 24 hr after an initial injection of endotoxin in the absence of humoral factors that transfer such acceleration (presumably a direct stimulating effect), as well as after passive transfer of tolerant-phase serum (presumably an opsonizing effect of antibodies to endotoxin [29, 37]. Thus, the same stimuli that evoke refractoriness of Kupffer cells to the release of endogenous pyro-

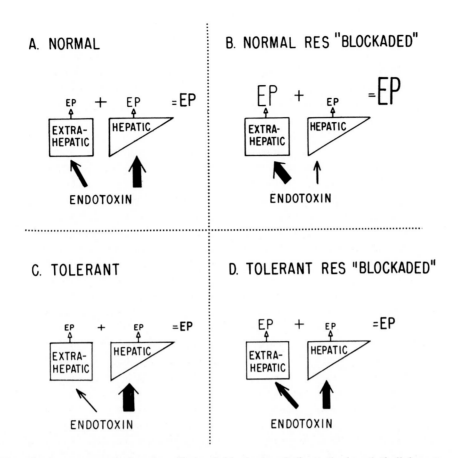

Figure 1. Hypothesis to account for the effect of blockade of the reticuloendothelial system (RES) on pyrogenic reactions to endotoxin in the normal and tolerant host. Concept is based on studies of rates of clearance from blood [33], hepatic uptake [34, 35] and pyrogenic reactions to ear vein [31] and portal vein [5] perfusion of endotoxin after RES blockade. This hypothesis is entirely comparable to that proposed by Dinarello et al. [14] based upon in-vitro studies of the release of endogenous pyrogen by endotoxin.

gen by endotoxin (direct cellular interaction with endotoxin and antibodies to endotoxin) are also those that evoke accelerated clearance of endotoxin. It is the former effect that permits the latter to become a significant protective mechanism.

The above concept implies that antibodies to endotoxin do not confer tolerance simply by accelerating RES clearance of endotoxin. This is strongly supported by the observations of Dinarello et al. [14] that tolerant-phase serum protects isolated hepatic Kupffer and spleen cells against the endogenous pyrogen-releasing activity of endotoxin. Since such protection did not extend to blood leukocytes or lung macrophages, it appears unlikely that the antiserum merely agglutinated the toxin. Rather, the findings suggest that antiserum

to endotoxin directly protects the RES against the endogenous pyrogen-releasing activity of endotoxin, perhaps by enhancing its intracellular detoxification or by interacting with specific receptors on RES membranes (cytophilic antibody). Studies in our laboratory support the latter possibility, since absorption of tolerant-phase rabbit serum at 4 C with packed normal rabbit spleen cells (in contrast to rabbit erythrocytes) removed significant protective activity. That accelerated RES clearance of endotoxin is not a requisite for induction of tolerance by antibodies to endotoxin was recently demonstrated by Braude and Douglas [29]. These investigators used an antiserum prepared against a heterologous endotoxin to circumvent the potent opsonizing activity of antibody to O antigen and demonstrated excellent protection against the local Shwartzman reaction in the absence of enhanced clearance of the toxin from the blood.

The concept that accelerated RES clearance of endotoxin confers minimal, if any, protection unless the RES is also refractory to the toxin would acount for the following observations.

(1) In 1955, Atkins and Wood reported that rabbits sensitized by one or two iv injections of typhoid vaccine three to five weeks previously cleared this pyrogen significantly more rapidly from the blood but exhibited febrile responses similar to those of nonsensitized controls [38].

(2) In 1962, Stuart and Cooper elicited enhanced RES clearance of endotoxin in mice by treatment with simple lipids and found that the animals became more susceptible to endotoxin. They concluded that "phagocytosis by itself is of relatively little importance in resistance to endotoxin" [39]. Di Luzio and Crafton confirmed these conclusions, using other reagents to modify RES clearance of endotoxin [40].

(3) In 1966, Chedid et al. reported that, in contrast to endotoxins from smooth gram-negative bacteria, toxic rough (R) endotoxins were cleared equally rapidly from the blood of nontolerant and tolerant mice. These investigators concluded: "It is, therefore, difficult to explain the increased resistance of the tolerant mouse against R endotoxin solely on the basis of its greater disappearance from the blood" [37].

(4) Recent studies in our laboratory have shown that endotoxin lethality cannot be prevented by an early exchange transfusion of normal rabbits 20 min after an iv LD_{80} dose of endotoxin, even though circulating levels of toxin are thereby rapidly reduced to levels equivalent to those seen in tolerant animals that survive this dose. It appeared probable from these studies that tolerance to endotoxin could not be based simply upon enhanced uptake by the RES of endotoxin but rather rested upon enhanced resistance of the RES to the injurious effects of the toxin [41]. The marked susceptibility of the normal RES to injury caused by endotoxin has already been well demonstrated by Heilman [42]. While man exhibits accelerated blood clearance of Cr^{51}-labeled endotoxin as pyrogenic tolerance is induced [33], it now appears likely from the above considerations that such tolerance is not achieved by this accelerated clearance per se, but by the concomitant development of refractoriness of the hepatic reticuloendothelial cells to the generation of endogenous pyrogen.

Duration of tolerance. Once tolerance to endotoxin has been actively induced by repetitive daily iv injections, its duration after discontinuance is a complex response. It was concluded during early studies in man that "the resistant state had disappeared in from four to five weeks. . . ." [23]. Careful analysis of the data presented, however, indicates that tolerance had declined, not disappeared, at this time. We are unaware of any systematic studies in man that define quantitatively the rate of tolerance loss. Studies in a few volunteers in our laboratory parallel the findings of Mulholland et al. in the rabbit [43], wherein the rate of tolerance decline is initially rapid after discontinuance of injections of endotoxin but then slows so that residual tolerance remains evident for several months. We postulate that the initial rapid rate of decline in tolerance reflects the transient nature of the cellular refractory state and that the subsequent slower-decline phase reflects the decline in protective antibody. The early, rapid-decline phase of pyrogenic tolerance was well recognized by Beeson, who initially emphasized that maximal pyrogenic tolerance requires repetitive, closely spaced injections of endotoxin [4].

Split-dose effect. Of great interest is the ability partially to overcome established tolerance to endotoxin in man by dividing the endotoxin dose in half and administering the second half-dose 2 hr after the first [2]. This maneuver was employed years ago to facilitate fever therapy with typhoid vaccines. The second dose was described as having the effect of "exploding the charge" supplied by the first [44]. The mechanism underlying this phenomenon is unknown but has been likened to the Danysz reaction, wherein the initial half-dose of endotoxin binds a high proportion of protective antibody, allowing the second half to act without inhibition by antibody [2]. Further tests of this hypothesis are required, but it conforms with observations in our laboratory that the accelerated blood clearance of endotoxin is slowed in tolerant volunteers 2 hr after the initial dose of toxin. This phenomenon of overcoming pyrogen tolerance unfortunately cannot be duplicated in the rabbit; the initial phase of the biphasic fever response to endotoxin is not abolished during tolerance in this animal, and division and administration of the endotoxin dose at intervals of 2 (or 3, 4, or 5) hr generally result simply in duplication of the initial febrile response. It should be stressed that, when a constant dose of endotoxin continues to be administered to tolerant man every 2 hr, tolerance is again observed by the fourth or fifth dose; thus, fever cannot be sustained by this method.

Effect of illness on tolerance. The tolerance mechanisms in man continue to function during febrile illness. In two such illnesses, typhoid fever and tularemia, it could be demonstrated that while the baseline reactivity to endotoxin was sharply increased, the tolerance mechanisms could still be activated within this hyperreactive framework. Tolerance could be readily induced during overt illness both by daily single injections of endotoxin and by continuous iv infusions. Tolerance induced before illness could also be demonstrated during illness [33]. Accelerated blood clearance of endotoxin, evident during induction of tolerance before illness, remained evident during typhoid fever [33]. This ability of man to acquire tolerance to endotoxin during typhoidal and tularemic illness and the inability to mitigate these illnesses by deliberate induction of tolerance permitted the conclusion that circulating endotoxin could not constitute the major cause of the sustained pyrexia and toxemia during these illnesses [33].

As in typhoid fever and tularemia, so in brucellosis; man hyperreacts to endotoxin and readily acquires tolerance to its iv administration [45]. Tolerance to endotoxin has been demonstrated in man after convalescence from tularemia, typhoid and paratyphoid fevers, and during chronic pyelonephritis secondary to gram-negative but not gram-positive bacteria [46–48]; presumably, tolerance results from exposure to the endotoxin component of the infecting microbe. Pyrogenic tolerance is also seen in man after malaria [49], but its basis is unknown.

Local vs. systemic tolerance. When man is rendered tolerant to repetitive daily iv injections of endotoxin, tolerance does not develop to its dermal inflammatory reactivity. Indeed, dermal inflammation is enhanced, probably as a result

of elevated titers of antibody to endotoxin [2]. Thus, it is clear that systemic tolerance to endotoxin cannot be equated with resistance to the local inflammation-evoking activity of the toxin. It has been proposed that endotoxin contributes to the sustained pyrexia and toxemia of typhoid fever primarily by virtue of this local inflammatory (rather than systemic) activity [33]. This concept would be consonant with the inability to mitigate this illness by deliberate induction of systemic tolerance to iv-injected endotoxin [33].

Effect of immunosuppression on endotoxin tolerance in man. Two types of studies have been performed in man as tests of the effect of impaired production of antibody on tolerance to endotoxin. Good et al. [50, 51] studied patients with agammaglobulinemia and reported that such patients acquired tolerance to the pyrogenic activity of typhoid vaccine, as did healthy controls. However, proof that titers of antibody to endotoxin were impaired was not documented. Wolff has studied three such patients and confirmed normal acquisition of tolerance in the absence of detectable increments in antibodies to endotoxin (S. M. Wolff, personal communication). In all of these studies, the endotoxin was administered iv at daily intervals. Since cellular refractoriness can rapidly lead to pyrogenic tolerance as discussed above, deficits in tolerance based upon deficits in protective antibody formation might not be apparent by such a schedule of challenge.

In our laboratory, a second model of impaired immunoglobulin production was studied (i.e., splenectomy), and tolerance was tested at intervals spaced sufficiently to allow waning of the cellular refractory state (five, seven, and 10 days after initial immunization). Three healthy volunteers who were splenectomized seven to 11 years previously after abdominal trauma responded to an initial iv injection of endotoxin in a manner identical to six healthy controls. These splenectomized subjects failed to develop increments in antibodies to endotoxin as measured by bentonite-flocculation and bactericidal-antibody techniques on days 5–10 after initial challenge and acquired tolerance at a rate significantly slower than that of the control group. From these findings, it would appear that while the spleen is not important in man's defense against (or reactivity to) initial challenge with small doses of endotoxin, loss of the antibody-producing activity of the spleen does constitute a significant impairment to subsequently acquired tolerance to endotoxin. It should again be emphasized, however, that for achievement of these results, tests for tolerance must be conducted with appropriate spacing of the toxin to reduce the contribution of the cellular refractory state, thus amplifying the protective role of antibody.

Discussion

The mechanisms by which man acquires tolerance to bacterial endotoxins are complex and still incompletely defined. Certain of the required extrapolations from animal studies to man permit only a tentative hypothesis. Nevertheless, a concept emerges that can now be critically tested. This concept holds (1) that pyrogenic tolerance to endotoxin is based primarily on enhanced resistance of hepatic Kupffer cells to the release of endogenous pyrogen and (2) that such resistance can be achieved either by direct cellular interaction with endotoxin or by antibodies to endotoxin, which protect these cells against injury due to the toxin. The direct cellular effect responsible for tolerance is transient and requires closely spaced or continuous exposures to endotoxin for maintenance; antibody-mediated protection is delayed but more enduring. The protective antibodies comprise those that react with antigens associated with O-specific groupings as well as those that react with common core antigens. When intact smooth endotoxins are used for immunization, the former antibodies appear to be the more prominent humoral protective component; when rough mutants are administered, the latter antibodies appear to comprise the most important humoral protective component. Accelerated clearance by the RES appears to be an ancillary protective mechanism that brings the endotoxin more rapidly to the refractory Kupffer cells. The host thus has at its disposal several mechanisms for acquiring resistance to endotoxin, and the diverse, often conflicting observations on pyrogenic tolerance can be explained by analysis of the interplay of these protective mechanisms.

References

1. Greisman, S. E., Hornick, R. B. Comparative pyrogenic reactivity of rabbit and man to bacterial endotoxin. Proc. Soc. Exp. Biol. Med. 131:1154–1158, 1969.
2. Greisman, S. E., Wagner, H. N., Jr., Iio, M., Hornick, R. B. Mechanisms of endotoxin tolerance. II. Relationship between endotoxin tolerance and reticuloendothelial system phagocytic activity in man. J. Exp. Med. 119:241–264, 1964.
3. Greisman, S. E., Hornick, R. B., Woodward, T. E. The role of endotoxin during typhoid fever and tularemia in man. III. Hyperreactivity to endotoxin during infection. J. Clin. Invest. 43:1747–1757, 1964.
4. Beeson, P. B. Tolerance to bacterial pyrogens. I. Factors influencing its development. J. Exp. Med. 86:29–38, 1947.
5. Greisman, S. E., Woodward, C. L. Mechanisms of endotoxin tolerance. VII. The role of the liver. J. Immunol. 105:1468–1476, 1970.
6. Atkins, E. Pathogenesis of fever. Physiol. Rev. 40:580–646, 1960.
7. Snell, E. S. An examination of the blood of febrile subjects for pyrogenic properties. Clin. Sci. 21:115–124, 1961.
8. Snell, E. S., Goodale, F., Jr., Wendt, F., Cranston, W. I. Properties of human endogenous pyrogen. Clin. Sci. 16:615–626, 1957.
9. Greisman, S. E., Hornick, R. B. On the demonstration of circulating human endogenous pyrogen. Proc. Soc. Exp. Biol. Med. 139:690–697, 1972.
10. Beeson, P. B. Temperature-elevating effect of a substance obtained from polymorphonuclear leucocytes [abstract]. J. Clin. Invest. 27:524, 1948.
11. Herion, J. C., Walker, R. I., Palmer, J. G. Endotoxin fever in granulocytopenic animals. J. Exp. Med. 113:1115–1125, 1961.
12. Page, A. R., Good, R. A. Studies on cyclic neutropenia. Am. J. Dis. Child. 94:623–661, 1957.
13. Bodel, P., Atkins, E. Release of endogenous pyrogen by human monocytes. N. Engl. J. Med. 276:1002–1008, 1967.
14. Dinarello, C. A., Bodel, P. T., Atkins, E. The role of the liver in the production of fever and in pyrogenic tolerance. Trans. Assoc. Am. Physicians 81:334–344, 1968.
15. Hahn, H. H., Char, D. C., Postel, W. B., Wood, W. B., Jr. Studies on the pathogenesis of fever. XV. The production of endogenous pyrogen by peritoneal macrophages. J. Exp. Med. 126:385–394, 1967.
16. Atkins, E., Bodel, P., Francis, L. Release of an endogenous pyrogen in vitro from rabbit mononuclear cells. J. Exp. Med. 126:357–384, 1967.
17. Bennett, I. L., Jr., Petersdorf, R. G., Keene, W. R. Pathogenesis of fever: evidence for direct cerebral action of bacterial endotoxins. Trans. Assoc. Am. Physicians 70:64–73, 1957.
18. Du Buy, B. Role of the granulocyte in the pyrogenic response to intracisternal endotoxin. Proc. Soc. Exp. Biol. Med. 123:606–609, 1966.
19. Carey, F. J., Braude, A. I., Zalesky, M. Studies with radioactive endotoxin. III. The effect of

tolerance on the distribution of radioactivity after intravenous injection of *Escherichia coli* endotoxin labeled with Cr51. J. Clin. Invest. 37:441–457, 1958.

20. Greisman, S. E., Woodward, W. E. Mechanisms of endotoxin tolerance. III. The refractory state during continuous intravenous infusions of endotoxin. J. Exp. Med. 121:911–933, 1965.

21. Greisman, S. E., Young, E. J., Woodward, W. E. Mechanisms of endotoxin tolerance. IV. Specificity of the pyrogenic refractory state during continuous intravenous infusions of endotoxin. J. Exp. Med. 124:983–1000, 1966.

22. Janoff, A., Weissmann, G., Zweifach, B. W., Thomas, L. Pathogenesis of experimental shock. IV. Studies on lysosomes in normal and tolerant animals subjected to lethal trauma and endotoxemia. J. Exp. Med. 116:451–466, 1962.

23. Morgan, H. R. Resistance to the action of the endotoxins of enteric bacilli in man. J. Clin. Invest. 27:706–709, 1948.

24. Greisman, S. E., Young, E. J., Carozza, F. A., Jr. Mechanisms of endotoxin tolerance. V. Specificity of the early and late phases of pyrogenic tolerance. J. Immunol. 103:1223–1236, 1969.

25. Greisman, S. E., Young, E. J. Mechanisms of endotoxin tolerance. VI. Transfer of the "anamnestic" tolerant response with primed spleen cells. J. Immunol. 103:1237–1241, 1969.

26. Boivin, A., Mesrobeanu, L. Recherches sur les antigènes somatiques et sur les endotoxines des bactéries. IV. Sur l'action anti-endotoxique de l'anticorps O. Rev. Immunol. (Paris) 4:40–52, 1938.

27. Perlman, E., Goebel, W. F. Studies on the Flexner group of dysentery bacilli. IV. The serological and toxic properties of the somatic antigens. J. Exp. Med. 84:223–234, 1946.

28. Creech, H. J., Hamilton, M. A., Nishimura, E. T., Hankwitz, R. F., Jr. The influence of antibody-containing fractions on the lethal and tumor-necrotizing actions of polysaccharides from *Serratia marcescens* (*Bacillus prodigiosus*). Cancer Res. 8:330–336, 1948.

29. Braude, A. I., Douglas, H. Passive immunization against the local Shwartzman reaction. J. Immunol. 108:505–512, 1972.

30. Beeson, P. B. Tolerance to bacterial pyrogens. II. Role of the reticulo-endothelial system. J. Exp. Med. 86:39–44, 1947.

31. Greisman, S. E., Carozza, F. A., Jr., Hills, J. D. Mechanisms of endotoxin tolerance. I. Relationship between tolerance and reticuloendothelial system phagocytic activity in the rabbit. J. Exp. Med. 117:663–674, 1963.

32. Wolff, S. M., Mulholland, J. H., Ward, S. B. Quantitative aspects of the pyrogenic response of rabbits to endotoxin. J. Lab. Clin. Med. 65:268–275, 1965.

33. Greisman, S. E., Hornick, R. B., Wagner, H. N., Jr., Woodward, W. E., Woodward, T. E. The role of endotoxin during typhoid fever and tularemia in man. IV. The integrity of the endotoxin tolerance mechanisms during infection. J. Clin. Invest. 48:613–629, 1969.

34. Cremer, N., Watson, D. W. Influence of stress on distribution of endotoxin in RES determined by fluorescein antibody technic. Proc. Soc. Exp. Biol. Med. 95:510–513, 1957.

35. Howard, J. G., Rowley, D., Wardlaw, A. C. Investigations on the mechanism of stimulation of non-specific immunity by bacterial lipopolysaccharides. Immunology 1:181–203, 1958.

36. Freedman, H. H. Passive transfer of tolerance to pyrogenicity of bacterial endotoxin. J. Exp. Med. 111:453–463, 1960.

37. Chedid, L., Parant, F., Parant, M., Boyer, F. Localization and fate of 51Cr-labeled somatic antigens of smooth and rough salmonellae. Ann. N.Y. Acad. Sci. 133:712–726, 1966.

38. Atkins, E., Wood, W. B., Jr. Studies on the pathogenesis of fever. I. The presence of transferable pyrogen in the blood stream following the injection of typhoid vaccine. J. Exp. Med. 101:519–528, 1955.

39. Stuart, A. E., Cooper, G. N. Susceptibility of mice to bacterial endotoxin after modification of reticulo-endothelial function by simple lipids. J. Pathol. Bacteriol. 83:245–254, 1962.

40. Di Luzio, N. R., Crafton, C. G. Influence of altered reticuloendothelial function on vascular clearance and tissue distribution of *S. enteritidis* endotoxin. Proc. Soc. Exp. Biol. Med. 132:686–690, 1969.

41. Greisman, S. E. Mechanisms of endotoxin tolerance: effect of exchange transfusion [abstract]. J. Clin. Invest. 51:39a, 1972.

42. Heilman, D. H. The selective toxicity of endotoxin for phagocytic cells of the reticuloendothelial system. Int. Arch. Allerg. Appl. Immunol. 26:63–79, 1965.

43. Mulholland, J. H., Wolff, S. M., Jackson, A. L., Landy, M. Quantitative studies of febrile tolerance and levels of specific antibody evoked by bacterial endotoxin. J. Clin. Invest. 44:920–928, 1965.

44. Nelson, M. O. An improved method of protein fever treatment in neurosyphilis. Am. J. Syph. 15:185–189, 1931.

45. Abernathy, R. S., Spink, W. W. Studies with brucella endotoxin in humans: the significance of susceptibility to endotoxin in the pathogenesis of brucellosis. J. Clin. Invest. 37:219–231, 1958.

46. Greisman, S. E., Hornick, R. B., Carozza, F. A., Jr., Woodward, T. E. The role of endotoxin during typhoid fever and tularemia in man. I. Acquisition of tolerance to endotoxin. J. Clin. Invest. 42:1064–1075, 1963.

47. Neva, F. A., Morgan, H. R. Tolerance to the action of endotoxins of enteric bacilli in patients convalescent from typhoid and paratyphoid fevers. J. Lab. Clin. Med. 35:911–922, 1950.

48. McCabe, W. R. Endotoxin tolerance. II. Its occur-

rence in patients with pyelonephritis. J. Clin. Invest. 42:618–625, 1963.

49. Heyman, A., Beeson, P. B. Influence of various disease states upon the febrile response to intravenous injection of typhoid bacterial pyrogen. J. Lab. Clin. Med. 34:1400–1403, 1949.

50. Good, R. A., Varco, R. L. A clinical and experimental study of agammaglobulinemia. J. Lancet 75:245–271, 1955.

51. Good, R. A., Zak, S. J. Disturbances in gamma globulin synthesis as "experiments of nature." Pediatrics 18:109–149, 1956.

Role of Lymphocytes in the Pyrogenic Response: Comparison of Endotoxin with Specific Antigen and the Nonspecific Mitogen, Concanavalin A

Elisha Atkins and Lorraine Francis

From the Department of Internal Medicine, Yale University School of Medicine, New Haven, Connecticut

Our studies indicate that lymphocytes can activate blood leukocytes to release endogenous pyrogen after incubation in vitro with concanavalin A, a nonspecific activator of thymus-derived lymphocytes, or with sufficient doses of filtered typhoid vaccine, one of the well-known endotoxins derived from gram-negative bacteria. If, as has been postulated, endotoxins produce some of their toxic reactions by acting as antigens in hosts with a naturally acquired, preexisting delayed hypersensitivity, the ability of endotoxin to induce fever, like that of certain antigens or mitogens described here, may also involve the intermediary activation of lymphocytes.

Despite intensive investigation during the last 20 years, the basic mechanisms by which endotoxins of gram-negative bacteria cause injury to host cells remain largely unknown. In the pathogenesis of fever, it is now well established that these agents, like other microbial and nonmicrobial activators, stimulate production and release of an endogenous pyrogen (EP) from various types of cells, including blood and exudate granulocytes, alveolar macrophages, splenic monocytes, and Kupffer cells of the liver, a major component of the reticuloendothelial system [1]. Lymphocytes do not appear to release EP [2, 3], but, because of their importance in delayed hypersensitivity, it seems possible that they play a role in mediating certain reactions to endotoxins (such as fever) that occur in hosts that have been naturally sensitized to these ubiquitous agents. The purpose of the present experiments was to determine whether lymphocytes, incubated with endotoxin or other mitogenic stimuli, can serve as intermediates in the production of EP and, hence, of fever.

Materials and Methods

Most of the materials and procedures used have been described [4]. For most experiments, mesenteric lymph nodes were removed by sterile techniques from normal rabbits after iv injection of 10,000 units of heparin (Liquaemin® sodium, Organon) and exsanguination by cardiac puncture. Single cell preparations (in a concentration of 2–3 $\times$ 10^7 cells/ml) were suspended in Eagle's minimal essential medium (MEM, Auto-Pow®, Flow Laboratories, Rockville, Md.) adjusted to pH 7.25–7.40 with 10% fresh normal plasma or serum. In several experiments, before incubation overnight with normal blood cells (NBC) in doses of 1–2 $\times$ 10^8 lymphocytes per 4–6 $\times$ 10^7 blood leukocytes, cells were incubated at 37 C for various periods (15 min to 5 hr) with either of the following two agents: (*1*) typhoid vaccine (monovalent reference standard NRV-LS no. 1, Walter Reed Hospital, Washington, D.C.) diluted 1:10 in pyrogenfree saline (PFS) and filtered through a Swinnex Millipore filter just before use; (*2*) concanavalin A (Nutritional Biochemicals) diluted in PFS to a concentration of 1.0 mg/ml and stored at −20 C until use.

Details of the procedures are given below for each experiment (see Results). Each preparation was routinely set up in aliquots of two to four doses. Media were made up with 5,000 units of penicillin and 5,000 µg of streptomycin per liter (penicillin-streptomycin solution GIBCO). Additional heparin (1,000 units/flask) was added to all lymphocyte suspensions that were incubated with plasma [4].

This study was supported by grant no. AI-01564 from the U.S. Public Health Service.

We thank Dr. Phyllis Bodel for helpful criticism in the preparation of this paper.

Please address requests for reprints to Dr. Elisha Atkins, Department of Internal Medicine, Yale University School of Medicine, New Haven, Connecticut 06510.

270

Results and Discussion

Production of EP by NBC incubated with antigen and specifically sensitized lymphocytes in vitro.   In recent studies of delayed hypersensitivity in rabbits [4], we demonstrated that when specifically sensitized lymphocytes are incubated with antigen, they release an agent that activates NBC to produce EP in vitro. A representative experiment done subsequently is presented here.

Two rabbits were sensitized by foot-pad injection of 10 mg of bovine γ-globulin (BGG, Armour) incorporated in complete Freund's adjuvant [4]. Ten and 12 days later, respectively, they were sacrificed, and draining lymph nodes (DLN) were removed from axillary and popliteal regions and incubated overnight as single-cell suspensions in MEM with normal blood cells. For assessment of the role of viable lymphocytes, half of the cells were disrupted by freezing and thawing three times before they were added to the NBC. The antigen, BGG, was then added in 5-mg doses to one-half of both the viable and the frozen-thawed portions; the other half served as a control.

The results (table 1) indicate that the combination of antigen and viable lymphocytes clearly activated NBC to produce EP. DLN cells alone or disrupted cells incubated with and without antigen were only rarely effective in this system. Similarly, DLN cells incubated alone with BGG released little or no EP.

On the basis of more extensive data based on this model, in which NBC were activated by non-pyrogenic supernates of sensitized lymphocytes previously incubated with antigen [4], we have

Table 1. Release of endogenous pyrogen by normal blood cells mixed with a suspension of sensitized lymphocytes from draining lymph nodes and incubated for 18 hr with (+BGG) or without (C) antigen. Results with viable and frozen-thawed (FT) lymphocytes are compared.

Viable*		FT	
+BGG†	C	+BGG	C
0.72‡	0.24	0.14	0.14

NOTE. All figures are averages of four recipients (two experiments). Donors were sensitized 10 and 12 days before experiment.

* Values are expressed as ΔT (°C).

† BGG = bovine gamma globulin.

‡ Significant response (>0.35 C) is printed in bold type.

postulated that antigen causes lymphocytes from animals with delayed hypersensitivity to release a soluble agent, presumably a "lymphokine" [5], which is capable of activating NBC in vitro. These studies suggest that specifically sensitized lymphocytes play an essential intermediary role in producing fevers of delayed hypersensitivity.

Comparative pyrogenicity of endotoxin injected iv and added to NBC in vitro.   In addition to its role as a powerful adjuvant, endotoxin has been found to produce signs of toxicity that appear to be related to the previous exposure of animals to gram-negative flora in the gastrointestinal tract [6]. On the assumption that these effects (like those of BGG or of a microbial antigen such as tuberculin in specifically sensitized individuals) might be mediated by naturally acquired states of delayed hypersensitivity, it seemed appropriate to determine whether lymphocytes could be similarly implicated in fevers induced by endotoxins of gram-negative bacteria. Since the method would necessarily involve use of lymphocytes exposed to endotoxin (as a presumed antigen), for various periods, a preliminary experiment was devised for comparison of the pyrogenicity of graded doses of endotoxin as a direct stimulus in vivo with its pyrogenicity as an inducer of EP when added to NBC in vitro.

Increasing doses of filtered vaccine (in triplicate 10-ml aliquots diluted in MEM) were mixed and incubated overnight with centrifuged blood cells from two rabbits after removal of plasma. Supernates were injected iv for determination of the amount of EP released.

As shown in table 2, the ability to detect small doses of endotoxin was increased at least 15–20 times when endotoxin was incubated for 18 hr with blood cells in vitro, as compared with its fever-inducing activity when given iv. For example, fevers of about 0.5 C were produced by EP present in supernates of NBC incubated with 0.1 ml of typhoid vaccine in vitro, whereas an iv injection of 2 ml was required for production of the same degree of fever in vivo. Again, roughly equivalent pyrogenic responses (0.8 C and 0.9 C) were produced by incubation of 0.3 ml of vaccine with NBC in vitro and injection of 5.0 ml in vivo.

Attempted release of EP from NBC incubated with mesenteric lymph-node lymphocytes previously exposed to endotoxin.   Since the in-vitro release of EP by blood leukocytes appeared to

Table 2. Comparative pyrogenicity of endotoxin (typhoid vaccine or TV) given iv or incubated for 18 hr with normal blood cells in vitro.

Increase in ability to detect endotoxin	Dose (TV, 1:100, ml)*	Induction of fever	
		iv [ΔT(°C)]	EP† [ΔT(°C)]
	5.0	**0.90**‡	...
>15×	2.0	**0.55**	...
	1.0	0.24§	**1.60**§
20×	0.3	0.12	**0.80**
	0.1	0.08§	**0.51**§
	0.01	0.08	0.27
	0.001	0.10	0.10
	Control	...	0.09§

NOTE. All figures are averages of three recipients, except as indicated.

* Average dose was 8.65×10^7 leukocytes per recipient.

† EP = endogenous pyrogen.

‡ Significant responses (>0.35 C) are printed in bold type.

§ Average of six recipients (two experiments).

provide an extremely sensitive test for the presence of endotoxin, the technique chosen for assessment of the effects of endotoxin on lymphocytes was based on earlier studies with sensitized lymphocytes and antigen. It was reasoned that when lymphocytes were exposed to endotoxin and when viable and disrupted cells were then compared in their ability to activate NBC to produce EP in vitro, the release of a lymphokine from viable cells could be distinguished by its additive effect from any residual, cell-associated endotoxin, which would be equally active in both viable and disrupted cell preparations.

The steps in this procedure were: (1) Normal lymphocytes (in doses of $1-2 \times 10^8$ cells) derived from mesenteric lymph nodes (MLNs) were suspended in MEM and incubated for various periods (5 min to 5 hr) with a dilute solution (1 ml, 1:100) of filtered typhoid vaccine. (2) Cells were centrifuged, washed once, suspended again in MEM, and added to NBC that had been resuspended in MEM after removal of most of the plasma. (3) After incubation for 18 hr, the mixture of lymphocytes and NBC was centrifuged, and the supernates were injected into rabbits for

determination of the amount of EP released. In some experiments, a portion of the endotoxin-exposed lymphocytes, after being washed and resuspended in medium, were disrupted by freezing and thawing three times before being added to NBC. (4) The supernatant fluids were removed at different intervals during incubation and tested for the presence of residual endotoxin or newly produced lymphokine by further incubation with NBC.

Table 3 shows the results of these experiments. On the left are pyrogenic reactions induced by washed lymphocytes that had been incubated with endotoxin in MEM with normal serum for the intervals indicated and then added to NBC. It is evident that neither viable nor frozen-thawed cells liberated significant amounts of EP from blood leukocytes despite previous exposure of from 5 min to 5 hr with the vaccine. The lack of significant difference between the results with viable cells and those with disrupted cells suggests that there was no detectable release of an EP-producing lymphokine with this dose of endotoxin, in contrast with results seen previously when sensitized DLN lymphocytes were incubated with specific antigen. The supernatant fluids of these lym-

Table 3. Release of endogenous pyrogen in vitro by normal blood cells (NBC) incubated for 18 hr with various preparations.

Prein-cubated with TV	NBC + MLN cells*		NBC + supernatant	
	V	FT	With LNC†	Without LNC‡
5 min	0.32 (12)§	0.14 (4)	**0.75**‖(8)	**0.86** (8)
1 hr	0.21 (6)	0.13 (2)		
3 hr	0.28 (2)	0.15 (2)		
5 hr	0.30 (12)	0.10 (4)	**0.37** (8)	**0.44** (8)
Control	0.28 (12)	0.20 (4)		

* Lymphocytes from mesenteric lymph nodes (MLN) previously incubated for various intervals (5 min to 5 hr) with typhoid vaccine (TV), which had been washed and resuspended in Eagle's minimal essential medium before addition to NBC. Effects of viable (V) and frozen-thawed (FT) lymphocytes are compared.

† Supernates of the same lymphocyte suspensions [plus lymph-node cells (LNC)] incubated with media plus 10% normal serum.

‡ Media plus 10% normal serum incubated with TV alone (no LNC).

§ Numbers in parentheses represent numbers of recipients.

‖ Significant responses (>0.35 C) are printed in bold type.

272

phocyte preparations, on the other hand, clearly had detectable amounts of endotoxin initially, as was evidenced by amount of EP generated by the 5-min samples incubated with NBC. The 5-hr samples liberated only half as much EP from NBC, indicating, presumably, that the amount of vaccine in these supernates had fallen to about half of its initial value. However, lymphocytes did not appear to play a significant role in removing or inactivating the endotoxin, since the control preparations, containing endotoxin incubated in media alone without lymphocytes, had lost about the same amount of activity during the same period (compare the two columns under "Supernatant").[1]

In summary, it seems reasonable to draw the following inferences from these experiments. (1) Under these conditions, endotoxin in the form of filtered typhoid vaccine does not attach firmly to normal lymphocytes derived from MLNs. (2) Unlike antigen incubated with specifically sensitized lymphocytes, this small dose of endotoxin does not stimulate these cells to generate detectable amounts of a lymphokine-like agent that, in turn, activates NBC to produce EP. (3) These lymphocytes do not appear to play a significant role in inactivation or removal of soluble endotoxin.

Pyrogenic properties of concanavalin A. During the past few years, a number of mitogenic agents have been described that have the property of activating normal T (thymus-derived) lymphocytes [7]. The products of such stimulated cells, collectively known as lymphokines, have a diverse range of biologic activities on lymphocytes, as well as on phagocytic and other types of cells in vitro [8]. These endogenous agents appear to be identical with those released from specifically sensitized lymphocytes incubated with antigen. In view of the failure of the small dose of endotoxin, an agent known to act on B (bone-marrow-derived) lymphocytes [9], to activate normal MLN cells in this pyrogen-assay system, it was of interest whether such cells could be stimulated by mito-

gens known to act on T rather than B lymphocytes [10]. Concanavalin A (Con A) was selected and doses of 0.01–1.0 mg were given iv for determination of its pyrogenic properties. As is evident in figure 1, this agent proved to be a potent pyrogen, producing biphasic fevers of 1.5 C–2 C with the highest dose after a latent period of 15–30 min; these fevers were similar to those after an iv dose of endotoxin (compare responses to filtered typhoid vaccine in figure 2). To determine whether the fever-inducing effects of Con A were due to contaminating endotoxin, we did a cross-tolerance study with typhoid vaccine. A group of four rabbits was injected daily with 0.3 mg of Con A. The animals rapidly developed pyrogenic tolerance to this dose, and on the sixth day, when the residual fevers were about one-third of those initially induced, these rabbits, along with three previously

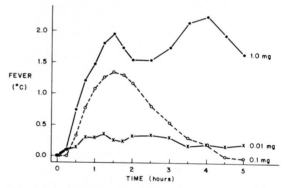

Figure 1. Mean febrile responses (two recipients each) to three 10-fold increments in dosage of concanavalin A given iv.

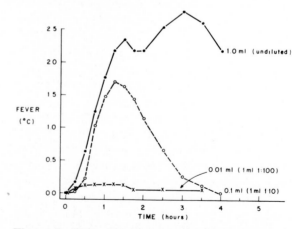

Figure 2. Mean febrile responses (two to eight recipients each) to three 10-fold increments in dosage of filtered typhoid vacine given iv.

---

[1] Endotoxin was incubated with both media alone and media with 10% normal sera. Since the results did not differ significantly, values in table 3 are for media with sera only. The comparable loss of pyrogenic activity of endotoxin incubated in media alone during the 5-hr period suggests that serum did not play a role in this process.

uninjected controls, were given a small dose of filtered typhoid vaccine (5 ml, 1:100 dilution). The febrile responses of the two groups to this small dose of vaccine were identical, indicating a lack of the kind of cross-tolerance ordinarily seen when endotoxins from unrelated species of gram-negative bacteria are tested in the same fashion [11]. Conversely, rabbits rendered tolerant to unfiltered typhoid vaccine (1.5 ml, 1:10 dilution) by seven daily iv inoculations showed only a slight degree of cross-tolerance to a 0.1-mg dose of Con A.

Release of EP from NBC incubated with MLN lymphocytes previously incubated with Con A or large doses of endotoxin. With this evidence that the pyrogenic action of Con A was probably not due to significant contamination with endotoxin, we devised the following experiment to determine whether this agent would activate normal lymphocytes to release mediators capable of causing blood leukocytes to generate EP in vitro.

The steps in this procedure, similar to the procedure described above, are shown in figure 3. Suspensions of normal MLN lymphocytes (in doses of $1–2 \times 10^8$ cells in 10 ml of MEM) were incubated for 2 hr with various doses (0.03–1.0 mg) of Con A. Cells were then separated by centrifugation and washed twice in an equal volume of MEM for removal of all soluble activator. One aliquot of resuspended cells was then disrupted by freezing and thawing three times. The viable and disrupted cell suspensions were then incubated for 18 hr with NBC (resuspended in MEM) as a test for the evolution of EP. As an assay for soluble pyrogen in the wash fluids, the second and final supernatant after washing was also added to a separate aliquot of NBC. Residual Con A was distinguished from EP by injection of all supernates into both normal rabbits and animals rendered refractory to the pyrogenic effects of exogenous pyrogens such as Con A and endotoxin but not EP [12].

The results of this experiment are shown on the left of table 4. MLN cells incubated with Con A clearly activated NBC to release EP.[2] The effects were dose-related and could not be attributed to activator still present in the supernatant fluid, since NBC, incubated with the second wash of the Con A-treated cells, released no EP. Most importantly, with the 0.3-ml dose of Con A (and with the 0.5-ml dose not presented in the table), viable cells appeared to be necessary for release of EP from NBC; this finding suggests that an EP-inducing lymphokine was liberated from activated MLN cells, as in the experiment discussed earlier with antigen and specifically sensitized cells.

[2] Cells incubated for 2 hr with either Con A or endotoxin in media with 10% normal sera consistently evoked greater release of EP from NBC than did those incubated in media with plasma. Results of both types of experiments have been averaged in table 4.

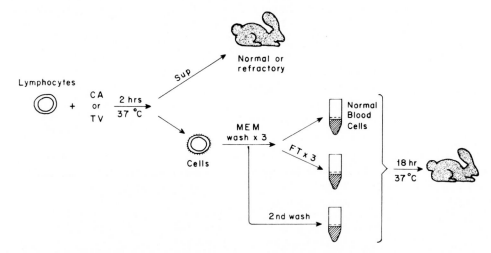

Figure 3. Schematic diagram of procedure to demonstrate release of endogenous pyrogen from normal blood cells incubated with mesenteric lymph-node lymphocytes preincubated with concanavalian A (CA) or filtered typhoid vaccine (TV). Sup = supernate; MEM = Eagle's minimal essential medium; FT = frozen-thawed (see text for details).

Table 4. Release of endogenous pyrogen in vitro by normal blood cells incubated for 18 hr with suspensions of lymphocytes from mesenteric lymph nodes (MLN).

NBC + MLN cells preincubated for 2 hr with					
Con A (1 mg/ml)			TV (1:10 dilution)		
Dose (ml)	ΔT(°C)	No. R*	Dose (ml)	ΔT(°C)	No. R*
0.03	0.35	3	0.1	0.21	6
0.10	**0.74**†	21	0.5	0.28	16
0.30	**0.94**	8	1.0	0.31	14
	0.13‡	8			
0.70	**1.23**	12	3.0	**0.84**	18
1.00	**1.58**	6	5.0	**1.13**	8
				0.13‡	8

* About half of the recipients given each dosage were rendered refractory (R) to the pyrogenic effects of either concanavalin A (Con A) or typhoid vaccine (TV) (see text). Febrile responses of this group did not differ significantly from those of normal recipients given the same material; hence, the responses of the two groups are averaged.

† Significant responses (>0.35) are printed in bold type.

‡ Cells frozen and thawed three times.

Similar experiments with higher doses of endotoxin are shown on the right of table 4. In contrast with results of the earlier experiment (table 3), the data here suggest that, in sufficient dosage, endotoxin may similarly activate lymphocytes in this system.[3] Since endotoxin appears to act directly on B rather than on T lymphocytes [9], it is possible that this effect may be mediated by products released by macrophages in the MLN population, since these products have recently been shown to activate T cells [13, 14].

The possibility exists that the disruption of previously exposed lymphocytes by freezing and thawing may simply prevent surface-bound Con A or endotoxin from directly activating blood leukocytes to produce EP. However, this possibility seems unlikely for several reasons. Both Con A and typhoid vaccine, frozen and thawed in the presence of intact lymphocytes or incubated with cell fragments, were still able to activate NBC (authors' unpublished observations). Fur-

[3] The second wash of endotoxin-treated cells failed to release EP when added to NBC, indicating a lack of soluble activator in the final suspension of lymphocytes, as in the case of Con A-treated cells.

ther studies are planned with lymphocytes subjected to irradiation or metabolic inhibitors for more precise analysis of the role of viable cells in this reaction.

In a final series of experiments, the mechanism for the pyrogenic effects of Con A and typhoid vaccine was examined by incubation of these agents with NBC or MLN cells (table 5). By use of normal and refractory recipient rabbits for distinguishing of EP from persisting exogenous pyrogen in the supernatant fluids [12], it could be seen that blood leukocytes, incubated alone with Con A or with typhoid vaccine, released an agent with the properties of EP (i.e., one that was equally pyrogenic in normal and refractory rabbits). On the other hand, when Con A or typhoid vaccine was incubated with suspensions of MLN cells alone, most, if not all, of the pyrogen in the supernates at either 2 or 18 hr was exogenous in its activity (producing little or no fever in refractory rabbits) and presumably represented activator that was still present in these cell suspensions. The small amount of EP from lymph-node cells could represent monocyte EP.

Preliminary studies have been done with NBC populations separated by centrifugation on a Ficoll-hypaque gradient [15]; the object is to determine whether Con A, like typhoid vaccine, can stimulate both granulocytes and monocytes

Table 5. Febrile responses of normal (NR) and refractory (RR) rabbits to supernates of normal blood-cell (NBC) or mesenteric lymph-node (MLN) cell suspensions incubated for various periods in vitro with concanavalin A (Con A) or typhoid vaccine (TV).

Agent	NBC (18 hr)		MLN			
			2 hr		18 hr	
	NR	RR	NR	RR	NR	RR
Con A (0.5–0.7 mg)	**1.48*** = **1.19**		**1.13** > 0.38		**1.00** > 0.40	
	(6)	(4)	(16)	(16)	(3)	(3)
TV Low dose (0.5/1.0 ml, 1:10 dilution)	**1.58** = **1.32**		**0.66** > 0.18		0.34	...
	(6)	(3)	(8)	(4)	(7)	
High dose (3.0/5.0 ml, 1:10 dilution)	**1.60**	...	**1.80** > 0.32		**1.22** > 0.32	
	(6)		(5)	(5)	(4)	(4)

* Values are given as ΔT(°C). Exogenous pyrogen: NR > RR; endogenous pyrogen: NR = RR. Numbers in parentheses = number of recipients. Significant responses (>0.35) are printed in bold type.

to produce EP. Preparations of rabbit-blood leukocytes containing lymphocytes and monocytes, but virtually devoid of granulocytes, were vigorously activated by Con A; to date, preparations of blood granulocytes free of monocytes have not been obtained. We have not yet tested Con A with preparations of either monocytes or granulocytes essentially devoid of lymphocytes to determine whether this mitogen can activate granulocytes or monocytes directly to generate EP, or whether it produces this effect solely via the release of an intermediary product from lymphocytes. The rapidity of onset of fever (within 15–30 min) after iv injection of Con A suggests, however, that this agent may directly stimulate EP-producing cells. The importance of EP release resulting from the intermediate role of lymphocytes, demonstrated in vitro by our experiments, remains to be determined.

References

1. Atkins, E., Bodel, P. Fever. N. Engl. J. Med. 286:27–34, 1972.
2. Dinarello, C. A., Bodel, P. T., Atkins, E. The role of the liver in the production of fever and in pyrogenic tolerance. Trans. Assoc. Am. Physicians 81:334–344, 1968.
3. Root, R. K., Nordlund, J. J., Wolff, S. M. Factors affecting the quantitative production and assay of human leukocytic pyrogen. J. Lab. Clin. Med. 75:679–693, 1970.
4. Atkins, E., Feldman, J. D., Francis, L., Hursh, E. Studies on the mechanism of fever accompanying delayed hypersensitivity. J. Exp. Med. 135:1113–1132, 1972.
5. Dumonde, D. C., Wolstencroft, R. A., Panayi, G. S., Matthew, M., Morley, J., Howson, W. T. "Lymphokines": non-antibody mediators of cellular immunity generated by lymphocyte activation. Nature (Lond.) 224:38–42, 1969.
6. Schaedler, R. W., Dubos, R. Relationship of intestinal flora to resistance. In M. Landy and W. Braun [ed.] Bacterial endotoxins. Rutgers University Press, New Brunswick, N.J., 1964, p. 390–396.
7. Andersson, J., Möller, G., Sjöberg, O. Selective induction of DNA synthesis in T and B lymphocytes. Cell. Immunol. 4:381–393, 1972.
8. Lawrence, H. S., Landy, M. [ed.] Mediators of cellular immunity. Academic Press, New York, 1969. 469 p.
9. Gery, I., Krüger, J., Spiesel, S. Z. Stimulation of B-lymphocytes by endotoxin. Reactions of thymus-deprived mice and karyotypic analysis of dividing cells in mice bearing T6T6 thymus grafts. J. Immunol. 108:1088–1091, 1972.
10. Stobo, J. D., Rosenthal, A. S., Paul, W. E. Functional heterogeneity of murine lymphoid cells. I. Responsiveness to and surface binding of concanavalin A and phytohemagglutinin. J. Immunol. 108:1–17, 1972.
11. Beeson, P. B. Tolerance to bacterial pyrogens. I. Factors influencing its development. J. Exp. Med. 86:29–38, 1947.
12. Snell, E. S., Atkins, E. Interactions of gram-negative bacterial endotoxin with rabbit blood in vitro. Am. J. Physiol. 212:1103–1112, 1967.
13. Gery, I., Gershon, R. K., Waksman, B. H. Potentiation of the T-lymphocyte response to mitogens. I. The responding cell. J. Exp. Med. 136:128–142, 1972.
14. Gery, I., Waksman, B. H. Potentiation of the T-lymphocyte response to mitogens. II. The cellular source of potentiating mediators. J. Exp. Med. 138:143–155, 1972.
15. Böyum, A. Isolation of mononuclear cells and granulocytes from human blood. Isolation of mononuclear cells by one centrifugation, and of granulocytes by combining centrifugation and sedimentation at 1 g. Scand. J. Clin. Lab. Invest. 21(Suppl. 97):77–89, 1968.

Humoral Immunity to Type-Specific and Cross-Reactive Antigens of Gram-Negative Bacilli

W. R. McCabe, A. Greely, T. DiGenio, and M. A. Johns

From the Evans Memorial Department of Clinical Research, University Hospital, Boston University Medical Center, Boston, Massachusetts

Active and passive immunization with antigens shared by most Enterobacteriaceae demonstrated that antibodies against the basal portion of core lipopolysaccharide (Re determinants) protected against challenge with heterologous, smooth gram-negative bacilli. Active immunization with Re induced a 13- to 100-fold increase in the number of *Escherichia coli* 107 and *Klebsiella pneumoniae* required for an LD_{50}. Other rough mutants and common enterobacterial antigen (CA) failed to induce consistent protection. In specimens of acute serum, antibody titers to O-specific antigen, CA, and Re were related to the frequency of shock or death in 206 episodes of gram-negative bacteremia. Frequency of shock and death could not be related to height of antibody titers to O antigen or CA. These complications occurred equally often in those with high and low antibody titers. In contrast, a significant correlation ($P < .01$) was observed between increasing titers of Re antibody and decreasing frequency of shock or death.

Despite their increasing frequency and clinical importance, relatively little is known about immunity to infections caused by Enterobacteriaceae and Pseudomonadaceae. Antibody to O-specific antigens and/or capsular antigens has been shown to protect experimental animals against challenge with some gram-negative bacilli [1–4]. There is no evidence, however, to indicate that humoral antibody exerts any protective effect in man against infections caused by enterobacteria, except possibly typhoid. So-called "natural antibodies" are present in most individuals but do not appear to be protective. In addition, persistence, and even acquisition, of new renal infections has occurred despite high titers of O-specific antibody [5, 6]. Similarly, lethal gram-negative bacteremia often develops despite high titers of O-specific antibody, as will be shown in this publication.

Since the core regions of the cell-wall lipopolysaccharides of most enterobacteria are identical or extremely similar chemically [7], gram-negative bacilli share several cross-reactive antigens.

These studies were supported by research grants no. AI 09584-02, AI 05901-08, and 5 P01-HE07299-11 and by training grant no. 5 T01-AI 213-10 from the U.S. Public Health Service.

Please address requests for reprints to Dr. W. R. McCabe, University Hospital, 750 Harrison Avenue, Boston, Massachusetts 02118.

Tate and Braude have demonstrated that antiserum to rough bacilli protected against challenge with heterologous endotoxin [8]. Similarly, Chedid et al. demonstrated that antiserum to rough salmonellae protected mice against lethal infections with *Klebsiella pneumoniae* [9]. In addition, it has been suggested that antibody to common enterobacterial antigen (CA), first described by Kunin [10], might have protective activity [11]. The progressively increasing frequency of nosocomial infections due to gram-negative bacilli [12–14] and the dearth of information concerning immunity to such infections in man prompted studies of the importance of antibodies to O-specific and cross-reactive antigens in both experimental and human bacteremia due to gram-negative bacteria.

Materials and Methods

Immunization and infection of animals. New Zealand white rabbits were immunized with CA (*Escherichia coli* 0:14) and *Salmonella minnesota* S218 and its chemically defined rough mutants, Ra 60, Rb 345, Rc 5, Rd_1 7, $Rd_2$3, and Re 595. In addition, these same strains, along with *K. pneumoniae* type II and *E. coli* 107, were used to immunize actively 20–22-g male CF1 mice. Two murine-virulent challenge strains, *K.*

pneumoniae type II and *E. coli* 107, were given intravenously to mice as a test of the efficacy of active and passive immunization with type-specific and shared cross-reactive antigens. The exact techniques used for immunization, for collection, preservation, and passive transfer of antiserum, and for challenge are described in detail elsewhere [15].

Gram-negative bacteremia. Specimens of serum were obtained from 206 patients during the acute phase of gram-negative bacteremia (when bacteremia was suspected clinically or immediately after growth was observed in blood cultures). Titers of antibody to O-specific antigen of the infecting strain, to determinants of Re mutants, and to CA were determined in these specimens and correlated with the severity and outcome of bacteremia. For these studies, it was assumed that the occurrence of shock and death reflected more severe bacteremia than bacteremia in which these complications did not occur. The frequency of these complications was determined for patients at each level of titer of antibody to O-specific antigen, Re antigen, and CA. Ideally, it would have been preferable to relate height of antibody titers to frequency of development of bacteremia. However, the importance of surgery and manipulative procedures in the development of bacteremia made selection of an appropriate group of controls impossible and thus precluded such an approach.

Earlier studies have clearly demonstrated that the severity of the host's underlying disease is the most important determinant of the outcome of gram-negative bacteremia. For this reason, all patients were classified into two categories of underlying disease, ultimately fatal and nonfatal underlying diseases, as previously described by McCabe and Jackson [12].

Antibody determination. Antibody titers to O-specific antigen of the patient's infecting organism, CA, and Re were measured in both experimental animals and specimens of acute serum from patients with gram-negative bacteremia by methods described in other publications [15–17].

Results

Studies with animals. Passive immunization. The protective activity induced in mice by passive transfer of 0.3 ml of antiserum from rabbits im-

munized with CA, smooth *S. minnesota,* and six chemically defined rough mutants of *S. minnesota* against challenge with 100 LD$_{50}$ of *K. pneumoniae* (10^4 bacilli) or 100 LD$_{50}$ of *E. coli* 107 (10^8 bacilli) (figure 1). Control animals received a similar quantity of nonimmune rabbit serum and were challenged with identical numbers of *K. pneumoniae* or *E. coli.* As shown in the upper portion of figure 1, passive transfer of antiserum from rabbits immunized with the Re mutant afforded significant protection ($\chi^2 = 45$; $P < .001$) against challenge with *E. coli* 107. In contrast, antiserum to CA, *S. minnesota* 218, and its rough mutants (Ra, Rb, Rc, Rd$_1$, and Rd$_2$) did not increase rates of survival after challenge with *E. coli.*

Passive transfer of antiserum to Re was also the most protective ($\chi^2 = 13.8$; $P < .001$) against challenge with *K. pneumoniae,* as shown in the lower portion of figure 1. Antiserum against the Rd$_2$3 mutant also afforded significant protection ($\chi^2 = 5.5$; $P < .02$) but was somewhat less effective than Re antiserum. Antiserum against CA, *S. minnesota* S218, and its Ra, Rb, Rc, and Rd$_1$ mutants afforded no protection

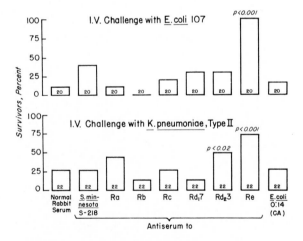

Figure 1. Survival rate in groups of mice given 0.3 ml of serum from rabbits immunized with *Salmonella minnesota* S218 and its Ra, Rb, Rc, Rd$_1$, Rd$_2$, and Re mutants or normal rabbit serum 60 min before intravenous challenge with *Escherichia coli* 107 or *Klebsiella pneumoniae* type II. In addition, the protective effect of antibody to common enterobacterial antigen (CA) was evaluated with antiserum from rabbits immunized with *E. coli* O:14. Only antiserum to the Re mutant afforded significant challenge against both strains, although Rd$_2$ antiserum provided significant protection with *K. pneumoniae.*

278

against challenge with *K. pneumoniae*. These studies of the effect of passive transfer of antiserum against various antigens shared by most gram-negative bacilli demonstrated that antibody to the basal portion of lipopolysaccharide core (Re determinants) enhanced survival after challenge with two strains of smooth heterologous bacilli by three- to 10-fold.

To insure that the protection observed after passive transfer of Re antiserum was actually mediated by antibody specific for determinants of Re mutants, we assayed the protective activity of unadsorbed Re antiserum and antiserum adsorbed with washed, killed Re bacilli. Adsorption of Re antiserum with dead Re bacilli completely removed hemagglutinating antibody to Re-coated erythrocytes and obviated the protective effect of this antiserum against challenge with *K. pneumoniae* and *E. coli*.

The effects of active intraperitoneal immunization with heat-killed *E. coli* O:14 (CA) and the Ra, Rb, Rc, Rd₂, and Re mutants of *S. minnesota* against challenge with heterologous gram-negative bacilli were then evaluated (figure 2). In addition to saline-injected controls, an equal number of animals were immunized intraperi-

toneally with *Pseudomonas aeruginosa* and served as controls for any nonspecific protective effect of endotoxin. As is illustrated in the upper portion of figure 2, active immunization with the Rc, Rd₂, and Re mutants of *S. minnesota* significantly enhanced survival after challenge with *E. coli* 107. Immunization with Re was most protective, producing an almost threefold increase in rate of survival over that in control animals.

The lower portion of figure 2 demonstrates that only the Re mutant provided a significant degree of protection when used for active immunization before challenge with *K. pneumoniae* type II. Active immunization with Ra, Rb, Rc, Rd₂, and CA failed to afford any significant increase in survival rates over those observed in controls injected with saline or pseudomonas after infection with *K. pneumoniae*. Thus, antibodies to Re mutants, induced by both active and passive immunization, were capable of affording significant protection against lethal infections caused by heterologous gram-negative bacilli.

The efficacy of immunization with the shared antigen, Re, was then compared with type-specific immunization. The number of *K. pneumoniae* and *E. coli* 107 required to produce an LD_{50} was determined in controls, mice immunized with Re, and mice immunized with the challenge strain. The LD_{50} of *E. coli* 107 was 4.7×10^6 in control animals, while a 13-fold increase (6.3×10^7) was required to produce an LD_{50} in mice immunized with Re. Type-specific immunization increased the LD_{50} by more than 100-fold, however. With *K. pneumoniae*, immunization with Re enhanced resistance 100-fold (LD_{50}, 4.4×10^4) over that in controls (LD_{50}, 2.7×10^2). However, type-specific immunization produced more than a 10,000-fold increase in the number of *K. pneumoniae* required to produce an LD_{50}.

Gram-negative bacteremia. Thus, either actively or passively transferred antibody to an antigen (determinants of Re mutants) shared by most gram-negative bacilli afforded significant protection against challenge with heterologous bacilli; this finding prompted investigation of the protective activity of antibody to type-specific antigens and to two shared cross-reactive antigens, CA and Re, in gram-negative bacteremia in man. Titers of antibody to O antigen of the patients' infecting organism, CA, and Re were determined in serum specimens obtained during the acute phase of 206

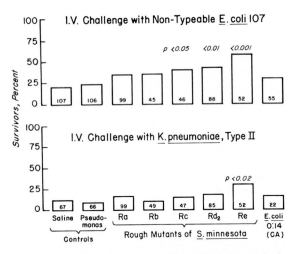

Figure 2. Survival rates in groups of mice actively immunized with saline, *Pseudomonas aeruginosa* (controls), the Ra, Rb, Rc, Rd₂, and Re mutants of *Salmonella minnesota* and CA (*Escherichia coli* O:14) and challenged with *E. coli* or *Klebsiella pneumoniae*. Although active immunization with Rc, Rd₂, and Re mutants provided significant protection against challenge with *E. coli* 107, only immunization with Re mutants protected against challenge with *K. pneumoniae*.

episodes of gram-negative bacteremia and correlated with the frequency of shock or death.

The proportion of patients at each level of O antibody titer who had either ultimately fatal or nonfatal underlying disease and who developed shock or succumbed to gram-negative bacteremia is shown in figure 3. As is readily apparent, both by examination of this figure and by statistical analysis, the height of antibody titer to the infecting organism at the onset of bacteremia bore no relation to the frequency with which these complications occurred. Neither shock nor death was any less frequent among patients with high titers of O antibody than among those with low titers, whether in patients with ultimately fatal disease, in those with nonfatal underlying disease, or in both groups combined.

Similarly, no relation could be detected between the frequency of shock and death and titers of antibody to CA (figure 4). The frequency of these complications did not differ significantly in patients with high titers or low titers of antibody to CA among those in either category of underlying host disease.

In contrast, high titers of antibody to Re were associated with amelioration of the severity of gram-negative bacteremia. There was an almost linear relation between increasing titers of antibody to Re and decreasing frequency of shock or death among patients with ultimately fatal underlying disease (figure 5). When this finding was examined by point biserial correlation (a statistical technique that allows correlation of titers, number of observations at each titer, and frequency of shock and death), a highly significant

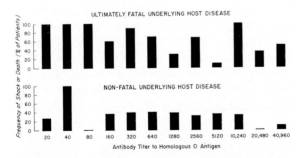

Figure 3. Frequency of shock or death in relation to titers of O antibody in specimens of serum taken during the acute phase of 206 episodes of gram-negative bacteremia. High titers of O antibody in these specimens were not associated with a decreased frequency of shock or death.

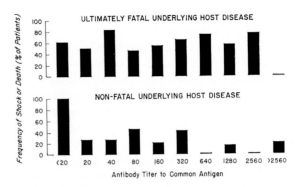

Figure 4. Frequency of shock or death in relation to antibody titers to Common Enterobacterial Antigen (CA) in gram-negative bacteremia. These complications were no less frequent among patients with high titers of antibody to CA than in those with low titers of antibody to CA in specimens of acute-phase serum.

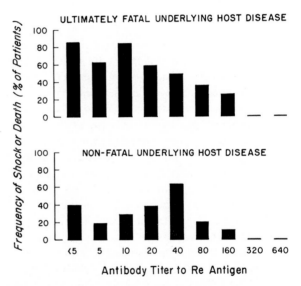

Figure 5. Relation of frequency of shock or death in gram-negative bacteremia to antibody titers to Re determinants in specimens of acute-phase serum. Although no correlation was apparent among patients with nonfatal underlying disease, a definite correlation between increasing titers of antibody to Re and decreasing frequency of shock or death was observed in patients with ultimately fatal underlying disease ($P < .001$) and when both groups of patients were combined ($P < .01$).

relation ($P < .001$) was found for patients with ultimately fatal underlying disease. This relation was not statistically significant ($P > .05$) for patients with nonfatal underlying disease. When both groups of patients were combined, the correlation between increased titers of antibody to

Re and decreased severity of bacteremia was statistically significant $(P < .01)$.

The protective effect of antibody to Re is most apparent when the frequency of shock and death is compared in patients with Re titers of less than 1:80 and with the frequency in those with titers $\geq 1:80$. Shock or death occurred in 69% of patients who had ultimately fatal underlying disease and titers of less than 1:80, in contrast to an incidence of 27% in patients with titers of $\geq 1:80$. Among patients with nonfatal underlying disease, 36% of those with Re titers of $<1:80$ experienced shock or death, while these complications occurred in only 11% of those with higher titers. In addition, the frequency of shock and death diminished progressively as Re titers increased above 1:80.

Discussion

The present investigations clearly demonstrate that antibodies to an antigen (Re) shared by most gram-negative bacilli afford significant protection to experimental animals against challenge with heterologous smooth enterobacteria. Although type-specific immunization of experimental animals was quantitatively superior to immunization with Re, antibody to Re was associated with a 13- to 100-fold increase in resistance to challenge with heterologous gram-negative bacilli. Studies of gram-negative bacteremia in man paralleled those in experimental animals and indicated that high titers of antibody to Re in specimens of acute-phase serum were associated with a threefold reduction in the frequency of shock and death. The failure to demonstrate any correlation between titers of antibody to O-specific antigen and severity of bacteremia is somewhat surprising in view of the paramount importance of type-specific immunity in streptococcal and pneumococcal disease. There is little evidence that antibody to O-specific antigens is associated with any protective activity in man, however. The present observations provide additional evidence for the relative unimportance of humoral antibody to O-specific antigens in resistance to infections caused by Enterobacteriaceae.

This demonstration that immunization with an antigen shared by most enterobacteria protects both humans and experimental animals against infections with heterologous bacilli suggests that it might be possible to develop an immunizing agent capable of enhancing resistance to a wide range of Enterobacteriaceae. There is a critical need for the development of more effective methods for the control of infections caused by gram-negative bacilli. Such infections are becoming increasingly prevalent and now constitute the most frequent type of nosocomial infection [12–14]. While the actual number of gram-negative infections in hospitals in the United States each year cannot be determined accurately, it is possible to provide some estimates of their frequency and thus to emphasize the magnitude of this problem. Gram-negative bacteremia has been reported with a frequency of at least one case per every 100 patients in several major medical centers [13, 14, 18]. If there is a similar prevalence for the 30,000,000 annual acute hospital admissions in the United States, there may be as many as 300,000 episodes of gram-negative bacteremia each year. The fatality rate from gram-negative bacteremia has approximated or exceeded 30% in most reports, and if this rate is extrapolated to these estimates of frequency, as many as 90,000 patients may die from this cause each year in this country. Although these estimates are based primarily on prevalence data from major medical centers rather than community hospitals, they do demonstrate that nosocomial gram-negative infections are a problem of considerable magnitude.

Attempts to control such infections have met with little success. Since most gram-negative bacilli have only a limited capacity to invade normal hosts, such infections tend to occur in debilitated patients or to result from direct introduction of bacteria into the vascular system or local sites by manipulative procedures, catheters, or intravascular devices. The universal presence of gram-negative bacilli in the gastrointestinal tract, their ubiquity within the hospital environment, and a high frequency of antibiotic resistance provide a ready explanation of their frequency as causes of nosocomial infections. The same factors also explain why preventive measures and antimicrobials have proved to be relatively ineffective methods of control. The development of immunologic methods to enhance resistance to enterobacterial infections appeared to be precluded by the large number of species and serotypes capable of producing such infections. The demonstration that antibody to a shared antigen does exert protective activity suggests that it might be feasible to develop a satisfactory immunizing agent against gram-negative bacteremia and that this should be evaluated further.

References

1. Markley, K., Smallman, E. Protection by vaccination against pseudomonas infection after thermal injury. J. Bacteriol. 96:867–874, 1968.
2. Wolberg, G., DeWitt, C. W. Mouse virulence of K (L) antigen-containing strains of *Escherichia coli*. J. Bacteriol. 100:730–737, 1969.
3. Sanford, J. P., Hunter, B. W., Souda, L. L. The role of immunity in the pathogenesis of experimental hematogenous pyelonephritis. J. Exp. Med. 115: 383–410, 1962.
4. Braude, A. I., Siemienski, J. The influence of endotoxin on resistance to infection. Bull. N.Y. Acad. Med. 37:448–467, 1961.
5. Vosti, K. L., Monto, A. S., Rantz, L. A. Host-parasite interaction in patients with infections due to *Escherichia coli* II. Serologic response of the host. J. Lab. Clin. Med. 66:613–626, 1965.
6. Williamson, J., Brainerd, H., Scaparone, M., Chueh, S. P. Antibacterial antibodies in coliform urinary tract infections. Arch. Intern. Med. 114:222–231, 1964.
7. Lüderitz, O., Staub, A. M., Westphal, O. Immunochemistry of O and R antigens of Salmonella and related Enterobacteriaceae. Bacteriol. Rev. 30: 192–255, 1966.
8. Tate, W. J., III, Douglas, H., Braude, A. I., Wells, W. W. Protection against lethality of E. coli endotoxin with "O" antiserum. Ann. N. Y. Acad. Sci. 133:746–762, 1966.
9. Chedid, L., Parant, M., Parant, F., Boyer, F. A

proposed mechanism for natural immunity to enterobacterial pathogens. J. Immunol. 100:292–301, 1968.
10. Kunin, C. M., Beard, M. V., Halmagyi, N. E. Evidence for a common hapten associated with endotoxin fractions of E. coli and other Enterobacteriaceae. Proc. Soc. Exp. Biol. Med. 111: 160–166, 1962.
11. Gorzynski, E. A., Ambrus, J. L., Neter, E. Effect of common enterobacterial antiserum on experimental *Salmonella typhimurium* infection of mice. Proc. Soc. Exp. Biol. Med. 137:1209–1212, 1971.
12. McCabe, W. R., Jackson, G. G. Gram-negative bacteremia I. Etiology and ecology. Arch. Intern. Med. 110:847–855, 1962.
13. Myerowitz, R. L., Medeiros, A. A., O'Brien, T. F. Recent experience with bacillemia due to gram-negative organisms. J. Infect. Dis. 124:239–246, 1971.
14. Finland, M. Changing ecology of bacterial infections as related to antibacterial therapy. J. Infect. Dis. 122:419–431, 1970.
15. McCabe, W. R. Immunization with R mutants of S. minnesota I. Protection against challenge with heterologous gram-negative bacilli. J. Immunol. 108:601–610, 1972.
16. McCabe, W. R., Kreger, B. E., Johns, M. Type specific and cross-reactive antibodies in gram-negative bacteremia. N. Engl. J. Med. 287:261–267, 1972.
17. Johns, M. A., Whiteside, R. E., Baker, E. E., McCabe, W. R. Common enterobacterial antigen. I. Isolation and purification from *Salmonella typhosa* O:901. J. Immunol. (in press).
18. DuPont, H. L., Spink, W. W. Infections due to gram-negative organisms. An analysis of 860 patients with bacteremia at the University of Minnesota Medical Center, 1958–1966. Medicine (Baltimore) 48:307–332, 1969.

Role of the Intestinal Flora in Major Trauma

P. W. H. Woodruff, D. I. O'Carroll,
S. Koizumi, and J. Fine

*From the Harvard Surgical Unit and the Department
of Pathology, Boston City Hospital,
Boston, Massachusetts*

The Limulus-lysate test revealed endotoxin in postmortem tissue homogenates of 20 of a random series of 35 patients. Data indicated that the finding of endotoxin in the livers of these patients signified an endotoxemia consequent on a failure of the endotoxin-detoxifying mechanisms. Endotoxemia was associated with a high incidence of gastrointestinal bleeding, hemorrhagic ulceration of the gastrointestinal mucosa, acute pulmonary pathology, and major injury to the liver. The absence of a septic focus as a source for endotoxemia in many of these patients is consistent with evidence that endotoxemia of intestinal origin commonly develops when the antibacterial defense mechanism has been depleted as a result of severe damage to the reticuloendothelial system.

Circulating endotoxin has been identified in various septic and nonseptic disorders in man and in experimental animals by the Limulus-lysate technique, a specific and highly sensitive quantitative assay for endotoxin [1]. In the experimental animal, persisting endotoxemia produces an increase in vascular permeability, plasma and red-cell extravasation into the lung and intestine, and progressive vascular collapse. In this communication we shall present data to show that a similar syndrome occurs in man, and that the intraintestinal gram-negative flora are frequently responsible for the development of this syndrome.

Clinical Studies

Table 1 lists a series of patients with septic and nonseptic disorders in whom circulating endotoxin was demonstrated [2]. In assembling the data, we encountered additional patients who had clinical features suggesting persisting endotoxemia

This study was aided by a grant from the National Institutes of Health, Bethesda, Maryland, and by a contract with the Office of the Surgeon General, United States Army.

We gratefully acknowledge the technical assistance of Miss Mary Kendrick, Channing Laboratory, and of Mr. Allan Kaufman, Department of Surgery, Harvard Surgical Unit.

Please address requests for reprints to Dr. J. Fine, Harvard Surgical Unit, Boston City Hospital, Boston, Massachusetts 02118.

but in whom the test was negative. Because of the extreme sensitivity of man to endotoxin [3, 4], we suspected that clinical endotoxemia and its consequences may exist even when the titer is below the capacity of the lysate to detect it. In view of this limitation, we undertook postmortem studies on a random series of patients, having observed that, in animals dying of endotoxic shock, the liver and spleen immediately after death contain considerably higher concentrations of endotoxin than the blood or other tissues.

The data summarized in table 2 are from such a series of 35 patients. The endotoxin content of liver, spleen, lung, and cardiac blood was determined from samples obtained under sterile conditions. Twenty of 35 patients had endotoxin in liver and other tissues. In 12 cases endotoxin was also found in plasma, in three endotoxin was not detected, and in five no sample of plasma was obtained. In 15 of the 35 cases, no endotoxin was found in blood or tissues.

Thirteen of the 20 endotoxin-positive cases suffered gastrointestinal bleeding, most of them within the week before death, while only one of the 15 negative cases suffered a comparable loss of blood. Twelve of the 13 positive cases that bled showed mucosal changes in the gastrointestinal tract, ranging from a congested edematous mucosa with petechiae to hemorrhagic infarction of the entire small bowel mucosa. Eight of these 12 patients exhibited superficial gastric erosions. Sixteen of the 20 positive cases showed severe

Table 1. Titers of circulating endotoxin in plasma of one series of patients studied. (Modified from [2] with permission of *The Lancet*.)

Septic disorder	Endotoxin (μg/ml of plasma)	Nonseptic disorder	Endotoxin (μg/ml of plasma)
Cystitis (renal transplant)	0.6	Heat stroke (hepatic failure)	0.5
Perforated appendicitis	0.008	Closed loop obstruction	0.008
Multiple trauma	0.128	Cirrhosis (hepatic failure)	0.004
Pneumonia, cystitis	0.128	Myeloblastic leukemia	0.25
Pneumonitis (renal transplant)	0.01	Splenectomy in advanced cirrhosis	0.03
Perforation of sigmoid (carcinoma)	0.02	Drug addiction (hepatic failure)	0.2
Wound sepsis (hip fracture)	0.008	Gangrene transverse colon	0.008
Ruptured bladder (pelvic fracture)	0.016	Acute pancreatitis	0.128
Cystitis (prostatectomy)	0.012	Mesenteric occlusion (gangrene)	0.008
Perforation of sigmoid (diverticulitis)	0.004	Burn 30% (C.V.A.)	0.016
Postoperative pericaecal abscess	0.08	Hepatectomy (right lobe)	0.02
Infected cardiac catheter	0.07	Volvulus sigmoid (portal vein)	0.004

acute pathology in the lungs, as compared to only three of the 15 cases without endotoxemia. Sepsis, in terms of a bacteremia or of an undrained gram-negative septic focus, was present in six of the 20 positive patients and in two of the 15 negative patients. Fifteen of the 20 positive cases had structural damage to the liver, which was severe in 10 cases (alcoholic hyaline, hepatitis, and cirrhosis) and mild in five (fatty metamorphosis, increased portal fibrous tissue, congestion, and bile stasis). Structural changes, all mild, were observed in seven of the 15 cases without endotoxemia.

The average interval between death and autopsy in both the endotoxin-positive and the endotoxin-negative groups was less than 20 hr. Studies were made for evaluation of the validity of endotoxin titers in livers obtained at various intervals after death. The amount present in rabbits that died from various causes was no greater up to 24 hr after death than at the time of death (table 3). The effect on these titers of bacterial growth after the patient's death was not significant, because the bacterial counts were usually no greater than 10^3/g of tissue in the rabbit and were less than 10^4/g in 80% of the human tissues ($N = 41$). Since proliferating bacteria do not produce a positive test until the count is 10^4/ml or higher (table 4), the amount detected, for the most part, represents killed or disintegrated bacteria present at the time of death. We excluded the possible effect of refrigeration on values after observing that three cycles of freezing and thawing of various cultures in the growth phase did not

Table 2. Incidence of disorders in liver, lung, and gastrointestinal tract (GI) of patients with and without terminal endotoxemia.

No. of patients	Endotoxemia	Liver injury (μg of endotoxin/g of liver)*	No. of patients with disorder			
			GI bleeding†	GI ulceration‡	Lung lesion§	Sepsis‖
15	No	7+ (0)	1	1	3	2
20	Yes	10++ (0.76)	9	6	10	6
		5+ (0.16)	2	4	3	0
		5° (0.31)	2	2	3	0

* Structural microscopic injury, mild (+) or severe (++). No injury = (°). Note presence of endotoxin in liver of five patients with endotoxemia, but without noticeable hepatic injury. (Endotoxin in liver signifies failure of reticuloendothelial detoxifying function.)

† Preterminal blood loss.

‡ Multiple mucosal ulcers in stomach and small intestine.

§ Edema and hemorrhagic extravasation, septic pneumonitis, or both.

‖ Gram-negative bacteremia or undrained gross gram-negative septic focus.

Table 3. Effect of time after death on endotoxin content and bacterial count/g of rabbit liver.

Cause of death	Endotoxin		Bacterial count	
	A*	B	A	B
Intestinal obstruction	0	0 (9)	...	...
Intestinal obstruction	0	0 (9)	...	...
Mucoid peritonitis	0	0 (20)	20	1×10^3 (20)
Pentobarbital	0	0 (24)	60	7.5×10^3 (24)
Pentobarbital	0	0 (4)	0	0 (4)
Pentobarbital	0	0 (4)	0	0 (4)
Occlusion of superior mesenteric artery (1 hr)	0.004	0.004 (18)	0	8 (18)
Occlusion of superior mesenteric artery (1 hr)	0.004	0.002 (18)	24	48 (18)
Occlusion of superior mesenteric artery (1 hr)	0.016	0.002 (24)	6×10^3	2×10^4 (24)
Occlusion of superior mesenteric artery (1 hr)	0	0 (24)	3×10^2	6×10^2 (24)

* A = data obtained at death. B = corresponding data obtained hours after death (number of hours specified in parentheses).

alter the amount of endotoxin. Since endotoxin was not detected in blood or tissues of normal animals immediately after they were killed and up to 24 hr afterward, we concluded that endotoxin in the liver within this interval after death signified an antemortem failure of detoxification. Loss of the ability to detoxify endotoxin in the 20 positive cases was not merely a terminal phenomenon, because endotoxemia did not occur in 15 of the 35 patients. Its presence may therefore be taken to indicate that it was responsible for the other disorders (found at postmortem examination) that are commonly associated with

Table 4. Endotoxin content of rabbit-liver cultures in early growth phase.

Source of endotoxin	Count/ml	Endotoxin (μg/ml)*
Klebsiella	1×10^4	0.008
Escherichia coli	2×10^4	0.016
Klebsiella	3×10^4	0.016
E. coli	4×10^5	0.064
Klebsiella	3×10^5	0.008
Klebsiella	1×10^5	0.128 (0.128)† (4)‡
Klebsiella	9×10^5	0.128 (0.128) (8)
Proteus mirabilis	6×10^5	0.25 (0.25) (8)
E. coli	1×10^4	0.5 (0.5) (8)

* Test was negative in all cultures (N = 11) with counts below 10^4.

† Values obtained after three cycles of freezing and thawing.

‡ Values obtained after boiling for 45 min.

endotoxemia in the experimental animal. Since a septic process was absent in a substantial number of patients in this series, we assume that endotoxin in blood and tissues of these patients was derived from the intestine. This may be the result of damage to the detoxifying system in the liver.

Recent data obtained by use of the Limulus-lysate technique in the rabbit show that the intestine is the source of endotoxemia in nonseptic disorders and also in many septic disorders. Thus, if the intestinal bacteria have been sufficiently suppressed by a nonabsorbable antibiotic injected directly into the ileum and caecum, the titer of circulating endotoxin is reduced, and the rate of survival rises substantially (tables 5 and 6) [6].

The quantitative significance of the role of intestinal flora is further demonstrated by the striking effect of prior treatment with intraintestinal kanamycin (table 7, groups A and B) on the amount of endotoxin in livers of rabbits subjected to an iv injection of endotoxin.

A telling piece of evidence on the critical role of intestinal flora is a recent change in the response of rabbits to massive injury that is related to a change in their intestinal flora. These rabbits (table 7, group C) survived an iv dose of endotoxin that is lethal in rabbits with a normal flora. Cultures of their blood and livers yielded gram-positive rather than the usual gram-negative organisms. Aerobic cultures of the intestinal contents yielded enterococci, staphylococci, and *Bacillus subtilis,*

Table 5. Relation of endotoxemia to lung lesion and survival rate in shock due to occlusion of the superior mesenteric artery. (Modified from [6] with permission of *Archives of Surgery*.)

Treatment and experiment	Circulating titer (µg/ml)	Duration of endotoxemia (hr)	Degree of lung injury*	Survival
None				
1	1.2	2	1	No
2	2.0	2	1	No
3	2.0	2	1	No
4	1.0	2	1	No
5	0.05	2	1	No
6	3.0	0.5	0	No
7	2.0	2	1	No
8	0.5	3	2	No
9	0.5	2	1	No
10	0.25	1	0	No
Rendered resistant				
11	0.03	1	0	Yes
12	0.06	1	0	Yes
13	0.1	1	0	Yes
14	0.06	1	0	Yes
15	0.03	1	0	Yes
16	0.2	2	1	No
17	0.03	1	0	Yes
Intraintestinal antibiotic				
18	0.5	2	1	No
19	0	0	0	Yes
20	1	1	1	No
21	0	0	0	Yes
22	0	0	0	Yes
23	0.01	1	0	Yes
24	0	0	0	Yes
25	1	1	1	No
26	0.01	1	0	Yes
27	0.25	2	1	Yes

* Intensity of hemorrhagic edema graded 0–3.

Table 6. Relation of endotoxemia to lung lesion and survival rate in shock due to *Escherichia coli* peritonitis. (Modified from [6] with permission of *Archives of Surgery*.)

Treatment and experiment	Circulating titer (µg/ml)	Duration of endotoxemia (hr)	Degree of lung injury*	Survival
None				
1	0.2	4+	2	No
2	0.1	4+	3	No
3	0.4	4+	3	No
4	0.2	4+	2	No
5	1.0	2	1	No
6	0.1	4+	3	No
7	1.0	4+	2	No
8	3.0	4+	2	No
9	0.01	4+	2	No
10	0.1	3	1	No
Rendered resistant				
11	0.01	4	2	No
12	0.01	4	2	Yes
13	0.01	2	1	Yes
14	0.01	1	0	Yes
15	0.01	1	0	Yes
16	0.02	4	3	No
Intraintestinal antibiotic				
17	0	0	0	No
18	0.1	4	3	No
19	0.1	3	2	No
20	0.1	2	1	No
21	0.01	4	2	No
22	0	0	0	Yes
23	0.4	2	1	Yes
24	0	0	0	Yes
25	0.05	2	1	Yes
26	0.01	1	0	Yes

* Intensity of hemorrhagic edema graded 0 to 3.

but only rarely a gram-negative organism. The counts showed a reduction from previous values of 10^9 or more per g to 10^5 or less. The concentration of intraluminal endotoxin showed a parallel decline from an average minimum of 3,000 to 125 µg/g of caecal content.

The difference that this change in the intestinal

Table 7. Mean content (µg) of endotoxin in various tissues of rabbits after iv injection of 100 µg/kg.

Type of recipient*	Plasma		Liver		Spleen		Lung		Total of means
	Per ml	Total	Per g	Total	Per g	Total	Per g	Total	
Group A (n = 4)	0.6	48.0	4.6	383.0	2.7	4.7	0.5	4.7	454.0
Group B (n = 5)	0	0	0.16	12.8	0.01	0.02	0	0	12.8
Group C (n = 5)	0	0	0	0	...	...	...	...	0

NOTE. Data were obtained after death or in survivors killed 24 hr after injection.

* All rabbits except those in group C had normal intestinal flora. (A) = no treatment; (B) = pretreated with intraintestinal kanamycin; (C) = no treatment; intestinal flora exclusively or predominantly gram-positive.

286

flora makes in response to major trauma is further demonstrated by the absence of endotoxemia and its consequences that regularly develop after a 30% immersion burn or a 1-hr occlusion of the superior mesenteric artery in rabbits with ordinary intestinal flora. Current studies indicate that prior instillation of endotoxin in the form of dead and living gram-negative bacteria restores the characteristic response to these types of trauma.

Discussion

The clinical records of the patients studied suggest that the syndrome of endotoxic shock in man, which is like that in the experimental animal, is frequently not recognized as such in patients without an identifiable septic focus. This absence of a septic focus as a source for endotoxemia in many of these patients is consistent with evidence that endotoxemia of intestinal origin commonly develops when the antibacterial defense mechanism has been depleted as a result of severe damage to the reticuloendothelial system [7].

The complicating effects of the intestinal flora in response to various types of major injury indicate the need to revise the rationale of therapy for disorders that deplete the antibacterial defenses.

References

1. Reinhold, R. B., Fine, J. A technique for quantitative measurement of endotoxin in human plasma. Proc. Soc. Exp. Biol. Med. 137:334–340, 1971.
2. Caridis, D. T., Reinhold, R. B., Woodruff, P. W. H., Fine, J. Endotoxemia in man. Lancet 1:1381–1386, 1972.
3. Greisman, S. E., Woodward, C. L. In vivo studies on the role of the liver in endotoxin fever and tolerance (abstract). J. Clin. Invest. 49:37a, 1970.
4. Kimball, H. R., Melmon, K. L., Wolff, S. M. Endotoxin-induced kinin production in man. Proc. Soc. Exp. Biol. Med. 139:1078–1082, 1972.
5. Artz, C. P., Moncrief, J. A. The treatment of burns. 2nd ed. Saunders, Philadelphia, 1969, p. 22–88.
6. Cuevas, P., de la Maza, L. M., Gilbert, J., Fine, J. The lung lesion in four different types of shock in rabbits. Arch. Surg. 104:319–322, 1972.
7. Cuevas, P., Fine, J. Demonstration of a lethal endotoxemia in experimental occlusion of the superior mesenteric artery. Surg. Gynecol. Obstet. 133:81–83, 1971.

Hemodynamic Measurements in Bacteremia and Septic Shock in Man

Rolf M. Gunnar, Henry S. Loeb,
Edward J. Winslow, Charlotte Blain,
and John Robinson

*From the Veterans Administration Hospital, Hines,
Illinois; the Loyola University Stritch School of Medicine,
Maywood, Illinois; and the Cook County Hospital and
the University of Illinois College of Medicine, Chicago,
Illinois*

Patients studied after genitourinary instrumentation have a high incidence of bacteremia. Of 31 such patients studied, eight developed gram-negative bacteremia, 10 developed gram-positive bacteremia, three developed mixed bacteremia, and 10 were abacteremic after instrumentation. Patients with gram-negative bacteremia had greater falls in central venous pressure and systemic vascular resistance and greater increases in cardiac index than either the abacteremic group or the patients with gram-positive bacteremia. These findings are consistent with vasodilatation during early gram-negative bacteremia due to release of a vasodilator substance. Fifty patients were studied during established septic shock, and the 31 with gram-negative infections had lower cardiac output and heart rates than the 19 with gram-positive infections. Vasoconstriction was more common in patients with gram-negative infections. In established shock due to gram-negative organisms, myocardial function appeared to be depressed, and vasodilatation was less marked.

To understand the acute hemodynamic effects of bacteremia in man, we thought it necessary to find a naturally occurring model in which the appearance of bacteremia is likely. Such a model is the patient undergoing genitourinary instrumentation [1]. In studying 31 such patients, we found eight who developed gram-negative bacteremia, 10 who developed gram-positive bacteremia, and three who developed mixed gram-negative and gram-positive bacteremia [2]. We performed hemodynamic measurements before and after surgical intervention and, therefore, could quantitate the effect of bacteremia on the hemodynamic parameters measured as change from the control measurement. Patients who did not develop bacteremia served as controls for the

operative procedures. None of the patients underwent general anesthesia. There were no significant differences in the ages or control hemodynamic values among groups of patients with gram-negative bacteremia, those with gram-positive bacteremia, and those with no bacteremia.

Table 1 lists the organisms cultured during or immediately after intervention. Fifteen of the bacteremic episodes were associated with positive urine cultures before instrumentation; in 11, the organism cultured from the blood was the same as that previously cultured from the urine. *Escherichia coli* and *Pseudomonas* were the most common of the gram-negative organisms found, while *Staphylococcus albus* was the most common gram-positive organism. In both groups, the most common surgical procedures were transurethral resection and cystoscopy.

Changes in systemic vascular resistance (SVR), cardiac output (CO), and central venous pressure (CVP) during bacteremia (as compared to the control measurement obtained the day before) are shown in table 2. Patients with gram-negative infections had a decrease in SVR of 10.3 ± 12.3 mm Hg/liter per min (mean $\pm$ SD), while the

This study was supported in part by NIH grant no. HE-08834 from the National Heart and Lung Institute and by grant no. N69-60 from the Chicago Heart Association.

Please address requests for reprints to Dr. Rolf M. Gunnar, Chief, Section of Cardiology, Loyola University Stritch School of Medicine, 2160 South First Avenue, Maywood, Illinois 60153.

Table 1. Organisms cultured from blood immediately after genitourinary instrumentation.

Gram-negative bacteremia*		Gram-positive bacteremia	
Organism			
Escherichia coli	1	Micrococcus	6
E. coli and		Enterococcus	1
Pseudomonas	2	Gamma Streptococcus	1
Pseudomonas	1	Beta Streptococcus	1
E. coli and Proteus	1	Diphtheroids	1
Aerobacter sp.	2		
Serratia	1		
Surgery			
Transurethral		Transurethral	
resection	3	resection	3
Cystoscopy	2	Cystoscopy	5
Bladder stone	1	Suprapubic cystect.	1
Suprapubic cyst	1	Evacuation of bladder	
Suprapubic cyst	1	clots	1

* Results are from eight patients with gram-negative bacteremia and 10 patients with gram-positive bacteremia.

patients with gram-positive infection had an insignificant mean change of 1.9 ± 3.4 mm Hg/liter per min. Patients without bacteremia also had insignificant change in SVR.

In seven of eight patients, there was a significant increase in CO associated with gram-negative bacteremia. In one patient, CO fell. Mean change was $+1.6 \pm 1.7$ liter/min in gram-negative infections, while there was essentially no change in CO during bacteremia in gram-positive infections or in the abacteremic group.

With gram-negative bacteremia, there was a decrease in CVP averaging 4.6 cm H_2O, while in the gram-negative group the CVP fell 0.8 ± 3.6 cm H_2O. It appears, therefore, that gram-negative bacteremia is associated with vasodilatation manifested by a decrease in CVP and SVR. In response to these changes, CO increases and, if this

increase is sufficient to overcome the fall in SVR, arterial pressure can be maintained.

In three patients (two with gram-negative bacteremia and one with gram-positive bacteremia), the shock syndrome developed. During shock in gram-negative bacteremia, SVR was markedly decreased (-35.4 and -6.5 mm Hg/liter per min), and CO increased ($+2.9$ and $+1.8$ liter/min). CO was increased further during volume infusion ($+0.8$ and $+5.1$ liter/min), and, even though there was little change in the lowered SVR, these two patients recovered without additional therapy. The increased CO was able to compensate for the decreased SVR, and the patients were relieved of the shock syndrome. These findings would indicate that there is no appreciable myocardial failure as part of the shock syndrome at this early stage of gram-negative bacteremia.

Before and during our study of acute bacteremia, we also studied patients with established septic shock [3, 4].

Table 3 presents clinical data on the 50 pa-

Table 3. Clinical characteristics of 50 patients with confirmed septic shock and positive blood cultures.

Patient data	Organism	
	Gram-positive	Gram-negative
Number	19	31
Average age (range)	60	64
	(40–70)	(48–88)
Sex		
Male	15	21
Female	4	10
Outcome		
Survived	2	2
Recovered, died later	5	9
Died in shock	12	20
Alcoholism	8	10
Diabetes	0	4

Table 2. Changes in hemodynamic values after genitourinary surgery in patients with and without bacteremia.

Parameter	Gram-positive	Gram-negative	P	Abacteremic
Cardiac output (liter/min)	$+.01 \pm 1.0*$	$+1.6 \pm 1.7$	$<.05$	$+0.6 \pm 0.8$
Systemic vascular resistance (mm Hg/liter per min)	-1.9 ± 3.4	-10.3 ± 12.3	$<.02$	-2.8 ± 5.2
Central venous pressure (cm H_2O)	-0.8 ± 3.6	-4.6 ± 2.5	$<.01$	-0.8 ± 2.7 cm H_2O
Stroke volume (ml)	$+1.1 \pm 12.3$	$+20.8 \pm 23.0$	$<.05$	$+13.0 \pm 19.4$

NOTE. Values represent change from measurements made the previous day. Significance is by Student's t-test. Values in the abacteremic group are not significantly different from those in the gram-positive group.

* Mean ± SE.

tients with confirmed positive blood cultures. Nineteen patients had gram-positive organisms, and 31 had gram-negative organisms. There were no differences by sex or incidence of alcoholism. The four diabetics were in the gram-negative group. It can be seen from data on survival that the patients had serious illnesses, and many of those who survived the shock episode succumbed later to their underlying illness.

In table 4 are listed the organisms cultured in patients with gram-positive infection. The majority of infections were due to *Pneumococcus,* and these patients had established pneumonia. The next most common organism cultured was *Staphylococcus aureus.*

In table 5, organisms cultured in patients with gram-negative infections are listed. *E. coli* and *Enterobacter* were the most common, with *Proteus* running a close third. Many of these patients had urinary-tract infections. Pneumonia was not uncommon; although it may not have been primary in these patients, this was their terminal presentation. There was one patient with enteritis due to *Shigella.*

Hemodynamic values before treatment are presented in table 6. At this phase of illness, cardiac output and heart rate were lower in patients with gram-negative infections than in patients with gram-positive infections. Twelve of the 13 patients with CO of < 2.5 were in the group with gram-negative bacteremia. Although the difference is not quite significant when means are compared, the SVR was higher in patients with gram-negative infections. Only two of 19 patients in the gram-positive group had SVR above 12 units, while 12 of 31 patients in the gram-negative group had vasoconstriction above this level; this difference is significant ($P < .02$).

The response to volume expansion demonstrates

Table 4. Organisms cultured and site of clinical infection in patients with gram-positive infections.

Organism	No. of patients	Infection
Diplococcus pneumoniae	9	Pneumonia
Staphylococcus aureus	5	Endocarditis (3) Pneumonia (1) Undetermined (1)
Beta hemolytic *Streptococcus*	1	Pneumonia
Microaerophilic *Streptococcus*	1	Undetermined
Anaerobic *Streptococcus*	1	Perforated diverticulum
Enterococcus	1	Urinary tract
Clostridium perfringens	1	Pneumonia
Total	19	

Table 5. Organisms cultured and site of clinical infection in 31 patients with gram-negative infections.

Organism	No. of patients	Infection (no.)
Escherichia coli	13 (5 mixed)	Urinary tract (7) Pneumonia (3) Bronchiectasis (1) Undetermined (2)
Enterobacter sp.	9 (3 mixed)	Urinary tract (5) Pneumonia (2) Cholangitis (1) Undetermined (1)
Proteus sp.	7 (4 mixed)	Urinary tract (35) Perf. appendix (1) Undetermined (1)
Klebsiella sp.	2	Pneumonia (1) Endocarditis (1)
Pseudomonas	2	Undetermined (2)
Paracolon	2	Urinary tract (1) Cholangitis (1)
Citrobacter	1	Urinary tract (1)
Shigella (Group B)	1	Acute enteritis (1)

Table 6. Comparison between hemodynamic values during septic shock in 19 patients with gram-positive organisms and 31 patients with gram-negative organisms cultured from blood.

Value for	MAP* (mm Hg)	CVP (mm Hg)	HR (per min)	CI (liter/min per M²)	SVR (mm Hg/ liter per min)	LVEDP (mm Hg)
Gram-positive	52 ± 2†	3.5 ± 0.6	119 ± 4	3.8 ± 0.3	8.5 ± 1.1	6.7 ± 1.3
Gram-negative	57 ± 3	5.1 ± 0.7	100 ± 4	2.9 ± 0.2	11.5 ± 0.9	9.8 ± 1.6
P (unpaired)	NS‡	NS	<.01	<.05	NS	NS

* MAP = mean arterial pressure; CVP = central venous pressure; HR = heart rate; CI = cardiac index; SVR = systemic vascular resistance; LVEDP = left ventricular end-diastolic pressure.

† Mean ± SE.

‡ NS = not significant.

impaired ventricular function in both groups. Thirteen patients with gram-positive infection received 646 ± 87 ml of fluid iv and increased their CO by 194 ± 55 ml/min per M² for each mm Hg increase in the CVP during infusion. Twenty-five patients with gram-negative infection received 590 ± 50 ml of fluid in a similar manner, and their CO increased only 79 ± 43 ml/min per M² for each mm Hg increase in CVP. These results suggest greater depression of ventricular function in the gram-negative than in the gram-positive group. Therefore, late in the history of gram-negative bacteremia, vasodilatation is no longer marked, and a major element in hemodynamic deterioration is myocardial depression.

In conclusion, the early changes of gram-negative bacteremia appear to be consistent with release of a vasodilator [5] and an increase in CO in response to this vasodilatation. In the late phase of gram-negative bacteremia, the SVR is no longer so markedly decreased, but CO and ventricular function are depressed. The latter could be explained by release of myocardial depressant factors [6], perhaps triggered by some direct or indirect effect of endotoxin on lysosomes [7]. The late peripheral vascular response or lack of response is probably not mediated by kinin, but rather is a resultant of various factors including release of catecholamine, acidosis, hypoxemia, and release of various vasoactive substances associated with activation of the complement system (see Mergenhagen's paper, Session 2).

References

1. Scott, W. W. Blood stream infection in urology. A report of 82 cases. J. Urol. 21:527–566, 1929.
2. Blain, C. M., Anderson, T. O., Pietras, R. J., Gunnar, R. M. Immediate hemodynamic effects of gram-negative vs gram-positive bacteremia in man. Arch. Intern. Med. 126:260–265, 1970.
3. Loeb, H. S., Cruz, A., Teng, C. Y., Boswell, J., Pietras, R. J., Tobin, J. R., Jr., Gunnar, R. M. Haemodynamic studies in shock associated with infection. Br. Heart J. 29:883–894, 1967.
4. Winslow, E. J., Loeb, H. S., Rahimtoola, S. H., Kamath, S., Gunnar, R. M. Hemodynamic studies and results of therapy in 50 patients with bacteremic shock. Am. J. Med. 1973 (in press).
5. Nies, A. S., Forsyth, R. P., Williams, H. E., Melmon, K. L. Contribution of kinins to endotoxin shock in unanesthetized rhesus monkeys. Circ. Res. 22:155–164, 1968.
6. Glenn, T. M., Lefer, A. M. Significance of splanchnic proteases in the production of a toxic factor in hemorrhagic shock. Circ. Res. 29:338–349, 1971.
7. Weissman, G. Lysosomal mechanism of tissue injury in arthritis. N. Engl. J. Med. 286:141–147, 1972.

Clinical and Experimental Observations on the Significance of Endotoxemia

Edward H. Kass, Philip J. Porter,
Michael W. McGill, and Ennio Vivaldi

From the Channing Laboratory, Thorndike Memorial Laboratory, Harvard Medical Unit, and the Department of Medical Microbiology, Boston City Hospital; and the Department of Medicine, Harvard Medical School, Boston, Massachusetts

Endotoxin-like materials, when sought in the plasma of septic patients, are found principally in association with infections due to gram-negative organisms, although positive tests are occasionally encountered under other circumstances. Many patients with infections due to gram-negative rods do not yield positive reactions, and positive reactions are not strong indicators of bacteremia or fatal outcome. Experiments in which hemorrhagic hypotension was induced in rabbits indicate that endotoxin-like materials are probably released from gram-negative gut flora but do not increase or decrease the rate of lethal outcome of controlled hemorrhagic hypotension. It is likely that factors other than endotoxin are more critical than endotoxin in determining lethal outcome of sepsis.

Our purpose in the present brief review is to indicate the clinical and experimental evidence for a point of view that has emerged in our laboratory. This point of view suggests that the presence of endotoxin-like materials in plasmas of patients and of experimental animals, although relatively indicative of an underlying infection due to gram-negative organisms, does not of itself constitute a particularly grave or ominous finding. It is hoped that by emphasis on this point, attention will be directed toward pathophysiologic features of severe sepsis other than those related to bacterial endotoxin.

The first substantial study of the presence of endotoxin-like materials in the blood of patients with severe infections was performed with use of the epinephrine skin reaction of Thomas [1]. Briefly, blood was drawn from patients with severe sepsis and was immediately placed in an ice bath; plasma was separated in the cold and either injected immediately into rabbits or frozen at —20

C until used. When small rabbits were given 2 ml of such plasma intravenously and appropriate dilutions of epinephrine intracutaneously, a localized hemorrhagic skin reaction at the site of injection of epinephrine sometimes occurred. A similar reaction was elicited by iv administration of 0.1–1.0 µg of purified lipopolysaccharide (LPS) instead of the plasma for delineation of the approximate sensitivity of this reaction. Initially, a small group of patients was recognized who had certain features in common. All of these patients had a severe localized infection due to a gram-negative organism, none had bacteremia, all had abnormalities of vasomotor and temperature-control mechanisms, and all died within 24 hr of the time when a positive test was detected.

In further studies [2], the hemorrhagic skin reaction in rabbits was made more sensitive by maintenance of the rabbits in a warm room and, thus, elevation of body temperature. Under these latter conditions, a hemorrhagic cutaneous reaction to epinephrine could be induced in rabbits that had received .001–0.1 µg of purified LPS iv. With the more sensitive test, a large number of plasmas from patients with diverse clinical disorders was examined. In general, the data indicated (table 1) that a positive skin test was elicited when gram-negative organisms but not gram-positive ones were causing the infection.

This work was supported by research grant no. HD-03693 from the National Institute of Child Health and Human Development and by training grant no. TO1 AI-00068 from the National Institute of Allergy and Infectious Diseases.

Please address requests for reprints to Dr. Edward H. Kass, Channing Laboratory, Boston City Hospital, Boston, Mass. 02118.

Table 1. Skin tests with plasma from patients with severe infection.

Type of organism in infection	No. of patients	Positive skin test	24-hr mortality	Death same admission	Positive blood culture
Gram-negative	23	10	4	10	8
Gram-positive	13	0	3	5	8
Gram-negative with positive test	10	10	1	6	1
Gram-negative with negative test	13	0	3	10	7

However, positive skin tests were sometimes elicited from the plasmas of individuals who had no apparent disease. Nevertheless, as can be seen from table 1, although the positive skin test was found in some but not all individuals who had infections due to gram-positive organisms, the presence of the test gave little insight into probability of death within 24 hr, death within the same hospital admission, or the presence of bacteremia. Thus, although there can be little doubt of the tendency of the test to be associated with infections due to gram-negative organisms, not all such infections, even fatal ones, are associated with a positive test, and there are many fatal infections due to other organisms that do not give rise to a positive test, as well as many nonfatal infections due to gram-negative organisms in which a positive test can be elicited. The presence or absence of bacteremia does not seem to be closely related to the probability of a positive skin test of the type that is elicited by purified LPS.

More recently, Levin [3, 4] has explored similar situations using the Limulus test. The Limulus test appears to be at least 10 times more sensitive than the hemorrhagic skin test in rabbits, although its specificity has not been completely explored as yet. In the first report by Levin, briefly summarized in table 2, there were 17 positive limulus reactions among 148 tests performed, and all 17 were associated with infections due to gram-negative rods. Of 22 patients with positive blood cultures, 12 had a positive Limulus test. In a later report (table 3), Levin et al. found positive Limulus tests in 20 of 34 patients with bacteremia due to a gram-negative organism and in none of 16 patients with bacteremia due to a gram-positive organism; they found 16 positive Limulus tests among 168 patients whose blood cultures were negative. There was some indication from the studies of Levin et al. of a tendency for positive tests to be associated with more severe infections, but the association was not invariable nor was it striking. Thus, in general, the Limulus test seems to have given results strikingly similar to those obtained with the hemorrhagic skin test, despite the increased sensitivity of the limulus test. Overall, the data indicate that endotoxin-like materials may appear in the plasma of patients in association with the presence of infections due to gram-negative rods, but rarely, if at all, in association with infections due to gram-positive organisms. However, despite this association, tests for the presence of endotoxin-like materials do not give an insight into the presence or absence of bacteremia or the likelihood of death, except perhaps in a relatively small number of unusually severe localized infections not associated with bacteremia.

We undertook experiments in our laboratory to explore this phenomenon further. These will not be presented in detail because they are being pub-

Table 2. Patients tested by Limulus test. (For details, see [3].)

Patients	Positive test	Negative test
Total no.	17	131
With bacteremia	12	10
With infection due to gram-negative rods	17	0

Table 3. Clinical results with Limulus test. (For details, see [4].)

Result	No. of patients	Gram-negative organism	Gram-positive organism
Blood culture positive	50	34	16
Skin test positive	20	20	0
Blood culture negative	168		
Skin test positive	16		

lished elsewhere; they have been mentioned in a preliminary report [5]. In brief, however, rabbits were exposed to sustained hypotension by controlled hemorrhage followed by spontaneous reinfusion of the previously removed blood. As is well established, the procedure can be so manipulated that a specified level of mortality in the animals can be obtained. In the experiments to be cited, the conditions of hemorrhage and spontaneous reinfusion were so adjusted that approximately 50% mortality from the hemorrhage was experienced (table 4). When this level of hemorrhagic collapse was induced in rabbits and plasmas of the rabbits were removed shortly before death, it was found that some, but not all, of the plasmas of the shocked rabbits contained endotoxin-like materials, as tested by injection into normal recipient rabbits that had received epinephrine intracutaneously and had been kept at ambient temperatures of 37 C. The rabbits whose plasmas elicited positive reactions were found to be those harboring facultative gram-negative rods in their intestinal contents; conversely, those rabbits that

did not harbor such facultative gram-negative rods did not produce positive reactions when they were exposed to fatal hemorrhagic collapse (table 5). To show that the positive cutaneous reaction was induced in some manner through the presence of facultative gram-negative rods in the fecal contents, we force-fed rabbits lacking such organisms by stomach tube with large numbers of *Escherichia coli,* sufficient to induce positive fecal cultures for *E. coli* when such cultures were previously absent. Rabbits colonized with *E. coli* under these conditions were then exposed to hemorrhagic manipulation, and their plasmas regularly elicited a positive reaction for endotoxin-like materials. Thus, it would appear to be confirmed that the endotoxin-like materials arise from organisms within the intestinal tract, and that the presence of facultative gram-negative rods is essential.

However, whether or not the animals harbored facultative gram-negative rods in their intestinal contents and whether or not *E. coli* was implanted, mortality from controlled hemorrhagic collapse was approximately the same in all groups.

We are led to the conclusion, therefore, that although it is correct to argue that endotoxin-like materials enter the blood stream in certain instances of hemorrhagic collapse, and that these materials reflect the presence of facultative gram-negative rods in the intestines of the animals, the presence or absence of such rods and the presence or absence of endotoxin-like materials in the blood of the shocked animals seems irrelevant to the ultimate fatal outcome. It would appear that both sides of the controversy have something in favor and something against their points of view.

If the effect of this analysis is to stimulate investigators to look for factors in severe sepsis other than those directly related to endotoxin (those that may be common to infections due to gram-positive as well as to gram-negative organisms), the presentation will have served its purpose.

Table 4. Mortality in hemorrhaged rabbits grouped according to volume of spontaneous reinfusion. (Reprinted from [5] with permission of the American Society for Microbiology.)

Spontaneous reinfusion (ml/kg)	Total no. of rabbits	No. dead	Mortality (%)
0	14	2	14
2	30	3	10
4	30	15	50
6	8	7	88

Table 5. Epinephrine skin-test results in survivors of groups of rabbits subjected to hemorrhagic shock of varying severity and divided according to presence or absence of fecal aerobic gram-negative rods. (Reprinted from [5] with permission of the American Society for Microbiology.)

Spontaneous reinfusion (ml/kg)	Positive epinephrine skin test/no. alive	
	Gram-negative rods present	Gram-negative rods absent
0	0/6	0/6
2	8/14	0/13
4	4/7	0/8

References

1. Porter, P. J., Spievack, A. R., Kass, E. H. Endotoxin-like activity of sera from patients with severe localized infections. N. Engl. J. Med. 271:445–447, 1964.
2. McGill, M. W., Porter, P. J., Kass, E. H. The use of

a bioassay for endotoxin in clinical infections. J. Infect. Dis. 121:103–112, 1970.

3. Levin, J., Poore, T. E., Young, N. S., et al. Gram negative sepsis; detection of endotoxemia with limulus tests. Ann. Intern. Med. 76:1–7, 1972.

4. Levin, J., Poore, T. E., Sauber, N. P., et al. Detection of endotoxin in the blood of patients with sepsis due to gram negative bacteria. N. Engl. J. Med. 283:1313–1316, 1970.

5. McGill, M. W., Porter, P. J., Vivaldi, E., Kass, E. H. Use of a bioassay for endotoxin in clinical infections and in experimental vasomotor collapse. Antimicrob. Agents Chemother. 1967:132–136, 1968.

Summary of Discussion

Edward H. Kass

A detailed discussion followed of the interrelationship between bioassay of endotoxin and clinical manifestation of disease. Several investigators indicated that, in agreement with data presented during the formal session, they were unable to relate shock and death to the presence or absence of a positive test for endotoxinlike materials by any of the several tests now available. Some investigators did not find that as many as 45%–50% of patients with positive blood cultures gave positive responses to the Limulus assay; others pointed out that even when such positive responses are found, they have relatively uncertain predictive value.

There was an extended discussion of the role of intestinal bacteria in releasing the substances found in the blood stream when there is a positive response by the skin test or the Limulus assay, and it was generally agreed that the present evidence favored the intestinal source of the materials. It was thought worthwhile to try to isolate and characterize substances found in the blood that account for the positive tests, since it is possible that the chemical form of these substances may not correspond precisely with the chemical form of the present lipopolysaccharide preparations.

The possibility was discussed that clinical testing for endotoxinlike materials would be facilitated by the use of pronase or related enzymes that would separate away serum proteins that might be complexing to the endotoxins. This would be an analogy to the experiments presented by Jon Rudbach earlier. However, it was pointed out by several investigators that the enzyme preparations used for this purpose were heavily contaminated with endotoxins and that the problem was an exceedingly difficult one to solve.

The implications of the presentations by Kass and by Gunnar were discussed in greater detail. A number of bacterial products are known to activate the complement system. This, in turn, can activate intravascular coagulation; thus it is quite reasonable to consider that both gram-positive and gram-negative infections may give similar pathologic changes and similar hemodynamic changes under appropriate conditions. It would be necessary to postulate only that endotoxins were not specific in terms of activating such systems.

Some disagreement that may have been due to differences in doses was expressed concerning the degree and timing of leukopenia in humans after the injection of endotoxin and concerning the presence or absence of thrombocytopenia after such injections. It appeared that the subject remains unresolved pending further controlled comparative studies.

The question of whether hemorrhagic collapse can be produced in a gnotobiotic animal was briefly discussed, but it was pointed out that, even under apparently germfree circumstances, the animals often received feeds contaminated with endotoxins. In general, however, the lethality of controlled hemorrhage is the same in germfree animals as in conventional animals. However, the problem of endotoxin contamination in the diet and its possible role in accounting for the lack of difference between germfree and conventional animals has not yet been fully settled.

There was a brief discussion of the possible role of prostaglandins in relation to endogenous pyrogen. Prostaglandins and endogenous pyrogens are quite different chemically, but the general subject is now being investigated.

Finally, several speakers reminded the audience that most studies of the reactivity of endotoxin were performed in healthy volunteers or in healthy laboratory animals. In fact, in the sick individual, responsivity to endotoxin may be greatly augmented, and amounts may become lethal that are otherwise relatively innocuous. Therefore, the search for a lethal effect may require that the recipient animal be in the same physiologic state that is characteristic of the severely ill host at the time when lipopolysaccharide might be released in substantial amounts.

Remarks of Dr. Robert C. Reisinger

DR. REISINGER: I would like first to show you some things that you probably have learned but may have forgotten. This first slide (figure 1) is

295

296

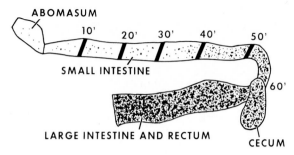

Figure 1. Distribution of *Escherichia coli* in the intestinal tract of the normal calf.

just to remind you that there are very few *Escherichia coli* in the more absorptive portion of the small intestine, the duodenum, jejunum, and proximal ileum. This is true of all mammalian species so far studied, including the cow and calf, the human infant and adult.

The next slide (figure 2) shows what was demonstrated by Theobald Smith and Marion Orcutt in 1922, i.e., that calves suffering diarrhea have tremendously increased numbers of *E. coli* in the intestinal tract. They described "a great increase in the number of *E. coli* in the lowest third of the small intestine with a spreading of the invasion towards the duodenum as the disease gains headway. Under these conditions, a general intoxication results . . . *E. coli* in the digestive tract has not been in general regarded as significant. This significance appears when the quantitative factor, obtained before natural death, is determined."

My work in Wisconsin, that of Gay in Canada, and that of Mebus in Nebraska confirmed that of Smith and Orcutt and further confirmed and demonstrated that the same mechanism of invasion of *E. coli* into the proximal small intestine may result in peracute deaths without septicemia

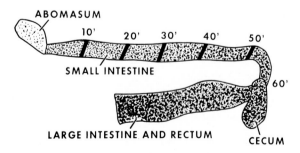

Figure 2. Distribution of *Escherichia coli* in the intestinal tract of the scouring calf.

and without diarrhea (SDS). Such deaths can occur within several hours after the feeding of an apparently healthy normal calf. Pathologic lesions at autopsy are absent or minimal, as in the crib-death syndrome in the human infant. Experimentally, the diarrhea syndrome can be precipitated by exposure to various viruses, cold and wet, avitaminous A, etc.—any "adverse contributing factor" that impairs optimal reticuloendothelial and/or gastrointestinal function. A classic feature of the calf-diarrhea syndrome is labored breathing (pneumonic signs without pneumonic lesions), but this disappears when, by appropriate antibiotic therapy, excessive numbers of *E. coli* are cleared from the digestive tract.

Bacterial endotoxins produced in the intestinal tract, which may be absorbed into the venous system, are transported directly into the liver. If for any reason these endotoxins are not inactivated in the liver, they obviously go via the inferior vena cava into the heart and lungs and, to a lesser extent (at least early in the syndrome), into the general arterial circulation.

Some factors common to many cases of SDS, the respiratory-distress syndrome, and endotoxin shock are hyperkalemia, hyponatremia, acidosis, thrombocytopenia, noncoagulability of blood, early respiratory signs without appropriate lesions, pulmonary edema, hemorrhage by diapesis, fast, weak pulse, and circulatory collapse. It is more illogical to consider these similarities as fortuitous than to realize the probability, or at least possibility, of a common cause or mechanism.

There are great differences between bacterial flora, *pH*, and physical characteristics of the intestinal contents of human infants fed human milk and these features in infants fed cow's milk. Due to its higher content of calcium and protein and its lower content of lactose, cow's milk fed to the human infant raises the *E. coli* count in the large intestine approximately 1,000-fold (from 10^6–10^7/g to 10^9–10^{10}/g), raises the *pH* from acid (4.5–5.6) to alkaline (7.0–8.0), makes curds hard and coarse instead of soft and fine, and makes bowel movements relatively infrequent. Characteristics of a healthy human infant on human milk are a relatively low coliform count, acid *pH*, soft, fine curds, and frequent bowel movements. Even if there were no SDS, these adverse changes wrought by a diet of cow's milk any time during the first six months of age are against all reason in producing the optimally healthy child and the healthy adult he should become.

I believe that crib death, or SDS, is only the tip of the iceberg with regard to effects of absorption of bacterial endotoxin in the human infant. The vasoconstriction, vasodilation, disseminated intravascular coagulation, and other characteristic effects of endotoxin may well account for much of the pathology associated with many cases of cerebral palsy, mental retardation, learning and speech defects, retrolental fibroplasia, hyaline-membrane disease, etc. These latter diseases appear predominantly in infants with immature or otherwise impaired hepatic function, i.e., the immature infants of diabetic mothers, infants anoxic due to placenta praevia, infants whose birth is long and difficult, etc. In long and difficult births, there is also a prolonged in-utero opportunity for seeding of the infants' digestive tracts should infected amniotic fluid be swallowed.

Since absorption of endotoxin from the digestive tract is a confirmed fact in various mammalian species, the probability, or at least the possibility, of similar absorption from the intestinal tract of the human infant should be considered and investigated.

Remarks of Dr. William R. McCabe

I would like to make a mini-mini comment about clinical disease in an exotic species, *Homo sapiens.* Our laboratory has been interested in determining whether the physiologic and biochemical changes induced by endotoxin in in-vitro systems and experimental animals also occur during human infections with gram-negative bacilli. The descriptions of consumption of complement components by endotoxin prompted measurements of C3 levels in 68 patients with gram-negative bacteremia. Specimens of serum were obtained from 68 patients with bacteremia as soon as this diagnosis was suspected and from 75 control patients admitted to the hospital for various reasons; C3 levels were determined by radial immunodiffusion. Mean C3 levels did not differ significantly in the 75 controls (141.4 ± 22 mg/100 ml) from those in the 68 patients with gram-negative bacteremia (132 ± 50.1 mg/100 ml). However, striking decreases in C3 concentration were observed in patients with gram-negative bacteremia in whom shock occurred or who succumbed. Mean C3 levels were 104.6 ± 31.4 in the 26 patients who experienced shock and 98.2 ± 37.6 in the 19 patients who died. In contrast, mean C3 levels were 150.7 ± 50.6 in the 41 patients with gram-negative bacteremia who did not experience shock or a fatal outcome. The decrease in mean C3 levels in bacteremic patients who were in shock or who succumbed was highly significant ($P < 0.001$) when compared with levels in patients with uncomplicated bacteremia or in controls.

If these values are viewed in another manner, C3 levels in specimens obtained shortly after initial suspicion of bacteremia appeared to have prognostic significance. Of the 24 patients with C3 levels of 100 or less, 15 (63%) developed shock, and 11 (46%) died. The initial C3 levels in 35 patients fell within two standard deviations above or below reported mean values in controls, 101 to 190. Eleven (31%) episodes of shock and six deaths (17%) occurred in these 35 patients. In contrast, one patient (11%) died, and no episodes of shock occurred in the nine patients with C3 levels greater than 190.

These findings indicate that, although decreases in C3 levels are not uniformly observed in gram-negative bacteremia, significant complement consumption occurs in lethal bacteremia or in bacteremia complicated by shock. These observations in human gram-negative bacteremia are quite similar to the demonstration by Gilbert and Braude that, while administration of lethal quantities of endotoxin induced complete consumption in rabbits, no decrease in complement occurred when nonlethal quantities of endotoxin were administered.

Participants

FRANK W. ADAIR
CIBA-Geigy Corporation
Pharmaceuticals Division
Summit, New Jersey 07901

JAMES C. ADYE
Evanston Hospital
2650 Ridge Avenue
Evanston, Illinois 60201

ANTHONY C. ALLISON
Clinical Research Center
Watford Road
Harrow, Middlesex HA1 3UJ
England

FRED ALLISON, JR.
Louisiana State University
  Medical Center
1542 Tulane Avenue
New Orleans, Louisiana 70118

LEONARD C. ALTMAN
National Institute of Dental
  Research
National Institutes of Health
Building 30, Room 323
Bethesda, Maryland 20014

DOUGLAS P. ANDERSON
Western Fish Disease Laboratory
Building 204, Sand Point NSA
Seattle, Washington 98115

DONALD ARMSTRONG
Memorial Hospital
444 East 68th Street
New York, New York 10021

MALCOLM S. ARTENSTEIN
Walter Reed Army Institute
  of Research
Washington, D.C. 20012

ELISHA ATKINS
Department of Internal Medicine
Yale University School of
  Medicine
New Haven, Connecticut 06510

G. AYME
Institut Francais d'Immunologie
69-Marcy l'Etoile
France

JOHN A. BACH
Unit 7831, Building 41-2
The Upjohn Company
Kalamazoo, Michigan 49001

PASQUALE F. BARTELLO
College of Medicine and
  Dentistry of New Jersey
100 Bergen Street
Newark, New Jersey 07103

WERNER F. BARTH
7705 Grandad Drive
Bethesda, Maryland 20034

FRANK M. BERGER
Wallace Laboratories
Cranbury, New Jersey 08512

L. JOE BERRY
Department of Microbiology
University of Texas
Austin, Texas 78712

JOSEPH G. BIRD
CIBA-Geigy Corporation
Summit, New Jersey 07901

HILDUR BLYTHMAN
Department of Microbiology
Yale University
New Haven, Connecticut 06510

PHYLLIS BODEL
Yale-New Haven Hospital
333 Cedar Street
New Haven, Connecticut 06510

CONSTANTIN A. BONA
Batiment INSERM
184 Faubourg Saint-Antoine
Paris XIIeme
France

S. GAYLEN BRADLEY
MCV Station, P. O. Box 847
Virginia Commonwealth
  University
Richmond, Virginia 23219

ABRAHAM BRAUDE
University Hospital, San Diego
San Diego, California 92037

J. WERNER BRAUN
Rutgers State University
New Brunswick, New Jersey
  08903

VOLKMAR BRAUN
Max-Planck-Institut Molekulare
  Genetik
1 Berlin 33
Germany

S. A. BROITMAN
Boston University School of
  Medicine
80 East Concord Street
Boston, Massachusetts 02118

THOMAS BUTLER
Johns Hopkins School of Medicine
Baltimore, Maryland 21205

J. A. CAMERON
Microbiology, Biological Sciences
University of Connecticut
Storrs, Connecticut 06268

LOUIS CHEDID
Institut Pasteur
Paris
France

CHEN-LO H. CHEN
Temple University School of
  Medicine
3400 North Broad Street
Philadelphia, Pennsylvania 19140

JOSEPH K. CHEN
Department of Microbiology
University of Pittsburgh School
  of Medicine
Pittsburgh, Pennsylvania 15213

FRANCIS V. CHISARI
10903 Raleigh Avenue
Garrett Park, Maryland

RITA R. COLWELL
Department of Biology
Georgetown University
Washington, D.C. 20007

JAMES F. COOPER
Bureau of Radiologic Health,
  Federal Drug Administration
5600 Fishers Lane
Rockville, Maryland 20852

WILLIAM A. CRAIG
Veterans Hospital
2500 Overlook Terrace
Madison, Wisconsin 53706

PETER CUEVAS
Sears Surgical Laboratory
Boston City Hospital
Boston, Massachusetts 02118

ARTHUR M. DANNENBERG
Johns Hopkins School of Hygiene
Baltimore, Maryland 21205

CHARLES A. DINARELLO
National Institutes of Health
Building 10, Room 11 S242
Bethesda, Maryland 20014

RAIMUND DIPAULI
University Department of
  Biology
775 Konstanz
Germany

SAM T. DONTA
University Hospitals
Iowa City, Iowa 52240

EUGENE L. DULANEY
Merck Institute of Therapeutic
  Research
Scott Avenue
Rahway, New Jersey 07065

C. R. DUNN
Department of Medicine, K-4
Cincinnati General Hospital
Cincinnati, Ohio 45229

HANS EIBL
Immuno AG, 1220 Wien
Industriestrasse 72
Vienna
Austria

THEODORE C. EICKHOFF
University of Colorado Medical
  Center
Denver, Colorado 80220

RONALD ELIN
National Institutes of Health
Bethesda, Maryland 20014

W. EDMUND FARRAR, JR.
Medical University Hospital of
  South Carolina
80 Barre Street
Charleston, South Carolina

DAVID S. FEINGOLD
University of Pittsburgh School
  of Medicine
Room 720, Scaife Hall
Pittsburgh, Pennsylvania 15213

JAMES P. FILKINS
Loyola University Medical Center
2160 South First Avenue
Maywood, Illinois 60153

JACOB FINE
Harvard Surgical Unit
Sears Surgical Laboratory
Boston City Hospital
Boston, Massachusetts 02118

RICHARD A. FINKELSTEIN
University of Texas
Southwestern Medical School
Dallas, Texas 75235

M. W. FISHER
Parke, Davis & Company
Joseph Campau at the River
Detroit, Michigan 48232

JACK S. C. FONG
Mayo Box 494
University of Minnesota Hospital
Minneapolis, Minnesota 55455

JOHN W. FOSTER
Neisseria Research Unit
Center for Disease Control
Atlanta, Georgia 30333

MICHAEL M. FRANK
National Institutes of Health
Building 10, Room 11N-104
Bethesda, Maryland 20014

HENRY H. FREEDMAN
Warner-Lambert Research
  Institute
170 Tabor Road
Morris Plains, New Jersey 07950

EARL H. FREIMER
Medical College of Ohio
P. O. Box 6190
Toledo, Ohio 43614

FRANK E. FRERMAN
Medical College of Wisconsin
561 North 15th Street
Milwaukee, Wisconsin 53233

HERMAN FRIEDMAN
Albert Einstein Medical Center
Department of Microbiology
Philadelphia, Pennsylvania 19141

GEORGE M. FUKUI
329 Prospect Avenue
Princeton, New Jersey 08540

CHRISTOS GALANOS
Max-Planck-Institut für
  Immunbiologie
78 Freiburg
Germany

MANUEL M. GARCIA
Animal Diseases Research
  Institute
P. O. Box 1400
Hull, Quebec
Canada

R. GERMANIER
Swiss Serum and Vaccine
  Institute
P. O. Box 2707, 3001 Berne
Switzerland

HENRY GEWURZ
Rush-Presbyterian-St. Luke's
  Medical Center
1753 West Congress Parkway
Chicago, Illinois 60612

DAVID GILBERT
Providence Hospital
700 N. E. 47th Avenue
Portland, Oregon 97235

EDWARD J. GOETZL
Robert B. Brigham Hospital
125 Parker Hill Avenue
Boston, Massachusetts 02120

JOHN N. GOLDMAN
Robert B. Brigham Hospital
125 Parker Hill Avenue
Boston, Massachusetts 02120

GERALD GOLDSTEIN
Department of Medicine
University of Virginia Hospital
Charlottesville, Virginia 22901

FRANCIS B. GORDON
Naval Medical Research Institute
Bethesda, Maryland 20014

EMIL C. GOTSCHLICH
Rockefeller University
New York, New York 10021

OTTO GOTZE
Scripps Clinic and Research
  Foundation
476 Prospect Street
La Jolla, California 92037

JOSEPH E. GRADY
The Upjohn Company
Infectious Diseases Research
Kalamazoo, Michigan 49001

SHELDON GREISMAN
University of Maryland Medical
  School
Department of Medicine
Baltimore, Maryland 21201

Rolf M. Gunnar
Veterans Administration
Edwards Hines Jr. Hospital
Hines, Illinois 69141

M. Carolyn Hardegree
Division of Biologics Standards
National Institutes of Health
Bethesda, Maryland 20014

Edward C. Heath
Department of Biochemistry
University of Pittsburgh School
  of Medicine
Pittsburgh, Pennsylvania 15213

James C. Hill
Naval Medical Research Institute
National Naval Medical Center
Bethesda, Maryland 20015

Monto Ho
School of Public Health
University of Pittsburgh
Pittsburgh, Pennsylvania 15213

Horace L. Hodes
Mount Sinai Hospital
100th Street and 5th Avenue
New York, New York 10029

J. Yuzuru Homma
P. O. Takanawa
University of Tokyo
Tokyo
Japan

William A. Hook
National Institute of Dental
  Research
National Institutes of Health
Bethesda, Maryland 20014

Robert Horn
School of Medicine
Vanderbilt University
21st Avenue, South
Nashville, Tennessee 37205

Richard B. Hornick
Division of Infectious Diseases
School of Medicine
University of Maryland
Baltimore, Maryland 21201

C. R. Jenkin
University of Adelaide
Adelaide
South Australia

Margaret A. Johns
Adult Infectious Disease
  Department
University Hospital
Boston, Massachusetts 02118

Arthur G. Johnson
University of Michigan Medical
  School
Ann Arbor, Michigan 48104

Robert Jones
Division of Experimental Surgery
Naval Medical Research Institute
National Naval Medical Center
Bethesda, Maryland 20014

John W. Jutila
Department of Botany and Micro-
  biology
Montana State University
Bozeman, Montana 59715

H. Ronald Kaback
Roche Institute of Molecular
  Biology
Nutley, New Jersey 07110

Michael A. Kane
4611 Glenbrook Parkway
Bethesda, Maryland

Dennis L. Kasper
Walter Reed Army Medical Center
Institute of Research
Washington, D.C. 20012

Edward H. Kass
Channing Laboratory
Boston City Hospital
774 Albany Street
Boston, Massachusetts 02118

Chris P. Katsampes
Warner-Lambert Company
170 Tabor Road
Morris Plains, New Jersey 07950

Masaya Kawakami
Gunma University, Showa-machi
Maebashi, Gunma-Ken
Japan

David Keast
Department of Microbiology
School of Medicine
University of Western Australia
Victoria Square
Perth
Western Australia

Lutz Kiesow
Experimental Medical Division
Naval Medical Research Institute
National Naval Medical Center
Bethesda, Maryland 20014

Mary Lidia Klodnycky
1516 North Harlem Avenue
River Forest, Illinois 60305

Avram R. Kraft
Division of Experimental Surgery
Naval Medical Research Institute
National Naval Medical Center
Bethesda, Maryland 20014

Arthur E. Krikszens
Department of Microbiology
New Jersey Medical School
100 Bergen Street
Newark, New Jersey 07103

Charles Kulpa
National Institutes of Health
Building 4, Room 111
Bethesda, Maryland 20014

Volker Lehmann
Department of Microbiology
University of Connecticut Health
  Center
Farmington, Connecticut

Loretta Leive
National Institutes of Health
Building 4, Room 111
Bethesda, Maryland 20014

Alf A. Lindberg
Statens Bakteriologiska
  Laboratory
S10521 Stockholm
Sweden

Robert B. Lindberg
U. S. Army Institute of Surgical
  Research
Brooke Army Medical Center
Fort Sam Houston, Texas 78234

Gerald A. LoGrippo
Henry Ford Hospital
2799 West Grand Boulevard
Detroit, Michigan 48202

Otto Lüderitz
Max-Planck-Institut für
  Immunbiologie
D78 Freiburg
Germany

P. HELENA MÄKELÄ
Central Public Health Laboratory
State Serum Institute
Helsinki
Finland

GERALD L. MANDELL
Box 251
University of Virginia School of
  Medicine
Charlottesville, Virginia 22901

ROBERT Q. MARSTON
National Institutes of Health
Building 1, Room 124
Bethesda, Maryland 20014

LUIS A. MARTINEZ
University of Connecticut Health
  Center
2 Holcomb Street
Hartford, Connecticut 06112

M. G. MAXIE
Ontario Veterinary College
University of Guelph
Guelph, Ontario
Canada

JOSEPH E. MAY
National Institutes of Health
Building 10, Room 11B-13
Bethesda, Maryland 20014

WILLIAM R. McCABE
University Hospital
750 Harrison Avenue
Boston, Massachusetts 02118

R. E. McCALLUM
Department of Microbiology
University of Texas
Austin, Texas 78712

JERRY McGHEE
University of Alabama,
  Birmingham
1919 Seventh Avenue South
Birmingham, Alabama 35233

FLOYD C. McINTIRE
29 North Elmwood Avenue
Waukegan, Illinois 60085

KENNETH L. MELMON
University of California
San Francisco, California

STEPHAN E. MERGENHAGEN
Laboratory of Microbiology and
  Immunology

National Institute of Dental
  Research
National Institutes of Health
Bethesda, Maryland 20014

J. G. MICHAEL
University of Cincinnati Medical
  Center
Cincinnati, Ohio 45219

RUSSELL L. MILLER
1562 47th Avenue
San Francisco, California 94122

KELSEY C. MILNER
Rocky Mountain Laboratory
U. S. Public Health Service
Hamilton, Montana 59840

GÖRAN MÖLLER
Karolinska Institutet Medical
  School
Stockholm
Sweden

ROBERT J. MOON
Department of Microbiology and
  Public Health
Michigan State University
East Lansing, Michigan 48823

SYLVIANE MOREAU
Worcester Foundation for
  Experimental Biology
222 Maple Avenue
Shrewsbury, Massachusetts 01545

KATHERINE MORRIS
Department of Microbiology and
  Public Health
Michigan State University
East Lansing, Michigan 48823

R. G. E. MURRAY
University of Western Ontario
London, Ontario
Canada

J. RAMA MURTHY
Evanston Hospital
2650 Ridge Avenue
Evanston, Illinois 60201

ERWIN NETER
Children's Hospital
219 Bryant Street
Buffalo, New York 14222

HAROLD C. NEU
Columbia Presbyterian Medical
  Center
New York, New York

HIROSHI NIKAIDO
Department of Bacteriology and
  Immunology
University of California
Berkeley, California 94720

SIGURD J. NORMANN
Department of Pathology
University of Florida
Gainesville, Florida 32601

ROBERT S. NORTHRUP
National Institutes of Health
Building 29, Room 431
Bethesda, Maryland 20014

WILLIAM M. O'LEARY
Cornell University Medical
  College
1300 York Avenue
New York, New York 10021

MARY JANE OSBORN
Department of Microbiology
University of Connecticut Health
  Center
Farmington, Connecticut

GEORGE A. PANKEY
Ochsner Clinic
1514 Jefferson Highway
New Orleans, Louisiana 70121

OLGERTS R. PAVLOVSKIS
Naval Medical Research Institute
National Naval Medical Center
Bethesda, Maryland 20014

ROBERT T. PFEIFER
The Upjohn Company
Infectious Diseases Research
Kalamazoo, Michigan 49001

NATHANIEL F. PIERCE
Department of Medicine
Baltimore City Hospitals
Baltimore, Maryland 21224

CARL PIERSON
University of Michigan Medical
  Center
3050 Kresgie 2 Medical Building
Ann Arbor, Michigan 48104

THOMAS G. PISTOLE
Spaulding Life Science Building
University of New Hampshire
Durham, New Hampshire 03824

DONALD PLANTEROSE
Research Laboratories
Beecham Laboratories
Betchworth, Surrey
England

PHILIP J. PORTER
Cambridge Hospital
Cambridge Street
Cambridge, Massachusetts

SAMUEL J. PRIGAL
77 Park Avenue
New York, New York 10016

PETER QUESENBERRY
St. Elizabeth's Hospital
736 Cambridge Street
Brighton, Massachusetts 02135

RICHARD QUINN
Burroughs-Wellcome Company
3030 Cornwallis Road
Research Triangle Park
North Carolina 27709

MARCEL RAYNAUD
Institut Pasteur
Garches
France

NORMAN D. REED
Department of Botany and
    Microbiology
Montana State University
Bozeman, Montana 59715

MICHAEL J. REICHGOTT
University of California Medical
    Center
1089 Moffett Hospital
San Francisco, California 94122

ROBERT C. REISINGER
National Institutes of Health
Building 37, Room 1D-05
Bethesda, Maryland 20014

GERARD RENOUX
Laboratoire d'Immunologie
Faculté de Médecin
Tours 37
France

HERBERT Y. REYNOLDS
National Institutes of Health
Building 10, Room 11B-13
Bethesda, Maryland 20014

STEPHEN H. RICHARDSON
Department of Microbiology

Bowman Gray Medical School
Winston-Salem, North Carolina
    27103

ERNST T. RIETSCHEL
Max-Planck-Institut für
    Immunbiologie
D78 Freiburg
Germany

R. J. ROANTREE
Department of Medical
    Microbiology
Stanford University
Stanford, California 94305

JOHN B. ROBBINS
National Institutes of Health
Building 10, Room 13N240
Bethesda, Maryland 20014

ROBERT R. ROBERTS
1C29 Wymount Terrace
Provo, Utah 84601

JOHN A. ROBINSON
Hines Veterans Administration
    Hospital
12 West
Hines, Illinois 60141

H. G. ROBSON
3775 University Street
Montreal, Quebec
Canada

VICTORIO RODRIGUEZ
M. D. Anderson Hospital
6723 Bertner
Houston, Texas 77025

RICHARD K. ROOT
University of Pennsylvania School
    of Medicine
552 Johnson Pavilion
Philadelphia, Pennsylvania 19104

DERRICK ROWLEY
University of Adelaide
Adelaide
South Australia

JON A. RUDBACH
Department of Microbiology
University of Montana
Missoula, Montana 59801

ALEXANDER RUTENBURG
Boston University School of
    Medicine
80 East Concord Street
Boston, Massachusetts 02118

MERLE A. SANDE
Department of Internal Medicine
University of Virginia School of
    Medicine
Charlottesville, Virginia 22901

JAY P. SANFORD
University of Texas
Southwestern Medical School
Dallas, Texas 75235

SAMUEL SASLAW
University Hospital
410 West 10th Avenue
Columbus, Ohio 43210

J. W. SHANDS, JR.
Department of Immunology and
    Medical Microbiology
University of Florida College of
    Medicine
Gainesville, Florida 32601

JOHN N. SHEAGREN
Howard University College of
    Medicine
Freedman's Hospital
Washington, D.C. 20301

N. RAPHAEL SHULMAN
National Institutes of Health
Building 10, Room 9N250
Bethesda, Maryland 20014

C. W. SHUSTER
Department of Microbiology
Case Western Reserve University
Cleveland, Ohio 44106

ROBERT C. SKARNES
Worcester Foundation for
    Experimental Biology
222 Maple Avenue
Shrewsbury, Massachusetts 01545

J. D. SMALL
National Institute of Dental
    Research
National Institutes of Health
Bethesda, Maryland 20014

RICHARD T. SMITH
Department of Pathology
University of Florida School of
    Medicine
Gainesville, Florida 32601

M. SOCHARD
Department of Biology
Georgetown University
Washington, D.C. 20007

EUGENE L. SPECK
14120 Heritage Lane
Silver Spring, Maryland 20906

GEORG F. SPRINGER
Evanston Hospital
2650 Ridge Avenue
Evanston, Illinois 60201

CHANDLER A. STETSON, JR.
New York University School of
   Medicine
550 First Avenue
New York, New York 10016

B. A. D. STOCKER
Department of Medical
   Microbiology
Stanford University Medical
   School
Stanford, California 94305

THOMAS P. STOSSEL
Children's Hospital Medical
   Center
300 Longwood Avenue
Boston, Massachusetts 02115

INDER JIT SUD
Beth Israel Hospital
Boston, Massachusetts 02115

BARNET M. SULTZER
Karolinska Institutet
104 05 Stockholm 50
Sweden .

L. SZABO
Institut de Biochimie
95-Orsay
France

BERNHARD URBASCHEK
Institute of Hygiene
University of Heidelberg
Klinikum Mannheim
68 Mannheim
West Germany

V. E. VALLI
Ontario Veterinary College
University of Guelph
Guelph, Ontario
Canada

STEPHEN I. VAS
McGill University
3775 University Street
Montreal, Quebec
Canada

WESLEY A. VOLK
Department of Microbiology
University of Virginia School of
   Medicine
Charlottesville, Virginia 22901

RICHARD V. WALKER
11 West Hills
Athens, Ohio 45701

PETER A. WARD
Department of Pathology
University of Connecticut Health
   Center
Farmington, Connecticut 06032

STANLEY B. WARD
National Institutes of Health
Building 10, Room 11-B-15
Bethesda, Maryland 20014

DENNIS W. WATSON
University of Minnesota
Box 196 Mayo
Minneapolis, Minnesota 55455

P. WERNET
Rockefeller University
New York, New York 10021

MARTIN WESTENFELDER
Veterans Administration Hospital
2500 Overlook Terrace
Madison, Wisconsin 53705

SILVIA WESTENFELDER
Veterans Administration Hospital
2500 Overlook Terrace
Madison, Wisconsin 53705

OTTO WESTPHAL
Max-Planck-Institut für
   Immunbiologie
Freiburg
West Germany

ROBERT W. WHEAT
P. O. Box 3020
Duke University Medical Center
Durham, North Carolina

GEORGE B. WHITFIELD, JR.
The Upjohn Company
Infectious Diseases Research
Kalamazoo, Michigan 49001

BRUCE WINTROUB
Robert B. Brigham Hospital
125 Parker Hill Avenue
Boston, Massachusetts 02120

SHELDON M. WOLFF
National Institute of Allergy and
   Infectious Diseases
National Institutes of Health
Bethesda, Maryland 20014

K. H. WONG
Division of Biologics Standards
National Institutes of Health
Bethesda, Maryland 20014

PETER WOODRUFF
Sears Surgical Laboratory
Boston City Hospital
Boston, Massachusetts 02118

ANDREW WRIGHT
Department of Molecular Biology
Tufts University School of
   Medicine
Boston, Massachusetts

E. THYE YIN
Jewish Hospital of St. Louis
216 South Kingshighway
St. Louis, Missouri 63110

LOWELL S. YOUNG
Memorial Hospital
444 East 68th Street
New York, New York 10021

WENDELL D. ZOLLINGER
Department of Bacterial Diseases
Walter Reed Army Institute of
   Research
Washington, D.C. 20012

Arrangements for the conference were made by:

PHYLLIS COHEN
Channing Laboratory
Boston City Hospital
774 Albany Street
Boston, Massachusetts 02118

EMILY DONOVAN
Channing Laboratory
Boston City Hospital
774 Albany Street
Boston, Massachusetts 02118

GERMAINE LEFTWICH
c/o Sheldon M. Wolff
National Institute of Allergy and
  Infectious Diseases
National Institutes of Health
Bethesda, Maryland 20014